USING MULTIVARIATE STATISTICS

Barbara G. Tabachnick
Linda S. Fidell

California State University, Northridge

1817

HARPER & ROW, PUBLISHERS, New York
Cambridge, Philadelphia, San Francisco,
London, Mexico City, São Paulo, Sydney

To Deba, who inspired the cover of this book and much more.

Sponsoring Editor: George A. Middendorf
Project Editor: Nora Helfgott
Production Manager: Willie Lane
Compositor: Kingsport Press
Printer and Binder: R. R. Donnelley & Sons Company
Art Studio: Fine Line, Inc.

Using Multivariate Statistics

Library of Congress Cataloging in Publication Data

Tabachnick, Barbara G.
  Using multivariate statistics.

  Includes bibliographical references and index.
  1. Multivariate analysis.   I. Fidell, Linda S.   II. Title.
QA278.T3   1983   519.5'35   82–11767
ISBN 0–06–042045–6

Contents

Preface

The primary goal of this book is to provide a summary of several of the more widely used multivariate procedures that is pitched at a level intelligible to the person who has completed a routine undergraduate course in inferential statistics. Such a summary should prove helpful both to the student who finds that an ever-increasing number of research papers report multivariate analyses and to the researcher who must prepare them.

We feel that many of our colleagues would also appreciate a usable summary of the multivariate procedures, particularly those whose major focus is on substantive research rather than on research design and analysis per se. Although the many books from which we learned these procedures are excellent, they do not address the problems encountered by the researcher who needs the techniques but has neither the time, energy, nor interest to master complicated mathematics. Another goal, then, is to share what we have learned about the use and interpretation of the techniques with colleagues who need to use them but do not wish to rummage through detailed mathematical treatments to try to discover what a given result might mean.

We believe that the researcher, as distinct from the statistician, is interested in the following major questions: Which analytic technique is most useful for answering my research question? How do my data fit the technique? How do I interpret the results in terms of the research question? Secondarily, the researcher, as well as the student, may be interested in two other questions: Why is one technique better than others in answering research questions? How do the various techniques fit together within a general model? We suspect that the researcher is probably *not* interested in reinventing multivariate statistics.

We assume that the user of the book (but not necessarily the reader) has access to a computer equipped with either the SPSS Batch System[1] or BMDP computer package.[2] Most multivariate analyses cannot be calculated by hand in the time available to researchers who are working on realistically sized problems (although we do show what is involved in the calculations for a tiny data set for each procedure). We encourage the reader to follow the steps of each analysis with either the small data set provided in Sections 4 of Chapters 5 through 10 or another available set.

Several features of the book may prove especially helpful. We have attempted both to identify and list the major research questions that can be answered through each analysis and to clarify theoretical and practical limitations to applying a technique to a data set. We indicate which underlying assumptions are critical to an analysis, how to test them, and what to do if they are violated.

In an effort to alleviate the anxiety some readers may feel about these procedures and in the belief that the procedures really are not, after all, so mysterious, we have provided a verbal statement of most equations below each equation. So a reader may elect to ignore the details of much of the mathematics (except during calculations) and still gain a sense of the calculations involved by reading the verbal statement rather than the equation. We have kept the number of equations to a minimum while attempting to retain the logic of the underlying mathematical processes.

As occasional reviewers for professional journals, both of us are concerned with inconsistency in reporting multivariate results. There seems to be a need among researchers and readers alike for standardization in regard to which part of the computerized outputs to include in Results sections. In other words, how should results be reported and which results are critical to report? We have tried to meet a portion of that need. Included in the section labeled "Complete Examples" for each technique is a Results section. Written in American Psychological Association (APA) journal format, including tables, graphs, and the like, it represents our decision regarding the minimum results that should be reported and the most concise way of reporting them. At the very least, we hope this attempt sparks debate and eventual resolution of problems involved in reporting multivariate results.

Two other potentially helpful features are the flow chart on page 63 around which the book was organized, which leads you from your major research question to the most appropriate technique to use, through consideration of numbers of dependent and independent variables; and a chapter discussing missing data, outliers, and the like.

But the book has distinct limitations as well. It includes only the most popular multivariate techniques out of the vast array currently available. Further, we have not shown derivations of mathematical equations, nor have we made

---

[1] SPSS is a trademark of SPSS Inc. of Chicago, Illinois, for its proprietary computer software. No materials describing such software may be produced or distributed without the written permission of SPSS Inc.

[2] Another increasingly popular package, SAS, is not reviewed in this book because it was unavailable to us at the time of writing.

original contributions to the theory of multivariate analyses. This book was written by users of the techniques for other users of them.

Perhaps a more serious limitation to the book is that some of it will become obsolete as the BMDP and SPSS computer packages are revised, expanded, and corrected. During the 4-year development of the book, several revisions in the programs necessitated revisions in the book. It is important, therefore, for both the reader and researcher to stay abreast of changes in canned programs as they are implemented in the local computing center, perhaps regularly updating the tables in each chapter that compare the programs.

This book is thoroughly student-tested, and it is to our students that we owe the greatest measure of thanks. They suffered through semi-intelligible early versions with enthusiasm and good grace. Although several classes in both our School Psychology and Clinical/Community Master's programs contributed their comments and observations to this final product, one class in particular had to learn from manuscript pages that were sometimes still warm, and they asked to be remembered as individuals. Thank you, Barry Beith, Larry Condelli, Russell Goto, Vaughn Inman, David Kaplan, Tom Keifer, Dean Kitnik, Marjo Kostama, Lew Kovner, Greg Krogh, Andrew Lanto, Sharon Sassé, Charles Smythe, Bruce Stein, Linda Vaughn, Robert Whaley, Hugh Whitehurst, and Morris Zlotnik. The valuable suggestions of this group and others helped us make several major decisions about the book, notably to err on the side of redundancy when listing the practical limitations (assumptions) of each procedure and explaining how to test for them.

Many of our colleagues have been more than helpful as well. George Middendorf, sponsoring editor, Nora Helfgott, project editor, and the staff of Harper & Row, along with Lieselotte Hofmann, copy editor, have been patient and encouraging. Jim Frane, a member of the BMDP (Biomedical Computer Programs, P-series) staff, shared his wealth of statistical knowledge plus a good deal of juicy gossip about multivariate analyses-gone-awry in numerous lengthy telephone conversations. Much that is right about the book is directly attributable to his careful reading of it. Jim Fleming, Edward Cobb, Schyler Huck, and Howard Sandler also kept us from making numerous silly errors, but none is responsible for other errors that have doubtless crept in during final revision of the manuscript. Colleagues cornered in halls and asked esoteric questions include Ralph Sabella, Don Butler, and Sam Pinneau. Charles Hofacker was also particularly helpful and generous with his time and expertise. The facilities and staff of the CSUN Computer Center and the CSU statewide timesharing system were invaluable in providing access to computer packages and ever-cheerful assistance. And we owe a debt of gratitude to our typists—Arlene Kaminsky, who typed most of the first draft, and most especially Joan Evans, a lady with a very fine brain behind her fingers, who corrected our grammar and spelling, produced several beautifully clean manuscripts, and, we suspect, learned some multivariate statistics, somewhat to her dismay, in the process. Neither of our husbands divorced us although we must have tried their patience severely from time to time.

In an effort to stay sane (or at least to be crazy in a different way) during preparation of the book, we took up *la danse orientale* as an avocation. Occasionally, after inverting too many matrices, a bit of this other interest crept into the manuscript. Please indulge us for the resulting silliness or, if you prefer, substitute other variable names for those we offer in a few of the examples. But we hope you will agree that the learning of statistics does not always have to be deadly serious.

And so we begin.

BARBARA G. TABACHNICK
LINDA S. FIDELL

Chapter 1

Introduction

1.1 MULTIVARIATE STATISTICS: WHY?

Multivariate statistics are a set of increasingly popular statistical techniques that are useful for analyzing more complicated data. They provide a system for analysis under conditions in which there may be several independent variables (IVs) and many dependent variables (DVs) all correlated with one another to varying degrees. Both because of the difficulty of addressing many modern research questions with restrictions imposed by univariate analyses and because of the availability of computing facilities with canned software for performing multivariate analyses, multivariate statistics are becoming widely used in social science and other research. Indeed, the day may be near when a standard undergraduate univariate statistics course ill prepares a student to read research literature or, for that matter, a researcher to produce it.

But how much harder are the multivariate techniques? Compared with the multivariate methods, univariate statistics seem so straightforward and neatly structured it is difficult to believe they once took so much effort to master. Yet most researchers can apply and validly interpret results of even the more intricate analysis of variance designs before the grand structure becomes apparent to them. While we would be delighted if some readers of this book gain insights into the full multivariate general linear model (Chapter 11 is an attempt to foster such insights), we will have accomplished our major goal if the reader feels comfortable attacking problems in multivariate statistics by computer and interpreting the resultant output.

Multivariate methods are more complex than univariate by at least an order of magnitude. But for the most part the greater complexity requires few conceptual leaps. The same old familiar concepts—such as sampling distributions and homogeneity of variance—simply become more elaborate.

The more complex multivariate models have not gained popularity by accident—or even by sinister design. Their growing popularity reflects the greater complexity of contemporary research in the behavioral and social sciences. In psychology, for example, we seem increasingly less enamored of the simple, clean, laboratory study where pliant, first-year college students each provide us with a single behavioral measure on cue.

1.1.1 The Domain of Multivariate Statistics: Numbers of IVs and DVs

Multivariate statistics represent a direct expansion of the more familiar univariate and bivariate statistics. Multivariate statistics can be seen as the "complete" or general case, with univariate and bivariate statistics as special cases or simplifications of the more general multivariate model. With many variables, multivariate techniques let you deal with all of your data in a single analysis, as an alternative to subjecting your data to a series of univariate or bivariate analyses.

Variables, it will be recalled, are measures taken on research units (people, companies, rats, stocks, airplanes, or whatever is considered the unit of analysis). Variables can roughly be dichotomized into two major types—independent and dependent. One type consists of input variables with respect to the research. In the context of your research, these are the variables that you consider to be prior. They are the IVs, which may consist of differing conditions to which you expose your research units (in which case they can be considered stimulus variables). Or, the IVs may reflect characteristics of your research units that the units themselves bring into the research situation. IVs can also be considered predictor variables within your research, because they predict the DVs—the other major type of variables. DVs can be considered the output variables in the context of your research. That is, they are the response, or outcome, variables or criteria. Note that an IV and a DV are defined within the research context. A DV in one research setting might be an IV in another.

Additional terms for variables are sometimes used synonymously with the independent-dependent dichotomy. Some of these terms are predictor-criterion, stimulus-response, task-performance, or simply input-output.

The term *univariate statistics* typically refers to analyses in which there is a single DV. There may, however, be more than one IV. For example, social behavior of undergraduates (the DV) could be studied as a function of geographical area (one IV) and type of training in social skills to which students are exposed (another IV). Analysis of variance is the prime example of univariate statistics (with t test as a special case).

The term *bivariate statistics* frequently refers to analyses of two DVs. For example, we might study the relationship between income (one DV) and the amount of education (another DV). The prototypical example of a bivariate statistic is the Pearson product-moment correlation coefficient. (Chapter 2 reviews univariate and bivariate statistics.)

With multivariate statistics, you can simultaneously analyze multiple dependent as well as multiple independent variables. This capability is important in both nonexperimental (correlational or survey) and experimental research.

1.1.2 Experimental and Nonexperimental Research

Recall that the critical distinction between experimental and nonexperimental research is whether or not the researcher has strict control over the IVs. Each IV in turn must be evaluated in judging whether or not a study is an experiment. Generally, if at least one IV is under experimental control, the research design as a whole is considered experimental.

If an IV is to be experimental, or manipulated, the researcher must have control over the levels (or conditions) of the IV to which a research unit is exposed. Further, the experimenter must ensure that the research units assigned to the various conditions differ *only* in exposure to the IV. That is, they could be expected to be alike (within random variation) on the DV were it not for differential exposure to the IV. When the manipulation has been carried out properly, the IV can be said to be experimental or manipulated (cf. Campbell & Stanley, 1966). If there are then systematic differences in the DV associated with the IV after exposure, these differences can be attributed to the IV. For example, if groups of undergraduates are exposed to different types of teaching techniques, and afterwards some groups of undergraduates perform better than others, the differences in performance can be said to be caused by the differences in teaching technique (with some stated degree of confidence). It is in this type of research that the terms *independent* and *dependent* have obvious meaning. The value of the DV *depends* on the manipulation of the IV. The IV is under the direct control of the experimenter; the DV is not.

In nonexperimental (correlational or survey) studies, neither the IVs nor the DVs[1] are under the control of the researcher. If an IV is *not* under experimenter control, it is considered nonexperimental or nonmanipulated. It is in this type of research that the distinction between IVs and DVs becomes more arbitrary. The distinction depends only on the research context and the goals of analysis. The researcher may assign labels to the categories of the IV in which the research units fall, but has no control over the assignment of the units to categories. For example, groups of people may be categorized in terms of geographic area (Northeast, Far West, etc.), but it is only the definition of the variable that is under researcher control. The choice of geographic area in which to live is hardly subject to manipulation by the researcher (except perhaps in unusual situations, such as the military). Nevertheless, naturally occurring differences such as those in geographic area may be considered an IV in a given research context if it is used, say, to predict some other nonexperimental (dependent) variable, such as income.

In this type of research, it is very difficult to unambiguously attribute

[1] In nonexperimental work, many researchers prefer the terms *predictor* to IV and *criterion* to DV.

causality to the IV. If there is a systematic difference in the DV associated with differences in the IV, the two variables can be said to be related (again with some degree of confidence). The cause of the relationship, however, remains unclear.

The use of the term *correlational* to describe this type of research is common but can be misleading. Though it is true that the correlation coefficient was traditionally developed and used in the context of nonexperimental research, its use is not limited to that type of research. Nor is it inappropriate to apply the so-called experimental statistics, such as analysis of variance, to data collected nonexperimentally. Statistics will work whether or not the researcher took control over the variables. One of the few considerations not relevant to choice of an optimal statistical technique is whether or not the data were collected experimentally. Only the interpretation, in terms of attributing causality, is crucially affected by the experimental-nonexperimental distinction.

1.1.2.1 Multivariate Statistics in Nonexperimental Research   Nonexperimental research can take many forms, some of which look very much like experiments, with only lack of appropriate controls to distinguish the two types of research. A more common example of nonexperimental research, however, is the survey. Typically, a large number of research units will be employed. For example, many people may be surveyed in a variety of geographic areas. Each research unit usually provides answers to many queries, producing data on a large number of variables.[2] Usually these variables will be interrelated in highly complex ways, and univariate or bivariate statistics are simply not sensitive to this complexity. A series of pairwise bivariate correlations, for example, may fail to reveal a fairly simple structure among variables. Thus it may be that the 20 or 25 variables tested really represent only two or three "supervariables," which by themselves can distinguish among research units almost as well as the original 20 or 25.

Another research goal might be to distinguish among subgroups in our survey sample on the basis of a variety of variables. We could, using univariate *t* tests or several analyses of variance, test group differences with respect to each variable separately. But if the variables are themselves interrelated, which is highly likely, the results of the multiple analyses of variance will be misleading as well as statistically suspect.

All tests of statistical significance are made with a margin of error, say 5%.[3] If several tests are done on the same data, the margin of error increases as the number of tests increases. If statistical tests are performed on 100 samples from the same population, for example, five of them would be expected to show statistical "significance" at the 95% level of confidence even if there is no true relationship between variables in the population being sampled. If the

[2] Having too many variables can cause a multitude of problems. Many procedures in this book are designed to help you choose a limited number of "good" variables and to eliminate the "bad" ones.

[3] Statistical inference is reviewed in Chapter 2.

variables being statistically tested are not only from the same sample, but also are themselves interrelated, the probability of error increases at an unknown rate.

With the use of multivariate statistical techniques, these complex interrelationships among variables not only can be revealed but also can be taken into account in statistical inference. It is possible to keep the overall error rate at some stated level, say 5%, no matter how many variables are being tested.

1.1.2.2 Multivariate Statistics in Experimental Research
For the most part, multivariate techniques were originally developed for use in correlational research. But once there are multiple DVs in experimental research, multivariate statistical techniques frequently become helpful as well. The problem of inflated error rate with multiple variables directly generalizes to experimental research. With multiple IVs, at least, the research can be designed so that the IVs are not correlated (see Chapter 2). With multiple DVs, however, it is highly unlikely that they will be uncorrelated.[4] Separate univariate analyses, then, are bound to lead to unacceptably high levels of error.

In the past, experimental research designs using multiple DVs were unusual. With attempts to make experimental designs more realistic, and with the availability of computer programs to analyze results of experiments, it is becoming increasingly rare to find an experiment with a single DV. When IVs are manipulated, the researcher wants to discover any differences in behavior that may result. It is simply wasteful to go through all the trouble of running an experiment only to miss the impact of the manipulation of the IV because the one measured DV wasn't the best possible one. By allowing measurement of a variety of behaviors within a single experiment without violating acceptable levels of error, multivariate statistics can help the experimenter design more efficient as well as more realistic experiments.

1.1.3 Computers and Multivariate Statistics

One answer to the question "Why multivariate statistics?" is that the techniques are now accessible to the nonstatistician. Only the most dedicated number cruncher would consider doing real-life problems in multivariate statistics without computers. Fortunately, there are excellent multivariate programs available in a number of computer packages. In this book, two packages have been chosen as representative. Discussion and examples throughout the book are based on programs in these packages. The packages are SPSS (Statistical Package for the Social Sciences) (Nie et al., 1975) and BMDP Statistical Software (Dixon, 1981).[5] The older BMD programs (Dixon, 1974) that contain information unavailable in the newer BMDP series are also discussed where appropriate. Each package contains a variety of specialized programs for univariate as well as

---

[4] Multiple uncorrelated DVs, however, can be *created* by the use of one type of multivariate technique—structural analysis (Chapter 10).

[5] Earlier editions (1977, 1979) were titled BMDP Biomedical Computer Programs—P Series.

multivariate statistics. Not only are the manuals excellent guides to both these packages, but additional documentation appears in various newsletters, pocket guides, technical reports, and other materials that are printed as the programs are continuously corrected, revised, and updated. The user of these programs should work closely with the local computing facility to stay abreast of changes in the programs as they occur.

SPSS and BMDP were selected primarily because of their popularity and ready availability at most large and medium-sized computer facilities with which a researcher is likely to be associated. If you are affiliated with a university or with a research organization that has access to a computer network, it is highly probable that at least one of the two packages will be available to you.

In the discussions and examples in Chapters 5 through 10 (the chapters that cover the specialized multivariate techniques in turn) an attempt is made to illustrate and explain output from a variety of programs within the two packages.[6] Our hope is that once the techniques are understood, it will be possible for you to generalize to virtually any multivariate program, so that you should not be limited to programs within the SPSS and BMDP packages.

If you have access to both programs, you are indeed fortunate. The programs within the packages are not completely overlapping, and some specialized analyses are better handled through one package than another. For example, if you want to do several versions of the same basic analysis on the same set of data, this is particularly easy to do with the SPSS package. That package is also especially well suited to management of data files. On the other hand, BMDP is well designed to do preliminary analysis of data and to test assumptions. Both packages have default options that can be relied on unless there is some reason to choose another option.

The packages also differ in ease of use. In the SPSS package, all programs use the same basic setup language. And once you learn the language, you can easily specify the description of the data set, and need refer to the manual only to specify the statistical analysis desired. Further, the manual itself (Nie et al., 1975) is highly informative. The statistical techniques associated with the various programs are discussed verbally. Many examples are given for setup of problems using several different forms of each type of analysis. Portions of computer output are also illustrated for each of the programs. (Discussion of the output, however, tends to be minimal.) There is an index that is helpful in locating information within the manual. The manual itself does not provide the statistical algorithms (equations) used in any detail. If a technique has alternative algorithms, then, it is often impossible to tell from the manual itself which one is in use. However, SPSS has recently provided a supplementary manual of algorithms (Norušis, 1978). SPSS output tends to be beautifully formatted and easy to follow. Everything is nicely labeled and can be located without much difficulty.

With respect to multivariate analysis, the SPSS package can be considered

[6] Each chapter on technique also includes a small hypothetical data set for which some hand calculations are illustrated. When using this book for reference, you can skip those sections.

in two parts. First there is the set of original programs included in the major manual (Nie et al., 1975). These programs were primarily designed for social sciences research and were deficient in programs for analysis of experimental as compared with survey research. For example, there was no provision for multivariate analysis of variance or for repeated-measures analysis of variance. This lack has been remedied in a newer program, SPSS MANOVA, included in the SPSS 7–9 update manual (Hull & Nie, 1981). Also included in the Release 9.0 update is an expanded program for multiple regression (SPSS NEW REGRESSION). The update manual is highly detailed and gives many examples. Interpretation of output is still deficient, however, and statistical algorithms are not provided in detail.

The BMDP package was developed by the originators of the BMD programs, but is far easier to set up. Some conventions for setup are standard over all of the programs, and verbal rather than numeric codes are used. The manual is a vast improvement over that for the BMD package, and in some ways surpasses that of SPSS. Many examples are given, both of setup and of output, and the output is annotated to facilitate interpretation. Like that of SPSS, the output tends to be well formatted and labeled. Handy explanatory notes are printed out for nonstandard statistics. For those who become familiar with the package and no longer need detailed guidelines for setup, each chapter concludes with a checklist containing critical information for setup in abbreviated form. BMDP programs tend to be highly sophisticated statistically but present the information at the level that most researchers can deal with. The package is also more comprehensive than SPSS.

In each of the chapters on technique (5 through 10), programs from SPSS, BMDP, and, occasionally, BMD are compared.

1.1.3.1 Program Updates In using commercial computer packages, it is essential that you know which version of the package is available to you. Programs in the packages are continually being changed, and not all changes are immediately implemented at each facility. This means that many versions of the various programs are in use simultaneously at different institutions. And even at one institution, more than one version of a package may be available.

Usually program updates are trivial; they simply correct errors that were discovered in earlier versions. Occasionally, though, major revisions in one or more programs are reflected in a new version, or new programs may be added to the package. Be sure to check with your computer facility to find out which version of each package is being used. Then be sure that the manual you are using is consistent with the package in use at your facility. Also check manual revisions and updates. Except where noted otherwise, information in this book is consistent with SPSS Release 9.0 and the 1981 update of BMDP.

1.1.3.2 Garbage In, Roses Out? The trick in multivariate statistics is not in the computation. That is easily done by computer. What is required of you as a researcher is that you choose the appropriate programs, use them correctly, and know how to interpret the output. Output from a commercial computer program, with its beautifully formatted tables, graphs, and matrices, can sometimes make garbage look like roses. Throughout this book we try to suggest clues that will indicate when the true message in the output more closely resembles fertilizer than flowers.

When you are dealing with multivariate statistics on large samples, you rarely get as close to the raw data as you do when applying univariate statistics to a relatively few cases. But the computer packages have programs designed to describe your data in the simplest univariate terms and to display graphically simple pairwise relationships among your variables. These programs provide preliminary analyses that are absolutely necessary if the later application of multivariate programs is to provide interpretable results.

1.1.4 Why Not?

There is, as usual, "no free lunch." The researcher will find that there are certain costs associated with the benefits of using multivariate procedures. Benefits from increased flexibility in research design, for instance, are sometimes negated by increased ambiguity in interpretation of results. In addition, multivariate results are quite sensitive to the analytic strategy chosen by the researcher (cf. Section 1.2.5) and do not always provide better protection against statistical errors than their univariate counterparts (cf. Chapter 8, Section 8.2.6). Add to this the fact that occasionally you *still* can't get a firm statistical answer to your research question, and you may wonder if the increase in complexity and difficulty is warranted.

Frankly, we think it is. Slippery as some of the concepts and procedures are, we think they provide alternative insights into relationships among variables that may more closely model social "reality." We find we can sometimes get at least partial answers to questions that couldn't be asked at all in the univariate framework. For a complete analysis, making sense out of your data usually requires a judicious mix of multivariate and univariate statistics.

And we also think the addition of multivariate statistics to your repertoire makes data analysis a lot more fun. If you liked univariate statistics, you'll love multivariate statistics!

1.2 SOME USEFUL DEFINITIONS

In order to describe multivariate statistics in a way that can be directly applied to real-world research, it is useful to review some common terms in research design and basic statistics. In the preceding section, distinctions were made between independent and dependent variables, and between experimental and nonexperimental research. A few additional terms that will be encountered in this book are described in this section.

1.2.1 Continuous, Discrete, and Dichotomous Data[7]

In applying multivariate techniques, it is important to consider the nature of the relationship between the numbers that comprise the data set and the properties or behaviors that they represent. For our purposes, the most useful distinction is between continuous, discrete, and dichotomous variables.

Continuous variables are those that are measured on some scale that changes values smoothly rather than in steps. Continuous variables can take on any value within the range of the scale. Precision is limited only by the measuring instrument, not by the nature of the scale itself. Some examples of continuous variables are time as measured on an old-fashioned analog clock face, annual income, age, temperature, distance, and scores on the Graduate Record Exam. Continuous variables, particularly when normally distributed, are appropriately analyzed by multivariate techniques.

Discrete variables are those that can take on a finite number of values (usually a fairly small number) and there is no smooth transition from one value or category to the next. Examples include time as measured by a digital clock, geographical area (e.g., categories based on continent), categories of religious affiliation, and type of community (rural or urban). Discrete variables may be used in multivariate analyses if there are a large number of categories and the categories represent some attribute that is changing in a quantitative way. For instance, a variable that represents numerous age categories by letting, say, 1 stand for 0–4 years, 2 stand for 5–9 years, 3 stand for 10–14 years, and so on up through the normal age span, would be useful because there are a lot of categories and the numbers designate a quantitative attribute (increasing age). But if the same numbers are used to designate categories of religious affiliation, they are probably not in appropriate form for analysis, because religions ordinarily do not fall along a quantitative continuum.

In the latter case, the discrete variable can still be analyzed by changing it into a number of dichotomous or two-level variables (e.g., Catholic vs. non-Catholic, Protestant vs. non-Protestant, Jewish vs. non-Jewish). Recategorization of a discrete variable into a series of dichotomous ones is called dummy variable coding, a process that is discussed and illustrated in Chapters 4 and 7. In general, dichotomous variables are analyzed with much more flexibility than those with more than two discrete levels.

It should be apparent that the distinction between continuous and discrete variables may be impossible to make in the limit. If you add enough digits to the digital clock, it becomes for all practical purposes a continuous measurement, while time as measured by the analog device can also be read in discrete categories, say, hours. Continuous measurements may be rendered discrete simply by specifying cutoffs on the continuous scale.

The property of variables that is crucial to application of multivariate

---

[7] Alternative labels (e.g., *interval* and *nominal* to replace *continuous* and *discrete*) might have been used. The terms used here were chosen as most consistent with common usage.

procedures is not kind of measurement, but rather normality, as discussed in Chapter 4 and elsewhere. Variables with badly skewed distributions, be they continuous, discrete, or dichotomous, present problems to multivariate analyses. Thus efforts to obtain symmetrically distributed variables at the outset of a research effort will be rewarded at the analytic stage.

In a frequent misapplication (cf. Gaito, 1980) of Stevens' (1946) classification of scales to statistical theory rather than measurement theory, a rank order or ordinal scale has been defined. This scale assigns a different number to each research unit to indicate its relative position along some quantitative dimension. Ranks assigned to contestants (first place, second place, third place, etc.) provide an example of an ordinal scale where one learns who was best, but not by how much. The problem with ordinal measures is that their distributions are rectangular (one frequency per number) instead of normal, unless tied ranks are permitted that pile up in the middle of the distribution. However, as with discrete scales, we see no problem with analyzing these data using multivariate techniques so long as there are a reasonably large number of ranks assigned, say 20 or more for each subsample to be analyzed.

1.2.2 Samples and Populations

Samples are measured in order to make generalizations about populations. Therefore we must be sure that our samples come from the populations of interest. Further, we must be sure that our samples are truly representative of the populations we want to generalize about. This requires that samples be chosen without bias from the population—or randomly. However, populations are frequently best defined in terms of samples, rather than vice versa. "The" population is that group from which you have randomly sampled.

Sampling has somewhat different connotations in experimental versus survey (or other nonexperimental) research. In nonexperimental research, you are searching for relationships among variables in some predefined population. Typically you will take elaborate precautions to ensure that you are achieving a representative sample of that population; you will first define your population, then do your best to randomly sample from it.[8]

In experimental research, the population to which you are generalizing is one that has been subjected to your experimental manipulation. You are not so much interested in populations per se as in differences between populations. Have you "created" different populations by treating them differently? Or, after treatment, can all of your research units still be described as coming from the same population, despite differences in treatment? The sampling objective here is to assure that all of your research units come from the same population *before* you treat them differently. Your total sample can be conceptualized as "the population." (In practice, of course, differences that are found are generalized to the larger populations from which the groups were sampled.) Random

[8] Strategies for random sampling in survey research are discussed in detail in such texts as Kish (1965) and Moser and Kalton (1972).

sampling consists of randomly assigning research units from that population to treatment groups, or levels of the IV. This ensures that before differential treatment all samples come from the same population. Statistical tests provide evidence as to whether, after treatment, all samples still come from the same population or whether they now represent different populations.

1.2.3 Descriptive and Inferential Statistics

Inferential statistics are designed to test hypotheses about populations by measuring samples of research units. Descriptive statistics are designed to describe samples of research units in terms of variables or combinations of variables. Depending on the outcome of inferential statistics, descriptive statistics may also be used to provide a best-guess description of populations. When descriptive statistics are used in this way, they are typically called parameter estimates rather than sample statistics.

Descriptive statistics can also be used to describe relationships among variables. In addition to means and standard deviations, then, descriptive statistics include correlations. In inferential statistics, we test our ability to generalize. Given statistical "significance," we assure ourselves that the sample characteristics can be generalized to the population of interest with a reasonable degree of confidence.

Inferential versus descriptive statistics is rarely an either-or proposition. With any set of data and array of research questions, we are usually interested in both the descriptive and the inferential aspects of the research. We want to describe our data, make generalizations from them, and estimate population values. It should be noted, however, that restrictions on generalization (inference) are far stronger than on description. Many of the assumptions required for multivariate statistics really are necessary only for inference. If simple description of the sample is the major goal, some of these assumptions (such as normality and homogeneity of variance) may be relaxed.

1.2.4 Orthogonality

Orthogonality can perhaps best be described as perfect *non*association. Two variables are orthogonal if their vectors are at right angles to each other so that the correlation between them is zero. Knowing the value of one variable gives no clue as to the value of the other.

Orthogonality is frequently desirable in statistical applications. It can be recalled from univariate statistics that when factorial designs (designs with more than one IV) are orthogonal (fully crossed with equal sample sizes in each combination of levels of each IV), hypotheses about main effects and interactions are independent of each other. Except for some relationships among statistical tests created by use of a common error term in the denominator of the F ratio, the outcome of each test gives no hint as to the outcome of the others. In experimental designs that are orthogonal, causality can unambiguously be attributed to various main effects and interactions.

Similarly, in multivariate analyses, sets of IVs or DVs may be orthogonal. That is, a set of independent or predictor variables may be orthogonal so that each adds, in a simple fashion, to prediction of the criteria or dependent variables. For example, among children, 80% of the variance in reading and spatial ability may be predicted from sex and age, with 20% predicted from sex and 60% predicted from age. Or a set of DVs may be orthogonal so that the overall effect of an IV can be partitioned into effects on each DV in an additive fashion.

1.2.5 Standard and Hierarchical Analyses

More usually, however, in multivariate situations the variables are correlated with each other (nonorthogonal). IVs become correlated, for instance, when unequal numbers of research units are measured in different cells. When dealing with human behavior, for instance, DVs are usually correlated by individual differences, alertness, or motivation if nothing else. When variables are correlated, they have shared or overlapping variance. A major decision for the multivariate analyst is determining how to handle this overlapping variance. Many of the techniques permit at least two strategies for handling it, some more.

In standard analysis, the overlapping variance is simply disregarded in assessing the contribution of each variable to the solution. Because it cannot be unambiguously assigned to either (or any) variable, the overlapping variance may contribute to the size of some summary statistic (such as R^2) but not to the variance associated with a particular variable. Figure 1.1(a) depicts a standard analysis as a Venn diagram, where the circles represent variance associated with Y, X_1, and X_2, overlapping variance is shown as overlapping areas in the circles, and the unique contributions of X_1 and X_2 to prediction of Y are shown as shaded areas. Note that total relationship between Y and the combination of X_1 and X_2 is the area bounded by the heavy line. For example, occupational prestige and sex are not orthogonal. Therefore there is overlapping variance between occupational prestige (X_1) and sex (X_2) in their relationship with income (Y).

Hierarchical analyses differ in that, after the researcher has assigned priority for entry of variables into equations, the first one to enter is assigned both unique variance and any overlapping variance it may have with other variables. Subsequent variables also take with them on entry both unique and any remaining overlapping variance. Figure 1.1(b) shows a hierarchical analysis for the same Y, X_1, and X_2 in which X_1 has been given priority over X_2. Note that the total variance explained by the relationship is the same as in Figure 1.1(a) but that the relative contributions of X_1 and X_2 have changed. For example, occupational prestige shows a stronger relationship with income than in the standard analysis, while the relation between sex and income remains the same.

The choice in strategy for dealing with overlapping variance is not trivial. The solution may look very different depending on whether a standard or a hierarchical strategy is employed. If the multivariate procedures sometimes have "a bad name," it is probably because of their reputation for unreliability—solutions change, sometimes dramatically, depending on which strategy is cho-

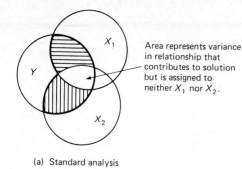
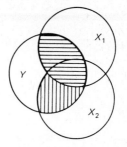

(a) Standard analysis

(b) Hierarchical analysis in which X_1 is given priority over X_2

Figure 1.1 Standard (a) and hierarchical (b) analyses of the relationship between Y, X_1, and X_2. Area indicated by darkened line is variance in Y associated with X_1 and X_2 in both analyses. Horizontal shading depicts variance assigned to X_1. Vertical shading depicts variance assigned to X_2.

sen. However, the strategies also ask different questions of the data and it is incumbent on the researcher to determine exactly which question is to be asked. We try to make the choices clear in the chapters that follow.

1.3 KINDS OF RESEARCH QUESTIONS

All statistical analyses consist of two major components: assessment of whether or not variables are related (inference); and, if so, description of the nature of the relationship. However, computer programs available for multivariate statistics, as for univariate statistics, tend to stress one component or the other. Further, programs tend to vary depending on the nature of the variables—particularly whether the variables are continuous or represent discrete groups.

One way to categorize the specialized multivariate programs is on the basis of the major research question to be asked. In Chapter 3, a flow diagram is developed so that the appropriate analysis can be chosen on the basis of a few decisions, the first of which is the major research question. A brief overview of these types of questions is given in the remainder of this section.

1.3.1 Measuring Degree of Relationship Among Variables

In the familiar bivariate correlation and regression analysis, the emphasis is on the strength of the relationship between the two variables. To what degree can one variable be predicted from the other? The greater the relationship, of course, the greater the predictability. In the multivariate generalization, the amount of relationship is assessed among multiple independent and dependent variables.

The simplest generalization is to the case in which there are multiple IVs but only a single DV. Here the major goal is to assess the degree to which the DV can be predicted from the set of IVs. This question and others are answered by multiple regression and multiple correlation analysis. For example, one might wonder how well graduate school grade point average (GPA) can be predicted by undergraduate grade point average, score on the Graduate Record Examination (GRE), ratings of letters of recommendation, and so on. For some statisticians, multiple regression techniques are not considered to be multivariate because they include only one DV. In this book, however, the usual tradition is followed in including the technique among the multivariate strategies.

In the full multivariate generalization, we can ask how well a *set* of DVs is related to a *set* of IVs. For example, we might have a number of measures of graduate school performance—grades, length of time to complete the program, comprehensive exams, ratings by major adviser, and so on. We could then ask how well this set of measures is predicted by undergraduate GPA, GRE, letters of recommendation, and the rest, using canonical correlation.

All computer analyses that assess degree of relationship also evaluate statistical significance; that is, output will include tests of whether or not the relationship is reliable, as well as descriptions of the amount of relationship. In addition, prediction afforded by the various IVs can be ascertained.

1.3.2 Measuring the Significance of Group Differences

If the IV(s) consist of groups or discrete levels, we can ask whether these groups differ on a set of DVs. Here, the emphasis is on inferring population differences among groups by testing sample differences. The IV(s) (groups) may be experimentally manipulated or not. For example, we might assign children to a number of different types of classroom structure (the IV) and assess the outcome in terms of a variety of DVs: a child's reading skill, behavior at school, behavior at home, expressed satisfaction with school, and so on. This would be a problem in multivariate analysis of variance that conceptually differs from the familiar univariate analysis of variance only in that there is more than one DV. As with ANOVA, there may be more than one IV, variables may be measured within or between research units, IVs may be hierarchically arranged, and so forth. (See Chapter 2 for a review of univariate analysis of variance.)

Included in this book are techniques for analysis of covariance, which is an analysis of variance that includes one or more extraneous or control variables in addition to the IV(s) and the single DV. Again, since there is only one DV, this is not strictly a multivariate technique but is treated here as such. In the example of type of classroom as the IV, and reading skill as the DV, a logical control variable might be the child's reading skill before the experimental manipulation (assignment to classroom) is made. After controlling for prior individual differences in reading skill, we could then ask whether assignment to classroom affects reading skill.

Finally, of course, there can be one or more IVs, one or more control variables, and one or more DVs. This is the problem handled by multivariate analysis of covariance.

Here again, although the major emphasis is on inferences regarding group differences, description of the nature and strength of the relationships is just as important in multivariate as in univariate applications of analysis of variance.

1.3.3 Predicting Group Membership

Sometimes, when there are many measures on research units that form natural groups (such as gender), the major emphasis is on the degree to which group membership can be predicted from the various measures. The groups act as discrete levels of the DV (the variable to be predicted) and the measures act as the predictors (the IVs). For example, we might want to know how well we can predict group membership in terms of medical diagnosis from various bodily measures (blood pressure, heart rate, temperature, etc.). This type of research question is handled by discriminant function analysis. Conceptually, the analysis differs from multivariate analysis of variance only in emphasis. If groups differ significantly on a set of measures, those measures can be turned around and used to predict group membership.

Although the emphasis in discriminant function analysis is on degree of prediction of group membership, analysis also includes tests of statistical significance (whether or not group membership can be reliably predicted) and evaluation of various predictor variables.

1.3.4 Determining Structure

A somewhat different perspective is required in asking questions about structure, for here there is no univariate or bivariate analog. With a single set of measures (which can be considered either IVs or DVs), the analysis attempts to uncover an underlying simple structure inherent in the set. Are the variables interrelated enough so that all the information in them can be summarized by a few newly created variables? This type of question is answered by principal components analysis or factor analysis. For example, given a host of measures on children with learning disabilities (e.g., several measures of achievement, ability, personality, behavior), can distinctions among these children be made on the basis of just a couple of dimensions (e.g., the degree to which they show language-communication difficulties and the degree to which they show perceptual-motor handicaps)?

1.4 COMBINING VARIABLES

Multivariate relationships are examined by studying the way in which variables are combined to do some useful work. The work that is done depends on the goal of analysis (e.g., to predict a single score from many others—regression; to predict group membership from many variables—discriminant function analysis; to group variables according to their interrelationship—factor analysis). The combination that is formed depends on the relationships between the variables, both IVs and DV(s). But in all cases the combination that is formed is a *linear* combination.[9]

A linear combination is one in which variables are assigned weights, and then products of weights and variable scores are added together. Equation 1.1 shows a linear combination between Y, the DV, and X_1 and X_2, considered IVs.

$$Y = W_1X_1 + W_2X_2 \tag{1.1}$$

For example, Y might be reading ability, X_1 age, and X_2 sex. The equation states that the highest relationship between reading ability and age and sex can be obtained by weighting age (X_1) by W_1, and sex (X_2) by W_2, and then adding them together. No other values of W_1 and W_2 would produce as strong a relationship between DV and IVs.

Notice that Eq. 1.1 includes neither X_1 or X_2 raised to powers (exponents) nor a product of X_1 and X_2. This would seem to place severe restrictions on the nature of multivariate solutions until one realizes that X_1 could itself be a product of two different variables or a single variable raised to a power. For example, X_1 might be age squared. A multivariate solution will not produce exponents or cross products of IVs to improve a solution, but a researcher can include them by creating X's that are cross products of IVs or are IVs raised to powers.

Inspection of the size of the W values (or some function of them) may reveal a great deal about the relationship between DV and IVs. If, for instance, the W value for some IV is zero, it means that the particular IV should be multiplied by zero to form the combination, or that the IV should be excluded for the best DV-IV relationship. Alternatively, if some IV has a large W value, then the IV is important to the relationship. Although some complications (to be explained later) prevent the whole interpretation of the relationship from being made on the basis of the sizes of the W values, they are nonetheless important in most multivariate procedures.

If one is lucky, the way in which the IVs as a group are best combined for some DV may be interpretable. The IVs, in particular combination, may represent a kind of supervariable, not directly measured but implicit in the combination that represents the best DV-IVs relationship. The supervariable may be thought of as an underlying dimension that performs the work desired from the multivariate analysis. Therefore attempts to understand the meaning of the IVs in combination are also critical to most multivariate procedures.

In the search for the best weights to apply in combining variables, computers do not try out all possible sets of weights. Various algorithms have been developed in order to replace a complete trial-and-error strategy. All algorithms involve manipulation of some form of a correlation matrix, a variance-covariance matrix, or a sum-of-squares and cross-products matrix. Section 1.6 describes

[9] Nonlinear multivariate statistics are also being developed but are not discussed in this book.

these matrices in very simple terms and shows their development from an appropriate data set. Appendix A describes some terms and manipulations appropriate to matrices. In Chapters 5 through 10, a small hypothetical sample of data is analyzed by hand to show how the combining weights can be derived for each analysis. Though this information is useful for a basic understanding of the multivariate statistics, it is *not* necessary for applying multivariate techniques fruitfully to your research questions.

1.5 NUMBER AND NATURE OF VARIABLES TO INCLUDE

Attention to the *number* of variables included in analysis is important. A general rule is to get the best solution with the fewest variables possible. As more and more variables are included, the solution usually improves, but only slightly. Sometimes the improvement does not compensate for the cost, in degrees of freedom, of including more variables, so that the overall power of the test diminishes. At an extreme, one might include so many variables relative to sample size that the solution provides an artificially good fit to the sample, but doesn't generalize to the population, a condition termed *overfitting*. To avoid overfitting, include only a limited number of variables in each analysis.

There are several important considerations regarding the *nature* of the variables to include in a multivariate analysis—cost, availability, and meaning among them. Depending on the analysis, one may want either a large number of related variables or a small number of unrelated ones. Further considerations for variable selection for each analysis are mentioned as they become appropriate.

But one of the most important considerations in variable selection for all analyses is reliability. How stable is the position of a given score in a distribution of scores when measured at different times or in different ways? Unreliable variables degrade an analysis while reliable ones enhance it. A few reliable variables will give a more meaningful solution than a large number of unreliable variables. Indeed, if variables are sufficiently unreliable, the entire solution may reflect only measurement error.

1.6 DATA APPROPRIATE FOR MULTIVARIATE STATISTICS

For multivariate statistics, an appropriate data set consists of values on a number of variables for some number of research units. For continuous variables, the values will be scores on those variables. For example, a continuous variable might be a score on the GRE, with values for the various research units (students) on the order of 500, 650, 420, and so on. For discrete variables, values are number codes representing group membership or treatment. For example, if there were three teaching techniques being evaluated, students assigned to one technique would be coded "1" on the teaching technique variable, those assigned to another technique would be coded "2," and so on.

TABLE 1.1   A DATA MATRIX OF HYPOTHETICAL
               SCORES

Student	X_1	X_2	X_3	X_4
1	1	500	3.20	1
2	1	420	2.50	2
3	2	650	3.90	1
4	2	550	3.50	2
5	3	480	3.30	1
6	3	600	3.25	2

1.6.1 The Data Matrix

The data matrix is an organization of values in which rows represent research
units, and columns represent variables. An example of a data matrix in which
there are six research units and four variables appears in Table 1.1. (Normally,
of course, there would be many more than six units.) For example, variable 1
(X_1) might be type of teaching technique, X_2 might be score on the GRE, X_3
might be GPA, and X_4 might be gender, with female coded as 1 and male
coded as 2.

In order to apply computer techniques to these data, they must either
be punched on cards or otherwise entered into some type of long-term storage
accessible by computer (e.g., internal computer files, paper tape, or some type
of external disk). Each experimental unit starts a new card or line of data.
Information identifying the unit is typically entered first, followed by the value
on each variable for that unit. On each card or line, values for a given variable
appear in the same place. Any computer package manual provides information
on setting up a data matrix.

Note that in this example, there are values for every variable for each
research unit. This is not always the case in research in the real world. With
large numbers of research units and variables, there are frequently data missing
in the matrix. For instance, some respondents may refuse to answer questions
about ethnicity, some students may be absent the day that one of the tests is
given. This will create blanks in the data matrix that must be dealt with. Chapter
4 covers this messy (but often unavoidable) problem.

1.6.2 The Correlation Matrix

Most readers are familiar with a correlation matrix, or **R**. The **R** matrix is a
square, symmetrical matrix. Each row (and each column) represents a different
variable, while the value at the intersection of each row and column is the
correlation between the variables in question. For instance, the value in the
second row, third column, is the correlation between the second variable and
the third variable. The same correlation, of course, also appears at the intersection
of the third row and the second column. Thus correlation matrices are said

to be symmetrical about the main diagonal, which means they are mirror images of themselves above and below the diagonal going from top left to bottom right. For this reason it is common practice to show only the bottom half or the top half of an **R** matrix. The entries in the main diagonal are often omitted as well, since they are all ones—correlations of variables with themselves.

Table 1.2 shows the correlation matrix for the three continuous variables, X_2, X_3, and X_4 of Table 1.1. The .85 is the correlation between X_2 and X_3 that appears twice in the matrix (as do the others). Other correlations are as indicated in the table.

TABLE 1.2 CORRELATION MATRIX FOR PART OF HYPOTHETICAL DATA FOR TABLE 1.1

		X_2	X_3	X_4
$\mathbf{R} =$	X_2	1.00	.85	−.13
	X_3	.85	1.00	−.46
	X_4	−.13	−.46	1.00

Many programs allow the researcher a choice between analysis of a correlation matrix and analysis of a variance-covariance matrix. If the correlation matrix is analyzed, a unit-free result is produced. That is, the solution reflects the relationships among the variables but not in the metric in which they were originally measured. If the metric of the scores was somewhat arbitrary to begin with, analysis of **R** is appropriate.

1.6.3 The Variance-Covariance Matrix

If, on the other hand, the scores were measured along a meaningful scale, it may be more appropriate to analyze a variance-covariance matrix. A variance-covariance matrix, $\mathbf{\Sigma}$, is also square and symmetrical, but the elements in the main diagonal are the variances of each variable, while the off-diagonal elements are covariances between different variables.

Variances, as you recall, are averaged squared deviations of each score from the mean of the scores. Because the deviations are averaged, the *number* of scores composing the variance is no longer relevant, but the metric in which the scores were measured is. Scores measured in hundreds tend to have largish numbers as variances; scores measured in units tend to have smaller variances.

Covariances are averaged cross products (the deviation between one variable and its mean times the deviation between a second variable and its mean). Covariances are similar to correlations except that they, like variances, retain information concerning the scales in which the variables were originally measured. The variance-covariance matrix for the continuous data of Table 1.1 appears in Table 1.3.

TABLE 1.3 VARIANCE-COVARIANCE MATRIX FOR PART OF HYPOTHETICAL DATA OF TABLE 1.1

	X_2	X_3	X_4
X_2	7026.66	32.80	−6.00
$\Sigma = X_3$	32.80	.21	−.12
X_4	−6.00	−.12	.30

1.6.4 The Sum-of-Squares and Cross-Products Matrix

This matrix, **S**, is a precursor to the variance-covariance matrix in which the deviations have not yet been averaged. Thus their sizes depend on the number of cases of which the elements are composed as well as on the metric in which they were measured (and also the sizes of the deviations, the amount of relationship among variables, etc.). The sum-of-squares and cross-products matrix for the continuous data in Table 1.1 appear in Table 1.4.

TABLE 1.4 SUM-OF-SQUARES AND CROSS-PRODUCTS MATRIX FOR PART OF HYPOTHETICAL DATA OF TABLE 1.1

	X_2	X_3	X_4
X_2	35133.33	164.00	−30.00
$S = X_3$	164.00	1.05	−0.59
X_4	−30.00	−0.59	1.50

In the major diagonal of the **S** matrix, the entry is the sum of squared deviations of scores around the mean for that variable, hence "sum of squares." That is, for each variable, the value in the major diagonal is

$$\text{Sum of squares } (X_j) = \sum_{i=1}^{N} (X_{ij} - \overline{X}_j)^2 \tag{1.2}$$

where $i = 1, 2, \ldots, N$. N is the number of research units. $j = 1, 2, \ldots P$. P is the number of variables. X_{ij} is the score on variable j by research unit i. $\overline{X}_j$ is the mean of all scores on the jth variable.

For example, for the fourth variable, X_4, the mean is 1.5. The sum of squared deviations around the mean is

$$\sum_{i=1}^{6} (X_{i4} - \overline{X}_4)^2 = (1 - 1.5)^2 + (2 - 1.5)^2 + (1 - 1.5)^2$$
$$+ (2 - 1.5)^2 + (1 - 1.5)^2 + (2 - 1.5)^2$$
$$= 1.50$$

The off-diagonal elements of the sum-of-squares and cross-products matrix represent the cross products of the variables. For each pair of variables, represented by row and column labels in Table 1.4, the entry is composed of the sum of the product of the deviation of one variable around its mean times the deviation of the other variable around its mean.

$$\text{Cross-product } (X_j X_k) = \sum_{i=1}^{N} (X_{ij} - \overline{X}_j)(X_{ik} - \overline{X}_k) \qquad (1.3)$$

where $k = 1, 2, \ldots P,$ and all other terms are as defined in Eq. 1.1, and $j \neq k.$ (Note that if $j = k,$ Eq. 1.3 becomes identical to Eq. 1.2.)

For example, the cross-product term for variables X_2 and X_3 (which appears in the middle of the first row and the middle of the first column) is

$$
\begin{aligned}
\sum_{i=1}^{N} (X_{i2} - \overline{X}_2)(X_{i3} - \overline{X}_3) &= (500 - 533.33)(3.20 - 3.275) \\
&\quad + (420 - 533.33)(2.50 - 3.275) \\
&\quad + \cdot \ \cdot \ \cdot \ + (600 - 533.33)(3.25 - 3.275) \\
&= 164.00
\end{aligned}
$$

Most computations start with S and proceed to Σ or $R.$ The progression from a sum-of-squares and cross-products matrix to a variance-covariance matrix is simple.

$$\Sigma = \frac{1}{N-1} S \qquad (1.4)$$

The variance-covariance matrix is produced by dividing every element in the sum-of-squares and cross-products matrix by $N - 1,$ where N is the number of research units.

Similarly, a correlation matrix may be derived from an S matrix by dividing each sum of squares by itself (to give the 1s in the main diagonal of R) and each cross product of the S matrix by the square root of the product of the sum of squared deviations around the mean for each of the variables in the pair. That is, each cross product is divided by

$$\text{Denominator } (X_j X_k) = \sqrt{\sum_{i=1}^{N} (X_{ij} - \overline{X}_j)^2 \sum_{i=1}^{N} (X_{ik} - \overline{X}_k)^2} \qquad (1.5)$$

where terms are defined as in Eq. 1.3.

For some multivariate operations, it may not be necessary to feed the data matrix to a computer program. Instead, a Σ or an R matrix can be entered, with each row (representing a variable) starting a new punched card or line of computer entry. Considerable computing time and expense can be saved by entering one or the other of these matrices rather than raw data into most programs.

1.6.5 Residuals

Often one goal of analysis, or test of its efficiency, is to reproduce the values of a variable or of a correlation matrix. For example, we might want to predict scores on the GRE (X_2 of Table 1.1) from knowledge of GPA (X_3) and gender (X_4). After applying the proper statistical operations, in this case a multiple regression analysis, it would be possible to derive a predicted score for each student. By applying the proper weights to GPA and gender for a student, we can make a best guess as to what that student's GRE would be. Since we already know GRE values for our sample of students, we can compare our best guess with the actual obtained GRE score. The difference between these two values is known as the residual and is a measure of error of prediction.

It is the nature of multivariate procedures that the residuals for the entire sample will sum to zero. That is, sometimes the best guess will be too large and sometimes it will be too small, but the average of all the errors will be zero. The squared value of all the residuals, however, is important as a measure of how good the prediction is. The way that the residuals are distributed is of further interest in evaluating the degree to which the data meet the assumptions of multivariate statistics. For example, do residuals diverge more from zero with high or low scores on the variable to be predicted? Residuals and their analysis will be discussed in Chapter 4 and elsewhere in Chapters 5 through 10.

1.7 ORGANIZATION OF THE BOOK

Chapter 2 begins with a review of univariate and bivariate statistics, while Chapter 3 operates as a guide to the multivariate techniques that are covered in this book and places them in context with the more familiar univariate and bivariate statistics. Included in Chapter 3 is a flow chart that organizes statistical techniques on the basis of the major research questions to be asked, as briefly described in Section 1.3.

Chapter 4 deals with the assumptions and limitations of multivariate statistics. More importantly, unavoidable violations of assumptions are discussed, with techniques presented for tackling the kind of dirty data that is inevitable in research in the real world. Chapter 4 is meant to be referred to whenever any of the techniques are pursued, and the reader is frequently guided back to it in the technique chapters.

Chapters 5 through 10 are the chapters covering specific multivariate techniques. They include descriptive, conceptual sections as well as a guided tour through a real-world data set for which the analysis is appropriate. The tour includes an example of a Results section, appropriate for submission to a professional journal, describing the outcome of the statistical analysis. Each technique chapter includes a comparison of computer programs available in the three packages described in Section 1.1.3.

Chapter 11 is an attempt to integrate univariate, bivariate, and multivariate statistics through the multivariate general linear model. Here the common elements underlying all of the techniques are emphasized, rather than the distinctions between them. Chapter 11 is meant to pull together the material in the remainder of the book, with a conceptual rather than pragmatic emphasis. Some may wish to consider this material earlier, for instance, immediately after Chapter 3, instead of waiting until the end.

Chapter 2

Review of Univariate and Bivariate Statistics

2.1 HYPOTHESIS TESTING

The fundamental goal of statistics, univariate or multivariate, is to make rational decisions under conditions of uncertainty. We need to make inferences (decisions) about populations based on samples of data taken from those populations. Since samples contain less than complete information—samples from the same population differ from one another and from the population—decisions regarding the population cannot be made with certainty.

The traditional solution to this problem is the application of statistical decision theory. Two hypothetical states of reality are set up, each represented by a probability distribution. Each distribution represents one of two alternative hypotheses about the true nature of events. Given sample results, we make a best guess as to which distribution the sample was taken from using formalized statistical rules to define "best."

2.1.1 One-Sample z Test as Prototype

Statistical decision theory is most easily illustrated through a one-sample z test, using the standard normal distribution as our model for the two hypothetical states of reality. Suppose we have a sample of 25 IQ scores and would like to decide whether this sample of scores was likely to have come from a random sample of a "normal" population (i.e., with $\mu = 100$ and $\sigma = 15$) or a population in which the average score (μ) is 108 and $\sigma = 15$.

First, note that we are testing hypotheses about *means*, not individual scores. Therefore the distributions representing hypothetical states of reality will have to reflect populations of means rather than populations of individual scores. Populations of means produce "sampling distributions of means" that differ systematically from distributions of individual scores in that, while the mean of a population distribution, μ, is equal to the mean of a sampling distribution, $\mu_{\bar{Y}}$, the standard deviation of a population, σ, is *not* equal to the standard deviation of a sampling distribution, $\sigma_{\bar{Y}}$. Sampling distributions have smaller standard deviations than populations of scores, and the decrease is related to N, the sample size.

$$\sigma_{\bar{Y}} = \frac{\sigma}{\sqrt{N}} \qquad (2.1)$$

For our sample, then,

$$\sigma_{\bar{Y}} = \frac{15}{\sqrt{25}} = 3$$

The question we are really asking is, Did our mean, taken from a sample of size 25, come from a sampling distribution with $\mu_{\bar{Y}} = 100$ and $\sigma_{\bar{Y}} = 3$ or did it come from a sampling distribution with $\mu_{\bar{Y}} = 108$ and $\sigma_{\bar{Y}} = 3$? Figure 2.1(a) illustrates the first of these hypothetical states of reality, defined for our purposes as the null hypothesis, H_0, a distribution formed from a very large number of means calculated from all possible samples of size 25 taken from a population in which $\mu = 100$ and $\sigma = 15$.

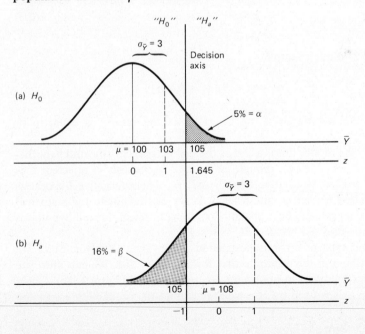

Figure 2.1 Sampling distributions for means with $N = 25$ and $\sigma = 15$ under two hypotheses: (a) H_0: $\mu = 100$ and (b) H_a: $\mu = 108$.

The null hypothesis has a special, fond place in statistical decision theory because it alone is used to define "best guess." The object is to position a decision axis between deciding in favor of H_0 or rejecting it, so that the probability of rejecting H_0 by mistake is small. "Small" is defined probabilistically as α, and an error in rejecting the null hypothesis is referred to as α, or Type I, error. We have little choice in picking α. Tradition and journal editors have decreed that it be .05 or .01, meaning that no more than 5% (or 1%) of the time would we reject the null hypothesis when it is true.

Using a table of areas under the standard normal distribution (the familiar table of standard normal deviates), we determine how many standard deviations from the mean of H_0 the decision axis must be so that the probability of obtaining a sample mean above that point is 5% or less. Looking up 5% in Table C.1, we find that the z (standard normal deviate) corresponding to the 5% cutoff is 1.645. Notice that the z scale is represented along one of two abscissas in Figure 2.1(a). If the decision axis is placed where $z = 1.645$, we must translate from the z scale to the $\overline{Y}$ scale to locate the decision axis along $\overline{Y}$. The transformation equation is

$$\overline{Y} = \mu + z\sigma_{\overline{Y}} \tag{2.2}$$

Equation 2.2 is a simple rearrangement of terms from the z test for a single sample:[1]

$$z = \frac{\overline{Y} - \mu}{\sigma_{\overline{Y}}} \tag{2.3}$$

Applying Eq. 2.2 to our example,

$$\overline{Y} = 100 + (1.645)(3) = 104.935$$

We reject the null hypothesis if the mean IQ of the sample is equal to or greater than 105.

Frequently in practice this is as far as the model is taken—the null hypothesis is rejected or not. Always, though, if the null hypothesis is rejected, it is rejected in favor of an alternative hypothesis, H_a. The alternative hypothesis is not always stated explicitly, but when it is we can complete the statistical decision model and evaluate the probability of failing to reject the null hypothesis when it *should* be rejected because H_a is true.

This second type of error is designated β, or Type II, error. Since we have explicitly stated in our example that for H_a, $\mu = 108$, we can illustrate the sampling distribution of means under H_a, as is done in Figure 2.1(b). Given that the decision axis is fixed by H_0, we need to find the probability associated

---

[1] Note that the more usual procedure in testing hypotheses with a single sample mean would be to solve for z on the basis of $\overline{Y}$ and then see whether the proportion above the obtained z were less than or equal to 5% (i.e., $z \geq 1.645$). If so, H_0 would be rejected.

with the place it crosses H_a. We first find z corresponding to an IQ score of 105 in a distribution with $\mu_{\bar{Y}} = 108$ and $\sigma_{\bar{Y}} = 3$. Applying Eq. 2.3, we find that

$$z = \frac{105 - 108}{3} = -1.00$$

Looking up $z = -1.00$ in an appropriate table, we find that about 16% of the time sample means would be equal to 105 or less even if the alternative hypothesis were true, i.e., $\mu = 108$. Therefore our $\beta = .16$.

Note that these distributions represent alternative realities, only one of which is true. If the null hypothesis is true, one of two things can occur: A wrong decision can be made by rejecting H_0 or a correct decision can be made by retaining it. If the probability of making the wrong decision is α, the probability of making the right decision is $1 - \alpha$. If, on the other hand, the alternative hypothesis is true, the probability of making the wrong decision is β, and the probability of making the right decision is $1 - \beta$. This information can be summarized in a "confusion matrix" (aptly named, according to beginning statistics students) showing the probability of each of these four outcomes:

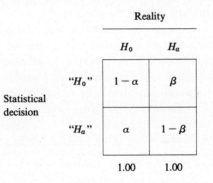

	Reality	
	H_0	H_a
"H_0"	$1 - \alpha$	β
"H_a"	α	$1 - \beta$
	1.00	1.00

(Statistical decision)

For our example, the probabilities are

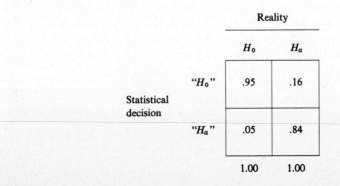

	Reality	
	H_0	H_a
"H_0"	.95	.16
"H_a"	.05	.84
	1.00	1.00

(Statistical decision)

2.1.2 Power

The lower right-hand cell of the confusion matrix has special meaning and represents the most desirable outcome. In using inferential statistics, it is the alternative hypothesis we are usually hoping to support with our research. We hope that it is true, and that our sample data will allow us to decide in favor of it. It has a special name, *power*. Notice that in Figure 2.1(b), power is represented by that portion of the H_a distribution that falls above the decision axis. Many of the choices in designing research are made with an eye toward increasing power.

Figure 2.1 and Eqs. 2.1 and 2.2 suggest some ways that power can be enhanced. One obvious way to increase power is to move the decision axis to the left. The problem with this strategy is that it may increase Type I error to an unacceptable level. Given the choice between .05 and .01 for α error, though, a decision in favor of .05 will increase power. A second strategy is to move the curves farther apart by applying a stronger treatment. Other strategies involve decreasing the variability in the sampling distributions either by decreasing variability (e.g., exerting greater experimental control) or by increasing sample size, N.

This model, and these strategies for increasing power, generalize both to other sampling distributions and to tests of hypotheses other than a single sample mean against hypothesized population means.

There is also the danger of *too much power* in hypothesis testing. The null hypothesis is probably never exactly true in nature. Any sample is likely to show some slight difference in μ from the population value. Given a large enough sample, rejection of H_0 becomes virtually certain. For that reason, the difference between μ_{H_a} and μ_{H_0} should be carefully considered in terms of a "minimal meaningful difference." Therefore with given levels of Type I and Type II error, the sample size chosen should be adequate but not excessive. Then rejection of H_0 is nontrivial. This issue is further considered in Section 2.4, "Strength of Association."

2.1.3 Extensions of the Model

As you may recall, the z test for differences between a sample mean and a population mean readily extends to an evaluation of the likelihood that the difference between two sample means is zero. A sampling distribution of the difference between means under the null hypothesis that $\mu_1 = \mu_2$ is generated and used for positioning the decision axis. Power of an alternative hypothesis is calculated with reference to the decision axis, just as before.

In the event that population variances are unknown, it is desirable to evaluate the probabilities by using Student's t rather than z, even for large samples. Numerical examples of use of t and of testing differences between two means are available in most univariate statistics books and will not be presented here. The logic of the process, however, is identical to that described in Section 2.1.1.

2.2 ANALYSIS OF VARIANCE

When two or more means are to be compared, we need to know whether there are any reliable differences among them. An illustration of distributions of scores from three samples is provided in Figure 2.2. Analysis of variance evaluates the differences among the mean scores on the DV (Y) for the samples relative to the overlap of the sample distributions. The null hypothesis is that $\mu = \mu_1 = \mu_2 = \cdots = \mu_k$ as estimated from $\overline{Y}_1, \overline{Y}_2, \ldots, \overline{Y}_k$, with k equal to the number of means being compared. A test of this hypothesis is provided by analysis of variance (ANOVA)—really a set of analytic procedures based on a common theme—in which two estimates of population variability are compared. One estimate is based on differences among scores within each group for which a sample mean is to be evaluated; it can be considered random or error variation. Another estimate comes from knowledge about the sampling distribution of means, plus the availability of k samples from that distribution (if the null hypothesis is true). If these two estimates of variability do not differ appreciably, it is reasonable to suppose that all of the sample means did indeed come from the same sampling distribution of means and that differences among them are due to random error. If, on the other hand, the estimate of population variability from sample means is much greater than the estimate of population variability from difference among scores within groups, it is unlikely that the k means derive from the same sampling distribution. The null hypothesis is rejected in favor of the alternative that sample means do *not* come from the same sampling distribution of means.

Differences among variances are evaluated as ratios, where the variance associated with difference among sample means is placed in the numerator, and the variance associated with differences within groups (error variance) is placed in the denominator. A ratio between two variances forms an F distribution. F distributions change shape (and variability) depending on degrees of freedom in both the numerator and the denominator of an F ratio. Tables of critical F, for testing the null hypothesis, depend on two degree-of-freedom parameters (cf. Appendix C, Table C.3).

The many varieties of analysis of variance can conveniently be summarized in terms of the partition of *sums of squares,* that is, sums of squared differences

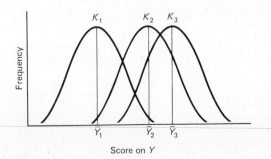

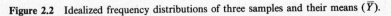

Figure 2.2   Idealized frequency distributions of three samples and their means ($\overline{Y}$).

between individual values and their means. A sum of squares (SS), it can be recalled, is simply the numerator of a variance, S^2, which in turn is the square of a standard deviation, S.

$$S^2 = \frac{\Sigma(Y - \overline{Y})^2}{N - 1} \qquad (2.4)$$

$$S = \sqrt{S^2} \qquad (2.5)$$

$$SS = \Sigma(Y - \overline{Y})^2 \qquad (2.6)$$

2.2.1 One-Way Between-Subjects ANOVA

Scores (representing measurement on the DV) appropriate to one-way between-subjects ANOVA with equal n can be represented in a two-dimensional table, with k columns representing groups (levels of the IV) and n scores within each group.[2] Table 2.1 shows how subjects are assigned to groups within this design.

Each column has an associated mean, $\overline{Y}_j$, with $j = 1, 2, \ldots k$. Each score within the table can be designated Y_{ij}, where $i = 1, 2, \ldots n$. We use the symbol GM to represent the grand mean of all scores over all groups.

It is possible to look at the difference between each score and the grand mean ($Y_{ij} - GM$) as the sum of two component differences:

$$Y_{ij} - GM = (Y_{ij} - \overline{Y}_j) + (\overline{Y}_j - GM) \qquad (2.7)$$

Algebraically, this is easily verifiable—we are simply subtracting out a component, $\overline{Y}_j$, and then adding it back in. We can then sum and square each term. The basic partition still holds because, conveniently, the cross-product terms

TABLE 2.1 ASSIGNMENT OF SUBJECTS IN A ONE-WAY BETWEEN-SUBJECTS ANOVA	Treatment		
	K_1	K_2	K_3
	S_1	S_4	S_7
	S_2	S_5	S_8
	S_3	S_6	S_9

[2] Throughout the book n will be used to represent size of a single combination of IV levels or groups; N will be reserved for total (or combined) sample size.

produced by squaring and summing cancel each other out. The partition then becomes

$$\sum_i \sum_j (Y_{ij} - GM)^2 = \sum_i \sum_j (Y_{ij} - \overline{Y}_j)^2 + n \sum_j (\overline{Y}_j - GM)^2 \qquad (2.8)$$

Each of these terms can be recognized as a sum of squares (SS)—that is, a sum of squared differences between a set of "scores" (sometimes means are treated as scores) and their associated "means." That is, each term is a special case of Eq. 2.6.

The term on the left of the equation is the total sum of squared differences between scores and the grand mean, ignoring groups with which scores are associated. This, then, can be designated SS_{total}. The first term on the right is the familiar sum of squares associated with a particular $\overline{Y}$ (or single group). If summed over all groups, it represents error, or the sum of squares associated with differences within groups, SS_{wg}. The final term of the equation is based on differences between each group mean and the grand mean. It can be designated SS_{bg}, the sum of squares associated with differences among groups. Equation 2.8, then, can be symbolized as

$$SS_{total} = SS_{wg} + SS_{bg} \qquad (2.9)$$

Degrees of freedom in ANOVA can be partitioned in much the same way as sums of squares:

$$df_{total} = df_{wg} + df_{bg} \qquad (2.10)$$

The total number of degrees of freedom in an ANOVA source table is the total number of scores in the table minus the number of means estimated from that set of scores. Since in this case only one mean, the grand mean, is being estimated, only 1 degree of freedom is used. Therefore

$$df_{total} = N - 1 \qquad (2.11)$$

For the within-groups term, there are still the same number of scores (N) but now k means are estimated, one for each of the k groups. Therefore k degrees of freedom are used:

$$df_{wg} = N - k \qquad (2.12)$$

For the between-groups term, there are only k "scores" (each group mean is now treated as a score) and a single estimate of the grand mean, so that

$$df_{bg} = k - 1 \qquad (2.13)$$

Verifying the equality proposed in Eq. 2.10, we get

$$N - 1 = N - k + k - 1$$

As in the partition of sums of squares, the term associated with group means is subtracted out of the equation and then added back in.

Another common notation for the partition in Eq. 2.7 is

$$SS_{total} = SS_K + SS_{S(K)} \tag{2.14}$$

as shown in Table 2.2(a). In this notation, the total sum of squares is partitioned into a sum of squares due to the k groups, SS_K, and a sum of squares due to subjects within the groups, $SS_{S(K)}$. Notice that the order of terms on the right side of the equation has been reversed from that of Eq. 2.9.

TABLE 2.2   PARTITION OF SUMS OF SQUARES AND DEGREES
OF FREEDOM FOR SEVERAL ANOVA DESIGNS

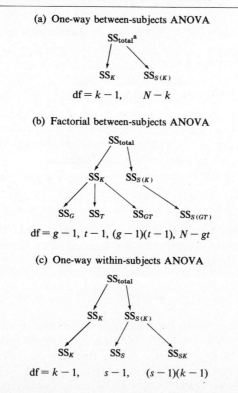

(a) One-way between-subjects ANOVA

SS_{total}^{a}

SS_K $SS_{S(K)}$

$df = k - 1,$ $N - k$

(b) Factorial between-subjects ANOVA

SS_{total}

SS_K $SS_{S(K)}$

SS_G SS_T SS_{GT} $SS_{S(GT)}$

$df = g - 1,\ t - 1,\ (g - 1)(t - 1),\ N - gt$

(c) One-way within-subjects ANOVA

SS_{total}

SS_K $SS_{S(K)}$

SS_K SS_S SS_{SK}

$df = k - 1,$ $s - 1,$ $(s - 1)(k - 1)$

TABLE 2.2 (*Continued*)

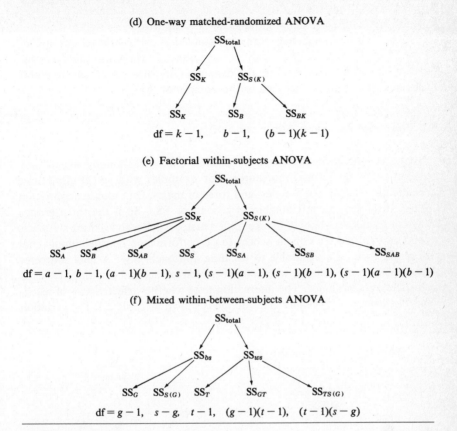

(d) One-way matched-randomized ANOVA

(e) Factorial within-subjects ANOVA

(f) Mixed within-between-subjects ANOVA

ᵃ For all SS_{total}, df $= N - 1$.

The division of each sum of squares by its associated degrees of freedom produces three variances: one associated with total variability among scores, MS_{total}; one associated with variability within groups, MS_{wg}; and one with variability between groups, MS_{bg}. In each case, MS stands for *mean square*, another term for variance. That is, a variance is an average sum of squares.

Two of these terms, then, provide the necessary variances to use an F ratio to test the null hypothesis, $\mu = \mu_1 = \mu_2 = \cdots = \mu_k$.

$$F = \frac{MS_K}{MS_{S(K)}}, \text{df} = (k - 1), N - k \tag{2.15}$$

An F table, such as Table C.3, can be consulted with numerator df $= k - 1$ and denominator df $= N - k$ for the desired level of alpha. If the obtained value of F from Eq. 2.15 exceeds the critical (tabled) value of F, then the null hypothesis is rejected and the hypothesis that there is a difference between population means among the k groups can be supported.

Notice that anything that increases obtained F also increases power (i.e., decreases Type II error). Power can be increased by decreasing the estimated population variability (reflected in the denominator, $MS_{S(K)}$) or by increasing differences among means (reflected in the numerator, MS_K).

2.2.2 Factorial Between-Subjects ANOVA

If groups differ on more than one dimension, differences among means may be attributed to more than one source. For example, with six groups, three may be composed solely of females, and three of males. One source of variation in means, then, may be attributable to gender, or SS_G. So if the three groups within each gender are exposed to three different approaches to teaching statistics, a second source of variation in scores (e.g., performance on a common final exam) might then be attributable to teaching approach, SS_T. A final source of variation, SS_{GT}, represents the interaction between gender and teaching approach in influencing performance. The interaction tests whether effectiveness of teaching approach varies with gender. This design is shown in Table 2.3. The variation

TABLE 2.3 ASSIGNMENT OF SUBJECTS IN A FACTORIAL BETWEEN-SUBJECTS ANOVA

		Teaching techniques		
		T_1	T_2	T_3
Gender	G_1	S_1 S_2	S_5 S_6	S_9 S_{10}
	G_2	S_3 S_4	S_7 S_8	S_{11} S_{12}

among groups and degrees of freedom are therefore partitioned as in Table 2.2(b). Error, or population variance, is now estimated by variation in scores within each of the six groups, $SS_{S(GT)}$. Three null hypotheses are now available for statistical test using the F distribution.

First, ignoring teaching method, do the male and female means represent the same sampling distribution of means? This is tested in the F ratio:

$$F = \frac{MS_G}{MS_{S(GT)}}, \ df = (g-1), N - gt \qquad (2.16)$$

Rejection of the null hypothesis supports an interpretation of differences between female and male in performance on the joint final exam.

Second, ignoring gender, do the means from the three teaching approaches represent the same sampling distribution of means? This is tested as:

$$F = \frac{MS_T}{MS_{S(GT)}}, df = (t - 1), N - gt \tag{2.17}$$

Rejection of the null hypothesis supports an interpretation of differential effectiveness of the three approaches.

Third, do the differences between gender means for each teaching approach represent the same sampling distribution of differences between means?

$$F = \frac{MS_{GT}}{MS_{S(GT)}}, df = (g - 1)(t - 1), N - gt \tag{2.18}$$

Rejection of the null hypothesis supports an interpretation that different teaching approaches are effective for females and males.

Notice that in each case, the population distribution of performance scores (error) is represented by $MS_{S(GT)}$, or within-group variance. In each case, the F table is consulted with appropriate degrees of freedom and desired alpha, and if the calculated F ratio (Eq. 2.16, 2.17, or 2.18) is greater than the associated tabled F ratio, the null hypothesis is rejected. The three hypotheses are *independent* when the number of scores within each group is equal. That is, the presence or absence of either main effect (gender or teaching approach) is in no way related to the presence or absence of the other main effect, or to the presence or absence of an interaction.

Recall from Section 2.2.1 that $k - 1$ degrees of freedom were used to test the null hypothesis regarding differences among groups. If the two-way partition of scores is made so that $k = gt$, then the three tests, G, T, and GT, will use up the $k - 1$ degrees of freedom. With proper partitioning, then, in a two-way factorial design you get three tests for the price of one.

With higher-order factorial designs, variation due to differences among groups can be further partitioned into main effects for each IV or dimension, two-way interactions between each pair of IVs, three-way interactions among each trio of IVs, and so on. In any factorial design, error sum of squares is represented by the sum of squared differences within each group or "cell" of the design.

2.2.3 Within-Subjects ANOVA

In some applications of ANOVA, the different levels of the IV for which means are to be compared represent not different groups but the same experimental units (subjects) measured on different occasions, as shown in Table 2.4.[3] In these situations, sources of differences among means for the various levels of the IV are the same as for the between-subject designs. Here, however, the error term can be further partitioned. Individual differences in performance can be viewed as a systematic source of variance in scores. If individuals are measured repeatedly, these individual differences can be examined. Since this variation is systematic, it can be subtracted from the error term. With a smaller error term, greater sensitivity is possible for test of the IV measured within

TABLE 2.4 ASSIGNMENT OF SUBJECTS IN A ONE-WAY WITHIN-SUBJECTS ANOVA

		Treatment		
		K_1	K_2	K_3
	S_1	S_1	S_1	S_1
Subjects	S_2	S_2	S_2	S_2
	S_3	S_3	S_3	S_3

subjects. The partition of sums of squares for a one-way design with k levels is as shown in Table 2.2(c), where s is the number of subjects.

Notice what has happened in this partition. Once individual differences between scores (SS_S) are subtracted out, there is no variation left within cells of the design. There is now left, in effect, one score per cell. The only source of variation remaining within the score matrix is the interaction between IV levels and subjects. That is, variation due to differences in the effect of the IV on different individuals is all that remains to serve as the best estimate of population (error) variance. If individuals react similarly to the IV, the interaction variance is a good estimate of error. But if different individuals are indeed affected differently by the IV, this error term may lower the calculated F ratio:

$$F = \frac{MS_K}{MS_{SK}}, df = (k-1), (k-1)(s-1) \tag{2.19}$$

In such a case, this becomes a conservative statistical test, making it difficult to reject the null hypothesis of no difference between means. It becomes a low-power test in which the Type II error rate is inflated. Since Type I error is unaffected, the test is not in disrepute for violation of the assumption of no interaction between subjects and levels of the IV.

There is another source of conservatism in this test. Notice that the degrees of freedom for error, $(k-1)(s-1)$, are reduced from those in the between-subjects design. Unless the reduction in error variance (by subtracting out indi-

---

[3] This design is sometimes also known as repeated measures, one-score-per-cell, randomized blocks, matched-randomized changeover, or crossover.

vidual differences, SS_S) is substantial, the loss in power from diminished degrees of freedom may overcome the power gained by reducing error variance.

In this design, there is another null hypothesis available for test—that different subject means, ignoring levels of the IV, represent the same sampling distribution of means. That is, the MS_S (variance due to individual differences) could be tested for statistical significance. Usually, however, this is not of research interest since if individuals do not maintain consistency over treatments, this would be a terrible analytic choice. Such a test would only serve warning that in future research with this DV, another design should be considered.

As suggested by the term *randomized blocks* this design has an application beyond that implied by the label "within-subjects." The design is also useful in a matched-randomized experimental design, as seen in Table 2.5. Here subjects

TABLE 2.5 ASSIGNMENT OF SUBJECTS IN A MATCHED-RANDOMIZED ANOVA[a]

		Treatment		
		A_1	A_2	A_3
	B_1	S_1	S_2	S_3
Blocks	B_2	S_4	S_5	S_6
	B_3	S_7	S_8	S_9

[a] Where subjects in the same block have been matched on some relevant variable.

are matched on the basis of variables thought to be highly related to the DV. They are divided into blocks, with as many subjects within each block as there are levels of the IV. Members of each block are randomly assigned to levels of the IV and statistically treated as if they were the same person. As with repeated measurement on individual subjects, matching should assure consistency in performance within blocks over levels of the IV. Variation associated with this consistency—sum of squares due to blocks, SS_B— can be subtracted from the error term just as for a true within-subjects design, as shown in Table 2.2(d). With b now representing the number of scores per level of IV, or the number of blocks, the statistical test proceeds just as for repeated measurement, following Eq. 2.19.

In factorial within-subjects (or randomized blocks) designs (Table 2.6), partition of the error sum of squares continues along with partition of the systematic sum of squares. As the design grows, a number of "subjects by effects" interactions are possible. Since violation of the assumption of no interaction may occur at several points, it is best to limit such violation in statistical tests. It is therefore common (though not universal) to develop a separate error term for each F test in factorial within-subjects designs. For the two-way design, this results in the partition shown in Table 2.2(e). Each effect to be tested in an F ratio has its own error term. The test for the main effect of IV A is

TABLE 2.6 ASSIGNMENT OF SUBJECTS IN A FACTORIAL WITHIN-SUBJECTS ANOVA

		Treatment A		
		A_1	A_2	A_3
Treatment B	B_1	S_1 S_2	S_1 S_2	S_1 S_2
	B_2	S_1 S_2	S_1 S_2	S_1 S_2

$$F = \frac{\text{MS}_A}{\text{MS}_{SA}}, \, df = (a-1), (s-1)(a-1) \qquad (2.20)$$

For the main effect of B, the test is

$$F = \frac{\text{MS}_B}{\text{MS}_{SB}}, \, df = (b-1), (s-1)(b-1) \qquad (2.21)$$

and for the interaction,

$$F = \frac{\text{MS}_{AB}}{\text{MS}_{SAB}}, \, df = (a-1)(b-1), (s-1)(a-1)(b-1) \qquad (2.22)$$

For higher-order factorial designs, the partition into sources of variation grows prodigiously, with an error term developed to go along with each effect (main or interaction) to be tested.

It should be noted that within-subject designs incur some controversy concerning conservatism of the F tests and whether or not separate error terms should be used. Additionally, if there are repeated measurements on single subjects, there may well be carry-over effects limiting generalizability to situations in which subjects are tested repeatedly. Finally, there is the assumption of homogeneity of covariance as well as homogeneity of variance. This means that consistency in rankings of subjects over each pair of IV levels is the same for all pairs of levels. If some pairs of levels represent measurement occasions close in time, while other pairs represent measures distant in time, this assumption may well be violated. Such violation can be serious because it may affect Type I error rate. Issues associated with the univariate approach to repeated measures are discussed by Frane (1980).

For some of these reasons, within-subjects ANOVA is sometimes replaced by multivariate analysis of variance, in which each repetition of the DV measure is treated as a separate DV (see Chapter 8).

2.2.4 Mixed Between-Within-Subjects ANOVA[4]

In some factorial analyses of variance, one or more IVs may be measured between subjects while one or more IVs are measured within subjects. These are the

[4] Also called split plot, repeated measures, and randomized block factorial design.

mixed between-within designs, the simplest of which involves one of each kind of IV, as seen in Table 2.7. Frequently the between-subject variables are based on observed differences among research units (e.g., age, sex) and it is said that the design is "blocked" on the subject variables. Notice that this is a different use of the term *blocks* from that of the preceding section. In a mixed design, both kinds of blocking can occur.

In order to show the partition, it is convenient to divide the total SS into that attributable to the between-subjects part of the design, and that attributable to the within-subjects part, as shown in Table 2.2(f). Each of these two sources can then be further partitioned into systematic and error components. The interaction between groups and trials (GT) is included in the within-subjects part of the design.

TABLE 2.7 ASSIGNMENT OF SUBJECTS IN A MIXED BETWEEN-WITHIN-SUBJECTS ANOVA

		Trials		
		T_1	T_2	T_3
Groups	G_1	S_1	S_1	S_1
		S_2	S_2	S_2
	G_2	S_3	S_3	S_3
		S_4	S_4	S_4

As more between-subjects IVs are added, they simply expand within the between-subjects part of the partition just as in the factorial between-subjects ANOVA, with a single error term consisting of subjects confined to each combination of the between-subjects IVs. Interactions among between-subjects IVs are included in this partition. As more within-subjects variables are added, the within-subjects portion of the design is expanded. Here the systematic sources of variance will include each of the within-subjects IVs, all interactions among within-subjects IVs, and all mixed interactions. Multiple error terms will also be developed, as they were for the pure within-subjects designs. Separate error terms are necessary for each within-subjects main effect and each within-subjects interaction.

Notice that "subjects" is no longer available as a separate, systematic source of variance for analysis. This is because subjects are confined to levels of the between-subjects variable(s); variations between subjects in each group are used as an estimate of error for testing variance associated with between-subjects variables.

Mixed designs can be used with "blocks" rather than repeated measures on individual subjects, as for within-subjects designs. In any event, problems associated with within-subjects designs (e.g., homogeneity of covariance) carry over to mixed designs, and MANOVA is sometimes used to circumvent these problems.

2.2.5 Design Complexity

To this point, discussion of analysis of variance has been limited to factorial designs in which numbers of scores in each cell are equal (equal n) and levels of each IV were purposely chosen by the researcher. Several deviations from these straightforward designs can occur. A few of the more common types of design complexity will be briefly reviewed here, but the reader actually faced with use of these designs is referred to one of the more complete analysis of variance texts such as Winer (1971), Keppel (1973), or Myers (1979).

2.2.5.1 Nesting   In the previously described ANOVA designs, it can be said that subjects are confined, or nested, within between-subjects IVs. That is, any given subject is exposed to only one level of each between-subjects IV. Nesting can also be carried out to higher levels of analysis. In particular, when it is impossible to accomplish random assignment to all levels of an IV within a certain setting, such assignment can be made across many settings.

Take the example of a variety of teaching techniques, in which children within a classroom cannot be randomly assigned to technique but whole classrooms can be so assigned. If teaching technique is the only IV of interest, this can be seen as a simple one-way between-subjects design where classrooms serve as the experimental units (i.e., subjects). For each classroom, an average score on the DV is obtained, and these averages now serve as scores in the one-way ANOVA.

If the effect of classroom is also to be assessed, this becomes a nested or hierarchical design, as seen in Table 2.8(a). Let's say that classrooms represent the same grade level at different schools, and we want to know whether performance differs in that grade at the various schools. The error term for the effect of classroom is represented by subjects within classrooms and teaching technique, while the error term for test of teaching technique is represented by classrooms within teaching technique. The latter test is precisely the one we would use if classroom effects were to be ignored and a one-way design used.

TABLE 2.8   SOME COMPLEX ANOVA DESIGNS

(a) Nested designs				(b) Latin square designs[a]			
Teaching techniques					Order		
T_1	T_2	T_3			1	2	3
Classroom 1	Classroom 2	Classroom 3		S_1	A_2	A_1	A_3
Classroom 4	Classroom 5	Classroom 6	Subjects	S_2	A_1	A_3	A_2
Classroom 7	Classroom 8	Classroom 9		S_3	A_3	A_2	A_1

[a] Where the three levels of treatment A are experienced by different subjects in different orders, as indicated.

2.2.5.2 Latin Square Designs   In true within-subjects designs, effects of order on the DV are frequently handled by having each subject exposed to the various levels of the IV in a different order. In that way, any differences among levels of the IV will be due to the IV rather than to the order in which levels were administered to the subjects. If there truly is an effect of order on the DV, this will simply elevate the error term, resulting in a conservative test of the IV. Another alternative is to treat order as another IV, as seen in Table 2.8(b), so that its effects can be partitioned out of the error term. In addition, the effects of order can be evaluated as a separate source of systematic variance. The designs can be extended to factorial arrangements of within-subjects variables as well as mixed within-between-subjects designs.

These designs share the basic problems of repeated measurement. Especially troublesome is the use of interactions (e.g., order by subjects, treatment by order) as error terms, which can create highly conservative tests of IVs. In addition, some of the interactions of interest may not be retrievable in some applications of Latin or Greco-Latin square designs.

2.2.5.3 Unequal *n* and Nonorthogonality   In a simple one-way between-subjects ANOVA, problems presented by unequal group sizes are relatively minor. Computation becomes more difficult, but that presents no real disaster if computer programs are available. If sample sizes are widely disparate, there is no inherent ambiguity in the design, but assumptions regarding homogeneity of variance become more critical. In particular, if the cell with the smaller n has the larger variance, the F test becomes too liberal, leading to a false rejection of a correct null hypothesis at an alpha level greater than that indicated in the tables.

In factorial designs with more than one between-subjects IV, unequal sample sizes for each combination of levels of IVs create ambiguity in results as well as difficulty in computation. With unequal n, a factorial design becomes nonorthogonal. Hypotheses about the various main effects and interactions are no longer independent, nor are sums of squares additive. This is because the same variance can be attributed to more than one source. That is, the various sources contain overlapping variance, as discussed in Chapter 1. If tests of effects were made without taking this overlap into account, the probability of a Type I error would increase, because any systematic variance could contribute to more than one test. For this reason a variety of strategies has been developed to deal with the problem, none of them completely satisfactory.

The simplest strategy is just to eliminate data randomly from cells of the design with larger sample sizes until all cells are equal. If unequal n is due to random loss of a few subjects here and there in the design originally set up for equal n, this may be a good choice. An alternative strategy with random loss of subjects producing unequal n is an unweighted-means analysis, described in Chapter 7 and most ANOVA textbooks, such as Winer (1971) and Keppel (1973). Since the unweighted-means approach has greater power than eliminating data to equalize cells, it is the preferred approach as long as computational aids are available.

But sometimes unequal n results from the nature of the population. Especially with nonexperimental research, differences in sample sizes may reflect true differences in nature. To artificially equalize the n's may be to distort these differences and lose generalizability. In these latter situations, decisions must be made as to how tests of effects are to be adjusted for overlapping variance. Standard methods for adjusting tests of effects for unequal n are discussed in Chapter 7, in the context of analysis of covariance.

2.2.5.4 Fixed and Random Effects In all of the ANOVA designs discussed so far, levels of each IV are chosen on an a priori basis, and only those levels are of interest in testing significance of the IV. This is the usual *fixed* effects IV. Sometimes, however, there is a desire to generalize to a population of levels for a particular IV. For example, in examining word familiarity (frequency of usage of words in the English language), the interest may be in the functional relationship between the DV (e.g., recall) and all levels of word familiarity. For purposes of analysis, some finite set of familiarity levels would be randomly chosen from a population of word frequency levels. Word frequency would then be considered a *random* effects IV.

The analysis could then be set up so that the results may be generalized to levels other than those chosen for test—i.e., to the population of levels from which the sample was chosen. This is similar to the notion of generalizing ANOVA results beyond the sample of subjects chosen—to the population of subjects from which the sample was randomly selected. During analysis, alternative error terms for evaluating the statistical significance of random effects IVs are required.

Although some readily available computer programs allow specification of random effects IVs, use of them is relatively rare. The interested reader is referred to one of the more sophisticated ANOVA texts, such as Winer (1971), for a full discussion of the random effects model.

2.2.6 Specific Comparisons

Numerator degrees of freedom, as defined in previous sections, can be thought of as a nonrenewable resource: They are used up in an analysis of variance. Altogether, in any ANOVA there are $k - 1$ of these degrees of freedom for the statistical tests of means representing treatment effects (where k is the number of levels of the first IV times the number of levels of the second IV, etc.). The typical way of using up these degrees of freedom is to test (via an F ratio) each main effect and each interaction. Each of these traditional F tests is called an omnibus F test.

2.2.6.1 Post Hoc Comparisons After finding a significant omnibus F ratio, you may want to snoop your data. For example, if $k = 4$ in a one-way design, you may want to know just *which* of your treatment means differs significantly from which other means (or which combinations of means are different from each other). At this point, however, you have used up your degrees of freedom and are capitalizing on chance differences among means, so any tests you perform are not at the stated level of alpha. Therefore, some form of adjustment of alpha is necessary. Many procedures for dealing with inflated Type I error rate are available and can be found in standard ANOVA texts. The one described here is that developed by Scheffé (1953) and is the most conservative of the popular methods. In part, the Scheffé method involves an adjusted level of critical F as follows:

$$F' = (k - 1) F_c \qquad\qquad (2.23)$$

where F' is adjusted critical F, and F_c is tabled F (with 1 degree of freedom in the numerator and degrees of freedom associated with the error term in the denominator) multiplied by degrees of freedom for effect being tested.

2.2.6.2 Planned Comparisons　In some research applications, you are in the enviable position of being able to specify, prior to data collection, highly explicit alternative hypotheses with respect to differences or patterns among cell means. In such happy circumstances, planned comparisons provide a powerful alternative to the practice of omnibus F tests followed by post hoc comparisons. As long as the entire analysis uses up no more than the available degrees of freedom, and all comparisons are mutually orthogonal (to be described in Section 2.2.6.4), no adjustment of tabled critical F is necessary, and all tests are done within the stated level of α error. If these conditions are violated, planned comparisons can still be done by using procedures less conservative than Scheffé's, for example, Newman-Keuls test for all pairwise comparisons.

2.2.6.3 Weighting Coefficients　Comparison among treatment means, whether planned or post hoc, begins by assigning a weighting factor (w) to each of the cell or marginal means in such a way that the weights reflect your null hypotheses. Suppose you have a set of k means (one-way design) and want to make comparisons among individual combinations. Assign a weight w_j to each mean $\overline{Y}_j$ (i.e., multiply each mean by a weight):

$$w_j \overline{Y}_j \qquad \text{where } j = 1, 2, \ldots, k$$

You may assign weights in any manner you choose, so long as (1) at least two of the means are assigned nonzero weights, and (2) the weights themselves sum to zero, that is,

$$\sum_{1}^{k} w_j = 0$$

reflecting your null hypothesis. For example, with two means ($k = 2$) you could assign $w_1 = 1$ and $w_2 = -1$, so that

$$(1)\,\overline{Y}_1 + (-1)\,\overline{Y}_2 = \overline{Y}_1 - \overline{Y}_2$$

Testing the null hypothesis: H_0: $\mu_1 - \mu_2 = 0$.

The idea behind the test is that the sum of the weighted means will be equal to zero if the null hypothesis is true. The more the sum diverges from zero, the greater the confidence with which the null hypothesis can be rejected.

With $k = 3$, a large number of comparisons could be developed, for example,

$$(1)\,\overline{Y}_1 + (-1)\,\overline{Y}_2 + (0)\,\overline{Y}_3 \quad \text{or} \quad (\tfrac{1}{2})\,\overline{Y}_1 + (\tfrac{1}{2})\,\overline{Y}_2 + (-1)\,\overline{Y}_3$$

The first would be a simple comparison of the first two means, ignoring the third, that is, H_0: $\mu_1 - \mu_2 = 0$. The second tests whether the average of means for levels 1 and 2 is equal to the mean for level 3, H_0: $(\mu_1 + \mu_2)/2 - \mu_3 = 0$. There are also special sets of weights that have been developed for testing trends over levels of a quantitative IV.

2.2.6.4 Orthogonality Any pair of comparisons is orthogonal if the sum of the cross products of the weights for the two comparisons is equal to zero. For example, in the two comparisons above:

	w_1	w_2	w_3
Comparison 1	1	-1	0
Comparison 2	$\tfrac{1}{2}$	$\tfrac{1}{2}$	-1

The sum of the cross products of the weights is

$$(1)(\tfrac{1}{2}) + (-1)(\tfrac{1}{2}) + (0)(-1) = 0$$

Therefore the two comparisons are orthogonal. If there were another comparison, say weights equal to 1, 0, and -1 for the three levels, it would be orthogonal with neither of the above two comparisons. As a set of three, then, the comparisons would not be mutually orthogonal. The special set of weights developed for trend analysis *are* mutually orthogonal.

2.2.6.5 *F* Tests for Comparisons To do a comparison on a set of means, the following general F test can be used as long as sample sizes are equal and the comparison is based on between-subjects differences.

$$F = \frac{n_c (\Sigma w_j \overline{Y}_j)^2 / \Sigma w_j^{\,2}}{\text{MS}_{\text{error}}} \tag{2.24}$$

The obtained F ratio to test the significance of a comparison is based on (1) the number of scores, n_c, in each of the means to be compared;

(2) a squared sum of weighted means, $(\Sigma w_j \overline{Y}_j)^2$; (3) the sum of the squared coefficients, Σw_j^2; and (4) the mean square for error in the ANOVA design.

For factorial designs, comparisons can be done on marginal or cell means, corresponding to contrasts involving main effects or interactions, respectively. The number of scores per mean and the error term will follow from the ANOVA design used. If any comparisons are to be made within subjects, a complication arises in that separate error terms must be developed for each comparison (just as separate error terms are developed for all omnibus tests of within-subjects IVs).

If the comparison is "planned" (i.e., prior development, available degrees of freedom, and mutual orthogonality with all other tests), the obtained F is tested against tabled F with 1 degree of freedom in the numerator and degrees of freedom associated with the MS_{error} in the denominator. If obtained F is greater than critical F, the null hypothesis represented by the weighting coefficients is rejected.

If the comparison is post hoc, obtained F is compared with F', as defined in Eq. 2.23. For tests of marginal means the adjustment $(k - 1)$ is based on degrees of freedom for the IV. For tests of cell means the adjustment is based on degrees of freedom for the interaction. For example, with two IVs, A and B, a comparison involving the cell means would have as an adjustment for critical F:

$$F' = (a - 1)(b - 1) \, F_c$$

A test on marginal means for A would have as adjustment:

$$F' = (a - 1) \, F_c$$

2.2.6.6 Type IV Errors There is some controversy regarding the appropriateness of certain types of post hoc comparisons following significant omnibus F tests, with inappropriate comparisons considered Type IV errors. As defined by Levin and Marascuilo (1972, p. 368), "a Type IV error is made whenever a researcher offers an incorrect interpretation to a correctly rejected statistical hypothesis." One question, for example, is whether testing of simple comparisons among marginal means (by any method other than Scheffé, 1953) is legitimate after finding a significant main effect. Another example is the practice of following a significant interaction in a two-way design with separate post hoc analyses of rows and/or columns or pairwise comparisons of all means. Without going into the argument (interested readers should refer to Levin and Marascuilo, 1972, 1973; Games, 1973), we find the following position reasonable. *All* post hoc comparisons are suspect by the very fact of testing hypotheses on the same data that generated the hypotheses. On the other hand, data snooping can aid immensely in what is often a difficult job of interpreting significant omnibus effects, if one has not been fortunate enough to be able to plan comparisons. Therefore as long as conservative procedures are used (e.g., as in Section 2.2.6.5) and results are interpreted cautiously, we see no serious harm in "snooping at will." That is, do the post hoc tests that will provide answers to the interesting questions generated by the data.

2.3 PARAMETER ESTIMATION

After rejecting a null hypothesis about differences among population means, we are probably also interested in actual values of population means. Having gathered data only on a sample from the population(s) of interest, there is no way to know exactly the population parameters. We do, fortunately, have unbiased estimators of these population means in the form of sample means. That is, our best guess of a population mean (μ) is the mean of the sample ($\overline{Y}$) randomly selected from that population. It would be an incomplete report of results, then, to state simply that the null hypothesis had been rejected without also indicating the size and direction of the differences in means. In most reports of research, therefore, sample mean values as well as statistical results are reported.

Notice that sample values (statistics) are only approximations of population parameters. There is error in this estimation, which is sometimes reported in addition to the estimates themselves. The familiar confidence intervals of introductory statistics reflect this error of estimation. Error of estimation depends on sample size and the estimation of population variability (both of which determine the variability of the sampling distribution), plus the degree of confidence one wishes to have in estimating μ.

2.4 STRENGTH OF ASSOCIATION

Although significance testing, specific comparisons, and parameter estimation illuminate the nature of group differences in the context of the DV, they do not convey in a straightforward way the degree to which the IV(s) and DV are related. It is important to assess this degree of relationship in order to avoid publicizing trivial results as though they had practical utility. As discussed in Section 2.1.2, powerful research can produce results that are statistically significant but realistically meaningless.

If we think again in terms of total variance in the DV (MS_{total}), it makes sense to ask what proportion of that variance is shared with, or predictable from, the IV. What proportion of variance (i.e., differences among scores) is attributable to differences among treatment levels for an IV? Note that strength of association is different from statistical significance as assessed during hypothesis testing. Probability of chance variation depends on sample size as well as amount of association. If the null hypothesis is rejected, it can be concluded that there is *some* association between the IV and DV. Strength of association measures indicate *how much* association there is.

A general, rough estimate of strength of association is available for any ANOVA in the form of η^2 (eta squared).

$$\eta^2 = \frac{SS_{effect}}{SS_{total}} \tag{2.25}$$

This is the simple, defining, equation for η^2 in ANOVA. It is equivalent to a squared point biserial correlation, in bivariate terms, between a continuous variable and a dichotomous variable. Note that for all these strength of association measures, the value depends on the particular choice of fixed levels of the IV and therefore cannot be generalized to research in which other levels were chosen. Given a significant effect (main effect or interaction), η^2 shows the proportion of score deviation in the DV (SS_{total}) attributable to the effect (SS_{effect}). In a balanced, equal-n design, if there is more than one statistically significant effect, the η^2's are additive. The sum of η^2 for all significant effects is the proportion of variation in the DV that is systematic, or predictable from knowledge of the IVs.

Although this is a simple, popular measure of strength of association in analysis of variance, it has two basic flaws. The first is that strength of association for a particular IV depends on the size of the design (in terms of number of IVs) in which the particular IV was tested. If a particular IV, say treatment groups, is tested in a one-way design, the computed value of η^2 will be larger than if treatment groups are tested in a two-way design in which the other IV and the interaction add to the total variance—especially if either one or both of these additional effects is large. That is, the denominator of η^2 contains not only error variance and systematic variance of interest but also irrelevant systematic variance. The larger the design (the more IVs), the greater the discrepancy becomes.

For this reason, an alternative form of η^2 has been developed, in which the denominator contains only that variance attributable to the effect of interest plus error, and eliminates any other systematic variance that may be operating on the DV.

$$\eta^2_{alt} = \frac{SS_{effect}}{SS_{effect} + SS_{error}} \tag{2.26}$$

The error term to be used, of course, is the one used in the F test for that effect, following the designs in previous sections of this chapter. Note that with this alternative formula, η^2's for all significant effects in the design *do not* sum to proportion of systematic variance in the DV. Indeed, the sum may be greater than 1.00. It is therefore imperative that the form of η^2 be reported if the version of Eq. 2.26 is used.

A second flaw in η^2 is that it computes proportion of systematic variance in terms of sample data, with no attempt to estimate proportion of systematic variance in the populations of interest. A statistic developed to estimate this strength of association between IV and DV in the population is ω^2 (omega squared).

$$\omega^2 = \frac{SS_{effect} - (df_{effect})(MS_{error})}{MS_{error} + SS_{total}} \tag{2.27}$$

This is the additive form of ω^2, in which the denominator represents total variance, not just that due to effect of interest plus error. As for η^2, a separate value of ω^2 is appropriately reported for each statistically significant main effect and interaction. *It is limited, however, to between-subjects analysis of variance designs with equal n.* It is inappropriate in ANOVAs in which there are any within-subjects IVs. There are forms of ω^2 available for designs containing repeated measures (or randomized blocks) but these are beyond the scope of this book. A set of procedures for finding ω^2 in a variety of designs is provided by Vaughn and Corballis (1969).

2.5 BIVARIATE STATISTICS: CORRELATION AND REGRESSION

Strength of association, as described in Section 2.4, is measured between a (usually) continuous DV and an IV that represents different levels of some treatment. Frequently, however, a researcher wants to measure the strength of association between variables where the IV-DV distinction is blurred. For instance, the association between years of education and income may be of interest even though neither was manipulated and inferences regarding causality are not required. In such a setting, correlation provides a measure of association, with squared correlation indicating strength of the association or proportion of shared variance.

Correlation differs from regression in that correlation measures the association between variables, whereas regression attempts to predict one variable from another (or many others).

2.5.1 Correlation

Because the Pearson product-moment correlation coefficient, r, is easily the most frequently used and because it is the basis of many multivariate calculations, we restrict ourselves to description of it. Perhaps the most interpretable form of the Pearson r is

$$r = \frac{\Sigma Z_X Z_Y}{N - 1} \tag{2.28}$$

The Pearson correlation coefficient is the adjusted (for degrees of freedom) average cross product of standardized X and Y variable scores.

where S is defined as in Eq. 2.5 and

$$Z_Y = \frac{Y - \overline{Y}}{S} \quad \text{and} \quad Z_X = \frac{X - \overline{X}}{S}$$

This measure of association is independent of scale of measurement (because both X and Y scores have been converted to standard scores) and independent of sample size (because of division by $N - 1$). Pearson r ranges between -1.00 and $+1.00$ in size where 0.00 represents no relationship between the X and Y variables; that is, a Pearson r of 0.00 indicates no improvement in predicting a person's score on X when the score on Y is known. An r value of -1.00 or $+1.00$ indicates perfect predictability of one score when the other is known. When correlation is perfect (i.e., $+1.00$ or -1.00), scores in the X distribution have the same relative positions as corresponding scores in the Y distribution. In that case, $Z_X = Z_Y$ for each pair and the numerator of Eq. 2.28 becomes, in effect, $\Sigma Z_X Z_X$. Because $\Sigma Z_X{}^2 = N - 1$, Eq. 2.28 reduces to $(N - 1)/(N - 1)$, or 1.00.

The raw score form of Eq. 2.28 also sheds light on the meaning of r:

$$r = \frac{N\Sigma XY - (\Sigma X)(\Sigma Y)}{\sqrt{[N\Sigma X^2 - (\Sigma X)^2][N\Sigma Y^2 - (\Sigma Y)^2]}} \tag{2.29}$$

Pearson r is the ratio of the covariance between X and Y to the square root of the product of the X variance times the Y variance.

Pearson r measures the covariance between X and Y relative to the square root of the product of the X and Y variances. (Only the numerators of variance and covariance equations appear in Eq. 2.29 because the denominators have canceled each other out.) The correlations are the entries in the correlation or **R** matrix as described in Chapter 1.

2.5.2 Regression

As already noted, whereas correlation measures the amount of relationship between two variables, regression attempts to predict one from the other. In bivariate (two-variable) regression, one attempts to predict a score on Y from a score on X, which involves finding a best-fitting straight line between two variables. Procedures for fitting the best line can be described by reference to Figure 2.3. Suppose you have a clear straight edge or thread. Place it so that it intersects the point marked $\overline{X}, \overline{Y}$. That is, the thread goes through the point on the scatterplot representing the mean of both variables. Now, pivot the thread so it minimizes the distance of all points around itself, with equal overall (squared) distance of points above and below the thread. Unsuccessful trial pivots are

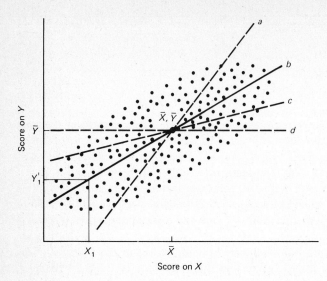

Figure 2.3 Predicting scores on Y from scores on X from a scatterplot.

illustrated by lines labeled *a, c,* and *d.* Line *b* most closely meets the requirements and represents the regression line as best fit by eye.

With scores known on X but unknown on Y, we can predict Y' from the regression line. For example, with a score of X_1, we would use the regression line to predict the Y_1' score.

To do this mathematically, an equation of the form

$$Y' = A + BX \tag{2.30}$$

is calculated where Y' is the predicted score, A is the value of Y when $X = 0.00$, B is the slope of the line (change in Y divided by change in X), and X is the value from which Y is to be predicted. The difference between the predicted and the obtained values of Y at each value of X represents error of prediction. The best-fitting straight line is operationally defined as that line which minimizes the squared errors of prediction. To solve Eq. 2.30 both B and A must be found.

$$B = \frac{N\Sigma XY - (\Sigma X)(\Sigma Y)}{N\Sigma X^2 - (\Sigma X)^2} \tag{2.31}$$

The bivariate regression coefficient, B, is the ratio between the covariance of the variables and the variance in the one from which predictions are to be made.

Note the differences and similarities between Eq. 2.29 (for correlation), and Eq. 2.31 (for the regression coefficient). Both have the covariance between the variables as a numerator but differ in denominator. In correlation, the variances of both are used: The square root of the product of the two variances serves as the denominator. In regression, the variance of the predictor variable serves as the denominator. If Y is predicted from X, X variance is the denominator. If X is predicted from Y, Y variance is the denominator.

To complete the solution, the value of A, the intercept, must also be calculated.

$$A = \overline{Y} - B\overline{X} \qquad (2.32)$$

The intercept is the mean of the predicted variable minus the product of the regression coefficient times the mean of the predictor variable.

The techniques discussed in this chapter for making decisions about differences, estimating population values, measuring association between variables, and predicting a score on one variable from a score on another are important to and widely used in the social and behavioral sciences. They form the basis for most undergraduate—and some graduate—statistics courses. It is hoped that this brief review will remind you of material already mastered so that, with common background and language, we can begin the study of multivariate statistics.

Chapter 3

A Guide to Statistical Techniques: Using the Book

3.1 RESEARCH QUESTIONS AND ASSOCIATED TECHNIQUES

All parametric statistics—univariate, bivariate, and multivariate—can be seen as applications of the general linear model. This integrated view of statistical techniques, however, will be reserved for Chapter 11, after examples of several techniques have been shown in detail. For the purpose of deciding on the most appropriate technique to use in any given research situation, it is more useful to emphasize differences among statistical techniques than similarities.

At the end of this chapter a flow diagram will appear, presenting some distinctions in research data in the form of a decision tree. On the basis of these distinctions, it should be possible to determine which statistical technique(s) can best be used to answer research questions of interest. But first, the questions and associated techniques will be briefly described.

The single most important criterion in choosing a technique is, of course, the major research question to be answered by application of the statistical analysis. Research questions can be categorized into four major types: degree of relationship among variables, significance of group differences, prediction of group membership, and structure.

3.1.1 Degree of Relationship Among Variables

If the major purpose of analysis is to assess the association among a group of variables, some form of correlation or regression is appropriate. By determining the number of independent and dependent variables, and whether or not any of the IVs can best be conceptualized as covariates,[1] a choice among four categories of statistical techniques is available.

3.1.1.1 Bivariate *r* Bivariate correlation, as reviewed in Chapter 2, assesses the degree of relationship between two variables, for example, belly dancing skill and years of musical training. In bivariate correlation there is no inherent distinction between the IV and the DV. Bivariate regression is a technique designed to predict the value of one variable from knowledge of the value of another, related, variable. This technique could be used to predict success in some skill as yet unlearned (e.g., belly dancing) on the basis of training already acquired (e.g., years of musical training). The variable to be predicted is the DV, while the predictor is considered the IV. Since bivariate correlation/regression is not considered multivariate, it will not be dealt with in greater detail until Chapter 11, in which it is considered in terms of the general linear model.

3.1.1.2 Multiple *R* Multiple correlation assesses the degree to which one variable (the DV) is related to a composite set of other variables (the IVs). That is, a set of IVs can be combined in such a way as to create a new, composite variable. Multiple correlation can then be seen as a bivariate correlation between the original DV and the newly created composite variable. For example, how large is the association between belly dancing skill and a number of measures taken prior to belly dancing training, such as years of musical training, body flexibility, and age?

Multiple regression is the technique by which the value of the DV is predicted from knowledge of the values of the IVs. The variables used to predict may or may not themselves be related. An example would be the prediction of belly dancing ability on the basis of age, body flexibility, and years of musical training. Another example might be the prediction of success in a program of education designed for gifted children on the basis of scores on a number of achievement tests. Or we might predict outcome in psychotherapy from scores on various personality measures and/or demographic characteristics. Or a measure of stock market behavior might be predicted from a variety of political and economic variables.

Multiple regression techniques also allow assessment of the relative contribution of each of the IVs toward predicting the DV. Chapter 5 deals with multiple regression/correlation.

3.1.1.3 Hierarchical *R* Hierarchical multiple regression is a form of multiple regression in which the contribution toward prediction of each IV is assessed in some predetermined hierarchical order.[2] For example, belly dancing ability

---

[1] If some IVs are to be assessed after the effects of other IVs are adjusted for, the latter are called covariates.

[2] Note that the term *hierarchical,* as used here, simply reflects the sequence in which variables are analyzed. There is no implication of a structural hierarchy, in which some variable (e.g., wards) is nested within a broader variable (e.g., hospitals).

(the DV) might first be predicted from years of musical training. Then it might be useful to see if age adds to the predictability of belly dancing over and above the contribution made by knowledge of prior musical training. Finally, it could be determined whether scores on a measure of body flexibility added to prediction of belly dancing ability after age and years of musical training were accounted for.

In an example related to psychotherapy, success of outcome might first be evaluated on the basis of demographic variables such as age, sex, and marital status. Then it might be useful to see if scores on personality tests added to predictability of outcome success after adjusting for the demographic variables.

In general, then, lower-priority IVs are assessed as predictors after adjusting for effects of higher-priority predictors (covariates). So for each IV in a hierarchical multiple regression, higher-priority predictors act as covariates. Hierarchical multiple regression can be useful in developing a reduced set of IVs (if that is desired) by indicating which variables no longer add to predictability.

The degree of relation between the DV and the IVs can be reassessed at each step of the hierarchy. That is, the growth of the hierarchical multiple correlation can be evaluated as new IVs are added to predict the DV. Hierarchical multiple regression is discussed in Chapter 5.

3.1.1.4 Canonical _R_ In canonical correlation, the goal is to assess the amount of relationship between two sets of variables. That is, there is a set of DVs as well as a set of IVs. For example, we might be interested in the relationship between a number of measures of belly dancing ability (knowledge of steps, ability to play finger cymbals, responsiveness to the music) and a number of variables measured prior to belly dancing training (body flexibility, musical training, and age). Or we might wonder whether there is a relationship between a selected set of achievement variables (arithmetic, reading, spelling as measured in elementary school) and a set of variables reflecting early childhood development (e.g., age at first speech, walking, toilet training).

In another example, it could be interesting to see if there is a relationship between demographic variables (religious affiliation, age, sex, marital status, socioeconomic status) on the one hand and a set of personality variables on the other (e.g., introversion-extraversion, degree of neuroticism, submissiveness-dominance). Or, if several measures of therapeutic outcome were available (e.g., ratings by psychotherapist, client, and client's family), it would be interesting to know the degree to which they could be predicted by a set of demographic and/or personality variables. Such research questions can be answered by canonical correlation, the subject of Chapter 6.

3.1.2 Significance of Group Differences

With discrete groups of research units (e.g., subjects) serving as the IV(s), the major research question frequently focuses on the extent to which DVs differ as a function of group membership. Especially in experimental research, this question is typically of major emphasis. But it should be noted that in all techniques for testing group differences, it is also possible to test strength of association between independent and dependent variables.

Here, again, choice among a variety of techniques hinges on the number of IVs and DVs, and whether or not some IVs are conceptualized as covariates.

3.1.2.1 One-Way ANOVA and *t* Test
These two statistics, reviewed in Chapter 2, are strictly univariate in nature and are adequately covered in most standard statistical texts.

3.1.2.2 One-Way ANCOVA
One-way analysis of covariance is designed to assess the effects of one IV (groups) on a single DV after the effects of one or more other potential IVs (covariates) are accounted for. Typically, covariates are chosen because of their known association with the DV, otherwise there is no point to their use. For example, it might well be expected that age and degree of reading disability would be related to outcome of a program of educational therapy. Groups are formed by randomly assigning children to different types of educational therapy. In assessing whether one type of therapy is better than other types (i.e., type of educational therapy is the IV), it would therefore be useful to first adjust for prior differences among children in age and reading disability, as covariates. The question in ANCOVA then becomes: Does type of educational therapy differentially affect reading ability (the dependent measure) after adjusting for the effects of age and amount of reading disability?

Analysis of covariance is designed to allow a more powerful look at the IV-DV relationship by minimizing error variance (cf. Chapter 2). The stronger the relationship between the DV and the covariate(s), the greater the power gained by ANCOVA over a simple ANOVA.

Another use of ANCOVA is to adjust for prior differences among groups when random assignment to groups is impossible, that is, where naturally formed groups are the IV.[3] For example, attitude toward abortion (as a DV) might well be expected to vary as a function of religious affiliation. However, it is not possible to randomly assign people to religious groups so that other differences between the groups, beyond religion, cannot be randomized out. In that situation, there might well be systematic differences between groups that are related to attitude toward abortion. As an example, age might vary as a function of religious affiliation. Since age is probably also related to abortion attitude, apparent differences between religious groups might well be due to difference in age, not religion per se. To get a "purer" measure of differences in attitude associated with religion, it might be worthwhile to first adjust attitude scores for age differences. That is, age could act as a covariate.

With more than two levels of the IV, specific comparisons among groups are possible. For example, if attitude is found to be associated with religion, one can evaluate which pairs of religious affiliation (e.g., Protestant compared with Catholic) differ from each other in attitude toward abortion after adjustment

[3] Some problems with this use are discussed in Chapter 7.

for differences associated with age. Analysis of covariance is covered in Chapter 7.

3.1.2.3 Factorial ANOVA

Factorial analysis of variance, reviewed in Chapter 2, is the subject of a large variety of statistics texts (e.g., Keppel, 1973; Winer, 1971; Myers, 1979). Further, it is frequently introduced in more elementary texts, such as Young and Veldman (1977). Therefore it will not be covered in greater detail in this book, although its place within the general linear model will be discussed in Chapter 11.

3.1.2.4 Factorial ANCOVA

Factorial analysis of covariance differs from one-way analysis of covariance only in that there is more than one IV. The desirability and use of covariates remains identical. For instance, in the example regarding educational therapy, in Section 3.1.2.2, another interesting IV might be a grouping based on sex of the child. The effect of type of educational therapy and of sex (and of the interaction between educational therapy and sex) on reading ability would then be assessed after adjusting for age and degree of reading disability. By adding sex as another IV, one can find out whether or not boys and girls differ as to which type of educational therapy is more effective. Again, specific comparisons among groups are possible. For example, for which types of educational therapy is there a differential effectiveness for boys and girls, and for which type of therapy are the effects similar regardless of sex?

For factorial ANCOVA, one or more IVs might be measured within subjects. In the above example, a third IV might be time of testing, with all children tested immediately after treatment, 6 months later, 1 year later, and so on. One could then see whether differences in effectiveness of therapy programs were maintained over time, after adjusting for initial age and degree of reading disability. Chapter 7 deals with factorial as well as one-way ANCOVA.

3.1.2.5 Hotelling's T^2

When the IV consists of only two groups, Hotelling's T^2 can be used to discover whether the groups differ on a set of DVs. For example, there might be two forms of educational therapy proposed for remedy of learning disabilities, emphasis on perceptual retraining (P) versus emphasis on academic training (A). And two DVs might be available to test the differential effectiveness of the two programs—for example, score on an academic achievement test and attention span in the classroom. Since these dependent measures involve multiple testing and might be intercorrelated as well, it would not be legitimate to test each of them using separate t tests. Instead, Hotelling's T^2 is available to test the hypothesis that groups differ on the composite set of measures. That is, do the centroids (combined "average" on both DVs) for the two groups differ, relative to the overlap between children in the two groups on the DVs? This example is illustrated in Figure 3.1. (A third DV would require a third axis.)

Hotelling's T^2 is a special case of multivariate analysis of variance, just as the t test is a special case of univariate analysis of variance, in which two

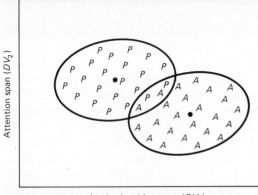

Figure 3.1 Simple Hotelling's T^2 situation with two DVs. Groups are perceptual emphasis (P) and academic emphasis (A). Filled circles within each ellipse indicate centroids; letters indicate individual children.

groups comprise the IV. Multivariate analysis of variance is discussed in detail in Chapter 8.

3.1.2.6 One-Way MANOVA

Multivariate analysis of variance is designed to investigate differences among two or more levels of an IV (groups) in terms of their effect on a set of DVs. The educational therapy example in the preceding section would be a candidate for MANOVA. Multivariate analysis of variance can also be used with more than two groups—for example, consider addition of a nontreatment control group to the two treatment groups, keeping the same set of two DVs.

With more than two groups, specific comparisons can be evaluated. For example, if assignment to treatment group is found to affect the DVs, it might be interesting to compare the two types of educational therapy, ignoring the control group. Do the two types of therapy differentially affect the outcome measures? And/or it might be interesting to see whether the performance of the control group differs from the average of the two educational therapy groups.

Additionally, techniques are available to assess which of the DVs are influenced by the IV. For example, it might be found that assignment to treatment group affects the academic DV but has no effect on attention span.

MANOVA can also be used with a within-subject IV. For example, in a series of photographs to be remembered, the IV might be serial position within the list: early, middle, or late. DVs might be rate of recall, number of words used to describe the photo in recalling it, rating of the vividness of the recall, and so on.

Any number of DVs may be used, and the procedure deals with intercorrelations among them. The entire analysis can be accomplished within a preset level of Type I error rate. One-way MANOVA is discussed in Chapters 8 and 9.

Discriminant function analysis is also available for those situations in which the IV comprises between-subjects groups (cf. Section 3.1.3.1).

3.1.2.7 One-Way MANCOVA In addition to dealing with multiple DVs, multivariate analysis of variance can be applied to problems in which there are one or more covariates (cf. Section 3.1.2.2). In that case, the technique becomes multivariate analysis of covariance. In the educational therapy example in Section 3.1.2.6, it might be worthwhile to adjust the DV scores for pretreatment baseline performance in academic achievement and attention span. Here the covariates are measured on the same scales as the DVs. This is a classic use of covariance analysis, in which pretest scores comprise the covariates. Since adjustment is made for pretreatment scores, any differences in the posttest scores (DVs) can more clearly be attributed to differences in assignment to treatment group (the two types of educational therapy plus control groups that make up the IV).

In the one-way ANCOVA example of religious groups in Section 3.1.2.2, it might be interesting to test a variety of attitudes in addition to that toward abortion. For example, additional dependent variables might be political liberalism versus conservatism, attitudes toward ecology issues, and so on. Here again, differences in attitudes might not only be reflecting differences in religion per se, but might be related to age (which in turn varies with religious affiliation). For the MANCOVA, then, age would be the covariate, religious affiliation the IV, and attitudes the DVs. It would thus be possible to test for differences among religious affiliation on this set of attitude measures, after adjustment for each respondent's age.

3.1.2.8 Factorial MANOVA Factorial multivariate analysis of variance is the extension of MANOVA to research with more than one IV—just as factorial ANOVA extends analysis of variance to multiple IVs. In the memory example in Section 3.1.2.6, a second IV might be the pictorial complexity of the photos to be recalled. The main effect of complexity on the set of DVs would then be available for test, as well as the main effect of serial position. It would also be possible to test hypotheses about differences in effects of serial position on the DVs as a function of pictorial complexity. The complexity IV (simple vs. complex) might be measured between subjects. This would produce a mixed within-between factorial MANOVA.

Specific comparisons can be made among cells in the design, and influence on individual DVs can be assessed. For example, an unexpected result might be a significant interaction between serial position and complexity. Recall rate might be the only DV influenced by the interaction, with no significant effect on number of words used in the recall or on rating of vividness. A trend analysis might then show that the nature of the interaction is such that there is no effect on recall of serial position for highly complex photos. For the less complex photos there is a quadratic trend, with recall relatively poor for items in the middle of the sequence, but better for items near the beginning or end of the list. Or, more likely, this result could have been anticipated, with a full a priori trend analysis replacing the test of the interaction, followed by the post hoc trend analysis.

For a between-subjects example, children assigned to educational therapy control groups in the previous examples might also be classified on the basis of sex as an additional IV. The effect of the interaction between treatment group and sex (whether effect of treatment is the same for boys and girls) on the set of DVs (academic achievement and attention span) could then be tested.

Virtually any complex ANOVA design (cf. Chapter 2) with multiple DVs can be analyzed through MANOVA procedures, given access to appropriate computer programs. Factorial MANOVA is covered in Chapter 8.

3.1.2.9 Factorial MANCOVA   It is sometimes desirable to incorporate one or more covariates into a factorial multivariate analysis of variance, producing factorial multivariate analysis of covariance. By combining features of both MANCOVA and a factorial arrangement into the educational therapy example, we could use pretest scores on academic achievement and attention span as covariates. That is, pretreatment scores on the two scales would serve as covariates. Posttreatment scores on the same two scales would then serve as the DVs. IVs would be assignment to one of the three treatment groups (two types of educational therapy and control) and sex. Each of the effects would then be tested after adjustment for covariates. The main effect of treatment assignment on the DVs would be assessed after adjustment for pretreatment (baseline) performance. The interaction would test whether treatment effects differed as a function of sex, again after adjustment for baseline performance. (The main effect of sex would probably not be of interest, strange as that statement may sound out of context.)

Here again, if significant main effects or interactions are found, procedures are available for making specific comparisons among groups and for evaluating IV and interaction influences on the various DVs. Factorial MANCOVA is discussed in Chapter 8.

3.1.3 Prediction of Group Membership

In survey or correlational research in which discrete groups can be identified, the emphasis is frequently placed on predicting group membership from a set of variables. Discriminant function analysis is designed to accomplish this.

3.1.3.1 One-Way Discriminant Function   In one-way discriminant function analysis, the goal is to predict membership in groups (with groups as the DV) on the basis of a set of IVs. With a little thought, it can be seen that this is the same question as is addressed by MANOVA, but turned around. If groups differ significantly on a set of variables, that set of variables will reliably discriminate among groups. As might be expected, the between-subjects example of Section 3.1.2.6 is amenable to analysis within the framework of discriminant functions. Discriminant function analysis could be used to evaluate prediction of a child's membership in one of the three groups (two types of educational therapy and a control) on the basis of posttreatment scores on academic achievement and attention span. That is, the analysis would tell us if group membership could be predicted reliably (which would be the case if MANOVA were to show a significant effect of group assignment on these measures).

Another example of one-way discriminant function analysis might be prediction of religious affiliation on the basis of scores on a number of attitude measures (e.g., attitude toward abortion, a liberal vs. conservative measure, a measure of ecological concern). Or an attempt could be made to discriminate belly dance students from trained belly dancers on the basis of previous dancing training, shimmy rate, and a measure of introversion-extraversion.

Procedures are available for assessing the relative contribution of IVs in predicting group membership. For example, it might be found that the major source of discrimination among religious groups was abortion attitude, with little predictability contributed by political and ecological attitudes.

Additionally, classification procedures have been developed, in which the effectiveness of IVs as a set is evaluated by determining how well individual cases are classified into their appropriate groups on the basis of knowledge of their scores on the IVs. One-way discriminant function analysis is covered in Chapter 9.

3.1.3.2 Hierarchical One-Way Discriminant Function

Frequently sets of IVs can be ordered in some way so that their effectiveness as predictors of group membership can be evaluated sequentially.[4] This is what hierarchical discriminant function analysis is designed to do. For example, in looking at attitudinal variables as predictors of religious affiliation, there might be some interest in a prior ordering of variables in terms of expected contribution to prediction. Abortion attitude might be given highest priority, political liberalism versus conservatism second highest priority, and ecology attitude lowest priority. Hierarchical discriminant function analysis would first assess the degree to which religious affiliation could be reliably predicted from abortion attitude. Then, gain in prediction could be assessed with the addition of political attitude, and finally with the further addition of ecology attitude.

This type of analysis could provide useful information in two ways. First, it might be desirable to reduce the size of a set of predictors. For example, if it were found that political and ecological attitudes did not add appreciably to abortion attitude in predicting religious affiliation, they could be dropped from further analysis. With a large set of IVs, this is frequently a preliminary step in analysis, sometimes with relatively inexpensive predictors given highest priority. Second, hierarchical discriminant function analysis fits within the framework of covariance analysis. At each step of the hierarchy, the analysis determines whether the variable being added significantly contributes toward predic-

---

[4] Here again, hierarchy refers to sequential entry of variables rather than structural hierarchy among them.

tion, after adjusting for the influence of higher-priority variables. At each step, then, higher-priority variables act as covariates.

Hierarchical discriminant function analysis can also be used to evaluate sets of variables. For example, by first entering a set of demographic variables with high priority and then entering a set of attitudinal variables with lower priority, one might see whether attitudes reliably predict religious affiliation after adjustment for demography. Hierarchical discriminant function analysis is covered in Chapter 9.

3.1.3.3 Factorial Discriminant Function
If groups are cross-classified on the basis of more than one attribute, prediction of group membership from a set of IVs is possible through factorial discriminant function analysis. For example, respondents might be classified on the basis of sex as well as religious affiliation. The DVs, then, are sex and religious groups. The three attitude measures listed in Section 3.1.3.1 (abortion, politics, and ecology) could be used to predict sex, ignoring religion; or religion, ignoring sex; or both sex and religion. It can be seen that this is the obverse of the questions answered in factorial MANOVA. For a number of reasons, procedures and computer programs designed for discriminant function analysis do not readily extend to factorial arrangements of groups. Unless some special conditions are met (cf. Chapter 9), it is usually expedient to rephrase the research question in such a way that factorial MANOVA can be used.

3.1.3.4 Hierarchical Factorial Discriminant Function
Difficulties inherent in factorial discriminant function analysis extend to hierarchical arrangements of predictors. Usually, however, you will find that questions of interest can readily be rephrased in terms of factorial MANCOVA.

3.1.4 Structure

A final set of questions is concerned with the latent structure underlying a set of variables. Depending on whether the search for structure is made on empirical or theoretical grounds, the choice is between principal components and factor analysis.

3.1.4.1 Principal Components
If scores on a number of variables are available from a group of subjects or other experimental units, it is frequently interesting to explore whether these variables reflect some smaller number of characteristics on which the subjects differ. That is, a number of DVs are available. The question is whether these DVs represent some underlying, or latent, structure of IVs. Or, perhaps more simply stated, how do variables cluster together? If no underlying structure is hypothesized prior to data collection, the technique typically is rotated principal components analysis. The underlying variables are called the principal components.

For example, suppose people are asked to rate various antidotes to stress or mechanisms for coping with it. Is it possible that these multitudinous antidotes really reflect a few basic global coping mechanisms? The underlying mechanisms might be degree of social contact, physical activity, and instrumental manipulation of stress producers. Or, suppose we have many potential measures of outcome of some treatment, such as psychotherapy. It might be worthwhile to develop a reduced set of components so that further research comparing types of psychotherapy could be enhanced. Principal components analysis is valuable in developing a set of components with particularly useful characteristics in future research (i.e., a small set of uncorrelated components based on the original dependent measures). Principal components analysis is discussed in Chapter 10.

3.1.4.2 Factor Analysis   When there is some prior hypothesis about underlying structure, rotated factor analysis can be used to assess the structure and the extent to which empirical structure conforms with hypothetical structure. In this case the underlying IVs are called factors. In the example in the preceding section involving mechanisms for coping with stress, there might be a prior expectation that a large set of coping mechanisms could be reflecting two major factors: nature of dealing with the problem (escape vs. direct confrontation) and degree of use of social supports (withdrawing from people vs. seeking them out).

It is also sometimes useful to explore differences between groups in terms of latent structure. For example, the two coping factors cited might be characteristic of young college students, whereas older adults might exhibit a substantially different structure of coping styles.

As implied in this discussion, factor analysis is useful in developing and testing theories. What is the structure of personality? Are there some basic dimensions of personality on which people differ? By collecting scores on a number of measures of personality, we can address questions about underlying structure through factor analysis. Factor analysis is covered in Chapter 10.

3.2 A DECISION TREE

A summary of these major research questions appears in Table 3.1. For each question, choice among techniques follows decisions about number of DVs and IVs and about whether or not some variables are usefully viewed as covariates. The table also briefly describes analytic goals associated with some techniques.

The paths in Table 3.1 are only recommendations concerning an analytic strategy. Researchers frequently discover that they need two or more of these procedures or, even more frequently, a judicious mix of univariate and multivariate procedures to answer fully their research questions.

In short, we recommend a flexible approach to data analysis in which both univariate and multivariate procedures are drawn on to clarify the results of research for both the researcher and the audience.

TABLE 3.1 CHOOSING AMONG STATISTICAL TECHNIQUES

Major research question	Number of dependent variables	Number of independent variables	Covariates	Analytic technique	Goal of analysis
Degree of relationship among variables	One	One		Bivariate r	
	One	Multiple	None	Multiple R	Create a linear combination of IVs to optimally predict DV.
	One	Multiple	Some	Hierarchical multiple R	
	Multiple	Multiple		Canonical R	Maximally correlate a linear combination of DVs with a linear combination of IVs.
Significance of group differences	One	One	None	One-way ANOVA or t test	
	One	One	Some	ANCOVA	
	One	Multiple	None	Factorial ANOVA	
	One	Multiple	Some	Factorial ANCOVA	
	Multiple	One	None	One-way MANOVA or Hotelling's T^2	Create a linear combination of DVs to maximize group differences.
	Multiple	One	Some	One-way MANCOVA	
	Multiple	Multiple	None	Factorial MANOVA	
	Multiple	Multiple	Some	Factorial MANCOVA	
Prediction of group membership	One (groups differ on one attribute)	Multiple	None	One-way discriminant function	Create a linear combination of IVs to maximize group differences (DVs).
	One (groups differ on one attribute)	Multiple	Some	Hierarchical one-way discriminant function	
	Multiple (groups differ on several attributes)	Multiple	None	Factorial discriminant function	
	Multiple (groups differ on several attributes)	Multiple	Some	Hierarchical factorial discriminant function	
Structure	Multiple (observed)	Multiple (latent)		Principal components (empirical)	Create linear combinations of observed variables to represent latent variables.
	Multiple (observed)	Multiple (latent)		Factor analysis (theoretical)	

3.3 TECHNIQUE CHAPTERS

Chapters 5 through 10 are the basic technique chapters and follow a common format. First, the technique is described and the general purpose briefly discussed. Then the kinds of questions that can be answered through application of that technique are suggested. These go beyond the "major research question" of Table 3.1. Next is a discussion of limitations associated with the technique on both theoretical and practical grounds. This section lists assumptions particularly associated with the technique. Checks of the assumptions are described, along with suggestions for dealing with violations. Then the statistical development of the procedure is briefly covered, along with a hand-worked small sample example, where possible.

The succeeding section explicates the major types of the technique, where appropriate. Then some major issues associated with the technique are covered. These include special statistical tests, data snooping, and the like. A direct comparison of computer programs is then made from among those available in the BMD and SPSS computer packages.

The next section then shows, step by step, application of the technique to a set of variables selected from an actual large sample survey, described in Appendix B. Assumptions are tested and violations are dealt with where necessary. Major hypotheses are evaluated and follow-up analyses are performed where indicated. Then a Results section is developed, as might be appropriate for submission to a professional journal. Where more than one major type of technique is covered in the chapter, there may be additional large sample examples. Finally, each technique chapter ends with a description of some applications of the technique found in the behavioral and social science literature. This is a near random selection of articles. The only criteria for selection of examples were that they were recent and covered a variety of fields. They are not meant to be taken as models, but rather as examples of the way techniques are currently being used.

In working with these technique chapters, it is suggested that the student/researcher apply the various analyses to some interesting large sample data set. Many data banks are readily accessible through computer installations.

Further, although we have tried to recommend methods of reporting multivariate results in Chapters 5 through 10, it may be inappropriate to report them fully in all publications. Certainly one would always want to mention that univariate results were supported and guided by multivariate inference. But the gory details associated with a full disclosure of multivariate results at a colloquium, for instance, may require more attention than one can reasonably expect from an audience. Likewise, a full multivariate analysis may be more than some journals are willing to reprint.

3.4 PRELIMINARY CHECK OF THE DATA

Before applying any technique, or sometimes even before choosing a technique, you should make some basic checks of the data set. The fit of the data with assumptions underlying parametric multivariate statistics should be checked. Though each technique chapter lists the assumptions specific to that technique, all require consideration of procedures in Chapter 4.

Chapter 4

Conditioning Matrices: Cleaning Up Your Act Before Analyzing Your Data

This chapter deals with a whole set of issues to be considered before analyzing data. Careful consideration of these issues can be time-consuming and tedious but is fundamental to an honest analysis of the data—or an analysis of honest data. Many of the multivariate procedures are known to be sensitive to rather slight changes in correlations among variables, so it is imperative that the correlations be as "clean" as possible.

Several sets of issues are relevant to a clean correlation matrix. Some of them involve obvious and not so obvious procedures that will artificially inflate or deflate the correlations themselves. Missing data, the bane of (almost) every researcher, *has* to be handled somehow. A variety of methods of treating missing data, each with its own implication for the results, is discussed. Outliers, cases that are extreme, create other headaches because the sizes of correlations are sometimes unduly influenced by them. Finally, the multivariate procedures are based on a whole set of assumptions. Some assumptions are critical to some procedures and some to others. Assumptions relevant to each procedure, and ways to test them, are discussed individually in each technique chapter. The rationale behind some of the more common assumptions is presented here together with data transformations that may bring distributions from aberrant variables into line with the assumptions.

You may find the material in this chapter difficult from time to time. It has been necessary to refer occasionally to material covered in technique chapters, material that will be more understandable after perusal of Chapters 5 through 10. You may find it helpful to read this chapter now to get an overview of tasks to be accomplished prior to actual data analysis and then read it more carefully once the remaining chapters have been mastered.

4.1 HONEST CORRELATIONS

In the best of all possible worlds, the variables measured in studies would be continuous, normally distributed, and linearly related to one another. Under such conditions, the Pearson product-moment correlation coefficient reflects accurately and completely the relationships among variables. However, in practice variables often do not meet these criteria and correlation can be inflated, deflated, or simply inaccurately calculated. Careful attention to construction of variables and to implications of their distributions can reduce the likelihood of artifactual correlation, as indicated next.

4.1.1 Inflated Correlation

Variables with skewness (Section 4.4.1) or variables with outlying values (Section 4.3) may produce artifactually high correlations. Transformation of the variables (Section 4.5) or various other methods for reducing skewness and eliminating outliers may be required.

 The manner in which a composite variable is constructed also may inflate correlation. Frequently a single variable is constructed by pooling responses to several different items. Scales on personality inventories, measures of socioeconomic status, health indices, and many other variables used in social and behavioral studies are so constructed. Correlations between variables are inflated if the same information is used in construction of more than one of them. Correlations differ in both magnitude and direction. The magnitude of correlation will be inflated in a positive direction if one response contributes to the size of two composite variables. The magnitude of r will be inflated in a negative direction if, for instance, a yes response increases the size of one variable while a no to the same item increases the size of another variable. If composite measures are being constructed, foreswear the joys of reusing items even when they could logically contribute to more than one variable.

4.1.2 Deflated Correlation

Relationships will appear lower than they should under several conditions. Presence of outliers, skewness, and failures of linearity and homoscedasticity are among them, as discussed in Sections 4.3 and 4.4. Data transformations (Section 4.5) may ameliorate some of these situations.

 A falsely small correlation can be obtained if the range of responses to a variable is restricted somehow in the sample. In a study of success in graduate school, for instance, quantitative ability would not emerge as highly correlated with other variables if all the students were selected from the mathematics department and all had about the same high scores in quantitative skills.

 If your correlation is too small because of restricted range in sampling, you can estimate its magnitude in a nonrestricted sample by using the following equation if you happen to know or are able to estimate the size of the standard deviation in the nonrestricted sample.

$$\tilde{r}_{xy} = \frac{r_{t(xy)}\left[\dfrac{S_x}{S_{t(x)}}\right]}{\sqrt{1 + r_{t(xy)}^2\left[\dfrac{S_x}{S_{t(x)}} - 1\right]}} \tag{4.1}$$

where $r_{t(xy)}$ is the correlation between X and Y with the range of X truncated, S_x is the unrestricted standard deviation of X, and $S_{t(x)}$ is the truncated standard deviation of X.

Correlations may also be deflated between a continuous and a dichotomous variable (or two dichotomous variables) if most of the responses to the dichotomous variable fall into one category. The highest correlation that such data could generate may be well below 1. Some recommend dividing the computed (but deflated) correlation by the maximum it could achieve given the split between the categories and then using the resulting figure in subsequent analyses, a procedure not without hazard, as discussed by Comrey (1973).

4.1.3 Inaccurately Computed R

The algorithms used by several of the multivariate programs may inaccurately compute the values in the correlation matrix when some of the variables have means that are large numbers and standard deviations that are small. The problem is that for some programs some computers may translate only the first six digits or so of a number but not the remaining digits. If all the variability in scores occurs in the digits that are dropped, then correlations between these variables and others are inaccurately computed. In effect, all the scores are the same for that variable even though different scores were entered into the computer. A statistic called the coefficient of variation, which is simply the standard deviation divided by the mean, alerts the researcher to this problem. When the coefficient of variation is 0.0001 (or less), computational inaccuracy is likely. The problem can be solved by subtracting the same large constant from every score for a variable so affected before calculating **R**. Subtracting (or adding) a constant from every score does not, of course, affect the size of r.

4.2 MISSING DATA

One of the nastiest problems in data analysis is that of missing data. The problem occurs when rats die, respondents become recalcitrant, or somebody goofs. Its seriousness depends on how much is missing and *why* it is missing.

If only a few units of data are missing from a large data set, the problems created are not so serious and almost any procedure for handling them should yield similar results. If, however, a goodly number of data points are missing, particularly from a small to moderate-sized data set, the problems can become very serious. Unfortunately, there are as yet no firm answers to questions regarding how much missing data can be tolerated for a sample of a given size.

The pattern of missing data may be more important than the amount missing. Randomly missing data points scattered throughout a data matrix rarely pose serious problems. Nonrandomly missing values, on the other hand, are always serious. If whole cases are missing in a nonrandom fashion (e.g., because of employment in a survey conducted mostly between 9 and 5 on weekdays), then inferences are limited to the sample from which data *were* obtained. Sometimes the problem is more subtle. Refusals to answer questions about income might, for instance, be correlated with certain attitudes so that exclusion of all incomplete cases from analysis on the basis of missing income data would distort inferences regarding attitudes in the population.

Although the temptation to assume randomness about missing data is nearly overwhelming, probably the safest thing to do is to *test it*. Use information that you *were* able to gather to test for relationships between critically important variables and the fact of missing data. For instance, from the preceding example, one could perform a *t* test of mean differences (or whatever statistic is appropriate given the data) in attitude between groups with and without data missing on income. A significant difference should lead to caution in generalizing results, whereas an insignificant one should produce more confidence in them. In any event, as a researcher you would then have more information about the relationship between income and attitude to report to your readers. If BMDPAM is available to you, you can learn more about your missing data than you might ever want to know.

A decision about how to handle missing data must be made by most researchers sooner or later (usually sooner) and the decision is important to the results obtained. At best, the decision is between several bad alternatives, five of which with some of their implications are listed in the five subsections that follow. For greater detail on these alternatives and others, consult Rummel (1970) and Cohen and Cohen (1975).

4.2.1 Treating Missing Data as Data

Although it might be embarrassing, it is possible that failure to respond to a question or to cooperate at all with research is a very good predictor of the behavior of interest. With the use of a dummy variable in your analysis, obtained by assigning cases 0 or 1 on a variable created for this purpose (depending on whether or not data are missing), the liability of missing data could become an asset. When coupled with insertion of the mean value on the variable with missing values, this procedure has numerous advantages over other strategies: The fact of missing data can be investigated as a predictor without distorting the central tendency of the variable itself (see Cohen & Cohen, 1975, pp. 265–290).

4.2.2 Deleting Cases or Variables

One of the more conservative procedures for handling missing values is simply to drop any cases or variables that contain them. If only a few cases have missing data and they seem to be a random subset of the whole sample, this may be a good procedure. Similarly, if missing values are concentrated in a few variables and they are not critical to the analysis, or are highly correlated with other, complete variables, the variable(s) with missing values might profitably be dropped.

But missing values are usually scattered throughout cases and variables so that deletion of them could mean a substantial loss of data. A researcher who has expended considerable time and energy collecting data is not likely to be eager to toss a lot of them out. Moreover, if cases with missing values are not randomly distributed through the data (if, for instance, failure to reveal income is systematically related to income level), then distortions of the sample occur by omitting them.

4.2.3 Using a Missing Data Correlation Matrix

Another option for handling missing data (called pairwise deletion in the SPSS framework or performed by BMDPAM or P8D) is to use all available pairs of values when calculating the correlations in **R**. A variable with 10 values missing would have all of its correlations with other variables based on 10 fewer pairs of numbers. If some of the other variables have missing data also, but among different cases, the N for those correlations will be further reduced. Because standard error of the sampling distribution for r is based on N, some correlations will be less stable than others in the same correlation matrix.

But that is not the only problem. In a correlation matrix based on complete data, the sizes of some correlations place constraints on the sizes of others. In particular,

$$r_{13}r_{23} - \sqrt{(1 - r_{13}^2)(1 - r_{23}^2)} \leq r_{12} \leq r_{13}r_{23} + \sqrt{(1 - r_{13}^2)(1 - r_{23}^2)} \quad (4.2)$$

The correlation between variables 1 and 2, r_{12}, cannot be smaller than the value on the left nor larger than the value on the right in a three-variable correlation matrix. If $r_{13} = .60$, and $r_{23} = .40$, then r_{12} may not be smaller than $-.49$ nor larger than $.97$. If, however, r_{12}, r_{23}, and r_{13} are all based on somewhat different cases because of missing data, the constraints are lifted.

Most multivariate statistics require calculation of eigenvalues and their corresponding eigenvectors from correlation matrices (see Appendix A). Under conditions of missing data and loosened constraints on size of correlations, eigenvalues may become negative. Because eigenvalues represent variance, negative eigenvalues represent something akin to negative variance. Moreover, because the total variance that is partitioned is constant (usually equal to the number of variables), the size of positive eigenvalues—variance—would have to be inflated by the size of negative eigenvalues—negative variance. The statistics derived under these conditions are very likely distorted. In some cases, the computer programs will tell you that derivation of eigenvalues from a correlation matrix is impossible; in others, not. You may check for yourself the size of all eigenvalues through SPSS FACTOR or BMDP4M.

However, for these problems to be created, the correlations have to move out of the range of their constraints with complete data. With fairly large data sets and fairly complete data, they may not do so even if some of them are based on slightly different pairs of cases. Under these conditions, use of a missing data correlation matrix may provide a reasonable multivariate solution and have the advantage of using all the data available. Use of this option for the missing data problem should not be rejected out of hand, but should be used cautiously with a wary eye to negative eigenvalues.

4.2.4 Estimating Missing Data

Another option is to eliminate empty slots in the data matrix by filling them. There are at least three popular schemes for doing so.

4.2.4.1 Using Prior Knowledge As a researcher in an area, one probably has an educated guess concerning the approximate size of values that are missing. For instance, one might be able to predict with certainty that a value would be above or below a median or, say, one, two, or three standard deviation units from the mean. In a largish data set with few values missing, insertion of well-educated guesses will not distort a multivariate solution.

Another possibility, if several data points are missing on a critical variable, is to dichotomize the variable: Convert all the points to "high" or "low" values (1s or 0s) and guess at the missing values. Often one can be certain that a case will fall above or below a median without being too certain just what value it would take in a continuous distribution. Dichotomization of a variable may, however, severely limit its predictive power, as well as misclassify some of the cases that fall near the median.

4.2.4.2 Inserting Mean Values Because, in the absence of all other information, the mean value is a best guess about a missing score on a variable, means are frequently inserted for missing values. Means are calculated using all available data and then placed in empty slots. Part of the attraction of this procedure is that it is conservative: The mean for the distribution as a whole remains the same and the researcher is not required to guess at missing values. If insertion of the mean is coupled with dummy variable coding for missing data, other advantages accrue, as well, as noted in Section 4.2.1.

The disadvantage to this procedure is that the correlations between a variable with mean inserted in several slots and other variables will be lowered (closer to zero). The amount of reduction in r depends on the amount of missing data. Because multivariate solutions are sometimes sensitive to slight changes in correlation coefficients, putting a variable at a disadvantage by inserting several means could change a solution considerably. Unless a researcher is completely in the dark concerning the value a missing datum might have, or is using the procedures described in Section 4.2.1, it is probably better to guess at the value than to insert the mean routinely.

A worthy compromise is to use certain demographic or classification attributes for the case to estimate the mean. For instance, one might substitute the mean value for women on a variable rather than the mean value for both sexes, if the case with a missing value were female. This procedure, which has a lot to recommend it, is not as conservative as inserting overall mean values and not as liberal as using prior knowledge.

4.2.4.3 Using Regression to Predict Missing Values　　An increasingly popular method for estimating missing values uses multiple regression (see Chapter 5). Variables that *are* available are used as IVs to write a regression equation for the variable with missing data (the DV). Cases with complete data generate the regression equation; the equation is then used to predict missing values for incomplete cases. Sometimes the predicted values from a first round of analyses are used along with complete cases to obtain a complete set of DV values for generating a second regression equation. The predicted values for missing data from round two are used for a third equation, and so forth until the predicted values from one step to the next are stable. The final predictions are the ones used to replace missing values. Using regression to estimate missing values is convenient through BMDPAM.

Advantages to regression for predicting missing values are that it is more objective than the researcher's guess but not as blind as simply inserting the grand mean. Disadvantages include risks of overfitting the data (as discussed in Chapter 1) and the requirement that good predictors for the variable with missing values be available in the data set. If none of the other variables is correlated with the one to be predicted, the regression solution will be tantamount to inserting the grand mean for those that are missing. Predictions regarding missing values using regression should be made only if the value predicted falls within the range of values included during development of the regression equation. Regression assumes a linear relationship between DV and IV that may hold just for the range of DV actually included. We recommend use of regression to estimate missing data only when sample size is substantial and there is not much missing data.

4.2.5 Repeating Analyses With and Without Missing Data

If you used a missing data correlation matrix or some method of estimating missing values, you might consider repeating your analyses using only complete cases. This is particularly important if the data set is small or the proportion of missing values high. If the results of analyses with and without missing data are similar, you can have confidence in them. If they are very different, however, you will need to investigate further the reasons for the change, with an eye to evaluating which result more nearly approximates "reality."

4.3 OUTLIERS

Outliers are cases with such extreme values on one or a combination of variables that they unduly influence the size of correlation coefficients, the average value

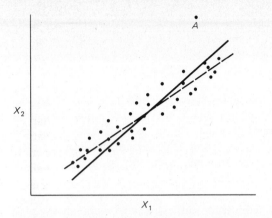

Figure 4.1 Bivariate scattergram illustrating the impact of outlying case (*A*) on selection of the best-fit regression line. To minimize squared errors of prediction with *A* included, the solid regression line rather than the dashed one would be selected.

for a group, or the variability of scores within a group. With respect to the correlation coefficient, outliers are influential because of their power to determine which one of a number of possible regression lines is chosen. Consider the oval-shaped bivariate scattergram of Figure 4.1. Any one of several regression lines, all at slightly different tilts, would provide a good fit to the data points inside the oval. But when the data point in the upper right-hand portion of the scattergram is included, the regression line that is chosen will be the one from among the several good alternatives that provides the best fit to the extreme case. That case is an outlier that had much more impact on the correlation coefficient than any of those inside the oval.

Outliers can be found in both univariate and multivariate situations, among both dichotomous and continuous variables, among both IVs and DVs, and in both the input and the output of analyses. They are a pervasive problem in the social, behavioral, biological, and medical sciences.

Cases can be extreme with respect to one variable (univariate outliers) or two or more variables in combination (multivariate outliers). For instance, a case may be perfectly within bounds regarding age (say, 15 years old) and more or less in bounds regarding income (say, $45,000 yearly) but extreme if you consider the two in combination. Multivariate outliers are cases with an unusual *combination* of two or more scores. Univariate outliers are easy to detect, multivariate outliers less so.

Cases can be extreme on continuous variables or on dichotomous ones if they are among the few cases on the distaff side of a bad split in proportions. A case can also be atypical regarding values on IVs (predictor variables), on DVs (outcome variables), or both. Lastly, a case can be unusual with respect to the input data (IVs and DVs) or with respect to the solution created by analysis: Sometimes a solution equation does not work very well for a few cases. Analysis of the cases producing a bad fit between predicted and obtained scores, for instance, may provide useful insights regarding the generalizability of the solution to the population.

4.3.1 Detecting Outliers

When outliers are present, one has either statistical problems if they are not eliminated, or moral problems if they are, due to alteration of either the sample or the variables. Some of the moral problems can be reduced if the researcher considers the search for outliers as a screening procedure for locating cases that were not part of the population to begin with. Purifying the sample is respectable; tossing out cases is suspicious.

4.3.1.1 Univariate Outliers   Univariate outliers may be found among both continuous and dichotomous variables and among both IVs and DVs by inspection of tables of univariate descriptive statistics.

Among dichotomous variables, those with very uneven splits between the two categories, say 90%–10%, may create outliers. Certainly the correlation coefficients between these variables and others are truncated. Probably the scores of the 10% of the cases are more influential in correlations than those in the 90% as well. Some (e.g., Rummel, 1970) suggest deleting dichotomous variables with 90 –10 splits between categories.

Among continuous variables, those cases with very deviant standardized scores on one (or more) variable(s) are outliers. Some recommend using a standardized score of ±3.00 as a cut for identifying outlying cases. Others point out that the proportion of cases one expects with extreme standardized scores depends on the size of the sample: With a large N, one would anticipate a few standardized scores in excess of 3.00. If a variable is approximately normally distributed, one can use the z tables to calculate the expected percentage of cases with standardized scores over 3.00 for a given sample size. This expected percentage can be compared with the obtained one to evaluate the extremeness of outlying cases. Z scores may be inspected by using BMDPID or produced by using SPSS CONDESCRIPTIVE (plus Option 3).

When one is trying to identify outliers in an analysis whose main goal is evaluating group differences, it is important to identify outliers among each group separately. If different treatments applied to groups are effective, they may shift scores around enough so that, with groups pooled, those most sensitive to treatment would appear to be outliers. Clearly, one would not want to throw out treatment effects by throwing out those cases. BMDP7D produces histograms of any variable separately for each group.

Once univariate outliers have been found and dealt with (Section 4.3.2), one should search for multivariate outliers.

4.3.1.2 Multivariate Outliers   Cases that have an unusual pattern of scores are harder to detect than those just described. These cases are multivariate outliers because they have unusual combinations of scores, but their univariate standardized scores on variables, considered singly, may be within expected ranges.

The first problem is to detect those cases that are suspected of being extreme; the second is to identify the variables involved. Detection of outlying cases can be accomplished through several of the programs in the BMDP series (4M, 7M, AM, and 9R) and through BMD10M. Multivariate outliers can also be analyzed in SPSS NEW REGRESSION (available beginning with Release 9.0) by requesting Mahalanobis distance within the CASEWISE subcommand. Interpretation of Mahalanobis distance is demonstrated in Section 5.8.1. Unfortunately, submitting the same data to different programs may produce a slightly different list of outlying cases. BMD10M uses a different algorithm to detect outliers and sometimes identifies up to 20% of the cases as outliers. Obviously, a very conservative probability estimate for a case being an outlier, say $p <$ 0.001, should be used with this program.

As with univariate outliers, multivariate outliers should be identified within each group if the goal of the analysis is evaluation of group differences. Tests for within-group multivariate outliers are provided in BMDPAM and BMDP7M. Use of other programs, including SPSS NEW REGRESSION, requires separate runs for each group.

It is also possible that, when a few cases identified as outliers are removed, other cases will become extreme with respect to the central tendency of the group. Sometimes you have to put data through the programs several times, each time deleting cases identified as outliers on the last pass, until finally no new outliers are identified.

Once the outliers are identified, you may want to discover the variables on which each case is extreme. BMDP1D may be used to show the scores on each variable for each outlying case (as illustrated in Chapter 5). Another way to find them is, for each case separately, to form two groups, a suspected outlier in one group and all nonoutliers in the other. Stepwise discriminant function analysis is then performed to identify variables that do and do not discriminate between the outlier and nonoutliers (see Chapter 9). Groups are formed by dummy variable coding: creating a new variable and assigning the case identified as an outlier the value 1 on the variable, and assigning a zero to nonoutlying cases. The variables studied in the discriminant analysis can be all variables or only some of them. Those variables that form the discriminant function, the regression equation that differentiates among groups, are the variables on which the case may be an outlier. Variables that do *not* enter the equation are not creating aberrant patterns.

One problem with discriminant function analysis, however, is that all the variables that do enter the equation may not be responsible for the unusual pattern of scores. If BMDP9R (see Chapter 5) is available to you, you can perform an "all possible subsets" regression analysis with a dichotomous DV (outlier vs. nonoutliers) instead of a discriminant function analysis. Variables on which a case is an outlier will turn up repeatedly in the regression equations, while others appear less regularly.

4.3.2 Reducing the Influence of Outliers

Once the offending cases have been identified, there are several strategies for lowering their influence on correlation.

4.3.2.1 Checking Accuracy of Data   A case just might be an outlier because data were entered incorrectly for it. Check the values for the case to make sure that they were entered correctly.

4.3.2.2 Transforming Distributions with Outliers   A good option for reducing the impact of outliers is to transform the data for the entire distribution of a variable so that extreme scores are moved nearer the central tendency of the distribution. This method preserves the unusualness of a case without allowing it undue influence on size of r. Data transformations are considered in Section 4.5.

4.3.2.3 Deleting Outlying Cases   A second alternative is to exclude cases identified as outliers from subsequent analyses. This alternative enjoys the advantage of expedience. But its disadvantage is that the sample is changed by exclusion of cases. Inferences can no longer be generalized to the population of cases from which the sample was originally drawn, although one could convincingly argue that the outlying case was not a member of the population to begin with. If this alternative is used, it is important to describe, if possible, the kinds of cases to which your results do not apply. An analysis of outlying versus nonoutlying cases on variables might reveal some interesting differences between the two groups.

4.3.2.4 Deleting Variables   If you are lucky, the same variable may be behaving badly in several outlying cases. Unless the variable is critical to the analysis, eliminating it could reduce the number of outliers.

4.3.2.5 Changing Scores on Variables Causing Outliers   Rather than simply dropping a case, it may be possible to manipulate the scores assigned to it on offending variables to retain it for analysis. Because measurement of variables is sometimes rather arbitrary anyway, this alternative may be attractive. There are several strategies for changing the scores.

One option is to change the standardized score—recode all standardized scores larger than 3.00 as 3.00, for instance. Another option is to assign the outlying case a raw score on the offending variable that is one unit larger (or smaller) than the next most extreme score in the distribution. Both these options preserve the deviancy of a case without allowing it to be so deviant that it perturbs correlation.

4.4 ASSUMPTIONS

Underlying some multivariate procedures and most statistical tests of their outcomes is a set of assumptions. Although blessed with different names, the assumptions of normality, linearity, and homoscedasticity are interrelated. However, for purposes of clarity they will be discussed separately. The fourth area of concern, multicollinearity and singularity, is not so much an assumption as a restriction on correlation matrices that, if violated, can render analyses meaningless. For want of a better location, it is included here.

Although the assumptions are important in derivation of the statistics, they are frequently less important in application of them to a data set. Just as univariate tests of significance are reasonably robust with respect to violations of assumptions (e.g., the F test is robust to violations of normality and homogeneity of variance, as long as sample sizes are relatively equal, but not to skewness), so also the multivariate statistics are undoubtedly robust to some violations of some of their assumptions sometimes. Unfortunately, the extent of robustness of various tests to various violations is not currently known. Probably, however, the researcher need only be concerned about flagrant violations of assumptions.

Although data transformations are offered as a remedy for some problems mentioned here, they are not universally recommended. One reason is that interpretation of an analysis is restricted to the variables that went into it. For instance, though IQ may be meaningfully interpreted, the logarithm of IQ scores may or may not be meaningful. Thus, our recommendation concerning data transformations is to decide before undertaking them whether potential gains in the outcome of analysis outweigh potential increases in difficulty of interpretation.

Lastly, these assumptions and others specific to various techniques do not apply with equal force to all multivariate statistics. Many apply only when one is performing signficance tests. The third section of each technique chapter indicates how seriously the various assumptions are to be regarded for that technique. The following is a discussion of the meaning of assumptions common to all the techniques.

4.4.1 Normality

The shape of the distribution of data points for each variable is as important to multivariate as to univariate solutions. The assumption of normality can refer either to the variables themselves or to sampling distributions of statistics calculated from samples of the same size drawn repeatedly from the same population.

In factor analysis, in canonical correlation, and in the DV in multiple regression with continuous IVs, the assumption refers to the variables themselves if significance tests are to be applied. In the other procedures—MANOVA, discriminant analysis, and ANOVA applications of multiple regression where inference regarding mean differences among groups is the goal—the assumption refers to sampling distributions of measures of central tendency. Actually, it is the sampling distributions at every level of the continuous IVs for which the DV is to be normally distributed in multiple regression, but the assumption is untestable. Therefore a conservative requirement is normality of the DV overall.

With sampling distributions, the central limit theorem protects against failures of normality when sample size is large and there are roughly the same number of cases in all groups. So evaluation of normality of variables is not as critical when inferences about group differences in central tendency are the goals of analysis.

Continuous variables are badly distributed, or skewed, if there is a pileup of scores at one end or the other of the distribution with a few scores thinly spaced along the opposite tail, as shown in Figure 4.2. Discrete and dichotomous variables are skewed if too many of the scores (say 80–90%) fall in the same category. Skewness in distributions may cause distortion of Type I error rate as well as instability in estimates of regression coefficients for variables (Fleming & Pinneau, 1980).

Tests of normality of variables can be conducted through the SPSS and BMDP packages by using descriptive programs in which measures of skewness

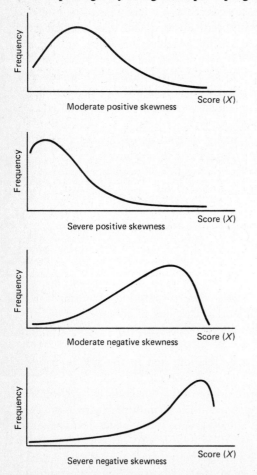

Figure 4.2   Moderate and severe positive and negative skewness.

are produced for distributions of variables. The value reported for skewness equals zero if the distribution is normal. To determine whether or not the value of skewness for a variable differs significantly from zero, you compare it against the standard error for skewness. The standard error for skewness is

$$s_s = \sqrt{\frac{6}{N}} \tag{4.3}$$

where N is the number of cases. The probability of obtaining that large a skewness value if data came from a normal distribution can then be evaluated using the z distribution, where

$$z = \frac{S - 0}{s_s} \tag{4.4}$$

and S is the value reported for skewness. A z value in excess of ± 2.58 would lead to rejection of the assumption of normality of the distribution at $p \leq$ .01. However, if sample size is sufficiently large, a variable may be significantly skewed but not enough to make a realistic difference in the analysis. Examine the actual shape of the distribution, using SPSS FREQUENCIES or one of the D (descriptive) programs in BMDP, before deciding that a transformation of the variable is required. If a variable is skewed, you should consider an appropriate transformation of it (see Section 4.5).

Another sense in which normality is important is in the assumption that the errors between predicted and obtained values are normally distributed about a mean of zero. Tests of this assumption involve inspection of residuals scatterplots and are discussed as appropriate in Chapters 5 through 10.

Most inferential statistics in multivariate analyses assume that the variables have a multivariate normal distribution. The assumption of multivariate normality is difficult to describe (but see, for instance, Tatsuoka, 1971) and difficult, if not impossible, to test if several variables are involved.

The relationship between univariate normality and multivariate normality is complicated. If a set of variables has a multivariate normal distribution, then the individual variables are univariate normal, but the reverse is not true: If variables are each univariate normal, they do not necessarily have a multivariate normal distribution.

Although univariate normality does not guarantee multivariate normality, the probability of multivariate normality for most real variables in social science is increased if all the variables have normal distributions. Univariate normality is desirable, anyway, because of the known and deleterious consequences of skewness and outliers on robustness of many significance tests. Therefore it is probably a good idea to transform variables to achieve normality, as discussed in Section 4.5, unless there are good reasons not to (e.g., the variables were measured in meaningful units).

4.4.2 Linearity

The assumption of linearity is that the relationship between two variables, between one variable and a combination of others, or between combinations of variables from each of two sets can be described using a straight line. Formulas for straight lines do not include terms with variables raised to powers or cross products of variables. (Frequently, however, nonlinear relationships can be converted to linear ones by mathematical manipulation, e.g., by taking the logarithm or some other transformation of the variables.)

Linearity is important to multivariate statistics in at least two ways. The first is that the model on which they are based is the general linear model. Chapter 11 describes the general linear model and the assumption of linearity in it.

The second reason for importance of this assumption is that the correlation coefficient, which forms the basis for most multivariate calculations, is sensitive only to the linear component of the relationship between two variables. Recall from trend analysis in ANOVA that the relationship between a DV and an ordered IV can be decomposed into several independent trends such as linear, quadratic, cubic, and so forth. Pearson r assesses only the first of these so

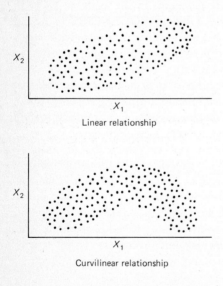

Linear relationship

Curvilinear relationship

Monotonic, nonlinear relationship

Figure 4.3   Bivariate scattergrams between X_1 and X_2 with linear and nonlinear relationships.

that a relationship that is strongly quadratic, for instance, may appear only weakly linear (or nonlinear) when evaluated using r. If only the linear components appear in the correlation matrix, and analysis proceeds exclusively from it, the true extent and pattern of the relationship among variables may not be revealed. Fortunately, in the behavioral sciences many relationships may be approximated by using straight lines even though there are significant nonlinear trends as well.

To detect gross departures from linearity among pairs of variables, bivariate scattergrams can be examined. If the scattergram is roughly oval-shaped, a correlation between the variables is indicated and the relationship is predominantly linear. If, however, the scattergram is shaped like a horseshoe, or has other curves and twists, nonlinear relationships are indicated, as shown in Figure 4.3. Data transformations of one or both variables may "linearize" a nonlinear relationship—but be sure to inspect a scattergram again after transformation.

4.4.3 Homoscedasticity

This assumption of homoscedasticity is that the variability in scores on one variable is roughly the same at all values of the other variable. Homoscedasticity is related to the assumption of normality as indicated in Figure 4.4. In the upper portion of the figure, both variables are normally distributed, as illustrated by the distributions along the axes. In this case, homoscedasticity is present. In the lower portion of the figure, one of the variables is skewed while the other is normal, and heteroscedasticity exists.

When heteroscedasticity is present, the relationship between the variables may be lawful, but it is not captured totally by the correlation coefficient. An analysis based on correlation will underestimate the extent of relationship between variables. Transformation of the skewed variables may restore normality and eliminate heteroscedasticity.

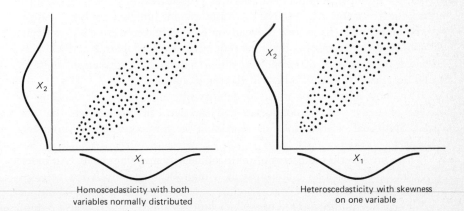

Homoscedasticity with both variables normally distributed

Heteroscedasticity with skewness on one variable

Figure 4.4 Bivariate scatterplots under conditions of homoscedasticity and heteroscedasticity.

4.4.4 Multicollinearity and Singularity

There are two different conditions of correlation matrices that can terminate an analysis or render portions of it unstable. Multicollinearity occurs when two variables in a matrix are perfectly (or nearly perfectly) correlated and when they show a similar pattern of correlations with the other variables. Scores on the WAIS and on the Stanford-Binet IQ scales, for instance, should be multicollinear. Singularity occurs when one score is a linear (or nearly linear) combination of others. Thus, the subscales that compose the WAIS would, in linear combination, produce the total WAIS score. Notice, however, that singularity does not imply high bivariate correlation, because the subscales of the WAIS are not perfectly correlated either with each other or with the total WAIS score. Nevertheless, singularity does imply a perfect multiple correlation, because all subscales of the WAIS, in linear combination, do produce the total WAIS score.

Although multicollinearity and singularity are conceptually distinguishable, they cause similar problems in multivariate analyses, namely in prohibiting or rendering unstable matrix inversion. Matrix inversion is the logical equivalent of division; calculations requiring division (and there are many of them—see the fourth sections of Chapters 5 through 10) cannot be performed on singular or perfectly multicollinear matrices because they would require division by the determinant of the matrix, which is zero (see Appendix A).

With nearly singular or highly multicollinear matrices, the determinant is not zero, but is nearly so. Division by a near-zero determinant produces very large numbers in the inverted matrix. Unfortunately, their exact sizes become unstable, fluctuating considerably with only minor changes in the sizes of the correlations in **R**. The portions of the multivariate solution that follow multiplication by an inverted matrix that is unstable are also unstable.

Although perfect singularity or multicollinearity will likely abort an analysis, the seriousness of near singularity or multicollinearity depends to some extent on which analysis is being performed. Factor analysis and canonical analysis, for example, might be described as procedures for collecting together multicollinear variables. In MANOVA, discriminant function analysis, and regression, on the other hand, one is usually interested in discovering a parsimonious set of predictors requiring the minimum number of variables. Because highly correlated pairs or sets of variables contain redundant information, there is usually no reason to include them in the same analysis.

Thus how much is too much singularity or multicollinearity depends on the type of analysis to be conducted and the parts of the analysis that are desired. Statistical entities, such as regression weights, that require inversion are suspect.

There are several indicators of multicollinearity and singularity. An index of bivariate (two-variable) collinearity is a high Pearson correlation between two variables. Inspection of the correlation matrix (standard or optional output for most programs) for values in excess of about .99 should reveal nearly redundant variables.

But sometimes the problem is more subtle, involving a combination of several variables instead of just two. Signs to look for include high SMCs or low tolerances.[1] SMCs are *squared multiple correlations* between one variable and a combination of others. To investigate multicollinearity and singularity in a data set, multiple regression is performed, with each variable in turn serving as DV and all others as IVs. If some combination of IVs has a high SMC with one of the DVs, those IVs are multicollinear or nearly singular; that is, there exists a linear combination of some variables that almost perfectly predicts another variable. The tolerance for a variable is 1 minus the SMC for that variable, so a high SMC, or a low tolerance, indicates trouble. SMCs and tolerances are available from many of the multivariate programs. Many, in fact, have default values that terminate analysis if tolerance for some variable is too low.

In factor analysis, a near-zero determinant may or may not indicate serious multicollinearity. The determinant can be seen to be the product of the eigenvalues. With many variables, a near-zero determinant can be produced when many eigenvalues are small but none dangerously so. For example, with 60 variables, several of which have eigenvalues around 0.1, the determinant may be nearly zero (to 15 or 20 decimal places) but the matrix is not seriously multicollinear. On the other hand, if any eigenvalues are negative, or are themselves nearly zero to a couple of decimal places, the matrix might be harmfully multicollinear.

Once multicollinearity or singularity is detected, there are several methods of dealing with it. Perhaps the simplest and best is to delete the offending variable(s). Because one variable is a combination of others, information is not lost by deleting it. Choice among variables to delete should include consideration of the ease of interpreting them, their relevance to other work, their cost, and the like.

A second method involves subjecting the variables to principal components analysis and then using scores on components (see Chapter 10) as the variables in subsequent analyses. Component scores are uncorrelated with one another even if composed of variables that are correlated. Of course, the principal component analysis itself must be interpretable before this procedure is helpful.

A third solution is to use stepwise, setwise, or hierarchical entry of variables into the analysis so that only one or a few of the variables that are multicollinear are used. With stepwise entry, if two variables are highly correlated with each other, the first to enter takes with it both its unique variance and the variance they share so that the second variable rarely has enough influence remaining to enter the equation.

Of these three methods, deletion of variables is probably the most straightforward and should be employed unless there are good reasons to do otherwise.

---

[1] Not all programs give tolerances or SMCs, although they could be obtained through use of regression analyses and are available in programs offering principal factor analysis (BMDP4M and SPSS FACTOR). In some cases (MANOVA), it is convenient, but occasionally misleading, to assess multicollinearity from near-zero values of determinants, as discussed in Chapters 8 and 9.

4.5 COMMON DATA TRANSFORMATIONS

Transformations are performed to change the distribution of a variable to reduce skewness, to reduce the influence of outlying cases, or to accomplish any of several other worthy statistical objectives (such as to create homogeneity of variance). It is important to check that the goal of transformation has been accomplished before incorporating a transformed variable into an analysis.

We have also observed cases in which transformed variables behaved no better (and occasionally worse) than the original ones did. So, although there are a number of good theoretical reasons for transformation, in practice the advantages may be slight.

It is also important to remember that interpretation of a multivariate solution is made on variables as they entered it. If transformed variables are used, they are the ones to be interpreted. Whether or not transformation increases difficulty of interpretation depends on the scale in which the variable was originally measured: If it was meaningful, transformation may hinder interpretation. Usually, however, measurement scales in the social sciences are somewhat arbitrary, so transformation does not increase the difficulty of interpretation.

If a square root transformation is employed, the test of mean differences may be interpreted as a test of differences between medians, since the now symmetrical distribution will have equivalent mean and median. If a logarithmic transformation is employed, the test of mean differences may be interpreted as a test of differences between geometric means. (The geometric mean is the mean of scores on the log scale transformed back to the original scale by taking the antilog of the log mean.) Transformation is undertaken because the distribution is skewed and the group mean is not a good indicator of the central tendency of the scores in the distribution. For skewed distributions, then, the median or geometric mean may be more meaningful than the mean itself, as may differences between medians or differences between geometric means.

Moderate and severe positive and negative skewness are shown in Figure 4.2. Distributions with positive skewness have a pileup of cases with very low scores. They also have significiant positive skewness values in Eq. 4.4. Distributions with negative skewness have a pileup of cases with large scores and significant negative skewness values in Eq. 4.4.

If a distribution has moderate positive skewness, the appropriate transformation is the square root. In SPSS, square root transformations are accomplished using COMPUTE; NEWX = SQRT(X). In BMDP the instruction is part of a paragraph, /TRANSFORM   NEWX = SQRT(X).

For severe positive skewness, a logarithmic transformation is appropriate. In SPSS one uses COMPUTE; NEWX = LG10(X). In BMDP, /TRANSFORM NEWX = LOG(X). If the LOG or SQRT transformation is to be applied to distributions in which there are values less than 1, add a constant to each score so that the smallest value on the variable becomes 1. If the smallest value on a variable were originally zero, for example, in SPSS use COMPUTE; NEWX = LG10(X + 1); in BMDP use /TRANSFORM   NEWX = LOG(X + 1).

If the variable has negative skewness, the best strategy is to "reflex" it and then apply the appropriate transformation for positive skewness. A variable is reflexed by subtracting each value from the largest score + 1 in the distribution. By this step, a variable with negative skewness is converted to one with positive skewness prior to transformation.

It should be clearly understood that this section merely scratches the surface of the topic of transformations, about which a great deal more is known. The interested reader is referred to Box and Cox (1964) or Mosteller and Tukey (1977) for a more flexible and challenging approach to the problem of transformation.

4.6 PRELIMINARY EVALUATION OF DATA

The checklist in Table 4.1 is offered as a guide to evaluation of data prior to analysis. The first two steps should be performed in order. The order of the next four steps is arbitrary because, for instance, outliers may cause skewness, and steps taken to eliminate the one problem may (or may not) solve the others.

TABLE 4.1 CHECKLIST FOR SCREENING DATA

1. Inspect univariate descriptive statistics for accuracy of input
   a. Out-of-range values
   b. Plausible means and dispersions
   c. Coefficient of variation
2. Evaluate number and distribution of missing data: deal with problem
3. Identify and deal with outliers
   a. Univariate outliers
   b. Multivariate outliers
4. Identify and deal with skewness
   a. Locate skewed variables
   b. Transform them (if desirable)
   c. Check result of transformation
5. Identify and deal with nonlinearity and heteroscedasticity
6. Evaluate variables for multicollinearity and singularity

Chapter 5

Multiple Regression

5.1 GENERAL PURPOSE AND DESCRIPTION

Regression analyses are a wonderfully powerful set of statistical techniques that allow one to assess the relationship between one DV and several IVs. For example, can reading ability in primary grades (the DV) be predicted from several IVs such as sex and preschool measures of perceptual and motor development? The set of techniques consists of multiple, hierarchical, and stepwise regression or correlation. Differences between these techniques involve the way variables enter the equation, that is, what happens to variance shared by variables and who determines the order in which they enter.

The terms regression and correlation are used more or less interchangeably to label these procedures, with regression generally used when the intent of the analysis is prediction, and correlation used when the intent is to measure degree of association.

Regression techniques are especially useful because they do not require that the IVs be uncorrelated with one another. This means that one can assess the relationship between several IVs (e.g., education, income, and socioeconomic status) and one DV (e.g., occupational prestige) when the IVs are correlated with one another and with the DV to varying degrees. Because regression techniques permit (but do not require) the IVs to be correlated, they are useful in experimental research when, for instance, nonindependence among treatments is created by the sudden death of an animal, or in observational and survey research in which nature has "manipulated" correlated variables. The flexibility of the regression techniques is, then, of special importance to the researcher who is interested in real-world or very complicated problems that cannot be meaningfully reduced to orthogonal designs in a laboratory setting.

The result of applying multiple regression techniques to a data set is an equation that represents a best-fit line between a (usually) continuous DV and several continuous or dichotomous IVs. Multiple regression is an extension of bivariate regression (see Chapter 2) in which several IVs, instead of just one, are used simultaneously to predict the outcome for each subject on some DV. The regression solution takes the form of the general linear model (see Chapter 11):

$$Y' = A + B_1 X_1 + B_2 X_2 + \cdots + B_k X_k$$

where Y' is the predicted value on the DV, the X's represent the various IVs (of which there are k), the A is a constant representing the Y intercept (the value of Y when all the X values are zero), and the B's are the weights assigned to each of the IVs by the regression solution.

The general problem for regression is to arrive at the set of B values, called regression coefficients, for the IVs that brings the Y values predicted from the equation as close as possible to the Y values obtained from measurement. The regression coefficients that are computed accomplish an intuitively appealing and highly desirable goal, namely, to minimize the sum of the squared deviations between the predicted Y values and the obtained Y values for the data set, or, to put it another way, to optimize the correlation between the predicted and obtained Y values for the data set. One of the important statistics derived from a regression analysis is the multiple correlation (or regression) coefficient that is a correlation coefficient between the obtained and predicted Y values (see Section 5.4.1). The predicted Y values may, in turn, be decomposed into the sum of weights and IVs that is the general linear equation.

5.2 KINDS OF RESEARCH QUESTIONS

The goal of research using regression is to illuminate the relationship between the DV of interest and a set of IVs. As a preliminary step, one can determine how strong the relationship is between DV and IVs and then, with some difficulty, assess the importance of various IVs to the relationship. With more theoretical interest, one can investigate the relationship between a DV and some IVs with the effect of other IVs statistically eliminated (e.g., the relationship between use of Valium and certain attitudinal variables, with effects of some health characteristics eliminated).

Sometimes researchers use regression to perform essentially a covariates analysis in which they ask if some critical variable adds anything to the prediction equation for some DV after other IVs—the covariates—have already entered (e.g., if biological sex adds to prediction of mathematical ability after effects of kinds of mathematical training are entered). Another strategy is to compare the ability of several competing sets of variables to predict the DV of interest (e.g., prediction of use of Valium by a set of health variables versus that by a set of attitudinal variables).

Occasionally a researcher is interested in finding the best prediction equation for some phenomenon regardless of the meaning of the equation, a goal met by stepwise regression.

The strength of regression analysis is in its ability to analyze the relationship between a DV and several IVs that can be correlated or uncorrelated, continuous or dichotomous. Discrete variables (e.g., religious affiliation) can be rendered dichotomous by converting them into a set of variables (Protestant vs. non-Protestant, Catholic vs. non-Catholic, Jewish vs. non-Jewish, and so on until the degrees of freedom have been used). When the set is entered as a group, the variance due to the discrete variable is analyzed, and, in addition, one can examine effects of the individual dichotomous components. This is called dummy variable coding and is covered in glorious detail by Cohen and Cohen (1975, pp. 173–188). ANOVA (Chapter 2) is a special case of regression in which uncorrelated IVs (treatments and their interactions) have been dummy variable coded. Most ANOVA problems can be handled through regression, but the reverse is not true.

Regression can also be used for a situation in which some IVs are dichotomous (or have been recoded to be dichotomous) while others are continuous. Many real-life problems include a mix of continuous and dichotomous predictors. If handled in the ANOVA framework, the continuous IVs have to be rendered discrete (e.g., high, medium, and low), which may impose arbitrary cuts that weaken real relationships. In regression, the full range of the IV can be maintained.

As a statistical tool, regression can be very helpful in answering a number of practical questions, as indicated by Sections 5.2.1 through 5.2.8.

5.2.1 Degree of Relationship

How good is the regression equation? That is, does the regression equation really provide a better-than-chance prediction, or, is the multiple correlation really any different from zero, when allowances for naturally occurring fluctuations in such correlations are made? For example, can one reliably predict reading ability given knowledge of perceptual and motor development and sex? The statistical procedures described in Section 5.6.2.1 will allow you to determine if your multiple correlation is probably different from zero.

5.2.2 Importance of IVs

If the multiple correlation is different from zero, then you may want to ask, Which of the IVs is important in the equation and which of the IVs can be deleted? For example, is knowledge of motor development helpful in predicting reading ability, or can we do just as well with knowledge of only sex and perceptual development? The methods in Section 5.6.1 will allow you to evaluate the relative importance of the various IVs to a regression solution.

5.2.3 Adding IVs

Suppose that you have just added a new IV (or new set of IVs) to your research design and you want to know whether or not you have improved your ability to predict. A test for improvement of the multiple correlation after addition of a new variable is given in Section 5.6.2.3. For example, can we enhance prediction of a child's reading ability by adding parental interest in reading to the three IVs already included in the equation?

5.2.4 Changing IVs

If you have already included in your equation all of the IVs that are known to be relevant and important, but prediction still is not perfect, you may be able to improve it by raising some of your IVs to powers greater than 1 (that is, by squaring, etc., some of them), or by including in the equation terms representing interactions between variables (by multiplying two—or more—IVs together).

Inspection of a scatterplot (known as residuals analysis—see Section 5.3.2.4) between the predicted and obtained Y values may reveal curvature, indicating that the relationship between the DV and the IVs is not strictly linear, but may better be described by some higher-order polynomial equation. To improve prediction, one may have to change some of the IVs as described above. For example, suppose a child's reading ability increases with increasing parental interest up to a point, and then levels off. Greater parental interest does not show greater reading ability. If the square of parental interest were added as an IV, better prediction of a child's reading ability could be achieved. There is danger, however, in too liberal a use of powers or cross products of IVs; the sample data may be overfit to the extent that results no longer generalize to a population. Procedures for using regression for polynomial curve fitting are discussed in Cohen and Cohen (1975) and McNeil, Kelly, and McNeil (1975).

5.2.5 Contingencies Among IVs

You may be interested in the way that one of the IVs behaves in the context of one, or a set, of the other IVs. Multiple regression can be used to hold several IVs statistically "constant" to examine the predictive relationship that remains between an interesting IV and the DV. For example, after adjustment for differences in perceptual and motor development, does sex predict reading ability? This procedure is described in Section 5.5.2.

5.2.6 Comparing Sets of IVs

Is one set of IVs a better predictor of a DV than another set of IVs? For example, one might calculate R (the multiple correlation coefficient) for reading ability based on perceptual and motor development and sex and another R

for reading ability based on family income and parental educational attainments. If one R is substantially larger than the other, a better set of predictor variables has been identified. Sometimes it is possible to argue that some phenomenon (e.g., use of Valium) is related to one set of IVs (e.g., health) but not another (e.g., attitudes) on the basis of this kind of comparison. Section 5.6.2.5 demonstrates a method for comparing the solutions given by two sets of predictors.

5.2.7 Predicting DV Scores for Members of a New Sample

One of the more important applications of regression involves predicting scores on a DV for subjects for whom only data on IVs are available. This application is fairly frequent in personnel selection for employment or graduate training and the like. Over a fairly long period, the researcher collects data on several IVs, say various aspects of an undergraduate record (overall GPA, GPA in major, motivation), and the DV, success in graduate school. If the IVs are significantly and strongly related to the DV, then for a new set of applicants to graduate school, regression coefficients are applied to IV scores to predict success in graduate school, and admission to the program is based partially on the outcome of solution of the regression equation.

5.2.8 Causal Modeling

Path analysis, or causal analysis, is a special application of regression analysis in which questions involving the minimum number of paths (causal influences) and their directions are asked by observing the sizes of regression coefficients with and without certain variables entered into the equation. For example, one can investigate the direct and indirect influence of intelligence on reading ability in the context of perceptual and motor development and sex. We have not included a description of path analysis in this book but direct the interested reader to Asher (1976) or Heise (1975).

    An increasingly popular set of new techniques, called structural analyses (Bentler, 1980), involves path analysis of a set of IVs and of latent variables (factors) all considered simultaneously. Analysis is facilitated by computer programs, most notably one called LISREL (Jöreskog & Sörbom, 1978). Although the method is not detailed here, you should be aware that many social scientists are finding a combination of the materials in this chapter (regression) and Chapter 10 (factor analysis) useful in inferring causal processes from nonexperimental or quasi-experimental research.

    Another loosely related development involving attempts at causal inference using both regression and quasi-experimental designs is time-series analysis. Simonton (1977) has provided a useful description of this procedure for the situation in which a relatively large number of cases is repeatedly measured for only a few periods. The same source provides a reference for other situations as well.

5.3 LIMITATIONS TO REGRESSION ANALYSES

5.3.1 Theoretical Issues

Perhaps the most important limitation to regression analyses concerns inference of causal relationships. Demonstration of causality is a logical and experimental, rather than statistical, problem. Statistics are helpful only in demonstrating that relationships occur reliably. A high multiple correlation indicates that a lot of variability is shared between one variable and a set of others, but not that the variables are causally related; shared variability could stem from many sources, including the influence of other, currently unmeasured variables. One can make an airtight case for causal relationship among a set of variables only by showing that manipulation of some of them is followed by change in others. But, as indicated in Section 5.2.8, work is proceeding in causal inference from less rigorous experimental designs.

Another problem for logic rather than statistics is that of inclusion of variables. Which DV should be measured (and how), and which IVs should be included (and how are they to be measured)? If one is striving for the highest possible multiple correlation, which IVs should be added to the equation for the most improvement in prediction? The answers to these questions can be provided by theory, astute observation, or good hunches, but they will not be provided by statistics. The regression solution itself is only as good as the selection and measurement of the variables that are used in it.

Some clues for selection of variables may be helpful. If the goal of research is *manipulation* of some DV (say, body weight), it would be strategic to include as IVs those variables that can be easily manipulated (e.g., caloric intake, physical activity) rather than those that cannot (e.g., genetic predisposition, patterns of early feeding). Similarly, if one is interested in *predicting* some variable (e.g., annoyance caused by noise) for the general population it would be strategic to use cheaply obtained IVs (e.g., neighborhood characteristics published by the Census Bureau) rather than expensively obtained ones (e.g., attitudes from in-depth interviews), provided, of course, that both work comparatively well.

5.3.2 Practical Issues

In addition to the theoretical considerations, use of multiple regression requires that several practical matters be considered, as described in Sections 5.3.2.1 through 5.3.2.4.

5.3.2.1 Number of Cases and Variables One must have more cases than variables or the regression solution will be perfect—and meaningless. With enough parameters, "one can fit an elephant." With enough variables, one can find a regression solution that completely predicts the DV for all cases, but only as an artifact of the case-to-variable ratio. Ideally one would have 20 times more cases than variables. If stepwise regression is to be used, a procedure that is notorious for capitalizing on chance, a case-to-variable ratio of 40 to 1 would be appropriate. *A suggested minimum requirement is to have at least 4 to 5 times more cases than IVs.* The lower the case-to-variable ratio, the more important it becomes that the residuals be normally distributed.

Although there is no firmly agreed-on minimum case-to-variable ratio, at least three considerations are relevant: skewness in DV, effect size, and measurement error. If the DV is skewed (and transformations are not undertaken), more cases are required. This recommendation is made for both continuous and dichotomous DVs. The size of anticipated effect is also relevant because more cases are needed to demonstrate a small effect than a large one. However, as the number of cases becomes quite large, almost any multiple correlation will depart significantly from zero, even one that predicts negligible variance in the DV. Finally, if substantial measurement error is expected from somewhat unreliable variables, more cases are needed.

If the minimum requirement is violated, there are some strategies that may help. You can reduce the number of IVs by deleting some IVs, by creating one (or more than one) IV that is a composite of several others (and deleting the others, of course), or by performing a principal components or factor analysis of the IVs (see Chapter 10) and using component or factor scores as IVs instead of the original IVs. The utility of the third technique depends on the interpretability of the principal components or factor analysis. Lastly, one could employ a "minimum subsets" regression strategy, as available in BMDP9R and as described in Section 5.5.3.

5.3.2.2. Outliers   Extreme cases will have the same deleterious effects on regression solutions as on other multivariate solutions and should be excluded from analysis, rescored, or changed in such a way that their influence on the components in the correlation matrix is reduced. Consult Chapter 4 for a summary of the procedures for detecting and eliminating univariate and multivariate outliers.

In regression, cases should be evaluated separately for univariate extremeness with respect to the DV and for both univariate and multivariate extremeness with respect to the IVs. In other words, keep the DV and the IVs separate when evaluating whether or not cases are outliers. A case that would be a multivariate outlier because the DV is unusual *in combination with the IVs* is a case that will have a large residual error for regression—and so it should. In other words, it would be "cheating" to eliminate a priori cases for which the regression solution is inadequate because the combination of DV and IVs is extreme.

5.3.2.3 Multicollinearity and Singularity   Calculation of regression coefficients requires inversion of the matrix of correlations among the IVs (Eq. 5.6), an inversion that is impossible if IVs are singular or *perfectly* correlated and is unstable if they are near singular or *highly* correlated. Review Chapter 4 on this issue. Recall that multicollinearity and singularity can be identified through high squared multiple correlations (SMCs, where each IV in turn serves as DV while the others are IVs), or *low tolerances* $(1 - SMC)$. Most multiple regression programs have default values for tolerance so that warnings are printed, variables are deleted, or the program terminates if default values are exceeded (see Section 5.7). *In regression, multicollinearity may also be detected by large standard errors of prediction of the regression coefficients.*

If multicollinearity is detected and cannot be eliminated by any of the procedures mentioned in Chapter 4 (e.g., by deletion of the least reliable variable that is involved), then ridge regression might be considered. Ridge regression is a controversial procedure that is available through BMDP2R (see Appendix C-7 of the BMDP manual). Ridge regression attempts to stabilize estimates of regression coefficients by inflating the variance that is analyzed. Although originally greeted with enthusiasm (cf. Price, 1977), serious questions about the procedure have been raised by Rozeboom (1979). The researcher who anticipates use of ridge regression should consult Rozeboom before conducting the analysis.

5.3.2.4 Normality, Linearity, and Homoscedasticity of Residuals   *Examination of residuals scatterplots provides a test of assumptions of normality, linearity, and homoscedasticity between predicted DV scores and errors of prediction.* Residuals scatterplots should be requested and examined as part of an initial regression analysis with selected but (probably) as-yet-untransformed variables. The focus of analysis is evidence of failures of normality, linearity, and homoscedasticity of residuals rather than the remainder of the regression solution. The decision regarding the desirability of variable transformation and the like includes the usual trade-off between improved fit to the assumptions of the regression model and potentially increased difficulty in interpreting variables.

Residuals scatterplots are provided by all four canned programs discussed in this chapter.[1] All provide a scatterplot in which one axis is predicted scores and the other axis is errors of prediction. In SPSS, both predicted scores and errors of prediction are standardized; in BMDP they are not. It is the overall shape of the scatterplot that is of interest. If all assumptions are met, it will be nearly rectangular in shape with a concentration of scores along the center. Figure 5.1(a) illustrates a distribution in which all assumptions are met. (Note that some programs show the standardized predicted DV scores as abscissa while others display it as ordinate.)

The assumption of normality is that the distribution of errors of prediction is independently and normally distributed at all levels of the predicted DV. Inspection of the residuals plot should reveal a pileup of scores in the center of the plot at each level of predicted score, with a normal distribution of residual errors around the center. Figure 5.1(b) illustrates a failure of normality, with a skewed distribution of residuals.

A second test for the normality of the distribution of residuals is available through SPSS NEW REGRESSION (starting with Release 9.0) and all the BMDP programs for regression. The test is a normal probability plot of residuals

---

[1] But they are not yet provided by SPSS REGRESSION (SPSS-6000) Release 8.0 as implemented and converted by Northwestern University, Vogelback Computing Center, for CYBER machines.

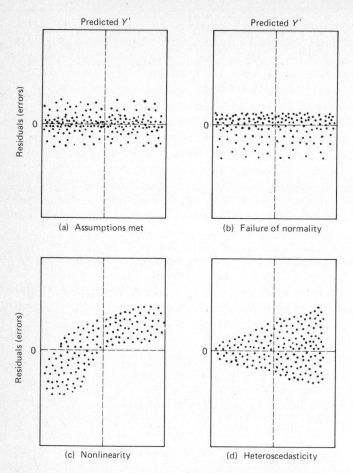

Figure 5.1   Plots of predicted values of the DV (Y') against residuals, showing (a) assumptions met, (b) failure of normality, (c) nonlinearity, and (d) heteroscedasticity.

in which their expected normal values are plotted against their actual sizes. Expected normal values are estimates of the z score of the residual a given raw score should have, given its rank in the original distribution. If the distribution of residuals is normal, the points will fall along a straight line running from the bottom left to the upper right corners of the graph, indicating that the lowest raw scores had the biggest normalized negative residuals, and so on systematically through the distribution. Distributions that are not normal will deviate from the straight line by curving above or below it in specific ways, depending on how the residuals are skewed. Bock (1975, pp. 156–160) illustrates the effects of several deviations from normality in the normal probability plot of residuals, as well as other details for the interested reader. The "fix" for the failure of normality of residuals is transformation of variables, as previously indicated.

Linearity of relationship between predicted DV scores and errors of prediction is also assumed. If nonlinearity is present, the overall shape of the scatterplot will be curved instead of rectangular, as seen in Figure 5.1(c). Typically, the relationship can be made linear by transforming variables or by adding terms to the regression equation, particularly terms involving interaction among the variables. Although adding these terms will increase R, the user must decide whether the price paid in overfitting with too many terms, or in difficulty of interpretation, is worth the gain in R.

The assumption of homoscedasticity is the assumption that the standard deviations of errors of prediction are approximately equal at all predicted DV levels. Heteroscedasticity does not invalidate the analysis so much as degrade it. Homoscedasticity means that the band enclosing the residuals should be approximately equal in width across all values of the predicted DV. A typical deviation is one in which the band becomes wider with increasing values of Y', as illustrated in Figure 5.1(d). That is, for high values of the DV, the error of prediction of the DV is greater. An example for a simple bivariate case might be the relation between income and age. For younger workers, salaries might be far more uniform than for middle-aged workers.

In some instances, heteroscedasticity may be suspected to be a function of something inherent in the order of cases. For example, subjects who are interviewed early in a survey might be expected to exhibit more variability of response owing to interviewer inexperience with the questionnaire. This can be checked in SPSS REGRESSION, BMDP1R, BMDP2R, or BMDP9R by entering cases in order of time of interview and requesting a plot of residuals against sequence of cases. The Durbin-Watson statistic, which provides a measure of autocorrelation of errors over the sequence of cases, can also be requested in either SPSS REGRESSION or BMDP9R. Details on the use of this statistic and a test for its significance are given by Wesolowsky (1976).

If failures of normality, linearity, and homoscedasticity are detected, variable transformation may be considered. Examine DV and IVs for departures from normality, using the procedures of Chapter 4. If, after transformation of nonnormal variables, the residuals scatterplot is still not rectangular, examine all DV-IV and IV-IV pairs for deviations from linearity. Inclusion of IVs raised to a power or cross products of IVs may be considered at this stage. In all cases, increased difficulty of interpretation must be balanced against an improved fit to the regression model.

5.4 FUNDAMENTAL EQUATIONS FOR MULTIPLE REGRESSION

A data set appropriate for multiple regression consists of a sample of research units (e.g., graduate students) for which scores are available on a number of IVs and on one DV. A small sample of hypothetical data with a few IVs is illustrated in Table 5.1.

Table 5.1 contains six sample units, with scores on three IVs: professional motivation (MOTIV), composite rating of qualifications for admission to gradu-

TABLE 5.1 SMALL SAMPLE OF HYPOTHETICAL DATA FOR ILLUSTRATION
OF MULTIPLE CORRELATION

Case No.	IVs			DV
	MOTIV (X_1)	QUAL (X_2)	GRADE (X_3)	COMPR (Y)
1	14	19	19	18
2	11	11	8	9
3	8	10	14	8
4	13	5	10	8
5	10	9	8	5
6	10	7	9	12
Mean	11.00	10.17	11.33	10.00
Standard deviation	2.191	4.834	4.367	4.517

ate training (QUAL), and composite rating of performance in graduate courses
(GRADE). The DV score is rating of performance on graduate comprehensive
exams (COMPR). We might now ask how well we can predict COMPR ratings,
knowing scores on MOTIV, QUAL, and GRADE. In actually addressing that
question, of course, we would consider a sample of six cases highly inadequate,
but this sample will serve to illustrate calculation of multiple correlation as
well as to demonstrate some analyses by canned computer programs. The reader
is encouraged to work problems involving these data by hand as well as by
available computer programs.

There are a variety of ways to develop the "basic" equation for multiple
correlation.

5.4.1 General Linear Equation

One way of developing multiple correlation involves the prediction equation,
yielding a predicted value of the DV (Y') that can be compared with obtained
Y.

$$Y' = A + B_1X_1 + B_2X_2 + \cdots + B_kX_k \tag{5.1}$$

where Y' is the predicted value of the DV, Y. B_1 to B_k represent best-fitting
weights, with A as the value of Y' when all X's are zero. X_i represents the
IVs, $i = 1, 2, \ldots, k$, available to predict Y.

Note that best-fitting regression weights are those that produce a prediction
equation for which squared differences between Y and Y' are at a minimum.
So this solution is called a least-squares solution.

In the sample problem, $k = 3$. That is, there are three IVs available to
predict the DV, COMPR.

$$(COMPR)' = A + B_M(MOTIV) + B_Q(QUAL) + B_G(GRADE)$$

To predict a student's COMPR score, the available IV scores (MOTIV,
QUAL, and GRADE) would be multiplied by their respective regression

weights. These weight-by-score products would be summed and added to the intercept, or baseline, value (A).

Differences among the observed values of the DV (Y), the mean of Y ($\overline{Y}$), and the predicted values of Y (Y') can be summed and squared, yielding estimates of variation attributable to different sources.

$$\text{SS}_y = \text{SS}_{\text{reg}} + \text{SS}_{\text{res}} \qquad (5.2)$$

The total sum of squares can be partitioned into two components: sum of squares predictable by regression, and sum of squares left over, or residual.

where

$$\text{SS}_y = \Sigma \, (Y - \overline{Y})^2$$

Total sum of squares is, as usual, the sum of squared differences between each individual's observed Y score and the mean of Y over all N cases. This represents total variation in the DV scores.

$$\text{SS}_{\text{reg}} = \Sigma \, (Y' - \overline{Y})^2$$

The sum of squares for regression is the predictable variation reflected by the sum of squared differences between predicted Y' and the mean of Y (which is the best guess of Y in the absence of any prior information).

$$\text{SS}_{\text{res}} = \Sigma \, (Y - Y')^2$$

The residual sum of squares is the sum of squared differences between observed Y and the predicted scores, Y', and represents a failure of prediction.

To find the squared multiple correlation, we write

$$R^2 = \frac{\text{SS}_y - \text{SS}_{\text{res}}}{\text{SS}_y} = \frac{\text{SS}_{\text{reg}}}{\text{SS}_y} \qquad (5.3)$$

The squared multiple correlation, R^2, is the proportion of sum of squares for regression in the total sum of squares for Y.

The squared multiple correlation is then the *proportion of variation in the DV that is predictable from the best linear combination of the IVs.* The multiple correlation itself is the correlation between the obtained and predicted Y values; that is, $R = r_{yy'}$.

The total sum of squares (SS_y) can be calculated directly from the observed values of the DV. For example, in the sample problem,

Mean $= 10.00$

$$\text{SS}_C = (18 - 10)^2 + (9 - 10)^2 + (8 - 10)^2 + (8 - 10)^2 + (5 - 10)^2 + (12 - 10)^2$$
$$= 102$$

To find the remaining sources of variation, it is necessary to solve the prediction equation (Eq. 5.1) for Y', which means that the best-fitting A and B_i must be found. For this purpose, however, a more direct method can be used, involving a slightly different perspective of multiple correlation in terms of individual correlations.

5.4.2 Matrix Equations

Another way of looking at R^2 is in terms of the correlations between each of the IVs and the DV. The summed products of these correlations, when multiplied by the regression weights associated with the respective IVs, yields the squared multiple correlation

$$R^2 = \sum_{i=1}^{k} r_{yi} \beta_i \tag{5.4}$$

where r_{yi} is the correlation between the DV and the ith IV, and β_i is the *standardized* regression coefficient, or beta weight (i.e., the regression weight that would be applied to the standardized X_i value—z scores—to predict the standardized Y').

Since the r_{yi} can be directly calculated from the data, resolution of the R^2 equation involves finding the standardized best-fitting regression coefficients (β_i) for the k IVs. Derivation of the k equations in k unknowns is beyond the scope of this book.

Solution of these derived equations, however, can best be illustrated by expressing them in matrix form:

$$R^2 = \mathbf{R}_{yi} \mathbf{B}_i \tag{5.5}$$

where $\mathbf{R}_{yi}$ is the row matrix of correlations between the DV and the k IVs, and $\mathbf{B}_i$ is a column matrix of standardized regression weights for the same k IVs.

The set of regression weights can be found by inverting the matrix of intercorrelations among IVs.

$$\mathbf{B}_i = \mathbf{R}_{ii}^{-1} \mathbf{R}_{iy} \tag{5.6}$$

The column matrix of standardized regression weights is the product of the inverse of the correlation matrix of IVs ($\mathbf{R}_{ii}^{-1}$) and the column matrix of correlations between IVs and the DV ($\mathbf{R}_{iy}$).

These equations,[2] then, can be used to calculate R^2 for the sample COMPR data from Table 5.1. All of the required intercorrelations are shown in Table 5.2.

Procedures for inverting a matrix are amply demonstrated elsewhere (e.g., Cooley & Lohnes, 1971; Harris, 1975) and are typically available in computer installations. Since the procedure is extremely tedious by hand, and becomes increasingly so as the matrix becomes larger, the inverted matrix for the sample data is presented without calculation in Table 5.3.

The $\mathbf{B}_i$ matrix is first found by postmultiplying the $\mathbf{R}_{ii}^{-1}$ matrix by the $\mathbf{R}_{iy}$ matrix.[3]

$$\mathbf{B}_i = \begin{bmatrix} 1.20255 & -0.31684 & -0.20435 \\ -0.31684 & 2.67113 & -1.97305 \\ -0.20435 & -1.97305 & 2.62238 \end{bmatrix} \begin{bmatrix} .58613 \\ .73284 \\ .75043 \end{bmatrix} = \begin{bmatrix} 0.31931 \\ 0.29117 \\ 0.40221 \end{bmatrix}$$

so that $\beta_M = 0.319$, $\beta_Q = 0.291$, and $\beta_G = 0.402$. Then, using Eq. 5.5, we obtain

TABLE 5.2 CORRELATIONS AMONG IVs AND THE DV FOR SAMPLE DATA IN TABLE 5.1

		$\mathbf{R}_{ii}$			$\mathbf{R}_{iy}$
		MOTIV	QUAL	GRADE	COMPR
	MOTIV	1.00000	.39658	.37631	.58613
	QUAL	.39658	1.00000	.78329	.73284
	GRADE	.37631	.78329	1.00000	.75043
$\mathbf{R}_{yi}$	COMPR	.58613	.73284	.75043	1.00000

TABLE 5.3 INVERSE OF MATRIX OF INTERCORRELATIONS AMONG IVs FOR SAMPLE DATA IN TABLE 5.1

	MOTIV	QUAL	GRADE
MOTIV	1.20255	-0.31684	-0.20435
QUAL	-0.31684	2.67113	-1.97305
GRADE	-0.20435	-1.97305	2.62238

[2] The equations can be solved in terms of Σ (variance-covariance) matrices or $\mathbf{S}$ (sum-of-squares and cross-product) matrices as well as correlation matrices. That is, Σ or $\mathbf{S}$ matrices can simply be substituted for corresponding $\mathbf{R}$ matrices. The regression coefficients in the case of $\mathbf{S}$ or Σ will be nonstandardized weights, as in Eq. 5.1.

[3] Procedures showing matrix inversion and matrix multiplication in detail appear in Appendix A.

$$R^2 = [.58613 \quad .73284 \quad .75043] \begin{bmatrix} 0.31931 \\ 0.29117 \\ 0.40221 \end{bmatrix} = .70237$$

so that 70% of the variance in graduate comprehensive exam scores is predictable from knowledge of motivation, admissions qualifications, and graduate course performance.

This solution for R^2 circumvents solution of the equation for the predicted values of COMPR (Y'). The beta weights (β_i) can be used to set up the prediction equation, but the coefficients must first be transformed to unstandardized B_i weights (unless all raw scores are converted to standardized z scores, and the prediction is desired in standardized form).

$$B_i = \beta_i \left(\frac{S_y}{S_i} \right) \tag{5.7}$$

Unstandardized weights (B_i) are found by multiplying beta weights (β_i) by the ratio of standard deviations of the DV to IV, where S_i is the standard deviation of the ith IV and S_y is the standard deviation of the DV.

$$A = \overline{Y} - \sum_{i=1}^{k} (B_i \overline{X}_i) \tag{5.8}$$

The intercept (A) is the mean of the DV less the sum of the means of the IVs multiplied by their respective unstandardized weights.

For the sample problem (refer to Table 5.1):

$$B_M = 0.319 \left(\frac{4.517}{2.191} \right) = 0.658$$

$$B_Q = 0.291 \left(\frac{4.517}{4.834} \right) = 0.272$$

$$B_G = 0.402 \left(\frac{4.517}{4.366} \right) = 0.416$$

$$A = 10 - [(0.658)(11.00) + (0.272)(10.17) + (0.416)(11.33)] = -4.72$$

The prediction equation for COMPR score, once scores on MOTIV, QUAL, and GRADE are known, is

$$(\text{COMPR})' = -4.72 + 0.658 \,(\text{MOTIV}) + 0.272 \,(\text{QUAL}) + 0.416 \,(\text{GRADE})$$

If a graduate student were to have ratings of 12, 14, and 15 respectively on MOTIV, QUAL, and GRADE, the predicted rating on COMPR would be

$$(\text{COMPR})' = -4.72 + 0.658(12) + 0.272(14) + 0.416(15) = 13.22$$

5.5 MAJOR TYPES OF MULTIPLE REGRESSION

There are three major analytic strategies in multiple regression, and the outcome of analysis can be drastically affected by choice of strategy.

5.5.1 Standard Multiple Regression

The simultaneous, or standard, strategy calls for entry of all IVs into the regression equation at once. This is the model reflected in the solution for the small sample graduate student data in Table 5.1. Each IV is assessed as if it had entered the regression after all other IVs had been entered. Each IV, then, can be evaluated in terms of what it adds to prediction of the DV, over and above the predictability afforded by all the other IVs.

Standard multiple regression is handled in the SPSS package by the REGRESSION program, as are all other types of multiple regression. A selected part of output is given in Table 5.4 for the sample problem illustrated in Table 5.1. (Full interpretation of program output will be reserved for the substantive examples to be presented later in this chapter.)

Within the BMD series, standard multiple regression is routinely handled by BMDP1R. A selected sample of output from that program is given in Table 5.5. Standard multiple regression can also be done through the more elaborate stepwise or setwise programs—BMDP2R or BMDP9R.

TABLE 5.4 DECK SETUP AND SELECTED SPSS OUTPUT FOR STANDARD MULTIPLE REGRESSION ON SAMPLE DATA IN TABLE 5.1 (RUN ON SPSS-6000)

```
                        REGRESSION    VARIABLE=MOTIV TO COMPR/
                                      REGRESSION=COMPR WITH MOTIV TO GRADE(2)/
                        OPTIONS       7
```

```
******************* MULTIPLE REGRESSION *************************
DEPENDENT VARIABLE..    COMPR

MEAN RESPONSE    10.00000    STD. DEV.    4.51664

VARIABLE(S) ENTERED ON STEP NUMBER   1..   MOTIV
                                           GRADE
                                           QUAL
```

		ANALYSIS OF VARIANCE	DF	SUM OF SQUARES	MEAN SQUARE	F	SIGNIFICANCE
MULTIPLE R	.83887					1.57313	.411
R SQUARE	.70235	REGRESSION	3.	71.64007	23.88002		
ADJUSTED R SQUARE	.25588	RESIDUAL	2.	30.35993	15.17997		
STD DEVIATION	3.89615	COEFF OF VARIABILITY	39.0 PCT				

		VARIABLES IN THE EQUATION					VARIABLES NOT IN THE EQUATION			
VARIABLE	B	STD ERROR B	F SIGNIFICANCE	BETA ELASTICITY		VARIABLE	PARTIAL	TOLERANCE	F SIGNIFICANCE
MOTIV	.65826582	.87213087	.56969898 .529	.3193058 .72409					
GRADE	.41603418	.64619051	.41451178 .586	.4022086 .47151					
QUAL	.27284787	.58911415	.21325133 .698	.2911579 .27658					
(CONSTANT)	-4.7217980	9.0656463	.27127968 .654						

TABLE 5.5   DECK SETUP AND SELECTED BMDP1R OUTPUT
FOR STANDARD MULTIPLE REGRESSION ON SAMPLE DATA
IN TABLE 5.1

```
              PROGRAM CONTROL INFORMATION

                   /PROBLEM   TITLE IS "SMALL SAMPLE STANDARD MULTIPLE REGRESSION".
                   /INPUT     VARIABLES ARE 4. FORMAT IS "(6X,4F4.0)". UNIT=50.
                   /VARIABLE  NAMES ARE MOTIV,QUAL,GRADE,COMPR.
                   /REGRESS   DEPENDENT IS COMPR.
                              INDEPENDENT ARE MOTIV,QUAL,GRADE.
                   /PRINT     DATA. CORR. COVA. RREG.
                   /PLOT      RESI. NORM. DNORM.
                   /END
```

```
  REGRESSION TITLE. . . . . . . . . . . . . . . .SMALL SAMPLE STANDARD MULTIPLE REGRESSION
  DEPENDENT VARIABLE. . . . . . . . . . . . . .        4 COMPR
  TOLERANCE . . . . . . . . . . . . . . . . . .        .0100
ALL DATA CONSIDERED AS A SINGLE GROUP
MULTIPLE R            .8381             STD. ERROR OF EST.        3.8961
MULTIPLE R-SQUARE     .7024

ANALYSIS OF VARIANCE
                    SUM OF SQUARES    DF    MEAN SQUARE    F RATIO    P(TAIL)
         REGRESSION        71.640      3      23.880        1.573     .41138
         RESIDUAL          30.360      2      15.180

                                              STD. REG
  VARIABLE      COEFFICIENT   STD. ERROR        COEFF       T    P(2 TAIL) TOLERANCE

INTERCEPT          -4.72180
MOTIV     1          .65827        .872          .319      .755     .529    .831565
QUAL      2          .27205        .589          .291      .462     .690    .374374
GRADE     3          .41603        .646          .402      .644     .586    .381333
```

5.5.2 Hierarchical Multiple Regression

In the hierarchical multiple regression model, IVs enter the regression in some
order specified by the researcher. Each IV is then assessed in terms of what it
adds to the equation at its own point of entry. Order of entry of variables
would normally be based on logical or theoretical considerations. For example,
IVs that are presumed (or manipulated) to be causally prior could be given
high priority of entry. That is, height might be considered prior to amount of
training in assessing success as a basketball player. Variables with greater theoretical importance could also be given early entry.

Or the opposite tack could be taken, by assessing the effect of some major
set of variables after a prior set of variables had been held constant. That is,
we could enter manipulated or other variables of major importance on later
steps, with "nuisance" variables given highest priority for entry. The lesser,
or nuisance, set would be entered first; then the major set could be evaluated
for what it adds to the prediction (and correlation) over and above the lesser
set. For example, we might want to see how well we can predict reading speed
(the DV) from intensity and length of a speed reading course (the major IVs)

after holding constant individual prior differences in reading speed (the nuisance IV). This is the basic analysis of covariance problem when the major set consists of, or can be coded into, dichotomous variables (i.e., dummy variables indicating group membership; cf. Cohen & Cohen, 1975).

IVs can be entered one at a time or in blocks. The analysis proceeds in stages, with information about variables both in and out of the equation given in computer output. Finally, after all variables are entered, a set of summary statistics is given, in addition to the information available at the last step of the progression.

In the SPSS package, hierarchical regression, like standard regression, is performed by the REGRESSION program. Because the program is set up to be highly flexible, some information in the output can be misleading. For example, the test for significance of contribution of individual variables must be recalculated (see Section 5.6.2.2). In Table 5.6, selected output is shown for the sample problem of Table 5.1, with higher priority given to admissions qualifications and course performance, and lower priority given to motivation.

In the BMDP package, hierarchical regression can be run on BMDP2R. Output from BMDP2R will be illustrated in the next section.

5.5.3 Stepwise and Setwise Regression

Stepwise regression is a rather controversial procedure, in which order of entry of variables is based on statistical rather than theoretical criteria. At each step the variable that adds most to the prediction equation, in terms of increasing R^2, is entered. The process continues until no more useful information can be gleaned from further addition of variables, with the researcher specifying statistical criteria for entry and deletion of variables.

The technique is typically used to develop a subset of IVs that is useful in predicting the DV, and to eliminate those IVs that do not provide additional prediction given this basic set. The procedure's controversy lies primarily in the capitalization on chance and overfitting of data inherent in this procedure. This problem seems most serious if explanatory interpretations of the data are being sought. The difficulty in using the procedure for exploratory research or simple prediction, where it is less controversial, lies in the variability of beta weights (and thus the contribution of variables) over samples from the same population that could produce a misleading subset of variables, and order of variables within the set, if decisions were to be based on a single sample. Further, a stepwise analysis might not lead to an optimum solution in terms of R^2. For example, several variables considered together might increase R^2, but any one alone would add no significant proportion of variance. In simple stepwise regression, none of the variables would enter.

An alternative to stepwise regression in choosing an optimal subset of variables is setwise regression. This is a variant in which the best possible subsets of variables are chosen, according to some criterion such as maximum R^2, from among all possible subsets of variables. One computer program in the BMD series, BMDP9R, is designed to do this. This alternative would avoid

TABLE 5.6 DECK SETUP AND SELECTED SPSS OUTPUT FOR
HIERARCHICAL MULTIPLE REGRESSION ON SAMPLE DATA
IN TABLE 5.1

```
                              REGRESSION    VARIABLE=MOTIV TO COMPR/
                                            REGRESSION=COMPR WITH QUAL,GRADE(4),MOTIV(2)/

* * * * * * * * * * * * * * * * * * * * * M U L T I P L E   R E G R E S S I O N * * * * * * * * * * * * * * * * * * * * * *
DEPENDENT VARIABLE..   COMPR

MEAN RESPONSE    10.00000    STD. DEV.    4.51664

VARIABLE(S) ENTERED ON STEP NUMBER   1..   QUAL
                                           GRADE


MULTIPLE R           .78586      ANALYSIS OF VARIANCE   DF    SUM OF SQUARES    MEAN SQUARE         F     SIGNIFICANCE
R SQUARE             .61757      REGRESSION             2.        62.99218       31.49609       2.42229       .236
ADJUSTED R SQUARE    .36262      RESIDUAL               3.        39.00782       13.00261
STD DEVIATION       3.60591      COEFF OF VARIABILITY   36.1 PCT

-------------------- VARIABLES IN THE EQUATION ----------------------      --------- VARIABLES NOT IN THE EQUATION -----------

VARIABLE        B        STD ERROR B      F           BETA             VARIABLE     PARTIAL    TOLERANCE       F
                                     ------------    ----------                                          ------------
                                     SIGNIFICANCE    ELASTICITY                                           SIGNIFICANCE

QUAL        .35065438    .53664200    .42696199      .3752862         MOTIV        .47085      .83157      .56969090
                                        .560           .35650                                                .529
GRADE       .47215984    .59408099    .63166494      .4564691
                                        .485           .53511
(CONSTANT)  1.0838690   4.4405959    .59575960E-01
                                        .823


- - - - - - - - - - - - - - - - - - - - - - - - - - - - - - - - - - - - - - - - - - - - - - - - - - - - - - - - - - - - -

* * * * * * * * * * * * * * * * * * * * * M U L T I P L E   R E G R E S S I O N * * * * * * * * * * * * * * * * * * * * * *
DEPENDENT VARIABLE..   COMPR

VARIABLE(S) ENTERED ON STEP NUMBER   2..   MOTIV


MULTIPLE R           .83807      ANALYSIS OF VARIANCE   DF    SUM OF SQUARES    MEAN SQUARE         F     SIGNIFICANCE
R SQUARE             .70235      REGRESSION             3.        71.64007       23.88002       1.57313       .411
ADJUSTED R SQUARE    .25588      RESIDUAL               2.        30.35993       15.17997
STD DEVIATION       3.89615      COEFF OF VARIABILITY   39.0 PCT

-------------------- VARIABLES IN THE EQUATION ----------------------      --------- VARIABLES NOT IN THE EQUATION -----------

VARIABLE        B        STD ERROR B      F           BETA             VARIABLE     PARTIAL    TOLERANCE       F
                                     ------------    ----------                                          ------------
                                     SIGNIFICANCE    ELASTICITY                                           SIGNIFICANCE

QUAL        .27204787    .58911415    .21325133      .2911579
                                        .690           .27658
GRADE       .41603418    .64619051    .41451178      .4022086
                                        .586           .47151
MOTIV       .65826582    .87213087    .56969090      .3193058
                                        .529           .72409
(CONSTANT) -4.7217980   9.8656463    .27127968
                                        .654


ALL VARIABLES ARE IN THE EQUATION.

- - - - - - - - - - - - - - - - - - - - - - - - - - - - - - - - - - - - - - - - - - - - - - - - - - - - - - - - - - - - -

DEPENDENT VARIABLE..   COMPR

                                  S U M M A R Y   T A B L E

STEP     VARIABLE          F TO        SIGNIFICANCE  MULTIPLE R  R SQUARE  R SQUARE  SIMPLE R    OVERALL F  SIGNIFICANCE
       ENTERED  REMOVED  ENTER OR REMOVE                                   CHANGE

 1     QUAL                .42696         .560        .73284      .53705    .53705    .73284      2.42229      .236
       GRADE               .63166         .485        .78586      .61757    .08052    .75043
 2     MOTIV               .56969         .529        .83807      .70235    .08478    .58613      1.57313      .411
```

the problem of finding variables that significantly improved R^2 only when considered as a group.

For stepwise regression, cross-validation with a second sample is highly recommended. At the very least, separate analyses of two halves of an available sample should be conducted, with conclusions limited to results that hold over the two analyses. This caution applies to setwise regression as well.

The SPSS REGRESSION program handles stepwise regression in a manner similar to that of hierarchical regression, and with the same need for recalculation of significance of contribution of IVs (see Section 5.6.2.2). Only one stepping method, "forward," is available, but three statistical criteria may be specified by the user, although default values are available. SPSS NEW REGRESSION includes additional stepping methods and statistical criteria.[4]

BMDP2R allows the user to specify the method for stepwise progression as well as the statistical criteria. There are four rules for moving from one step to the next, fully described in the BMDP manual (Dixon, 1981). For the data in Table 5.1, output from a computer run with default values is illustrated in Table 5.7.

By specifying block stepwise entry levels for IVs, one can set up combinations of hierarchical and stepwise regression. For example, a block of high priority IVs might be set up to compete for order of entry according to statistical criteria; then a lower priority block of predictors would compete for order of entry. The regression would thus be hierarchical over blocks, but stepwise within blocks. The SPSS manual (Nie et al., 1975) illustrates additional variations on entry of variables.

5.5.4 Choosing Among Regression Strategies

To simply assess relationships among variables, and answer the basic question of multiple correlation, the method of choice would be standard multiple regression. As a matter of fact, unless there is good reason to use some other technique, this would be the recommended one. Reasons for using other methods might be theoretical or for development of hypotheses.

Hierarchical regression allows the researcher to control the advancement of the regression process. Importance of variables in the prediction equation can be manipulated by the researcher according to logic or theory. Explicit hypotheses can be tested about proportion of variance attributable to some variables after others have been accounted for.

Although there is similarity in programs chosen and output produced for hierarchical and stepwise regression, there are fundamental differences in the way that variables enter the prediction equation and in the interpretations that can be made from the results. In hierarchical regression the researcher controls entry of variables, while in stepwise (including setwise) regression the sample data controls order of entry. Stepwise and setwise regression can therefore

---

[4] Some of the features of SPSS NEW REGRESSION are available in earlier releases of the CDC CYBER implementation of SPSS REGRESSION (SPSS-6000).

TABLE 5.7   INPUT AND SELECTED BMDP2R OUTPUT FOR STEPWISE
REGRESSION ON SAMPLE DATA IN TABLE 5.1

```
                    PROGRAM CONTROL INFORMATION

                         /PROBLEM    TITLE IS "SMALL SAMPLE STEPWISE REGRESSION".
                         /INPUT      VARIABLES ARE 4. FORMAT IS "(6X,4F4.0)". UNIT=50.
                         /VARIABLE   NAMES ARE MOTIV,QUAL,GRADE,COMPR.
                         /REGRESS    DEPENDENT IS COMPR.
                                     INDEPENDENT ARE MOTIV,QUAL,GRADE.

                         /END


     STEP NO.    0
    ISTD. ERROR OF EST.     4.5166

    ANALYSIS OF VARIANCE
                         SUM OF SQUARES     DF     MEAN SQUARE
              RESIDUAL      102.00000        5       20.40000

                         VARIABLES IN EQUATION                              .                    VARIABLES NOT IN EQUATION
                                       STD. ERROR  STD REG                  .                        PARTIAL             F TO
          VARIABLE     COEFFICIENT    OF COEFF    COEFF    TOLERANCE   F TO        LEVEL.  VARIABLE   CORR.   TOLERANCE  ENTER  LEVEL
        (Y-INTERCEPT      10.000 )                                    REMOVE       .
                                                                                   . MOTIV      1   .58613  1.00000     2.89    1
                                                                                   . QUAL       2   .73284  1.00000     4.64    1
                                                                                   . GRADE      3   .75043  1.00000     5.16    1



     STEP NO.    1
    VARIABLE ENTERED    3 GRADE
    IMULTIPLE R             .7504
    MULTIPLE R-SQUARE      .5631
    ADJUSTED R-SQUARE      .4539
    STD. ERROR OF EST.    3.3376

    ANALYSIS OF VARIANCE
                         SUM OF SQUARES     DF     MEAN SQUARE    F RATIO
           REGRESSION      57.440559        1      57.44056        5.16
           RESIDUAL        44.559441        4      11.13986

                         VARIABLES IN EQUATION                              .                    VARIABLES NOT IN EQUATION
                                       STD. ERROR  STD REG                  .                        PARTIAL             F TO
          VARIABLE     COEFFICIENT    OF COEFF    COEFF    TOLERANCE   F TO        LEVEL.  VARIABLE   CORR.   TOLERANCE  ENTER  LEVEL
        (Y-INTERCEPT       1.203 )                                    REMOVE       .
        GRADE       3         .776      .342      .750    1.00000      5.16     1  . MOTIV      1   .49600   .85839      .98    1
                                                                                   . QUAL       2   .35297   .38645      .43    1
    I* * * * * F-LEVELS(   4.000,    3.900) OR TOLERANCE INSUFFICIENT FOR FURTHER STEPPING

                               STEPWISE REGRESSION COEFFICIENTS
     I VARIABLES    0 Y-INTCPT  1 MOTIV    2 QUAL     3 GRADE
     STEP
        0           10.0000*    1.2083      .6847      .7762
        1            1.2028*     .7295      .3507      .7762*

    NOTE-
        1)  REGRESSION COEFFICIENTS FOR VARIABLES IN THE EQUATION ARE INDICATED BY AN ASTERISK
        2)  THE REMAINING COEFFICIENTS ARE THOSE WHICH WOULD BE OBTAINED IF THAT VARIABLE WERE TO ENTER IN THE NEXT STEP
    SUMMARY TABLE
    STEP          VARIABLE            MULTIPLE  INCREASE    F-TO-      F-TO-   NUMBER OF INDEPENDENT
    NO.     ENTERED    REMOVED          R      RSQ   IN RSQ   ENTER    REMOVE   VARIABLES INCLUDED
       1    3 GRADE                    .7504   .5631   .5631   5.1563                   1
```

be seen as model-building rather than model-testing procedures. As exploratory
techniques, they can be useful for such purposes as eliminating variables that
are clearly superfluous in order to tighten up future research. (When multicollinearity or singularity are present, setwise regression can be indispensable in
identifying multicollinear variables, as indicated in Chapter 4.)

For the example of Section 5.4, in which performance on graduate comprehensive exam (COMPR) was to be predicted from professional motivation
(MOTIV), qualifications for graduate training (QUAL), and performance in
graduate courses (GRADE), the differences among regression strategies might
be illustrated as follows. If standard multiple regression were used, two fundamental questions would be asked: (1) What is the size of the overall relationship

between COMPR and MOTIV, QUAL, and GRADE? and (2) How much of
the relationship is contributed uniquely by each IV? If hierarchical regression
were used, with QUAL and GRADE entered before MOTIV, the question
would be, Does MOTIV significantly add to prediction of COMPR after differences among students in QUAL and GRADE have been statistically eliminated?
If stepwise regression were used, one would ask, What is the best linear combination of IVs to predict the DV in this sample? And use of setwise regression
would lead to the query, What is the size of R^2 from each one of the IVs,
from all possible combinations of two IVs, and from all three IVs for this
sample?

5.6 SOME IMPORTANT ISSUES

5.6.1 Importance of IVs

The question of contribution of IVs to multiple correlation cannot be resolved
in an entirely straightforward manner. In the simplest case of uncorrelated
IVs, regression coefficients (standardized or unstandardized) or even simple correlations between IVs and the DV could be used to assess importance. When
IVs are intercorrelated, however, regression weights and correlations can carry
redundant or misleading information. For each IV, correlations reflect not only
variance shared with the DV, but also variance shared with other IVs. Regression
weights are misleading where there are suppressor variables. That is, some IVs
are heavily weighted not because they directly predict the DV, but because
they suppress irrelevant variance in other IVs (see Section 5.6.4). Unstandardized
weights are expressed in units of the DV, which may be unfamiliar to you.[5]
Neither of these statistics convey precise information about *unique* contribution
of IVs, which is what frequently would be of greatest interest.

   Two related statistics are commonly used for assessing this unique contribution: partial correlations and semipartial correlations (the latter are called part
correlations in the SPSS manuals, and contributions to R^2 by BMDP9R). For
partial correlations, the contribution of remaining IVs (other than the ith) is
partialled out of both the ith IV and the DV. For semipartial correlations,
the contribution of remaining IVs is partialled out of only the ith IV. Thus
semipartial correlation squared expresses the unique contribution of the IV as
a proportion of total variance of the DV, while partial correlation squared
expresses the unique contribution of the IV as a proportion of R^2.

   Of these statistics, *squared semipartial correlation* ($sr_i{}^2$) *is probably the
single most useful measure of importance of an IV.* The interpretation assigned
to $sr_i{}^2$ differs, however, depending on the type of multiple regression employed.

   [5] However, if one wishes to use regression weights to predict scores on a *new* sample, unstandardized (B) weights will be less biased than beta weights because B weights are not influenced
by the standard deviations as estimated by the original sample.

5.6.1.1 Standard Multiple and Setwise Regression   In standard multiple regression and setwise regression, sr_i^2 for an IV is the amount by which R^2 would be reduced if that IV were not included in the regression equation. Alternatively, sr_i^2 represents the unique contribution of the IV to R^2.

For SPSS REGRESSION, sr_i^2 can most easily be calculated using Eq. 5.9.[6]

$$sr_i^2 = \frac{F_i}{\text{df}_{\text{res}}}(1 - R^2) \tag{5.9}$$

The squared semipartial correlation (sr_i^2) for the ith IV is calculated from the F_i ratio given in the SPSS output section labeled VARIABLES IN THE EQUATION, the residual degrees of freedom (df$_{\text{res}}$) given in the analysis of variance table, and R^2, which appears as R SQUARE in the SPSS output in the section to the left of the analysis of variance table (see Table 5.4).

From the output shown in Table 5.4, then, the squared semipartial correlation for **GRADE** is

$$sr_G^2 = \frac{0.415}{2}(1 - .70235) = .0618$$

For BMDP1R and BMDP9R, T_i values (student's t) are given for each variable instead of F_i. Since $F = t^2$, semipartial correlations can be found with Eq. 5.9, using the squared value of T_i for F_i.

If partial correlations squared (pr_i^2) are desired, they also can be calculated from the F_i (or T_i^2) and df$_{\text{res}}$ as defined for Eq. 5.9.[7]

$$pr_i^2 = \frac{F_i}{F_i + \text{df}_{\text{res}}} \tag{5.10}$$

In all standard multiple regression programs, the T_i or F_i represents the significance test for sr_i^2 and pr_i^2, as well as B_i and β_i, as discussed in Section 5.6.2.

When the IVs are intercorrelated, the squared semipartial correlations in standard multiple regression will not necessarily sum to multiple R^2. The sum of sr_i^2 will usually be smaller than R^2 (although under some rather extreme circumstances, the sum can be larger than R^2). When the sum is smaller, the difference between R^2 and the sum of the sr_i^2 represents shared variance, variance

---

[6] This same procedure is appropriate if BMDP2R is used for standard multiple regression. The information used would be from the last step in the analysis. For SPSS NEW REGRESSION, sr_i is optionally available as PART COR by requesting STATISTICS = ZPP.

[7] Available as PARTIAL in SPSS NEW REGRESSION by requesting STATISTICS = ZPP.

that is contributed to R^2 by two or more IVs. It is rather common to find a substantial R^2 with all of the unique contributions of IVs quite small.

5.6.1.2 Hierarchical or Stepwise Regression

In these two forms of regression, sr_i^2 is interpreted as the amount of variance added to R^2 by each IV as it enters the equation at that particular point. The sr_i^2 *do* sum to R^2 in hierarchical and stepwise regression, but the apparent importance of an IV may depend critically on its point of entry into the equation. The research question then becomes, How much does this IV add to multiple R^2 after IVs with higher priority have contributed their share to prediction of the DV?

In SPSS as well as BMDP programs, the squared semipartial correlations are given as part of routine output for hierarchical and stepwise regression. For SPSS REGRESSION, the RSQ CHANGE or R SQUARE CHANGE (see Table 5.6) for each IV in the SUMMARY TABLE is the squared semipartial correlation. For BMDP2R, the squared semipartial correlation is given as INCREASE IN RSQ (see Table 5.7) for each IV.

If desired, partial correlations squared can be found from F ratios, as in Eq. 5.10. *The F ratios given in computer output, however, may be inappropriate.* Section 5.6.2.2 should be consulted for the method of calculating appropriate F_i values from squared semipartial correlations (Eq. 5.12) prior to using Eq. 5.10 to get squared partial correlation values for hierarchical or stepwise regression.

5.6.2 Statistical Inference

This section covers significance tests for overall multiple correlation and for the individual components (IVs) of the multiple regression. A test, F_{inc}, is also given for evaluating the statistical significance of a subset of variables added to a prediction equation in stepwise or hierarchical analysis. Calculation of confidence limits on unstandardized regression weights and procedures for comparing differences in predictive capacity of two sets of IVs conclude the section.

It should be noted that when the researcher is using either stepwise or setwise regression as an exploratory tool, inferential procedures of any kind may be inappropriate. Inferential procedures require that the researcher have a hypothesis to test. When stepwise or setwise regression is used to snoop data, there may be no hypothesis, even though the statistics themselves are available.

5.6.2.1 Test of Multiple *R*

The overall inferential test in multiple correlation is whether the sample of scores was drawn from a population in which the multiple R is zero. This is equivalent to the null hypothesis that all regression coefficients are zero, or that all correlations between DV and IVs are zero. With large N, the test of this hypothesis becomes trivial because it is almost certain to be rejected. Testing overall R is more interesting with smaller numbers of cases.

The test of this hypothesis is presented in all computer outputs for standard multiple regression as the analysis of variance. For standard regression performed through stepwise programs, or for hierarchical analysis, the analysis of variance table at the last step gives the relevant information. The F ratio for mean square regression over mean square residual (or deviation about regression) tests the significance of multiple R (and, of course, multiple R^2). Mean square regression is the sum of squares for regression in Eq. 5.2 divided by k degrees of freedom; and mean square residual (or deviation about regression) is the sum of squares for residual in the same equation divided by $(N - k - 1)$ degrees of freedom.

For stepwise regression in which all potential IVs do not enter the equation, the sample R^2 is not distributed as F. Therefore the analysis of variance table at the last step is misleading, and the reported F is biased so that an F ratio that appears to be statistically significant actually reflects a Type I error rate in excess of alpha. Wilkinson (1979) has developed tables for critical R^2 when forward selection procedures are used for stepwise addition of variables and set size is specified in advance.[8] He suggests that these tables are a fair approximation for other selection procedures. Appendix C contains Table C.5 for the 95 and 99 percentage points ($\alpha = .05$ and .01) for multiple R^2, given N, the sample size; k, the number of potential IVs; and M, the number of IVs selected in the final stepwise or setwise solution. Table C.5 thus shows how large multiple R^2 must be to be statistically significant at .05 or .01 levels, given N, M, and k.

For example, for a stepwise regression in which there were 100 subjects, 20 potential IVs, and 15 IVs chosen for the solution, a multiple R^2 of approximately .30 would be required to be considered significantly different from zero at $\alpha = .05$ (and approximately .34 at $\alpha = .01$). Wilkinson reports that linear interpolation on N and k works well. For values of M that do not appear in the table, the conservative strategy is to choose the next highest M in the table. Table C.5 could also be used to find critical R^2 values if a post hoc decision were made to terminate hierarchical regression analysis as soon as R^2 reached statistical significance.

After looking at the data, you may wish to test the significance of some subsets of IVs in predicting the DV. Subsets might even consist of single IVs. If several of these a posteriori tests are done, Type I errors become increasingly likely. Larzelere and Mulaik (1977) recommend the following conservative F test, which keeps Type I error rate below α for all linear combinations of IVs:

$$F = \frac{R_s^2/k}{(1 - R_s^2)/(N - k - 1)} \tag{5.11}$$

where R_s^2 is the square of the multiple (or bivariate) correlation to be tested for significance, and k is the *total* number of IVs (not just the one or more considered in the R_s^2 being tested). The obtained F is compared against tabled F, with k and $(N - k - 1)$ degrees of freedom. That is, the critical value of F for each subset is the same as the critical value for overall multiple R.

---

[8] Tables are under development for tests of statistical significance when set size is *not* specified in advance (Wilkinson & Dallal, 1980).

In the sample problem, the bivariate correlation (tested a posteriori) between MOTIV and COMPR can be evaluated as follows:

$$F = \frac{.58613^2/3}{(1 - .58613^2)/2} = 0.349 \qquad \text{with df} = 3, 2$$

5.6.2.2 Test of Regression Components For testing the individual regression components against the null hypothesis in *standard* multiple regression, the same test is used for all statistics related to contribution of individual IVs, including unstandardized B_i and standardized β_i, as well as sr_i and pr_i, discussed in Section 5.6.1. For this type of regression, the test is straightforward, and the values are given in computer output. In SPSS REGRESSION for standard multiple regression, the F_i values (significance of unique contribution of the ith regression coefficient) appear in the output section labeled VARIABLES IN THE EQUATION (see Table 5.4). Degrees of freedom are 1 and df$_{res}$, which appears in the accompanying analysis of variance table. In BMDP1R, T_i values are given (Student's t test) for each IV, and again are tested with df$_{res}$ (deviation about regression) from the analysis of variance table (see Table 5.5). If BMDP2R is used for standard multiple regression, F_i values are available at the last step of the analysis (F-TO-REMOVE) and are interpreted as for SPSS REGRESSION.

But regardless of the statistical form of the test of significance for regression coefficients in standard multiple correlation, it is important to recall its limitations during interpretation. The significance test is sensitive only to the unique variance an IV adds to R^2. A very important IV that happened to share variance with another IV in the analysis might be nonsignificant although the two IVs in combination were responsible in large part for the size of R^2. An IV that is highly correlated with the DV but has a nonsignificant regression coefficient might have suffered just such a fate. For this reason, we recommend reporting and, where desirable, interpreting r_{iy} in addition to F_i for each IV, as shown in Table 5.17.

For stepwise and hierarchical regression, assessment of contribution of variables is more complex, and *appropriate significance tests may not appear in the computer output.* First, there is an inherent ambiguity in the testing of each variable. In stepwise and hierarchical regression, tests for sr_i^2 and pr_i^2 will not produce the same result as tests of the regression weights (B_i and β_i). This is because regression weights are independent of order of entry of the IVs, whereas sr_i^2 and pr_i^2 depend directly on priority. However, since the sr_i^2 reflect "importance" as typically interpreted in stepwise-hierarchical regression, tests based on those values might be intuitively more appealing.[9]

Second, there is a choice to be made between two error models. Model I bases the error term and degrees of freedom for F on only those IVs entered

---

[9] For combined standard-hierarchical regression, it might be desirable to use the "standard" method for all IVs, simply to maintain consistency. If so, be sure to report that F test is for regression coefficients.

at that point in the hierarchy reached (up to and including the ith IV, being assessed). Model II, more commonly recommended, bases the error term and degrees of freedom on all IVs eventually entered into the equation. A full discussion of factors to consider in choosing error terms appears in Cohen and Cohen (1975). In general, the preference would be for Model II. But if the number of IVs approaches sample size, the increase in degrees of freedom with Model I might make it attractive; however, interpretation would be ambiguous (cf. Section 5.3.2.1). Therefore only Model II will be covered in detail here.

For SPSS, the F_i given for each IV is appropriate only for B_i and β_i, and *must be recalculated* entirely for $sr_i{}^2$ in hierarchical or stepwise regression:

$$F_i = \frac{sr_i{}^2}{(1 - R^2)/df_{res}} \tag{5.12}$$

For Model II, the F_i for each IV is based on $sr_i{}^2$ (the squared semipartial correlation, appearing in SPSS output as RSQ or R SQUARE CHANGE and explained in Section 5.6.1.2); multiple R^2 after all IVs are considered; and residual degrees of freedom from the analysis of variance table for the final step.

Degrees of freedom for this test are 1 and df_{res}. For the example shown in Table 5.6, the F ratio for GRADE would be

$$F_G = \frac{.08052}{(1 - .70235)/2} = 0.541 \qquad \text{with df} = 1, 2 \qquad \text{(Model II)}$$

For BMDP2R, F_i is given for hierarchical or stepwise regression (F value to enter or remove in SUMMARY TABLE), but it is based on a Model I error term. In order to find Model II error, the $sr_i{}^2$ (represented as INCREASE IN RSQ) can be entered in Eq. 5.12, along with overall multiple R^2 and residual degrees of freedom from the analysis of variance table at the final step of the progression. For GRADE in Table 5.7, using R^2 from Table 5.5, the Model II F would be

$$F_G = \frac{.5631}{(1 - .7024)/2} = 3.78 \qquad \text{with df} = 1, 2 \qquad \text{(Model II)}$$

For setwise regression (BMDP9R, all possible subsets) the individual IVs are best interpreted as if for standard multiple regression (i.e., each variable entering last). That is, the T_i values given for each variable in the "best" subset are interpreted as those for BMDP1R noted earlier.

5.6.2.3 Test of Added Subset of IVs

For hierarchical or stepwise regression, it is also possible to test whether a subset of one or more variables adds significantly to variance already explained by a prior subset of variables.[10]

---

[10] In SPSS NEW REGRESSION this can be done through the TEST subcommand.

$$F_{inc} = \frac{(R_{wi}^2 - R_{wo}^2)/M}{(1 - R_{wi}^2)/df_{res}} \qquad (5.13)$$

where F_{inc} is the incremental F ratio; R_{wi}^2 is the multiple R^2 achieved with the added subset of IVs in the equation; R_{wo}^2 is the multiple R^2 *without* the additional subset of IVs in the equation (both values may be found in the SUMMARY TABLE of any computer output); M is the number of IVs in the added subset; and $df_{res} = (N - k - 1)$ come from residual degrees of freedom in the final analysis of variance table.

The null hypothesis of no increase in R^2 is tested as F with M and df_{res} degrees of freedom. If the null hypothesis is rejected, then the subset of added variables *does* significantly increase the explained variance.

For the hierarchical example in Table 5.6, we can test whether MOTIV adds significantly to the variance contributed by the first two variables to enter the equation, QUAL and GRADE.

$$F_{inc} = \frac{(.70235 - .61757)/1}{(1 - .70235)/2} = 0.570 \qquad \text{with } df = 2, 2$$

5.6.2.4 Confidence Limits Around B In estimating population values for the parameters, confidence limits for the unstandardized regression coefficients (B_i) can be calculated on the basis of information available in all output. The standard error of the regression coefficients (SE_{B_i}) for the ith IV can be found in the section labeled VARIABLES IN THE EQUATION for the final step of an analysis, or in the SUMMARY TABLE. This will be listed as STD ERROR B or SE B (SPSS), STD ERROR OF COEFF (BMDP2R), or simply STD ERROR (BMDP1R and BMDP9R). The critical value of t for the desired confidence level is based on $(N - 2)$ degrees of freedom, where N is the sample size.

$$CL_{B_i} = B_i \pm SE_{B_i} (t_{\alpha/2}) \qquad (5.14)$$

The $1 - \alpha$ confidence limits for the unstandardized regression coefficient for the ith IV (CL_{B_i}) is the regression coefficient (B_i) plus or minus the standard error of the regression coefficient (SE_{B_i}) times the critical value of t, with $(N - 2)$ degrees of freedom at the desired level of α.

For the example illustrated in Table 5.4, the 95% confidence limits on the unstandardized regression coefficient for GRADE, with $df = 4$, would be

$$CL_{B_G} = 0.416 \pm 0.646(2.78) = 0.416 \pm 1.796 = -1.380 \leftrightarrow 2.212$$

If the confidence interval contains zero, one cannot reject the null hypothesis that the population regression coefficient is zero.

5.6.2.5 Comparing Two Sets of Predictors It is sometimes of interest to know whether one set of predictors correlates better with a criterion than another set of predictors. For example, can ratings of current belly dancing ability be better predicted by personality tests or by past dance and musical training? The procedure for finding out is fairly convoluted, but if you have access to BMDP or are willing to hand-punch a pair of predicted scores for each subject in your sample and if you have a large sample, a test for the significance of the difference between two "correlated correlations" (both correlations are based on the same sample and share a variable) is available (Steiger, 1980). If sample size is small, nonindependence among predicted scores for cases can result in serious violation of the assumptions underlying the test.

As suggested in Section 5.4.1, a multiple correlation can be thought of as a simple correlation between criterion scores such as belly dancing ability (Y) and the scores that are predicted from a set of variables (Y'), that is, $R = r_{yy'}$. If there are two sets of predictors, Y'_a and Y'_b (e.g., a = personality scores and b = past training), a comparison of their relative effectiveness in predicting Y can be made by testing for the significance of the difference between $r_{yy'_a}$ and $r_{yy'_b}$. For simplicity, let's call these r_{ya} and r_{yb}. The trick is that to test for the difference we need to know the correlation between the predicted scores from set A (personality) and those from set B (training), that is, $r_{y'_a y'_b}$ or, simplified, r_{ab}. This is where the BMDP file manipulation procedures or hand punching become necessary.

The Z test for the difference between r_{ya} and r_{yb} is

$$\bar{Z}^* = (z_{ya} - z_{yb}) \sqrt{\frac{N-3}{2 - 2\bar{s}_{ya,yb}}} \qquad (5.15)$$

where N is, as usual, the sample size,

$$z_{ya} = \frac{1}{2} \ln \left(\frac{1 + r_{ya}}{1 - r_{ya}} \right) \quad \text{and} \quad z_{yb} = \frac{1}{2} \ln \left(\frac{1 + r_{yb}}{1 - r_{yb}} \right)$$

And now the fun begins:

$$\bar{s}_{ya,yb} = \frac{(r_{ab})(1 - \bar{r}^2 - \bar{r}^2) - \frac{1}{2}(\bar{r}^2)(1 - \bar{r}^2 - \bar{r}^2 - r_{ab}{}^2)}{(1 - \bar{r}^2)^2}$$

where $\bar{r} = \frac{1}{2}(r_{ya} + r_{yb})$.

So, for the example, if $R_a = .40 = r_{ya}$ (i.e., the correlation between currently measured ability and ability as predicted from personality scores), $R_b = .50 = r_{yb}$ (i.e., the correlation between currently measured ability and ability as predicted from past training), $r_{ab} = .10$ (i.e., the correlation between ability as predicted from personality and ability as predicted from training), and $N = 103$,

$$\bar{r} = \frac{1}{2}(.40 + .50) = .45$$

$$\bar{s}_{ya,yb} = \frac{(.10)(1 - .45^2 - .45^2) - \frac{1}{2}(.45)^2(1 - .45^2 - .45^2 - .10^2)}{(1 - .45^2)^2}$$

$$= .0004226$$

$$z_{ya} = \frac{1}{2} \ln \left(\frac{1 + .40}{1 - .40} \right) = .42365$$

$$z_{yb} = \frac{1}{2} \ln \left(\frac{1 + .50}{1 - .50} \right) = .54931$$

and, finally,

$$\bar{Z}* = (.42365 - .54931) \sqrt{\frac{103 - 3}{2 - .000845}} = -0.88874$$

Since $\bar{Z}*$ is within the critical values of ± 1.96 for a two-tailed test, there is no statistically significant difference between multiple R when predicting Y from Y'_a or Y'_b. That is, there is no statistically significant difference in predicting current belly dancing ability from past training versus personality tests.

Steiger (1980) presents additional significance tests for situations where both the criteria and the predictors are different, but from the same sample, and for comparing the difference between any two correlations within a correlation matrix.

5.6.3 Adjustment of R^2

Just as simple r_{xy} from a sample can be expected to fluctuate around the value of the correlation in the population, sample R^2 can also be expected to fluctuate around the population value. The problem is that R^2 (and R) can never take on negative values, so that all chance fluctuations will be in the positive direction adding to the magnitude of R^2. As in any sampling distribution, the magnitude of chance fluctuations increases as the sample size decreases. Therefore, R^2 tends to be overestimated, and the smaller the sample the greater the overestimation. For this reason, in estimating the population value of R^2, adjustment should be made for expected inflation in sample R^2.

SPSS REGRESSION, BMDP2R, and BMDP9R provide an adjusted R^2 as part of routine output. Wherry (1931) provided a simple equation for this adjustment, which is called adjusted $\tilde{R}^2$:

$$\tilde{R}^2 = 1 - (1 - R^2)\left(\frac{N-1}{N-k-1}\right) \tag{5.16}$$

where N is the sample size, k is the number of IVs, and R^2 is the overall squared multiple correlation. For the same sample problem,

$$\tilde{R}^2 = 1 - (1 - .70235)\,(\tfrac{5}{2}) = .25588$$

as printed out for SPSS, Table 5.4.

    For stepwise regression, Cohen and Cohen (1975) recommend k based on the number of IVs considered for inclusion, rather than on the number of IVs selected by the program. They also suggest the convention of reporting $\tilde{R}^2 = 0$ when the value spuriously becomes negative.

    In cases where the number of subjects is 60 or fewer and there are numerous IVs (say, more than 20), Eq. 5.16 may provide inadequate adjustment for R^2. The adjusted value may be off by as much as .10 (Cattin, 1980). In these situations of small N and small $N - k$ values, Eq. 5.17 (Browne, 1975) provides further adjustment.

$$\tilde{R}_s^2 = \frac{(N-k-3)\tilde{R}^4 + \tilde{R}^2}{(N-2k-2)\tilde{R}^2 + k} \tag{5.17}$$

The adjusted R^2 for small samples is a function of the number of cases, N, the number of IVs, k, and the $\tilde{R}^2$ value as found from Eq. 5.16.

An equation producing even less bias when $N < 50$, but requiring far more computation, is shown in Cattin (1980).

5.6.4 Suppressor Variables

In some situations, there may be an IV that is virtually uncorrelated with the DV but is useful in predicting the DV and in increasing the multiple R^2 by virture of its correlations with other IVs. This suppressor variable receives its name because it "suppresses" some variance in the other IVs that is irrelevant to prediction of the DV. In a full discussion of suppressor variables, Cohen and Cohen (1975) describe several varieties of suppression and provide examples of situations where this might occur.

    Typical examples in psychology might involve the administration of two paper-and-pencil tests, only one of which predicts the DV (say, belly dancing ability). The nonpredicting IV (e.g., a vocabulary test) would serve to partial out variance due to ability in taking paper-and-pencil tests, thus enhancing the prediction of the DV by the combined IVs over that achieved by the single IV (e.g., ability to list dance patterns) that is most closely related to the DV.

    In output, suppressor variables can be recognized by their pattern of regression coefficients and correlation with the DV. To find suppressor variables, the simple correlations between the IVs and the DV can be checked from the correlation matrix available from output, and compared against the standardized regression weights (beta values) for the IVs. (Rearrangement of terms in Eq. 5.7 allows for calculation of beta weights if the values are not available from computer output.) Either one of the following two conditions signals the presence of a suppressor variable: (1) the absolute value of the simple correlation between the IV and DV is substantially smaller than the beta weight for that IV, or (2) the simple correlation and beta weight have opposite signs. In either case the beta weight would have to be significantly different from zero.

5.7 COMPARISON OF SPSS AND BMD PROGRAMS

The primary difference between the SPSS package and the BMDP series with respect to regression is that in SPSS a single program serves for all varieties of multiple regression, while in the BMDP packages there are separate programs specializing in various types of regression.

Direct comparisons of multiple and stepwise/hierarchical programs are summarized in Tables 5.8 (pages 118–119) and 5.9 (pages 120–123), respectively. Some of these features are elaborated on in Sections 5.7.1 and 5.7.2.

5.7.1 SPSS Package

The original SPSS program for regression is simply called SPSS REGRESSION. Beginning with Release 9.0, a more elaborate program became available, called NEW REGRESSION. The distinctive feature of both SPSS REGRESSION programs is their flexibility.

SPSS REGRESSION offers two options for treatment of missing data (fully described in the manual). Correlation matrices can be input instead of raw data, and as in all SPSS procedures, analysis can be limited to subsets of cases. A special option can be selected so that correlation matrices are printed only when there are bad elements in the matrix, that is, when one or more of the correlations cannot be calculated.

The stepwise procedure offers only one option regarding the progression of steps: forward stepping.[11] Variables are added to the equation in the order in which they contribute most to R^2, given that they meet the statistical criteria for entering. Variables once in the equation stay there. The three statistical criteria for stepwise regression specifiable for this program are (1) the maximum number of IVs to be entered into the equation, (2) the minimum F ratio for adding a variable to the equation, and (3) tolerance. This last value is the

---

[11] The CDC CYBER implementation, SPSS-6000, offers three options regarding progression of steps: forward stepping, backward stepping, and stepwise. The stepwise procedure uses forward stepping but eliminates variables that no longer meet criteria for staying in the equation.

TABLE 5.8 COMPARISON OF SPSS AND BMD PROGRAMS FOR STANDARD MULTIPLE REGRESSION

Feature	SPSS REGRESSION	SPSS NEW REGRESSION[a]	BMDP1R
Input			
Subsamples	Yes	Yes	Yes
Correlation matrix input	Yes	Yes	Yes
Covariance matrix input	No	Yes	Yes
Missing data handled	Yes	Yes	Yes
Regression through the origin	No[b]	Yes	Yes
Tolerance	Yes	Yes	Yes
Regression output			
Labeling of variables	Yes	Yes	Yes
Analysis of variance for regression	Yes	Yes	Yes
Multiple R	Yes	Yes	Yes
R^2	R SQUARE	R SQUARE	MULTIPLE R-SQUARE
Standard error for R	STANDARD ERROR	STANDARD ERROR	STD. ERROR OF ESTIMATE
Adjusted R^2	Yes	Yes	Yes
Correlation matrix	Yes	Yes	Yes
Sum-of-squares and cross-products matrix	No	Yes	No
Covariance matrix	No	Yes	Yes
Number of valid cases	Yes	Yes	Yes
Means and standard deviations	Yes	Yes	Yes
Minimum and maximum of variables	No	No	Yes
Coefficient of variation	No	No	Yes
Unstandardized regression coefficients	B	B	COEFFICIENT

	STD. ERROR B	SE B	STD. ERROR
Standard error of regression coefficient	STD. ERROR B	SE B	STD. ERROR
F or t test for regression coefficient	F	T(F optional)	T
Intercept	(CONSTANT)	(CONSTANT)	(CONSTANT)
Standardized regression coefficient	BETA	BETA	STD. REG. COEFF.
Proportion of variance cumulated	RSQ CHANGE	RSQCH	No
Variance-covariance matrix for unstandardized B coefficients	No	Yes	No
Correlation matrix for unstandardized B coefficients	No	No	Yes
Sweep matrix	No	Yes	No
Condition number bounds	No	Yes	No
Residuals			
Tables of predicted scores and residuals	Yes	Yes	Yes
Plot of residuals against predicted scores	Yes	Yes	Yes
Normal plot of residuals	No	Yes	Yes
Other plots available	Yes	Yes	Yes
Durbin-Watson statistic	Yes	Yes	Yes
Standardized residuals (cases)	No	Yes	No
Mahalanobis and/or Cook's distance	No	Yes	No[c]

Note: BMDP2R (cf. Table 5.9) can also be used for standard multiple regression.

[a] Available starting with Release 9.0 (Hull & Nie, 1981).

[b] Available in SPSS-6000 CYBER implementation of Release 8.0.

[c] Available through BMDPAM, BMDP4M, or BMD10M.

TABLE 5.9 COMPARISON OF SPSS AND BMD PROGRAMS FOR STEPWISE AND/OR HIERARCHICAL REGRESSION

Feature	SPSS REGRESSION	SPSS NEW REGRESSION[a]	BMDP2R	BMDP9R
Input				
Correlation matrix input	Yes	Yes	Yes	Yes
Covariance matrix input	No	Yes	Yes	Yes
Missing data handled	Yes	Yes	Yes	Yes
Specify stepping algorithm	No[b]	Yes	Yes	Yes
Specify tolerance	Yes	Yes	Yes	N.A.
Specify F to enter	Yes	Yes	Yes	N.A.
Specify F to remove	No	Yes	Yes	N.A.
Specify probability of F to enter and/or remove	No	Yes	No	N.A.
Specify maximum number of steps	No	Yes	Yes	N.A.
Specify maximum number of variables	Yes	No	No	Yes
Maximum number of inclusion levels	50	As many as IVs	As many as IVs	Yes
Regression through the origin	No[b]	Yes	Yes	Yes
Regression output				
Labeling of variables	Yes	Yes	Yes	Yes
Analysis of variance for regression, each step	Yes	Yes	Yes	Yes
Multiple R, each step or subset	Yes	Yes	Yes	No
R^2, each step	R SQUARE	R SQUARE	MULTIPLE R-SQUARE	Each best set
Standard error for R, each step	STANDARD ERROR	STANDARD ERROR	STD. ERROR OF EST.	No

Statistic				Each best set
Adjusted R^2	Yes	Yes	Yes	No
Sum-of-squares and cross-products matrix	No	Yes	No	Yes
Correlation matrix	Yes	Yes	Yes	Yes
Covariance matrix	No	Yes	Yes	Yes
Correlation matrix of regression coefficients	No	No	Yes	Yes
Covariance matrix of regression coefficients	No	Yes	No	Yes
Number of valid cases	Yes	Yes	Yes	Yes
Means and standard deviations	Yes	Yes	Yes	Yes
Coefficients of variation	No	No	Yes	Yes
Minimums and maximums	No	No	Yes	Yes
Smallest and largest z	No	No	Yes	Yes
Skewness and kurtosis	No	No	Yes	No
Sweep matrix	No	Yes	No	No
Condition number bounds	No	Yes	No	No
Variables in equation (each step)				(Each best set)
Unstandardized regression coefficients	B	B	COEFFICIENT	COEFFICIENT
Standard error of regression coefficient	STD. ERROR B	SE B	STD. ERROR OF COEFFICIENT	No
Standardized regression coefficient	BETA	BETA	STD. REG. COEFF.	No
Variables in equation (each step)				(Each best set)
F (or T) to remove from equation	F	T	F TO REMOVE LEVEL	N.A.
Intercept	(CONSTANT)	(CONSTANT)	(Y-INTERCEPT)	INTERCEPT

TABLE 5.9 (*Continued*)

Feature	SPSS REGRESSION	SPSS NEW REGRESSION[a]	BMDP2R	BMDP9R
Standardized regression coefficient for entering equation	BETA IN	BETA IN	No	No
Partial correlation coefficient	PARTIAL	PARTIAL	PARTIAL CORR.	No
Tolerance	Yes	Yes	Yes	Yes
F (or T) to enter equation	F	T	F TO ENTER LEVEL	No
Table of stepwise regression coefficients	No	No	Yes	No
Summary table				(Single best set)
Step number	No	Yes	Yes	N.A.
Multiple R	Yes	MULTR	Yes	N.A.
R^2	R SQUARE	RSQ	MULTIPLE RSQ	N.A.
Change in R^2 (squared semipartial correlation)	RSQ CHANGE	RSQCH	INCREASE IN RSQ	N.A.
Adjusted R^2	No	ADJRSQ	No	N.A.
F to enter equation	No	FCH	F-TO-ENTER	N.A.
F to remove from equation	No	FCH	F-TO-REMOVE	N.A.
Significance of T to enter or remove	No	Yes	No	N.A.
Unstandardized regression coefficient	B	No	No	Yes
Intercept	(CONSTANT)	No	No	Yes
Standardized regression coefficient	BETA	BETA IN	No	Yes
Number of independent variables included	No	No	Yes	Yes

Criterion value for best subset	N.A.	N.A.	N.A.	Yes
Residual mean square	N.A.	N.A.	N.A.	Yes
Standard error of estimate	N.A.	N.A.	N.A.	Yes
F statistic for equation	No	Yes[c]	No	Yes
Degrees of freedom for F	No	No	No	Yes
Significance	No	Yes[c]	No	Yes
Standard error of regression coefficient	No	No	No	Yes
t test and significance for regression coefficient	No	No	No	Yes
Residuals				
Table of predicted scores and residuals	Yes	Yes	Yes	Yes
Plot of residuals against predicted scores	Yes	Yes	Yes	Yes
Normal plot of residuals	No	Yes	Yes	Yes
Other plots available	Yes	Yes	Yes	Yes
Durbin-Watson statistic	Yes	Yes	No	Yes
Standardized residuals	No	Yes	No	Yes
Mahalanobis and/or Cook's distance	No	Yes	No[d]	Yes
Summary statistics for residuals	No	Yes	No	Yes

[a] Available starting with Release 9.0 (Hull & Nie, 1981).
[b] Available in SPSS-6000 CYBER implementation, Release 8.0.
[c] See Section 5.6.2.1 for interpretation.
[d] Available through BMDPAM, BMDP4M, or BMD10M.

proportion of variance of a potential IV that is not explained by IVs already in the equation. The lower the tolerance value, the less restriction is placed on entering variables. By disallowing entry of variables that add virtually nothing to predictability, one can avoid multicollinearity.

There can be as many as 50 inclusion levels in hierarchical or hierarchical-stepwise regression. This means that up to 50 variables can be entered in predetermined order, although with listing in blocks many more variables could be entered.

A table of predicted scores and residuals can be requested, and can be accompanied by a plot of standardized residuals against standardized predicted values of the DV (z scores of the Y' values). Plots of standardized residuals against sequenced cases are also available. For a sequenced file, one can request a Durbin-Watson statistic, which is used for a test of autocorrelation between adjacent cases.

The flexibility of input for the SPSS REGRESSION program does not carry over into output. The only difference between standard and stepwise or hierarchical regression is in the printing of a progression of steps. Otherwise the statistics and parameter estimates are identical. These values, however, have different meanings, depending on the type of analysis. As pointed out earlier, the RSQ CHANGE refers to squared semipartial correlations, but only for stepwise or hierarchical analysis. For standard multiple regression, the semipartial correlations must be calculated from other values (see Eq. 5.9), and RSQ CHANGE simply reflects proportion of added variance of variables in the order in which they were listed in requesting the program run. As a matter of fact, all useful information in the summary table for standard multiple regression appears also in the section labeled VARIABLES IN THE EQUATION. The summary table can and should, therefore, be suppressed by use of the appropriate option when running standard multiple regression.

In the case of stepwise or hierarchical regression, the test of significance of contribution of IVs to the equation is in error and must be recalculated (see Eq. 5.12).

Several other types of regression, in addition to the three major types listed in Section 5.5, can be analyzed through the SPSS REGRESSION program. The SPSS manual (Nie et al., 1975), in the chapter on general linear models, demonstrates procedures for examining nonlinear relationships (including polynomial regression) through proper transformation and addition of variables, for analysis of variance and covariance through dummy variable coding, and for path analysis.

As can be seen in Tables 5.8 and 5.9, SPSS NEW REGRESSION has retained most of the features of SPSS REGRESSION, and added a number of new ones, including those available in the SPSS-6000 CYBER implementation. Especially welcome is the availability of statistics for evaluating multivariate outliers, most notably the Mahalanobis distance for each case. This is the only SPSS program that currently offers such capability. Moreover, the user can specify the variables on which multivariate outliers are to be based, making the program useful for evaluating outliers prior to any other multivariate procedures. In addition, a useful new method for replacing missing data, substitution of mean value, has been added (cf. Chapter 4).

For stepwise regression, probability of F to enter or remove a variable can be specified as well as the F value itself. But the availability of F for the regression equation for each step (shown in the summary table as well as in step-by-step output) is misleading, as discussed in Section 5.6.2.1.

5.7.2 BMD Series

The BMDP packages offer separate programs for the various types of regression. All the programs do standard multiple regression, but some may give misleading information when used to do it. The applicable program in the BMDP series is BMDP1R. For stepwise and hierarchical regression, the program is BMDP2R. BMDP9R is also available as a variant of stepwise (but not hierarchical) program. In the P series, procedures have been instituted for handling missing data, a feature not available in many of the original BMD programs.

For standard multiple regression, BMDP1R offers analysis of "case combinations," or subsamples. Reduction of residuals due to grouping serves as a test of the equality of regression among the groups. The covariance and correlation matrices are available as output or can be used as input. In addition to a table of predicted values and residuals, the BMDP1R program allows a number of plots of residuals, all described in the manual (Dixon, 1981).

The stepwise/hierarchical regression program in the BMDP series, BMDP2R, combines several features of the other programs and adds some that are unavailable elsewhere in the two packages. The user can specify the stepping algorithm, that is, how the program adds or deletes variables in the equation as it progresses, in addition to specifying statistical criteria. These are fully described in the BMDP manual (Dixon, 1981). The four specifiable statistical criteria are tolerance, F to enter, F to remove, and maximum number of steps. At the end of the output for all steps, a table of stepwise regression coefficients is given. This indicates the full regression equation for each step. The intercept (A) is given, along with unstandardized regression coefficients (B_i) for each of the variables in the equation at the next step. There is no limit to the number of inclusion levels for hierarchical analysis. Input and output matrices and residuals tables and plots are the same as for BMDP1R.

BMDP9R is a setwise rather than stepwise program, which searches among all possible subsets of variables for that subset that is best. "Best" is defined by the user, with such criteria available as maximum R^2 (see Section 5.6.3). A number of best subsets can be identified, at the option of the researchers.

Cross-validation is simplified by a case-weighting procedure, in which cases to be omitted from the computation of the regression equation are given a weight of zero. Residuals analysis is extensive, providing information for evaluation of several assumptions, including outliers.

For regression analysis other than the three major types discussed in this chapter, BMDP offers a number of additional programs. These cover such procedures as nonlinear regression, analysis of variance and covariance, periodic regression, and asymptotic regression.

5.8 COMPLETE EXAMPLES OF REGRESSION ANALYSIS

To illustrate applications of regression analyses, variables were chosen from among those measured in the research described in Appendix B. Two analyses are reported here, both with number of visits to health professionals (TIMEDHS) as the DV and both using the SPSS REGRESSION program.

The first example is a standard multiple regression between the DV and other health variables: number of physical health symptoms (PHYHEALT), number of mental health symptoms (MENHEALT), and stress from acute life changes (RAHE). From this analysis, one can assess the degree of relationship between the DV and the IVs, the proportion of variance in the DV predicted by regression, and the relative importance of the various IVs to the solution.

The second example demonstrates hierarchical regression of the same IVs that were used in the first example. The first step of the analysis is entry of PHYHEALT to determine how much variance in number of visits to health professionals can be accounted for by differences in physical health. The second step is entry of RAHE to determine if there is a significant increase in R^2 when differences in stress are added to the equation. The final step is entry of MENHEALT to determine if differences in mental health are related to number of visits to health professionals after differences in physical health and stress have been statistically accounted for.

5.8.1 Evaluation of Assumptions

Because both analyses use the same variables, assumptions may be evaluated simultaneously.

5.8.1.1 Number of Cases and Variables With 465 respondents and 3 IVs, the cases to variables ratio was 155:1, well above the minimum requirements for regression.

5.8.1.2 Normality, Linearity, and Homoscedasticity of Residuals An initial run with standard multiple regression and untransformed variables was performed to examine the shape of the scatterplot of residuals against predicted DV scores. Because the scatterplot was unavailable through SPSS REGRESSION as implemented and converted for CYBER machines, BMDP1R was used to obtain it. The residuals scatterplot appears in Figure 5.2.

Notice the execrable overall shape of the scatterplot, which violates all of the assumptions regarding distribution of residuals. Comparison between Figure 5.2 and Figure 5.1 (in Section 5.3.2.4) suggests further analysis of the distributions of the variables. (It was noticed in passing, although we tried not to look, that R^2 for this analysis was significant, but only .22.)

SPSS CONDESCRIPTIVE was used to examine the distributions of the

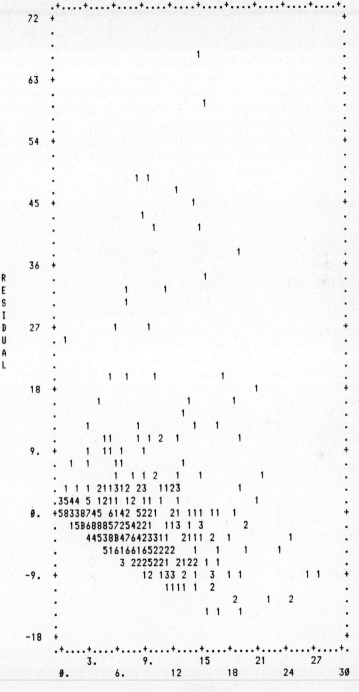

Figure 5.2 Residuals scatterplot using untransformed variables. Output from BMDP1R.

variables. All of the variables had significant positive skewness (see Chapter 4), which explained, at least in part, the problems in the residuals scatterplot. Logarithmic and square root transformations were applied as appropriate, and the transformed distributions checked once again for skewness. Thus TIMEDHS and PHYHEALT, with logarithmic transformations, became LTIMEDHS and LPHYHEAL. After square root transformation, RAHE became SRAHE. In the case of MENHEALT, which is itself positively skewed, application of the milder square root transformation made the variable significantly negatively skewed, so no transformation was ultimately undertaken. Table 5.10 shows output from CONDESCRIPTIVE for both untransformed and transformed variables.

TABLE 5.10 DECK SETUP AND SPSS CONDESCRIPTIVE OUTPUT
FOR UNTRANSFORMED AND TRANSFORMED VARIABLES

```
           COMPUTE        LTIMEDHS = LG10(TIMEDHS + 1)
           COMPUTE        LPHYHEAL = LG10(PHYHEALT + 1)
           COMPUTE        SRAHE = SQRT(RAHE)
           CONDESCRIPTIVE TIMEDHS TO SRAHE
           STATISTICS     ALL


VARIABLE   TIMEDHS

MEAN          7.901    STD ERR        .508    STD DEV      10.948
VARIANCE    119.870    KURTOSIS     13.101    SKEWNESS      3.248
MINIMUM           0    MAXIMUM      81.000    SUM        3674.000
C.V. PCT    138.570    .95 C.I.      6.903               TO
                                                  8.899

VALID CASES     465    MISSING CASES      0



VARIABLE   PHYHEALT

MEAN          4.972    STD ERR        .111    STD DEV       2.388
VARIANCE      5.704    KURTOSIS      1.124    SKEWNESS      1.031
MINIMUM       2.000    MAXIMUM      15.000    SUM        2312.000
C.V. PCT     48.035    .95 C.I.      4.754               TO
                                                  5.190

VALID CASES     465    MISSING CASES      0



VARIABLE   MENHEALT

MEAN          6.123    STD ERR        .194    STD DEV       4.194
VARIANCE     17.586    KURTOSIS      -.292    SKEWNESS       .602
MINIMUM           0    MAXIMUM      18.000    SUM        2847.000
C.V. PCT     68.494    .95 C.I.      5.740               TO
                                                  6.505

VALID CASES     465    MISSING CASES      0
```

TABLE 5.10   (Continued)

VARIABLE   RAHE

MEAN	204.217	STD ERR	6.297	STD DEV	135.793
VARIANCE	18439.662	KURTOSIS	1.801	SKEWNESS	1.043
MINIMUM	0	MAXIMUM	920.000	SUM	94961.000
C.V. PCT	66.494	.95 C.I.	191.843	TO	216.592

VALID CASES   465   MISSING CASES   0

VARIABLE   LTIMEDHS

MEAN	.741	STD ERR	.019	STD DEV	.415
VARIANCE	.172	KURTOSIS	-.177	SKEWNESS	.228
MINIMUM	0	MAXIMUM	1.914	SUM	344.698
C.V. PCT	56.018	.95 C.I.	.703	TO	.779

VALID CASES   465   MISSING CASES   0

VARIABLE   LPHYHEAL

MEAN	.744	STD ERR	.008	STD DEV	.167
VARIANCE	.028	KURTOSIS	-.639	SKEWNESS	.156
MINIMUM	.477	MAXIMUM	1.204	SUM	345.850
C.V. PCT	22.432	.95 C.I.	.729	TO	.759

VALID CASES   465   MISSING CASES   0

VARIABLE   SRAHE

MEAN	13.400	STD ERR	.231	STD DEV	4.972
VARIANCE	24.723	KURTOSIS	.117	SKEWNESS	-.091
MINIMUM	0	MAXIMUM	30.332	SUM	6230.789
C.V. PCT	37.107	.95 C.I.	12.946	TO	13.853

VALID CASES   465   MISSING CASES   0

The residuals scatterplot from BMDP1R following regression of the transformed variables (and deletion of outliers) appears as Figure 5.3. Notice that, although the scatterplot is still not perfectly rectangular, its shape is considerably improved over that in Figure 5.2.

5.8.1.3 Outliers Multivariate outliers were tested among the IVs using BMDP4M, which prints the Mahalanobis distance from each case to the centroid

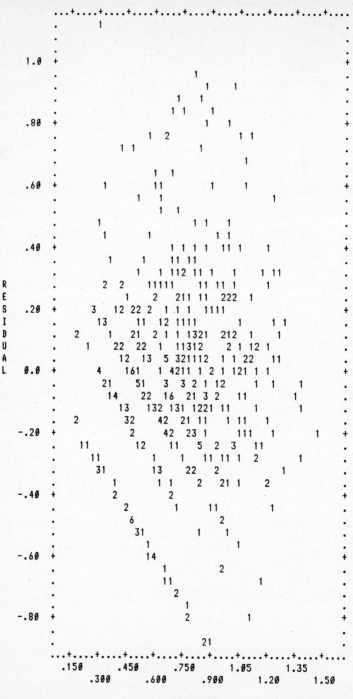

Figure 5.3 Residuals scatterplot following regression with transformed variables. Output from BMDP1R.

of all cases for original data, among other things.[12] The Mahalanobis distance is a chi square (χ^2) variable; in the BMDP4M output it has been divided by its degrees of freedom. To determine which cases are multivariate outliers, one must calculate a critical value by looking up critical χ^2 at the desired alpha level (Table C.4) and dividing the critical χ^2 by its degrees of freedom. In this case, critical χ^2 at $\alpha \leq .01$ for 3 df was 11.3449. The critical value, then, is $11.3449/3 = 3.78$. Any case with a tabled value larger than 3.78 is a multivariate outlier among the IVs.

Two cases, 125 and 403, had tabled values in excess of 3.78 (see Table 5.11). Because variables were already transformed and because N was substantial,

TABLE 5.11 DECK SETUP AND PARTIAL OUTPUT FOR BMDP4M TO ASSESS MULTIVARIATE OUTLIERS

```
        PROGRAM CONTROL INFORMATION

            /PROBLEM TITLE IS "MULTIVARIATE OUTLIERS FOR REGRESSION".
            /INPUT VARIABLES ARE 3. ADD=2. UNIT=55.
                   FORMAT IS "(20X,2F4.0,28X,F4.0,20X/)".
                   CASE=465.
            /VARIABLE NAMES ARE PHYHEALT,MENHEALT,RAHE,LPHYHEAL,SRAHE.
                   USE=2,4,5.
            /TRANSFORM LPHYHEAL=LOG(PHYHEALT+1).
                   SRAHE=SQRT(RAHE).
            /ROTATE METHOD=NONE.
            /PLOT FINAL=0. FSCORE=0.
            /END
```

ESTIMATED FACTOR SCORES AND MAHALANOBIS DISTANCES (CHI-SQUARES) FROM EACH CASE TO THE CENTROID OF ALL CASES FOR ORIGINAL DATA (3 D.F.) FACTOR SCORES (1 D.F.) AND THEIR DIFFERENCE (2 D.F.).
EACH CHI-SQUARE HAS BEEN DIVIDED BY ITS DEGREES OF FREEDOM.
CASE NUMBERS BELOW REFER TO DATA MATRIX BEFORE DELETION OF MISSING DATA.

CASE LABEL	NO.	CHISQ/DF 3	CHISQ/DF 1	CHISQ/DF 2	FACTOR 1
	101	.934	2.082	.359	-1.443
	102	1.817	3.320	1.066	1.822
	103	2.767	5.879	1.211	2.425
	104	1.054	.237	1.462	-.487
	105	.478	.015	.709	-.124
	106	.321	.355	.305	.596
	107	1.092	.723	1.277	.850
	108	1.225	2.499	.588	1.581
	109	.078	.001	.116	-.034
	110	1.818	2.268	1.593	1.506
	111	.956	1.361	.753	1.167
	112	.436	.009	.650	-.093
	113	2.787	4.419	1.971	2.102
	114	.180	.019	.260	.136
	115	1.236	.988	1.361	-.994
	116	.435	.229	.538	.479
	117	1.408	1.832	1.196	1.353
	118	.457	.852	.260	.923
	119	1.804	2.648	1.381	1.627
	120	1.533	2.552	1.024	1.597
	121	.988	.000	1.482	.006
	122	1.652	1.044	1.956	-1.022
	123	.917	1.500	.625	-1.225
	124	.073	.165	.027	.407
→	125	3.935	1.374	5.216	1.172
	126	1.792	.001	2.687	.025
	127	.676	.367	.830	.606
	128	2.118	.698	2.827	-.836

---

[12] SPSS NEW REGRESSION, unavailable to us at time of writing, also produces Mahalanobis distance to evaluate outliers.

these two cases were deleted from further analysis.

To identify the variables on which outliers occurred, BMDP9R was run separately for each of the two outlying cases, with the DV coded 1 for the outlier and 0 for the remaining cases (cf. Chapter 4). Input and selected output for one of these, case 403, appears in Table 5.12. As can be seen, only SRAHE discriminated case 403 from the remainder. A similar run for case 125 showed the discriminating variables to be MENHEALT and LPHYHEAL.

To describe the deleted cases, and search for any additional univariate outliers, BMDP1D was used, with data for deleted cases printed out. As seen in Table 5.13 (page 134), case 403 showed an extreme score ($z = 3.41$) on SRAHE. Case 125 scored relatively high on MENHEAL and relatively low on LPHYHEAL.

5.8.1.4 Multicollinearity and Singularity   The BMDP4M run of Table 5.11 resolved doubts about possible multicollinearity and singularity among the transformed IVs. It was found that the largest SMC between each IV and the other two was .31738 (see Table 5.14, page 134). Further, the highest correlation, between MENHEALT and LPHYHEAL, was .513.

5.8.2 Standard Multiple Regression

SPSS REGRESSION was used to compute a standard multiple regression between LTIMEDHS (the transformed DV), and MENHEALT, LPHYHEAL, and SRAHE (the IVs).

The output of REGRESSION appears as Table 5.15 (page 135).[13] Included are the values of R, R^2, and $\tilde{R}^2$, a summary of the analysis of variance for regression, and the portion labeled VARIABLES IN THE EQUATION, which prints unstandardized and standardized regression coefficients and their significance levels. The significance level for R is found in the analysis of variance source table, $F(3, 459) = 93.07$, $p < .01$.

The significance levels for the regression coefficients (see Table 5.15, VARIABLES IN THE EQUATION) were evaluated against 1 and 459 df. Only two of the IVs, SRAHE and LPHYHEAL, contributed significantly to regression; F values were 22.40 and 139.43, respectively. For these two IVs, confidence limits for the unstandardized coefficients could be calculated, if desired, using Eq. 5.14.

$$CL_{SRAHE} = 0.01611 \pm 0.00340(1.96)$$

$$0.0095 \leq B_{SRAHE} \leq 0.0228$$

$$CL_{LPHYHEAL} = 1.28314 \pm 0.10866(1.96)$$

$$1.0702 \leq B_{LPHYHEAL} \leq 1.4961$$

---

[13] Note that this output, produced by the CYBER (SPSS-6000) version of SPSS, differs somewhat from the standard SPSS.

TABLE 5.12 DECK SETUP AND PARTIAL OUTPUT FOR BMDP9R
TO DISTINGUISH OUTLYING VARIABLES FOR CASE NO. 403

```
PROGRAM CONTROL INFORMATION

      /PROBLEM   TITLE IS "LS REGRESSION, DESCRIBE OUTLIERS".
      /INPUT     VARIABLES ARE 3.  ADD=3.  UNIT=55.
                 FORMAT IS "(20X,2F4.0,28X,F4.0,20X/)".
      /VARIABLE  NAMES ARE PHYHEALT,MENHEALT,RAHE,
                 LPHYHEAL,SRAHE,DUMMY.  USE=2, 4 TO 6.
      /TRANSFORM LPHYHEAL=LOG(PHYHEALT).
                 SRAHE=SQRT(RAHE).
                 DUMMY=0.
                 IF (KASE EQ 403) THEN DUMMY=1.0.
      /REGRESS   DEPENDENT IS DUMMY.
                 INDEPENDENT ARE 2, 4, 5.
      /END


                              ****  SUBSETS WITH    1 VARIABLES  ****

                ADJUSTED
      R-SQUARED  R-SQUARED      CP

      .025046    .022940      2.61  VARIABLE      COEFFICIENT  T-STATISTIC
                                    5 SRAHE         .00147603      3.45
                                      INTERCEPT    -.0176276

      .004752    .002603     12.26  VARIABLE      COEFFICIENT  T-STATISTIC
                                    4 LPHYHEAL      .0155041       1.49
                                      INTERCEPT    -.00790194

      .000095   -.002065     14.47  MENHEALT


                              ****  SUBSETS WITH    2 VARIABLES  ****

                ADJUSTED
      R-SQUARED  R-SQUARED      CP

      .028080    .023872      3.16  VARIABLE      COEFFICIENT  T-STATISTIC
                                    2 MENHEALT     -.000659352    -1.20
                                    5 SRAHE         .00168899      3.65
                                      INTERCEPT    -.0164443

      .025438    .021219      4.42  VARIABLE      COEFFICIENT  T-STATISTIC
                                    4 LPHYHEAL      .00469593       .43
                                    5 SRAHE         .00141432      3.13
                                      INTERCEPT    -.0198454

      .005635    .001330     13.84  MENHEALT  LPHYHEAL


                              ****  SUBSETS WITH    3 VARIABLES  ****

                ADJUSTED
      R-SQUARED  R-SQUARED      CP

      .030530    .024221      4.00  VARIABLE      COEFFICIENT  T-STATISTIC
                                    2 MENHEALT     -.000953885    -1.56
                                    4 LPHYHEAL      .0131067       1.08
                                    5 SRAHE         .00161186      3.44
                                      INTERCEPT    -.0221055
STATISTICS FOR "BEST" SUBSET
MALLOWS" CP                          2.61
SQUARED MULTIPLE CORRELATION        .02505
MULTIPLE CORRELATION                .15826
ADJUSTED SQUARED MULT. CORR.        .02294
RESIDUAL MEAN SQUARE               .002101
STANDARD ERROR OF EST.             .045839
F-STATISTIC                         11.89
NUMERATOR DEGREES OF FREEDOM           1
DENOMINATOR DEGREES OF FREEDOM       463
SIGNIFICANCE                        .0006
```

VARIABLE NO. NAME	REGRESSION COEFFICIENT	STANDARD ERROR	STAND. COEF.	T- STAT.	2TAIL SIG.	TOL- ERANCE	CONTRIBUTION TO R-SQUARED
INTERCEPT	-.0176276	.00611610	-.380	-2.88	.004		
5 SRAHE	.00147603	.000427985	.158	3.45	.001	1.000000	.025046

TABLE 5.13   DECK SETUP AND PARTIAL OUTPUT FOR BMDP1D
TO ASSESS OUTLIERS

```
PROGRAM CONTROL INFORMATION

        /PROBLEM   TITLE IS "LS REGRESSION, DESCRIBE OUTLIERS".
        /INPUT     VARIABLES ARE 3.  ADD=3.  UNIT=55.
                   FORMAT IS '(20X,2F4.0,28X,F4.0,20X/)'.
        /VARIABLE NAMES ARE PHYHEALT,MENHEALT,RAHE,
                   LPHYHEAL,SRAHE,DUMMY. USE=2, 4 TO 6.
        /TRANSFORM  LPHYHEAL=LOG(PHYHEALT).
                    SRAHE=SQRT(RAHE).
                    DUMMY=0.
                    IF (KASE EQ 125 OR KASE EQ 403) THEN DUMMY=XMIS.
        /PRINT     MISSING.
        /END


PRINT CASES CONTAINING MISSING VALUES.

CASE LABEL NO.   2 MENHEALT  4 LPHYHEAL  5 SRAHE    6 DUMMY

        125        16.0000      .4771     19.4165    MISSING
        403         7.0000      .9542     30.3315    MISSING

NUMBER OF CASES READ. . . . . . . . . . . .     465


VARIABLE                    STANDARD    ST.ERR.   COEFF. OF   S M A L L E S T    L A R G E S T                TOTAL
NO. NAME            MEAN     DEVIATION   OF MEAN   VARIATION   VALUE   Z-SCORE   VALUE    Z-SCORE    RANGE     FREQUENCY

  2 MENHEALT        6.123     4.194       .1945     .68494     0.000    -1.46    18.000    2.83     18.000      465
  4 LPHYHEAL         .648      .206       .0096     .31802      .301    -1.68     1.176    2.56       .875      465
  5 SRAHE          13.400     4.972       .2386     .37107     0.000    -2.69    30.332    3.41     30.332      465
  6 DUMMY           0.000     0.000       0.0000    0.00000    0.000     0.00     0.000    0.00      0.000      463
```

TABLE 5.14   PARTIAL OUTPUT FROM BMDP4M TO ASSESS
MULTICOLLINEARITY; DECK SETUP ON TABLE 5.12

SQUARED MULTIPLE CORRELATIONS (SMC) OF EACH VARIABLE WITH
ALL OTHER VARIABLES

                   SMC

  2 MENHEALT   .31738
  4 LPHYHEAL   .28089
  5 SRAHE   .16660

That neither of the confidence limits includes zero reiterates the significance of the IVs.

Equation 5.9 was used to calculate the squared semipartial correlations for SRAHE and LPHYHEAL, as follows:

TABLE 5.15 STANDARD MULTIPLE REGRESSION ANALYSIS
OF LTIMEDHS (THE DV) WITH MENHEALT, SRAHE,
AND LPHYHEAL (THE IVS); DECK SETUP AND SELECTED
OUTPUT FROM SPSS REGRESSION

```
          COMPUTE      LTIMEDHS=LG10(TIMEDHS+1)
          COMPUTE      LPHYHEAL=LG10(PHYHEALT+1)
          COMPUTE      SRAHE=SQRT(RAHE)
          REGRESSION   VARIABLES=LTIMEDHS,MENHEALT,LPHYHEAL,SRAHE/
                       REGRESSION=LTIMEDHS WITH MENHEALT,LPHYHEAL,SRAHE(2)
          STATISTICS   1,2,3,6
```

DEPENDENT VARIABLE..   LTIMEDHS

MEAN RESPONSE   .74136   STD. DEV.   .41522

VARIABLE(S) ENTERED ON STEP NUMBER   1..   MENHEALT
                                            SRAHE
                                            LPHYHEAL

		ANALYSIS OF VARIANCE	DF	SUM OF SQUARES	MEAN SQUARE	F	SIGNIFICANCE
MULTIPLE R	.61500						
R SQUARE	.37823	REGRESSION	3.	30.12659	10.04220	93.06993	0
ADJUSTED R SQUARE	.37416	RESIDUAL	459.	49.52585	.10790		
STD DEVIATION	.32848	COEFF OF VARIABILITY	44.3 PCT				

```
-------------------- VARIABLES IN THE EQUATION ----------------------      ---------- VARIABLES NOT IN THE EQUATION ----------
```

VARIABLE	B	STD ERROR B	F	BETA	VARIABLE	PARTIAL	TOLERANCE	F
			SIGNIFICANCE	ELASTICITY				SIGNIFICANCE
MENHEALT	.21149342E-02	.44551124E-02	.22535977	.0212767				
			.635	.01740				
SRAHE	.16110740E-01	.34039826E-02	22.400429	.1905896				
			.000	.29011				
LPHYHEAL	1.2831395	.10866479	139.43442	.5149863				
			0	1.28687				
(CONSTANT)	-.44064995	.76286361E-01	33.365193					
			0					

$$sr^2_{\text{SRAHE}} = \frac{22.400}{459}(1 - .37823) = .0303$$

$$sr^2_{\text{LPHYHEAL}} = \frac{139.434}{459}(1 - .37823) = .1889$$

These values represent the amount by which R^2 would be reduced if SRAHE or LPHYHEAL were omitted from the equation. Their sum, .2192, is the amount of R^2 attributable to unique sources. The difference between R^2 and unique variance $(.37823 - .2192 = .1590)$ represents variance that SRAHE, LPHYHEAL, and MENHEALT jointly contribute to R^2.

It was noted from the correlation matrix that MENHEALT was correlated with LTIMEDHS $(r = .36308)$ but did not contribute significantly to regression. Equation 5.11 was used a posteriori to evaluate the significance of the correlation coefficient.

$$F = \frac{(.3631)^2/3}{(1 - .3631^2)/463 - 3 - 1} = 23.24$$

The correlation between MENHEALT and LTIMEDHS was significantly different from zero: $F(3, 459) = 23.24$, $p < .01$.

Thus, although the bivariate correlation between MENHEALT and

TABLE 5.16   CHECKLIST FOR STANDARD MULTIPLE
REGRESSION

1. Issues
   a. Number of cases and variables
   b. Outliers
   c. Multicollinearity and singularity
   d. Normality, linearity, and homoscedasticity of residuals
2. Major analyses
   a. Multiple R, F ratio
   b. Adjusted multiple R, overall proportion of variance accounted for
   c. Significance of regression coefficients
   d. Squared semipartial correlations
3. Additional analyses
   a. Post hoc significance of correlations
   b. Unstandardized (B) weights, confidence limits
   c. Standardized (β) weights
   d. Unique versus shared variability
   e. Suppressor variables

LTIMEDHS is reliably different from zero, the relationship seems to be mediated by, or redundant to, the relationship between LTIMEDHS and other IVs in the set. Had the researcher only measured MENHEALT and LTIMEDHS, however, the significant correlation might have led to stronger conclusions than are warranted about the relationship between mental health and number of visits to health professionals.

Table 5.16 contains a checklist of analyses performed in standard multiple regression. An example of a Results section in journal format appears next.

RESULTS

A standard multiple regression was performed between number of visits to health professionals as the dependent variable and physical health, mental health, and stress as independent variables. Analysis was performed using SPSS REGRESSION with an assist from several BMDP programs in evaluation of assumptions.

Results of evaluation of assumptions led to transformation of the variables to reduce skewness in their distributions, reduce the number of outliers, and improve the normality, linearity, and homoscedasticity of residuals. A square root transformation was used on the measure of stress (SRAHE). Logarithmic transformations were used on the number of visits to health professionals (LTIMEDHS) and on physical health (LPHYHEAL). One IV, MENHEALT, was positively skewed without transformation and negatively skewed with it; it was not transformed. With the use of a $p < .01$ criterion and BMDP4M, two outliers among the IVs were identified and deleted from analysis. Of the two deleted cases, one had an extremely high score on SRAHE ($z = 3.41$), while the other combined a relatively high score on MENHEALT with a relatively low score on LPHYHEAL. No cases had missing data and no suppressor variables were found.

Table 5.17 displays the correlations between the variables, the unstandardized regression coefficients (B) and intercept, the standardized regression coefficients (β), the semipartial correlations (sr^2) and R, R^2, and adjusted R^2. R for regression was significantly different from zero: $F(3, 459) = 93.07$, $p < .001$. For the two regression coefficients that differed significantly from zero, 95% confidence limits were calculated. The confidence limits for SRAHE were 0.0095 to 0.0228, and those for LPHYHEAL were 1.0702 to 1.4961. **[Means, standard deviations, and correlations are from a portion of SPSS REGRESSION not reproduced in Table 5.15.]**

Insert Table 5.17 about here

Only two of the IVs contributed significantly to prediction of number of visits to health professionals as logarithmically transformed (LTIMEDHS), log of physical health scores (LPHYHEAL, $sr^2 = .19$) and square root of acute stress scores (SRAHE, $sr^2 = .03$). The three IVs in combination contributed another .16 in shared variability. Altogether, 38% (37% adjusted) of the variability in visits to health professionals could be predicted by knowing scores on these three IVs.

Although the correlation between LTIMEDHS and MENHEALT was .36, MENHEALT did not contribute significantly to regression. Post hoc evaluation of the correlation revealed that it was significantly different from zero, $F(3, 459) = 23.87$, $p < .01$. Apparently the relationship between the number of visits to health professionals and mental health is an indirect result of the relationships between physical health, stress, and visits to health professionals.

TABLE 5.17 STANDARD MULTIPLE REGRESSION OF HEALTH AND STRESS VARIABLES ON NUMBER OF VISITS TO HEALTH PROFESSIONALS

Variables	LTIMEDHS (DV)	LPHYHEAL	SRAHE	MENHEALT	B	β	sr^2 (Unique)
LPHYHEAL	.59				1.283**	0.51	.19
SRAHE	.36	.31			0.016**	0.19	.03
MENHEALT	.36	.52	.38		0.002	0.02	
				Intercept = −0.441			
Means	0.74	0.74	13.35	6.10			
Standard deviations	0.42	0.17	4.91	4.18			R^2 = .38[a]
						Adjusted R^2 = .37	
						R = .62**	

** $p < .01$.

[a] Unique variability = .22; shared variability = .16.

138

5.8.3 Hierarchical Regression   The second example involves the same three IVs entered one at a time in an order determined by the researcher. LPHYHEAL was the first IV to enter, followed by SRAHE and then MEN-HEALT. The main research question asked then is whether or not information regarding differences in mental health can be used to predict visits to health professionals after differences in physical health and in acute stress have been statistically eliminated. In other words, do people go to health professionals for more numerous mental health symptoms if they have similar physical health symptoms and stress as other people?

Table 5.18 shows selected portions of the output from the SPSS REGRES-SION program. Notice that a complete regression solution is provided at the end of each of steps 1 through 3. The significance of the bivariate relationship between LTIMEDHS and LPHYHEAL may be assessed at the end of step 1, $F(1, 461) = 241.08$, $p < .001$. The bivariate correlation is .59, accounting for 34% of the variance. After step 2, with both LPHYHEAL and SRAHE in the equation, $F(2, 460) = 139.73$, $p < .01$, $R = .61$, and $R^2 = .38$. With the addition of MENHEALT, $F(3, 459) = 93.07$, $R = .62$, and $R^2 = .38$. Increments in R^2 at each step may be read directly from the R SQUARE CHANGE column of the SUMMARY TABLE. Thus, $sr^2_{\text{LPHYHEAL}} = .34$, $sr^2_{\text{SRAHE}} = .03$, and $sr^2_{\text{MENHEALT}} = .00$.

Significance of the additions of both SRAHE and MENHEALT may be tested by calculating F_{inc} with the use of Eq. 5.13. For the addition of SRAHE to LPHYHEAL:

$$F_{\text{inc}} = \frac{(.37792 - .34338)/1}{(1 - .37792)/459} = 25.49$$

with 1 and 459 df. Thus there is a reliable increase in R^2 with addition of SRAHE at $p < .01$.

For the addition of MENHEALT to LPHYHEAL and SRAHE:

$$F_{\text{inc}} = \frac{(.37823 - .37792)/1}{(1 - .37823)/459} = 0.23$$

with 1 and 459 df. Thus there is no reliable increase in prediction of LTIMEDHS by addition of MENHEALT if differences in LPHYHEAL and SRAHE are already accounted for. Apparently the answer is no to the question, Do people go to health professionals for more numerous mental health symptoms if they have similar physical health symptoms and stress to others?

Because regression was hierarchical, the significance levels of the Model II squared semipartial correlations must be calculated using Eq. 5.12. (Note that this F ratio is equal to F_{inc} if only one variable has been added.)

$$F_{\text{LPHYHEAL}} = \frac{.34338}{(1 - .37823)/459} = 253.49$$

$$F_{\text{SRAHE}} = \frac{.03454}{(1 - .37823)/459} = 25.50$$

TABLE 5.18 DECK SETUP AND PARTIAL OUTPUT FOR SPSS HIERARCHICAL REGRESSION

```
COMPUTE      LTIMEDHS=LG10(TIMEDHS+1)
COMPUTE      LPHYHEAL=LG10(PHYHEALT+1)
COMPUTE      SRAHE=SQRT(RAHE)
REGRESSION   VARIABLES=LTIMEDHS,MENHEALT,LPHYHEAL,SRAHE/
             REGRESSION=LTIMEDHS WITH LPHYHEAL(6) SRAHE(4) MENHEALT(2)
STATISTICS   1,2,3,6
```

DEPENDENT VARIABLE..   LTIMEDHS

MEAN RESPONSE   .74136   STD. DEV.   .41522

VARIABLE(S) ENTERED ON STEP NUMBER   1..   LPHYHEAL

		ANALYSIS OF VARIANCE	DF	SUM OF SQUARES	MEAN SQUARE	F	SIGNIFICANCE
MULTIPLE R	.58599					241.08479	0
R SQUARE	.34338	REGRESSION	1.	27.35139	27.35139		
ADJUSTED R SQUARE	.34196	RESIDUAL	461.	52.30105	.11345		
STD DEVIATION	.33683	COEFF OF VARIABILITY	45.4 PCT				

-------------------- VARIABLES IN THE EQUATION ----------------------   ---------- VARIABLES NOT IN THE EQUATION ----------

VARIABLE	B	STD ERROR B	F SIGNIFICANCE	BETA ELASTICITY	VARIABLE	PARTIAL	TOLERANCE	F SIGNIFICANCE
LPHYHEAL	1.4600520	.94033674E-01	241.08479 0	.5859899 1.46429	SRAHE	.22934	.90123	25.537891 .000
(CONSTANT)	-.34420926	.71646403E-01	23.081091 .000		MENHEALT	.08276	.72753	3.1722357 .076

VARIABLE(S) ENTERED ON STEP NUMBER   2..   SRAHE

		ANALYSIS OF VARIANCE	DF	SUM OF SQUARES	MEAN SQUARE	F	SIGNIFICANCE
MULTIPLE R	.61475					139.72752	.00
R SQUARE	.37792	REGRESSION	2.	30.10227	15.05114		
ADJUSTED R SQUARE	.37522	RESIDUAL	460.	49.55017	.10772		
STD DEVIATION	.32820	COEFF OF VARIABILITY	44.3 PCT				

-------------------- VARIABLES IN THE EQUATION ----------------------   ---------- VARIABLES NOT IN THE EQUATION ----------

VARIABLE	B	STD ERROR B	F SIGNIFICANCE	BETA ELASTICITY	VARIABLE	PARTIAL	TOLERANCE	F SIGNIFICANCE
LPHYHEAL	1.3067650	.96517052E-01	183.31032 0	.5244684 1.31056	MENHEALT	.02215	.67436	.22535977 .635
SRAHE	.16547587E-01	.32744785E-02	25.537891 .000	.1957575 .29798				
(CONSTANT)	-.45114808	.72949222E-01	38.246957 .000					

VARIABLE(S) ENTERED ON STEP NUMBER   3..   MENHEALT

		ANALYSIS OF VARIANCE	DF	SUM OF SQUARES	MEAN SQUARE	F	SIGNIFICANCE
MULTIPLE R	.61500					93.06993	0
R SQUARE	.37823	REGRESSION	3.	30.12659	10.04220		
ADJUSTED R SQUARE	.37416	RESIDUAL	459.	49.52585	.10790		
STD DEVIATION	.32848	COEFF OF VARIABILITY	44.3 PCT				

-------------------- VARIABLES IN THE EQUATION ----------------------   ---------- VARIABLES NOT IN THE EQUATION ----------

VARIABLE	B	STD ERROR B	F SIGNIFICANCE	BETA ELASTICITY	VARIABLE	PARTIAL	TOLERANCE	F SIGNIFICANCE
LPHYHEAL	1.2831395	.10866479	139.43442 0	.5149863 1.28687				
SRAHE	.16110740E-01	.34039826E-02	22.400429 .000	.1905896 .29011				
MENHEALT	.21149342E-02	.44551124E-02	.22535977 .635	.0212767 .01740				
(CONSTANT)	-.44064995	.76286361E-01	33.365193 0					

ALL VARIABLES ARE IN THE EQUATION.

* M U L T I P L E   R E G R E S S I O N *
DEPENDENT VARIABLE..   LTIMEDHS

S U M M A R Y   T A B L E

STEP	VARIABLE ENTERED REMOVED	F TO ENTER OR REMOVE	SIGNIFICANCE	MULTIPLE R	R SQUARE	R SQUARE CHANGE	SIMPLE R	OVERALL F	SIGNIFICANCE
1	LPHYHEAL	241.08479	0	.58599	.34338	.34338	.58599	241.08479	0
2	SRAHE	25.53789	.000	.61475	.37792	.03454	.36058	139.72752	.000
3	MENHEALT	.22536	.635	.61500	.37823	.00031	.36308	93.06993	0

TABLE 5.19 CHECKLIST OF HIERARCHICAL
REGRESSION ANALYSIS

1. Issues
   a. Number of cases and variables
   b. Outliers, missing values
   c. Multicollinearity and singularity
   d. Normality, linearity, and homoscedasticity of residuals
2. Major analyses
   a. Multiple R, F ratio
   b. Adjusted R, proportion of variance accounted for
   c. Squared semipartial correlations
   d. Significance of regression coefficients
   e. Incremental F
3. Additional analyses
   a. Unstandardized (B) weights, confidence limits
   b. Standardized (β) weights
   c. Prediction equation from stepwise analysis
   d. Post hoc significance of correlations
   e. Suppressor variables
   f. Cross-validation (stepwise and setwise)

F levels are evaluated with 1 and 459 df. Both are reliably different from zero at $p < .01$.

Table 5.19 is a checklist of items to consider in hierarchical regression. An example of a Results section in journal format appears next.

RESULTS

Hierarchical regression was employed to determine if addition of information regarding stress and mental health symptoms improved prediction of visits to health professionals beyond that afforded by differences in physical health symptoms. Analysis was performed using SPSS REGRESSION with an assist from several BMDP programs in evaluation of assumptions.

Results of evaluation of assumptions led to transformation of the variables to reduce skewness in their distributions, reduce the number of outliers, and improve the normality, linearity, and homoscedasticity of residuals. A square root transformation was used on the measure of stress (SRAHE). Logarithmic transformations were used on the number of visits to health professionals (LTIMEDHS) and physical health (LPHYHEAL). One IV, MENHEALT, was positively skewed without transformation and negatively skewed with it; it was not transformed. With the use of a $p < .01$ criterion and BMDP4M, two outliers among the IVs were identified and deleted from analysis. No cases had missing data and no suppressor variables were found.

Table 5.20 displays the correlations between the variables, the unstandardized regression coefficients (B) and intercept, the standardized regression coefficients (β), the semipartial correlations (sr^2), and R, R^2, and adjusted R^2 after entry of all three IVs. R was significantly different from zero at the end of each step. After step 3, with all IVs in the equation, $R = .62$, $F(3, 459) = 93.07$, $p < .01$. **[Means, standard** deviations, and correlations are from a portion of SPSS REGRESSION not reproduced in Table 5.18.]

Insert Table 5.20 about here

    After step 1, with LPHYHEAL in the equation, $R^2 = .34$, $F(1, 461) = 241.08$, $p < .01$. After step 2, with SRAHE added to prediction of LTIMEDHS by LPHYHEAL, $R^2 = .38$, $F_{inc}(1, 459) = 25.49$, $p < .01$. Thus addition of SRAHE to the equation results in a significant increment in R^2. After step 3, with MENHEALT added to predictions of LTIMEDHS by LPHYHEAL and SRAHE, $R^2 = .38$, $F_{inc}(1, 459) = 0.23$. Therefore addition of MENHEALT to the equation did not reliably improve R^2.

TABLE 5.20 HIERARCHICAL REGRESSION OF HEALTH AND STRESS VARIABLES ON NUMBER OF VISITS TO HEALTH PROFESSIONALS

Variables	LTIMEDHS (DV)	LPHYHEAL	SRAHE	MENHEALT	B	β	sr² (Incremental)
LPHYHEAL	.59				1.283	0.51	.34**
SRAHE	.36	.31			0.016	0.19	.03**
MENHEALT	.36	.52	.38		0.002	0.02	.00
					Intercept = −0.441		
Means	0.74	0.74	13.35	6.10			
Standard deviation	0.42	0.17	4.91	4.18			
						$R^2 = .38$	
						Adjusted $R^2 = .37$	
						$R = .62**$	

** $p < .01$.

5.9 SOME EXAMPLES FROM THE LITERATURE

The following examples have been included to give you a taste for the ways in which regression can be, and has been, used by researchers.

Multiple regression is the statistical tool used in Judgment Analysis, a procedure developed to identify importance of variables underlying policy decisions. In a demonstration involving diagnosis of learning disabilities, Beatty (1977) gave diagnosticians (judges) test profiles for a sample of children. Judges ranked the children in order of severity of learning disability, and these rankings served as the DV.[14] Beatty then used as IVs the age of the child and the set of five test scores on which judges had based their rankings. Judges were analyzed individually, and then in combination, so that judges using similar policies could be grouped. So for each group of judges with a distinct policy, it was possible to evaluate how successfully the rankings could be predicted and the relative importance of IVs in influencing the diagnostic rankings. In Beatty's demonstration, the three scores on the Wechsler Intelligence Scale for Children (Verbal, Performance, and Full Scale) were important for all judges, contributing most to the multiple correlation. Some judges seemed to have more predictable policies than others, and for one judge, 96% of the variance in rankings could be explained by knowledge of the IVs.

In evaluating a negative income tax project in New Jersey, Nicholson and Wright (1977) argue for inclusion of "participant's understanding of the treatment" as one of the IVs in policy experimentation. At each stage of their analysis, separate multiple regression equations were set up for six DVs—all related to a family's earnings and hours worked. (These six DVs could have been combined into a single canonical correlation; see Chapter 6.) Separate equations were also run on the basis of ethnicity. IVs for all equations included (1) pretreatment value of the DV; (2) variables related to family age, education, and size; (3) variables related to site and whether or not families received welfare; (4) a dummy variable indicating whether a family was in the treatment or the control group; and (5) variables related to specification of which experimental plan a family was in, with eight combinations of guaranteed income and tax levels. As the main focus of the evaluation, the regression weight for the difference between experimental and control groups was evaluated for statistical significance in each of the regression equations. Then Nicholson & Wright reran the regression equations with 11 binary variables related to participant's knowledge about the tax plan. Finally, a third set of equations was run that included terms related to interactions between knowledge levels and estimated effects of the guarantee and tax levels. It was found that the multiple correlation increased significantly with the addition of variables representing participant knowledge, and increased again when interaction variables were included. In general, estimates of disincentive effects for families with high understanding were larger

---

[14] Opinions differ regarding application of regression to rank order data. However, since rank order data produce rectangular distributions with neither skewness nor outliers, the application may be considered justified.

than for those who lacked understanding of the system. That is, among families who understood the system, those in the experimental group worked fewer hours and earned less than those in the control group. The difference between experimental and control was less, and in some case in the opposite direction, for families who showed little understanding of the policy.

S. Fidell (1978) applied stepwise regression procedures to prediction of the extensity of attitudes among groups of people. Respondents were interviewed at 24 sites across the United States, stratified by noise exposure and population density. The proportion of respondents at each site who had been highly annoyed by neighborhood noise exposure was predicted from both situational (demographic and physical) variables and other attitudinal variables. Since the prevalence of shared attitudes in the community was of greater interest than the intensity of individual beliefs, the unit of analysis was the site rather than the individual respondent. Primary interest in prediction also dictated stepwise entry of variables, many of which were available from census data. Excellent prediction was available from several combinations of variables, with one equation of 3 IVs accounting for 88% of the variance in community annoyance at exposure to noise.

Species density as a function of environmental measures in the United States was studied by Schall and Pianka (1978) in stepwise regression analysis. Separate analyses were run with birds and lizards as DVs. The 11 IVs were the same for both analyses, consisting of such environmental variables as highest and lowest elevation, sunfall, average annual precipitation, several temperature measures, and so on. Units of measurement (cases) were 895 quadrates, each 1° of latitude by 1° of longitude. Environmental measures were found to be much better predictors of lizard density (multiple $R^2 = .82$) than of bird density (multiple $R^2 = .34$). There is also suggestion that different environmental factors predict density for lizards than for birds. Sunfall, entering the regression first for lizards, by itself accounts for .67 of the variance in density. For birds, sunfall appears to be acting as a suppressor variable. Simple correlation with bird density is virtually zero, but entering fifth in the regression, sunfall adds 9% of predictable variance. Although no cross-validation was reported, discrepancies in the two analyses are large enough to be considered important. (Hierarchical regression would have produced results that could be interpreted with more confidence.)

Maki, Hoffman, and Berk (1978) used standard multiple regression to study the effectiveness of a water conservation campaign. Two parallel analyses were done on sales and production of water (DVs) over 126 months, with months acting as the unit of study. IVs were moratorium (whether or not it and/or conservation measures were in effect), rainfall, season, population, and dollars spent on public relations (as a measure of intensity of campaign). Results of the two analyses were similar, with both showing significant effects of the moratorium-conservation efforts. Money spent on the campaign was not statistically significant in either campaign. Rainfall was a significant predictor of production, but not of sales. Population and season predicted both sales and production.

Chapter 6

Canonical Correlation

6.1 GENERAL PURPOSE AND DESCRIPTION

The goal of canonical correlation is to analyze the relationships between two sets of variables. It may be useful to think of one set of variables as the IVs and the other set as the DVs, or it may not. In any event, canonical analysis provides a tool for more complicated, but perhaps more general, research settings where two sets of variables are measured and the experimenter wants to know how the two sets relate to each other.

To perform this task, canonical analysis generates pairs of linear combinations of variables, one linear combination from each of the two sets. Linear combinations of variables are called canonical variates. The first pair of canonical variates maximizes the correlation between a linear combination of one set and a linear combination of the other. A second pair of canonical variates, if calculated, is uncorrelated with the first pair and maximizes the correlation between linear combinations of variables after the variance due to the first pair of canonical variates has been removed. Discovery of pairs of canonical variates continues either until no significant linkages between sets remain in the residual correlation matrices or until as many pairs of canonical variates have been defined as there are variables in the smaller set.

Canonical analysis is the most general of the multivariate techniques. In fact, the other procedures—multiple regression, discriminant function analysis, and MANOVA—are all special cases of it. But it is also the least used and most impoverished of the techniques, as explained in the next section.

6.2 KINDS OF RESEARCH QUESTIONS

Although a large number of research questions may be asked of canonical analysis in its many specialized forms (such as discriminant analysis), relatively few intricate research questions are readily answerable through direct application of computer programs currently available for canonical analysis. In its present stage of development, canonical analysis is best considered a descriptive analysis or a screening procedure rather than a hypothesis-testing procedure. The following sections, however, contain questions that can be addressed with the aid of SPSS and BMDP programs.

6.2.1 Number of Canonical Variate Pairs

Along how many dimensions are the variables in one set related to the variables in the other? Canonical analysis calculates pairs of variates either until it runs out of variables on one side of the equation or until no significant relationships remain. The number of pairs of variates is the number of dimensions along which the sets of variables are related. Significance tests for canonical variates are reported in Sections 6.4 and 6.5.1.

Suppose that a researcher were interested in the relationship between a set of variables measuring compliance with medical personnel (willingness to use prescriptions, to restrict activity, and the like) and a set of demographic characteristics (educational level, religious affiliation, etc.). Canonical analysis between the sets of variables might reveal that two pairs of canonical variates accounted for all of the significant linkages in the correlation matrix. These two, if interpretable, represent the different dimensions along which compliance is related to demography. But as pointed out by James Fleming (personal communication), the number of statistically significant pairs of canonical variates will be greater than the number of interpretable pairs if N is at all large.

6.2.2 Nature of Canonical Variates

How might the dimensions that relate two sets of variables be interpreted? What is the meaning of the pattern of variables that compose one variate and the pattern composing the other in the same pair? Interpretation of pairs of canonical variates usually proceeds from matrices of correlations between variables and canonical variates, as described in Sections 6.4 and 6.5.2.

Once two significant pairs of canonical variates have been discovered, the next step would be to interpret them. The first pair, for instance, might reveal that forms of medical compliance requiring expenditures of funds (drugs, tests, etc.) were related to economic demographic characteristics; the second pair might show some relationship between a tendency to restrict activity and a conservative versus liberal dimension of demographic characteristics (political party affiliation, religiosity, etc.).

6.2.3 Importance of Canonical Variates

How much variance in each pair of canonical variates is overlapping variance?[1]
And, as a separate question, How much variance do the canonical variates
from one set of variables extract from the variance in the other set? These
questions are answered by examining the squared canonical correlation coefficients and the redundancy statistics, respectively. Sections 6.4 and 6.5.1 address
these issues.

From the example in Section 6.2.1 it is possible that the first pair of
canonical variates have 50% overlapping variability while the second pair overlap
only 15%. Moreover, the first canonical variate in the compliance set may
"explain" 10% of the variance in demography while the first canonical variate
from the demographic set explains 20% of the variance in compliance. Researchers are coming to understand that percent of variance is one of the best indices
of the importance of relationships in a variety of contexts. Statistical significance,
on the other hand, can be thought of as a necessary but not sufficient criterion
for importance.

6.2.4 Canonical Variate Scores

Had the canonical variates from both sets of variables been measured directly,
what scores would the research units (e.g., subjects) have received on them?
This question is similar to one asked of factor analysis (Section 10.2.7). Examination of canonical variate scores may reveal deviant cases as well as illuminate
the shape of the relationships between the canonical variates and the original
variables, as discussed briefly in Section 6.4. Were a pair of canonical variates
easily interpretable, scores on them might be useful as IVs or DVs in other
analyses as well. For instance, the researcher might use medical compliance
scores related to availability of funds from the demographic variables to examine
the effects of publicly supported medical facilities.

6.3 LIMITATIONS

6.3.1 Theoretical Limitations[2]

Canonical correlation has several important theoretical limitations that may
explain its rarity in the literature. Perhaps the most critical limitation involves
interpretability: Procedures that maximize correlation between linear combinations of variables do not necessarily facilitate interpretation of the underlying
dimension. Although in principal component analysis and factor analysis (Chapter 10) linear combinations are usually rotated to facilitate interpretation, rotation

---

[1] R^2 of Chapter 5 can be seen as a measure of overlapping variance between a single DV
and a set of IVs. In canonical analysis the overlap is between a *set* of DVs and a set of IVs.

[2] The authors are indebted to James Fleming for many of the insights of this section.

of canonical variates is not common practice or even available through SPSS or BMD. Therefore a canonical solution may be mathematically elegant but uninterpretable.

Not only are pairs of canonical variates restricted to linear relationships and limited in number, but also they are calculated to be independent of all other pairs. In factor analysis (Chapter 10) one has a choice between an oblique (correlated) and an orthogonal (uncorrelated) solution, but in canonical analysis only the orthogonal solution is available. A researcher who suspects, perhaps on the basis of a previous analysis, that groups of variables within a set are identifiable but correlated with one another, might reconsider use of canonical correlation with the set.

A last concern is the sensitivity of the solution in one set of variables to the inclusion of variables *in the other set.* In canonical analysis, the solution depends on the separate correlations among variables in each set and also on the correlations among variables between sets. Changing the variables in the second set may markedly alter the composition of canonical variates in the first set. This is to be expected, of course, given the goals of the analysis, yet the sensitivity of the procedure to minor alterations is a cause for concern.

6.3.2 Practical Issues

6.3.2.1 Normality
Although there is no requirement that the variables be normally distributed when canonical correlation is used descriptively, the analysis will be enhanced if they are. Inference regarding number of significant canonical variate pairs, however, proceeds on the assumption of multivariate normality. Multivariate normality is not itself a testable hypothesis, but the likelihood of it is increased if the IVs and DVs are all normally distributed. As indicated in Chapter 4, *the crucial aspect of normality of variables can be evaluated through skewness.* If SPSS MANOVA is used, the normal and residuals plots (cf. Section 5.3.2.4) are helpful in evaluating skewness.

With minimally skewed variables and large samples, the central limit theorem can be invoked to assuage the researcher's guilt with respect to multivariate normality. But with small samples or if it is severe, skewness can threaten a meaningful solution. In that case, the transformations suggested in Chapter 4 should be seriously considered.

6.3.2.2 Missing Data
Because missing data can create inconsistencies, they should be estimated carefully. Levine (1977) gives an example of dramatic changes in the canonical solution with a change in procedures for handling missing data. Consult Chapter 4 for methods of estimating or eliminating cases with missing data.

6.3.2.3 Outliers
Cases that are unusual can have untoward effects on canonical analysis just as they can on other multivariate techniques. The search for outliers should be conducted separately with each set of variables. Consult

Chapter 4 for methods of detecting and reducing the effects of both univariate and multivariate outliers.

6.3.2.4 Multicollinearity and Singularity   Because calculation of canonical variates requires inversion of correlation matrices (different ones, depending on whether SPSS or BMD is used), it is important that the matrices be of full rank. That is, none of the variables should be too highly correlated with, or near linear combinations of, others in the set. This restriction applies to R_{xx}, R_{yy}, and R_{xy} (see Eq. 6.1). Consult Chapter 4 for methods of identifying and eliminating multicollinearity and singularity in correlation matrices.

6.3.2.5 Linearity   Linearity is important to canonical analysis in at least two ways. The first is that the analysis is performed on a set of correlation matrices that contain Pearson product-moment correlations. Pearson r, as you recall from Chapters 2 and 4 and elsewhere, is sensitive to the straight-line component of relationships but not to other components. *Inspection of scatterplots between all pairs of variables may reveal gross departures from linearity that can be corrected through variable transformation.* If SPSS MANOVA is used, residuals plots can first be examined for violation of linearity between any variables. Pairwise scatterplots are unnecessary if residuals plots are acceptable (cf. Section 5.3.2.4).

Linearity is also important to canonical correlation in that the analysis maximizes Pearson r between a combination of variables from one set and a combination from the other. As discussed in Section 6.3.1, there may be no theoretical reason to suppose that sets of variables are linearly related, and they may not be. *Inspection of scatterplots between pairs of canonical variates may reveal departures from linearity that can be corrected through inclusion of higher orders of variables or cross products of variables.* Scatterplots are available directly, through BMDP6M or, in SPSS, by requesting output of scores on canonical variates and using them as input to the SCATTERGRAM subroutine.

6.4 FUNDAMENTAL EQUATIONS FOR CANONICAL CORRELATION

A data set that is appropriately analyzed through canonical correlation is a set with several subjects (or experimental units), each of whom has been measured on four or more variables. The variables themselves form two sets with at least two variables in the smaller set. In many situations it may be useful to think of one of the sets of variables as IVs and the other set as DVs. Although this distinction is not required, it will be useful to maintain it for our present purposes.

To illustrate some of the calculations in canonical correlation, a hypothetical data set is presented in Table 6.1. Five intermediate- and advanced-level belly dancers were rated on the quality of their "top" shimmies (TS), "top" circles (TC), "bottom" shimmies (BS), and "bottom" circles (BC). Each charac-

TABLE 6.1 SMALL SAMPLE OF HYPOTHETICAL DATA FOR ILLUSTRATION OF CANONICAL CORRELATION	Set 1 (X)		Set 2 (Y)	
Belly dancers	TS	TC	BS	BC
S_1	6.3	6.6	6.0	5.9
S_2	3.0	5.8	3.2	5.5
S_3	4.1	3.9	4.0	3.9
S_4	5.5	3.5	5.7	3.3
S_5	3.1	2.9	2.8	3.0
Mean	4.4	4.5	4.3	4.3
Standard deviation	1.46	1.58	1.45	1.31

teristic of the dance was rated by two judges on a 7-point scale (with larger numbers indicating higher quality) and the responses were averaged. The goal of analysis was to discover patterns, if any, between the quality of the movements on top and the quality of those on bottom.

The first step in a canonical analysis is generation of a correlation matrix. In this case, however, the matrix of correlations between all the variables can be subdivided into four parts: the correlations between the DVs ($\mathbf{R}_{yy}$), the correlations between the IVs ($\mathbf{R}_{xx}$), and two matrices of correlations between DVs and IVs ($\mathbf{R}_{xy}$ and $\mathbf{R}_{yx}$). Table 6.2 contains the correlation matrices for the data in the example.

There are several ways to write the fundamental equation for canonical correlation, one of them more intuitively appealing than the others. The equations that are used by SPSS and BMD are slightly different from the one to be given here and from each other. However, they are all variants on the following equation.

$$\mathbf{R} = \mathbf{R}_{yy}^{-1} \mathbf{R}_{yx} \mathbf{R}_{xx}^{-1} \mathbf{R}_{xy} \qquad (6.1)$$

The correlation matrix that is analyzed is a product of four matrices of correlation: between DVs, between IVs, and between DVs and IVs.

TABLE 6.2 CORRELATION MATRICES FOR THE DATA SET IN TABLE 6.1	$\mathbf{R}_{xx}$	$\mathbf{R}_{xy}$		
	$\mathbf{R}_{yx}$	$\mathbf{R}_{yy}$		
	TS	TC	BS	BC
TS	1.0000	.3597	.9852	.2704
TC	.3597	1.0000	.3613	.9947
BS	.9852	.3613	1.0000	.2725
BC	.2704	.9947	.2725	1.0000

It may be helpful conceptually to compare Eq. 6.1 with Eq. 5.6 for regression. Equation 5.6 indicates that regression coefficients for predicting Y from a set of X's are a product of the inverse of the matrix of correlations between the X's and the matrix of correlations between the X's and Y. Equation 6.1, although containing more than one Y, might be thought of as a product of regression weights for predicting X's from Y's ($\mathbf{R}_{yy}^{-1} \mathbf{R}_{yx}$) and those for predicting Y's from X's ($\mathbf{R}_{xx}^{-1} \mathbf{R}_{xy}$).

Canonical analysis proceeds by solving for the eigenvalues and eigenvectors of the matrix, $\mathbf{R}$ of Eq. 6.1. As discussed in Chapter 10 and in Appendix A, solving for the roots (eigenvalues) of a matrix is a process that redistributes the variance in the matrix, consolidating it in many cases into a few composite variables rather than many individual variables. The eigenvectors corresponding to each eigenvalue are transformed into the weights used to combine the original variables to accomplish the consolidation. As demonstrated in Appendix A, calculations of eigenvalues and corresponding eigenvectors are difficult and not particularly enlightening. For this example, the task was accomplished with an assist from SPSS MANOVA (see Section 6.6.1). The goal of the calculations, however, is to redistribute the variance in the original variables into a very few sets, each set capturing a large share of the variance and defined by linear combinations of IVs on one side and DVs on the other. Canonical correlation maximizes the correlation between each linear combination of IVs and the corresponding linear combination of DVs.

The relationship between eigenvalues and canonical correlation is, however, simple. Namely,

$$\lambda_i = r_{ci}^2 \tag{6.2}$$

Each eigenvalue, λ_i, is equal to the squared canonical correlation, r_{ci}, for the dimension ($i = 1, 2, \ldots, m$ where m is the smaller of either k_x or k_y, as defined in Eq. 6.3).

Once the eigenvalues are calculated, the correlation between the linear combination of DVs and the linear combination of IVs is found by taking the square root of the eigenvalue. The size of r_{ci} may be interpreted as an ordinary Pearson product-moment correlation coefficient. When r_{ci} is squared, it represents, as usual, overlapping variance between two variables. In this case, however, it represents overlapping variance between pairs of canonical variates. Because $r_{ci}^2 = \lambda_i$, the eigenvalues themselves represent overlapping variance between pairs of canonical variates. This relationship between eigenvalues and variance is seen elsewhere in multivariate statistics (e.g., principal components).

For the data set of Table 6.1, two eigenvalues were calculated (one for each variable in the smaller set—both sets in this case). The first eigenvalue was .99820, which corresponded to a canonical correlation of .99910. The second eigenvalue was .96747, so r_{c2} was .98360. That is, the first pair of canonical variates correlate .99910 and overlap 99.82% in variance, while the second pair correlate .98360 and overlap 96.75% in variance.

Significance tests (Bartlett, 1941) are available to test whether or not one or a set of r_c's differ from zero.

$$\chi^2 = -\left[N - 1 - \left(\frac{k_x + k_y + 1}{2}\right)\right] \ln \Lambda_m \qquad (6.3)$$

The significance of one or more canonical correlations is evaluated as a chi square variable where N is the number of cases, k_x is the number of variables in the IV set, and k_y is the number in the DV set. The natural logarithm of lambda, Λ, is defined in Eq. 6.4. This chi square has $(k_x)(k_y)$ df.

and

$$\Lambda_m = \prod_{i=1}^{m} (1 - \lambda_i) \qquad (6.4)$$

Lambda, Λ, is the product of differences between eigenvalues and unity, generated across m canonical correlations.

For the example, to test if the canonical correlations as a set differ from zero, we write

$$\Lambda_2 = (1 - \lambda_1)(1 - \lambda_2) = (1 - .99820)(1 - .96747) = .000058554$$

$$\chi^2 = -\left[5 - 1 - \left(\frac{2 + 2 + 1}{2}\right)\right] \ln .000058554$$

$$= -(1.5)(-9.74556)$$

$$= 14.618$$

This χ^2 is evaluated with $(k_x)(k_y) = 4$ df. The two canonical correlations differ from zero: $\chi^2(4) = 14.62$, $p < .006$. With the first canonical correlate removed, does the second still differ from zero?

$$\Lambda_1 = (1 - .96747) = .03253$$

$$\chi^2 = -\left[5 - 2 - \left(\frac{2 + 2 + 1}{2}\right)\right] \ln .03253$$

$$= -(1.5)(-3.42559)$$

$$= 5.138$$

This chi square has $(k_x - 1)(k_y - 1) = 1$ df and also differs significantly from zero: $\chi^2(1) = 5.14$, $p < .02$.

Significance of canonical correlations is sometimes evaluated using the F distribution as, for example, in SPSS MANOVA (Hull & Nie, 1981).

Two sets of canonical coefficients (analogous to regression coefficients) are required for each canonical correlation, one set to combine the DVs and the other to combine the IVs. The canonical weights for the DVs are found as follows:

$$\mathbf{B}_y = (\mathbf{R}_{yy}^{-1/2})'\hat{\mathbf{B}} \tag{6.5}$$

Canonical coefficients for the DVs are a product of the transpose of the inverse of the square root of the matrix of correlations between DVs and the matrix of eigenvectors, $\hat{\mathbf{B}}$, for the DVs.

For the example:[3]

$$\mathbf{B}_y = \begin{bmatrix} 1.02 & -0.14 \\ -0.14 & 1.02 \end{bmatrix} \begin{bmatrix} 0.18 & 0.99 \\ 0.99 & -0.19 \end{bmatrix} = \begin{bmatrix} 0.05 & 1.04 \\ 0.98 & -0.33 \end{bmatrix}$$

Once the canonical coefficients for the DVs are known, those for the IVs are found by using the following equation:

$$\mathbf{B}_x = \mathbf{L}\mathbf{R}_{xx}^{-1}\mathbf{R}_{xy}\mathbf{B}_y \tag{6.6}$$

Coefficients for the IVs are a product of four matrices: $\mathbf{L}$, a diagonal matrix of reciprocals of eigenvalues; $\mathbf{R}_{xx}^{-1}$, the inverse of the matrix of correlation between IVs; $\mathbf{R}_{xy}$, the matrix of correlations between IVs and DVs; and $\mathbf{B}_y$, the coefficients for the DVs.

For the example:

$$\mathbf{B}_x = \begin{bmatrix} \dfrac{1}{.99} & 0 \\ 0 & \dfrac{1}{.98} \end{bmatrix} \begin{bmatrix} 1.15 & -0.41 \\ -0.41 & 1.15 \end{bmatrix} \begin{bmatrix} .99 & .27 \\ .36 & .99 \end{bmatrix} \begin{bmatrix} 0.05 & 1.04 \\ 0.98 & -0.33 \end{bmatrix}$$

$$= \begin{bmatrix} -0.05 & 1.07 \\ 1.02 & -0.34 \end{bmatrix}$$

The two matrices of canonical coefficients can then be used to estimate scores on canonical variates.

$$\mathbf{X} = \mathbf{Z}_x\mathbf{B}_x \tag{6.7}$$

and

$$\mathbf{Y} = \mathbf{Z}_y\mathbf{B}_y \tag{6.8}$$

---

[3] These calculations, like the others in this section, were performed using five or six decimal places and then rounded back. The results agree with computer analyses of the same data but the rounded-off figures presented here do not themselves always check out to both decimals.

Scores on canonical variates are estimated as the product of the standardized scores on the original variables, $\mathbf{Z}_x$ and $\mathbf{Z}_y$, and the canonical coefficients used to weight them, $\mathbf{B}_x$ and $\mathbf{B}_y$.

For the example:

$$\mathbf{X} = \begin{bmatrix} 1.30 & 1.30 \\ -0.96 & 0.80 \\ -0.21 & -0.40 \\ 0.75 & -0.66 \\ -0.89 & -1.04 \end{bmatrix} \begin{bmatrix} -0.05 & 1.07 \\ 1.02 & -0.34 \end{bmatrix} = \begin{bmatrix} 1.26 & 0.95 \\ 0.86 & -1.30 \\ -0.40 & -0.09 \\ -0.71 & 1.03 \\ -1.02 & -0.60 \end{bmatrix}$$

$$\mathbf{Y} = \begin{bmatrix} 1.15 & 1.21 \\ -0.79 & 0.90 \\ -0.23 & -0.32 \\ 0.94 & -0.78 \\ -1.06 & -1.01 \end{bmatrix} \begin{bmatrix} 0.05 & 1.04 \\ 0.98 & -0.33 \end{bmatrix} = \begin{bmatrix} 1.24 & 0.80 \\ 0.84 & -1.12 \\ -0.33 & -0.13 \\ -0.72 & 1.23 \\ -1.04 & -0.77 \end{bmatrix}$$

The first belly dancer, in standardized scores (and appropriate costume), earned 1.30 on TS, 1.30 on TC, 1.15 on BS, and 1.21 on BC. When these scores were weighted by canonical coefficients, she had a score of 1.26 on the first canonical variate and a score of 0.95 on the second canonical variate for the IVs (the X's), and scores of 1.24 and 0.80 on the first and second canonical variates, respectively, for the DVs (the Y's). Notice that the sum across the belly dancers' scores on each canonical variate is zero, within rounding error. These scores, like factor scores (Chapter 10) represent estimates of scores the dancers would have received had they been judged on the canonical variates directly.

Although researchers occasionally interpret canonical variates from canonical coefficients, this procedure has the same difficulty as interpreting a regression equation from the regression coefficients (or factors from factor score coefficients, and so forth). Matrices of correlations between variables and canonical variates are usually interpreted instead.

$$\mathbf{A}_x = \mathbf{R}_{xx}\,\mathbf{B}_x \tag{6.9}$$

and

$$\mathbf{A}_y = \mathbf{R}_{yy}\,\mathbf{B}_y \tag{6.10}$$

Correlations between variables and canonical variates are found by multiplying the matrix of correlations between variables by the matrix of canonical coefficients.

For the example:

$$\mathbf{A}_x = \begin{bmatrix} 1.00 & .36 \\ .36 & 1.00 \end{bmatrix} \quad \begin{bmatrix} -0.05 & 1.07 \\ 1.02 & -0.34 \end{bmatrix} = \begin{bmatrix} .32 & .95 \\ 1.00 & .05 \end{bmatrix}$$

$$\mathbf{A}_y = \begin{bmatrix} 1.00 & .27 \\ .27 & 1.00 \end{bmatrix} \quad \begin{bmatrix} 0.05 & 1.04 \\ 0.98 & -0.33 \end{bmatrix} = \begin{bmatrix} .31 & .95 \\ .99 & -.05 \end{bmatrix}$$

Thus the first canonical variate for the IVs correlates .32 with TS and 1.00 with TC. The second canonical variate for the IVs correlates .95 with TS and .05 with TC. The first canonical variate for the DVs correlates .31 with BS and .99 with BC, while the second correlates .95 with BS and −.05 with BC. Recall that canonical variates are produced in pairs, each with a corresponding r_c. Thus the first pair (the first variate from each set of variables) is primarily a "circles" (top and bottom) pair that correlates .99910. The second pair is primarily a "shimmies" pair that correlates .98360. Ability to do circles well is apparently one dimension of belly dancing, while ability to do shimmies well is another independent dimension.

The proportion of variance extracted from the IVs by the canonical variates of the IVs is

$$pv_{xc} = \sum_{i=1}^{k_x} \frac{a_{ixc}^2}{k_x} \tag{6.11}$$

and

$$pv_{yc} = \sum_{i=1}^{k_y} \frac{a_{iyc}^2}{k_y} \tag{6.12}$$

The proportion of variance extracted from a set of variables by a canonical variate of the set (called *averaged squared loading* by BMDP6M) is the sum of the squared correlations divided by the number of variables in the set.

Thus for the first canonical variate in the set of IVs,

$$pv_{x1} = \frac{.32^2 + 1.00^2}{2} = .5512$$

and for the second canonical variate in x,

$$pv_{x2} = \frac{.95^2 + .05^2}{2} = .4525$$

In summing for the two variates, 100% of the variance in the IVs is extracted by both canonical variates.

For the DVs and the first canonical variate,

$$pv_{y1} = \frac{.31^2 + .99^2}{2} = .5381$$

and for the second canonical variate,

$$pv_{y2} = \frac{.95^2 + (-.05)^2}{2} = .4525$$

Together the two canonical variates extract 99.06% of the variance in the DVs.

Often, however, one is interested in knowing how much variance the canonical variates from the IVs extract from the DVs, and vice versa. In canonical analysis, this variance is called redundancy.

$$rd = (pv)(r_c{}^2) \tag{6.13}$$

The redundancy in a canonical variate is the percent of variance it extracts from its own set of variables times the canonical correlation squared for the pair.

Thus for the example:

$$rd_{x1} = \left(\frac{.32^2 + 1.00^2}{2}\right)(.99820) = .5502$$

$$rd_{x2} = \left(\frac{.95^2 + .05^2}{2}\right)(.96747) = .4378$$

$$rd_{y1} = \left(\frac{.31^2 + .99^2}{2}\right)(.99820) = .5371$$

and

$$rd_{y2} = \left(\frac{.95^2 + (-.05)^2}{2}\right)(.96747) = .4378$$

So, the first canonical variate from the IVs extracts 55.02% of the variance from the DVs, and the second extracts 43.78% of the variance. Together they extract 98.8% of the variance in the DVs. The first and second canonical variates for the DVs extract 53.71% and 43.78% of the variance in the IVs, respectively. Together they extract 97.49% of the variance in the IVs.

6.5 SOME IMPORTANT ISSUES

6.5.1 Importance of Canonical Variates

As in most statistical procedures, evaluation of significance is the first step in evaluating the importance of a solution. Relationships that fail the test of significance are not reliable and should not be interpreted. Conventional statistical procedures apply to use of results of Eqs. 6.3 and 6.4 (or a corresponding F test), which are reported directly by both SPSS and BMDP programs.

The only potential source of confusion is the meaning of the chain of significance tests: The first test is for all pairs taken together and essentially tests for independence between the two sets of variables; the second test is for all pairs of variates with the first and most important pair of canonical variates removed; the third is done with the first two pairs removed, and so forth. If the first test, but not the second, reaches significance, then only the first pair of canonical variates should be interpreted. If the first and second tests are significant but the third is not, then the first two pairs of variates are interpreted, and so on. Because canonical correlations are reported out in descending order of importance, usually only the first few pairs of variates are significant.

Once significance is established, amount of variance accounted for is of critical importance. Because there are two sets of variables, several assessments of variance are relevant. The first, and easiest, is the variance overlap between each pair of significant variates. As indicated in Eq. 6.2, overlapping variance for a pair is the eigenvalue, or the squared canonical correlation, for the pair. Because r_c values of .30 or less represent, squared, less than 10% of the variance, most researchers do not interpret pairs with a canonical correlation lower than .30 even if significant (significance depends, to a large extent, on N).

The next variance relationship to be examined is the variance a canonical variate extracts from its own set of variables. Equations 6.11 and 6.12 show that percent of variance, pv, is assessed as the sum of squared loadings on a variate divided by the number of variables in the set. This calculation is identical to the one used in factor analysis for the same purpose, as shown in Table 10.4. A pair of canonical variates may extract very different amounts of variance from their respective sets of variables. Because canonical variates are independent of one another (orthogonal), percents of variance can be summed across variates to arrive at the total variance extracted from the variables by all the significant variates of the set.

The last relationship to be considered is redundancy (Stewart & Love, 1968; Miller & Farr, 1971). Equation 6.13 shows that redundancy is simply the percent of variance extracted by a canonical variate times the canonical correlation for the pair. Although we have defined redundancy elsewhere, for ease of description, as the variance one variate extracts from the variables in the opposing set, that description is a bit misleading. More precisely, redundancy answers the question: If I knew the score on a canonical variate from the IVs, how much would my uncertainty regarding the DVs be reduced? It is possible, for instance, for a canonical variate from the IVs to be an important dimension in its own set of variables, but correlated with an unimportant dimension among the DVs (and vice versa). Therefore the redundancies for a pair of canonical variates are not usually equal. Similar to pv, because canonical variates are orthogonal, redundancies for a group of variables can be added across canonical variates to get a total for the variables and, thus, a total redundancy measure for the DVs relative to the IVs, and vice versa.

6.5.2 Interpretation of Canonical Variates

As with all multivariate procedures, canonical correlation creates new entities, called canonical variates, which represent mathematically viable dimensions of sets of observed variables. The dimensions are mathematically, but not necessarily logically, viable. A major task for the researcher is to discern the meaning of pairs of canonical variates.

Interpretation of canonical variates involves assessment of the correlations in the loading matrices, A_x and A_y (Eqs. 6.9 and 6.10, respectively), for all pairs of canonical variates deemed important when using the procedures of Section 6.5.1. Each pair of canonical variates is interpreted as a pair, with each variate representing a dimension from the IVs that is highly correlated with a dimension of the DVs.

One interprets the pattern of variables highly correlated (loaded) with the variate from the IVs as a dimension that is related to the dimension similarly interpreted from the DVs. Because the numbers in **A** are correlations, and because squared correlations measure overlapping variance, correlations of .30 (10% of variance) and above are usually considered part of a pattern, while variables with loadings below .30 are disregarded. Establishing the cutoff for correlations is, however, somewhat a matter of taste, although the guidelines in Section 10.6.5 may prove helpful.

Because the goal of canonical analysis is to maximize correlation among mathematically defined dimensions, results of canonical analysis are not always easily interpretable. This drawback may well account for its relative scarcity in research literature.

6.6 COMPARISON OF PROGRAMS

One program in the BMDP series and two in the SPSS package are available for canonical analyses. Table 6.3 provides a comparison of important features of the programs. If available, the program of choice is BMDP6M. A second choice, but with limitations, is SPSS MANOVA.

6.6.1 SPSS Package

The two programs in the SPSS package for canonical analysis are CANCORR (Nie et al., 1975) and MANOVA (Hull & Nie, 1981). Although the CANCORR program was designed specifically for this purpose, it provides less complete information than MANOVA. A major strength of CANCORR is its flexibility in handling missing data and multicollinearity. Weaknesses include unavailability of loading matrices, of percents of variance, and of redundancies. Although these statistics can be hand-calculated by methods shown in Section 6.4, a minimal amount of matrix multiplication (see Appendix A) is required.

A more complete canonical analysis may be conducted through SPSS MANOVA, the only program to output percents of variance and redundancies

TABLE 6.3 COMPARISON OF SPSS AND BMDP PROGRAMS
FOR CANONICAL CORRELATION

Feature	SPSS CANCORR	SPSS MANOVA	BMDP6M
Input			
Raw data	Yes	Yes	Yes
Correlation matrix	Yes	No	Yes
Covariance matrix	No	No	Yes
Number of canonical variates	Yes	No	Yes
Tolerance	Yes	No	Yes
Minimum canonical correlation	Specify significance	No	Yes
Data set printed	No	No	Yes
Output			
Univariate:			
Means	Yes	Yes	Yes
Standard deviations	Yes	Yes	Yes
Number of cases	Yes	Yes	Yes
Coefficient of variation	No	No	Yes
Smallest value	No	No	Yes
Largest value	No	No	Yes
Smallest standard score	No	No	Yes
Largest standard score	No	No	Yes
Skewness	No	No	Yes
Kurtosis	No	No	Yes
Normal plots	No	Yes	No
Multivariate:			
Canonical correlations	Yes	Yes	Yes
Eigenvalues (r_c^2)	Yes	Yes	Yes
Significance test	χ^2	F	χ^2
Lambda	Yes	Yes	No
Correlation matrix	Yes	No	Yes
Covariance matrix	Yes	No	Yes
Loading matrix	No	Yes	Yes
Raw canonical coefficients	No	Yes	Yes
Standardized canonical coefficients	Yes	Yes	Yes
Canonical variate scores	Yes	No	Yes
Percent of variance	No	Yes	No
Redundancy	No	Yes	No
Variable-variable plots	No	No	Yes
Variable-variate plots	No	No	Yes
Variate-variate plots	No	No	Yes
Numerical consistency	No	No	Yes
Multicollinearity (within-sets SMCs)	Yes	No	Yes
Multiple analyses	Yes	No	No
Residuals plots	No	Yes	No

directly. But problems can arise with reading the results, because MANOVA was not designed specifically for canonical analysis and some of the labels are misleading.

Canonical analysis is requested on the MANOVA card by calling one set of variables DVs and the other set covariates; no IVs are listed. An example of the MANOVA card is as follows: MANOVA; DV1, DV2, DV3 WITH CV1, CV2, CV3, CV4. Interpretation of the output is facilitated if parameter estimates are suppressed and DISCRIM statistics are requested, simply because there is less output and much of it pertains directly to canonical analysis.

Remembering that one set of variables is identified as DVs and the other set as covariates assists in reading the output. Immediately following the first section, labeled MULTIVARIATE TESTS OF SIGNIFICANCE, is output labeled EIGENVALUES AND CANONICAL CORRELATIONS. Both r_c (CANON. CORR.) and $r_c{}^2$ (SQUARED CORR.) are given for each pair of canonical variates (ROOT NO.). The DIMENSION REDUCTION ANALYSIS that follows gives significance tests for the canonical correlations. RAW CANONICAL COEFFICIENTS FOR DEPENDENT VARIABLES (Set 1), STANDARDIZED CANONICAL COEFFICIENTS (Set 1), and CORRELATIONS (loadings) BETWEEN DEPENDENT AND CANONICAL VARIABLES follow unambiguously. Under VARIANCE EXPLAINED BY CANONICAL VARIATES OF THE DEPENDENT VARIABLES one finds percent of variance (PCT. VAR. DEPENDENTS) and redundancy (PCT. VAR. COVARIATES). These statistics are repeated immediately for the second set of variables, labeled COVARIATES.

Although SPSS MANOVA provides a rather complete canonical analysis, it does not calculate canonical variate scores, nor does it test for multivariate outliers.

6.6.2 BMD Series

BMDP6M (Dixon, 1981) provides an almost complete canonical analysis in addition to checks on accuracy of input, univariate outliers (LARGEST and SMALLEST VALUES), multicollinearity (WITHIN SET SMCs), linearity (variable-variable, variable-variate, and variate-variate PLOTS), and normality (SKEWNESS). Flexibility of input, control over progression of the analysis, and flexibility of output are other desirable features of the program. Although percent of variance and redundancy are not given, they are easily calculated from the loading matrices that are included. BMDP6M is the program of choice, if available.

6.7 COMPLETE EXAMPLE OF CANONICAL CORRELATION

For an example of canonical correlation, variables were selected from among those made available by research described in Appendix B. The goal of analysis was to discover the dimensions, if any, along which certain attitudinal variables were related to certain health characteristics.

Selected attitudinal variables (Set 1) include attitudes toward the role of women (WOMENROL), toward locus of control (IESCALE), toward current marital status (HAPSTAT) and toward self (SELFESTE). Larger numbers indicate increasingly conservative attitudes about the proper role of women, increasing feelings of powerlessness to control one's fate (external as opposed to internal locus of control), increasing dissatisfaction with current marital status, and increasingly poor self-esteem.

Selected health variables (Set 2) include assessment of mental health (MENHEALT), assessment of physical health (PHYHEALT), number of visits to health professionals (TIMEDHS), attitude toward use of medication (ATTMED), and a frequency-duration measure of use of psychotropic drugs (DRUGUSE). Larger numbers reflect poorer mental and physical health, more visits, greater willingness to use drugs, and more use of them.

Canonical correlation provides a means for studying the relationships among these two sets of variables and for studying the number and nature of dimensions of correspondence. Analysis of these data was conducted through BMDP6M.

6.7.1 Evaluation of Assumptions

6.7.1.1 Normality Because one goal was a significance test of the number of reliable pairs of canonical variates, and because the metrics along which the measures were taken were arbitrary, distributions of variables were evaluated for skewness, as described in Section 4.4.1, using SPSS CONDESCRIPTIVE. Several variables were found to be quite skewed and were transformed by use of square roots or logarithms (see Section 4.5). Transformations were applied to each distribution until it no longer had significant skewness. A square root transformation was applied to HAPSTAT, now SHAPSTAT. Variables requiring logarithmic transformation were TIMEDHS, PHYHEALT, and DRUGUSE, to become LTIMEDHS, LPHYHEAL, and LDRUGUSE, respectively.

It should be reiterated that interpretation of the results is restricted to variables as transformed. Had conceptual difficulties with one or more of the variables arisen as a result of transformation, its use would have been reconsidered. If creating statistical "purity" produces uninterpretable results, the results are useless no matter how well the data fit some theoretical distribution.

6.7.1.2 Missing Data Fortunately (and almost unbelievably), no datum was missing on these variables among the 465 cases.

6.7.1.3 Outliers Transformed and untransformed variables were run through BMDP1D to identify univariate outliers. The output appears in Table 6.4. It was noted that the smallest and largest z-score values were sizable for some of the variables before transformation but within ± 3.00 standard score

TABLE 6.4 SELECTED BMDP1D OUTPUT FOR ANALYSIS OF UNIVARIATE OUTLIERS

```
PROGRAM CONTROL INFORMATION

/PROBLEM  TITLE IS "OUTLIERS AND SKEWNESS FOR CANONICAL".
/INPUT    VARIABLES ARE 9. ADD=4. UNIT=55.
          FORMAT IS "(8X,2F4.0,4X,2F4.0,8X,5F4.0,24X/)".
/VARIABLE NAMES ARE TIMEDHS,ATTMED,PHYHEALT,MENHEALT,
          SELFESTE,IESCALE,HAPSTAT,DRUGUSE,WOMENROL,
          LTIMEDHS,LPHYHEAL,SHAPSTAT,LDRUGUSE.
/TRANSFORM LTIMEDHS=LOG(TIMEDHS+1).
          LPHYHEAL=LOG(PHYHEALT+1).
          SHAPSTAT=SQRT(HAPSTAT).
          LDRUGUSE=LOG(DRUGUSE+1).

/END
```

VARIABLE NO. NAME	MEAN	STANDARD DEVIATION	ST.ERR. OF MEAN	COEFF. OF VARIATION	SMALLEST VALUE	Z-SCORE	LARGEST VALUE	Z-SCORE	RANGE	TOTAL FREQUENCY
1 TIMEDHS	7.901	10.948	.5077	1.38570	0.000	-.72	81.000	6.68	81.000	465
2 ATTMED	7.686	1.156	.0536	.15041	5.000	-2.32	10.000	2.00	5.000	465
3 PHYHEALT	4.972	2.388	.1108	.48035	2.000	-1.24	15.000	4.20	13.000	465
4 MENHEALT	6.123	4.194	.1945	.68494	0.000	-1.46	18.000	2.83	18.000	465
5 SELFESTE	15.834	3.943	.1828	.24899	8.000	-1.99	29.000	3.34	21.000	465
6 IESCALE	6.733	1.302	.0604	.19341	0.000	-5.17	10.000	2.51	10.000	465
7 HAPSTAT	22.733	8.832	.4096	.38850	0.000	-2.57	58.000	3.99	58.000	465
8 DRUGUSE	9.002	10.113	.4690	1.12340	0.000	-.89	66.000	5.64	66.000	465
9 WOMENROL	35.135	6.758	.3134	.19235	18.000	-2.54	55.000	2.94	37.000	465
10 LTIMEDHS	.741	.415	.0193	.56018	0.000	-1.79	1.914	2.82	1.914	465
11 LPHYHEAL	.744	.167	.0077	.22432	.477	-1.60	1.204	2.76	.727	465
12 SHAPSTAT	4.666	.980	.0454	.21002	0.000	-4.76	7.616	3.01	7.616	465
13 LDRUGUSE	.764	.487	.0226	.63783	0.000	-1.57	1.826	2.18	1.826	465

units after transformation (e.g., compare TIMEDHS with LTIMEDHS). It was also noted that some sizable standard scores were still present (cf. IESCALE, SHAPSTAT) that might be associated with outliers.

Analysis of multivariate outliers is not available through BMDP6M. Therefore the two variable sets were evaluated separately for outliers using BMDP4M. (BMD10M or BMDPAM could have been used.) Outliers appear in the BMDP4M program as cases with large Mahalanobis distances, shown as χ^2/df, from the centroid of the cases for original data, as demonstrated in Section 5.8.1. To determine the critical value for identifying outliers, one looks up critical χ^2 with number of variables in the set as degrees of freedom, and then divides critical χ^2 by degrees of freedom. Cases with Mahalanobis distances (χ^2/df) larger than the calculated value (critical χ^2/df) are outliers.

At $\alpha = .01$,[4] critical χ^2 for the set with four variables was $13.28/4 = 3.32$. For the set with five variables the cutoff value was $15.09/5 = 3.02$. With the use of these cutoff values, seven cases had values larger than 3.32 for the first set of variables, and three cases had values larger than 3.02 in the analysis of the second set. Because one of these cases was the same in both sets of variables, nine cases in all were deleted from canonical analysis. Cases were deleted because N was large. Had N been marginal, other methods of handling outliers (Section 4.3.2) were available.

Descriptive statistics produced by BMDP6M of the variables with outliers deleted appear in Table 6.5. Along with skewness, it was noted that the coefficients of variation were sufficiently large so as to correspond to accurately calculated correlation matrices (Section 4.1.3) and that the smallest and largest stan-

---

[4] Others might prefer a .001 criterion for deleting cases.

TABLE 6.5   SELECTED INPUT AND OUTPUT FROM BMDP6M;
DESCRIPTIVE STATISTICS FOR TRANSFORMED VARIABLES
WITH OUTLIERS DELETED

```
PROGRAM CONTROL INFORMATION

/PROBLEM TITLE IS "CANONICAL CORRELATION EXAMPLE".
/INPUT  VARIABLES ARE 9. ADD=4. UNIT=55.
        FORMAT IS "(8X,2F4.0,4X,2F4.0,8X,5F4.0,24X/)".
/VARIABLE NAMES ARE TIMEDHS,ATTMED,PHYHEALT,MENHEALT,
         SELFESTE,IESCALE,HAPSTAT,DRUGUSE,WOMENROL,
         LTIMEDHS,LPHYHEAL,SHAPSTAT,LDRUGUSE.
         USE=2,4,5,6,9,11 TO 13.
/TRANSFORM LTIMEDHS=LOG(TIMEDHS+1).
         LPHYHEAL=LOG(PHYHEALT+1).
         SHAPSTAT=SQRT(HAPSTAT).
         LDRUGUSE=LOG(DRUGUSE+1).
/CANONICAL FIRST ARE LTIMEDHS,ATTMED,LPHYHEAL,MENHEALT,LDRUGUSE.
          SECOND ARE SELFESTE,IESCALE,SHAPSTAT,WOMENROL.
/PRINT MATRICES ARE CORR,COEF,LOAD.
/PLOT XVARS ARE CNVRF1,CNVRF2.
     YVARS ARE CNVRS1,CNVRS2.
/END
```

UNIVARIATE SUMMARY STATISTICS

VARIABLE	MEAN	STANDARD DEVIATION	COEFFICIENT OF VARIATION	SMALLEST VALUE	LARGEST VALUE	SMALLEST STANDARD SCORE	LARGEST STANDARD SCORE	SKEWNESS	KURTOSIS
5 SELFESTE	15.79386	3.91857	.248107	8.00000	29.00000	-1.99	3.37	.45	.20
6 IESCALE	6.75439	1.27227	.188362	5.00000	10.00000	-1.38	2.55	.48	-.45
12 SHAPSTAT	4.71330	.84886	.180099	3.31662	7.61577	-1.65	3.42	.57	-.22
9 WOMENROL	35.22807	6.72085	.190781	18.00000	55.00000	-2.56	2.94	.07	-.46
10 LTIMEDHS	.73597	.41262	.560647	0.00000	1.91381	-1.78	2.85	.22	-.18
2 ATTMED	7.67982	1.15716	.150675	5.00000	10.00000	-2.32	2.01	-.11	-.45
11 LPHYHEAL	.74259	.16634	.224005	.47712	1.20412	-1.60	2.77	.16	-.63
4 MENHEALT	6.09649	4.13129	.677651	0.00000	18.00000	-1.48	2.88	.57	-.33
13 LDRUGUSE	.76585	.48695	.635828	0.00000	1.82607	-1.57	2.18	-.15	-1.09

dard scores were within a range anticipated in a sample of over 400 people.
The extremely large standard scores that were found in similar analyses before
scores were transformed and outliers deleted were no longer seen.

6.7.1.4 Multicollinearity and Singularity   BMDP6M provides a direct test
of multicollinearity/singularity, as shown in Table 6.6. SMCs with each variable
in a set serving as DV and the others in the set as IVs, in turn, are reported.
If R-SQUARED values become large (say, .99 or above), then one variable in
the set is nearly a linear combination of others and near singularity is present,
to be handled by procedures mentioned in Section 4.4.4. In this case, low
R-SQUARED values indicate absence of singularity or multicollinearity and,
indeed, considerable heterogeneity in the sets of variables, particularly in the
attitudinal set.

6.7.1.5 Linearity   BMDP6M provides a particularly flexible scheme for
assessing linearity between pairs of variables, between pairs of canonical variates,
and between variables and canonical variates. The PLOT paragraph allows one
to request any number of scatterplots of variables and canonical variates and
to control the size of the plot.

Figure 6.1 (page 166) shows two scatterplots produced by BMDP6M for
the example using default size values for the plots. The scatterplots are between
the first and second pairs of canonical variates, respectively. Their shapes reflect
the low canonical correlations for the solution (see the next section) but no
obvious departures from linearity.

TABLE 6.6   SELECTED BMDP6M OUTPUT FOR ASSESSMENT
OF MULTICOLLINEARITY

```
SQUARED MULTIPLE CORRELATIONS OF EACH VARIABLE IN
SECOND SET WITH ALL OTHER VARIABLES IN SECOND SET


             VARIABLE
   NUMBER      NAME      R-SQUARED


      10  LTIMEDHS       .37835
       2  ATTMED         .08134
      11  LPHYHEAL       .47412
       4  MENHEALT       .29087
      13  LDRUGUSE       .31696


SQUARED MULTIPLE CORRELATIONS OF EACH VARIABLE IN
FIRST SET WITH ALL OTHER VARIABLES IN FIRST SET


             VARIABLE
   NUMBER      NAME      R-SQUARED


       5  SELFESTE       .22406
       6  IESCALE        .14026
      12  SHAPSTAT       .12581
       9  WOMENROL       .06608
```

6.7.2 Canonical Correlation

The number and importance of canonical variates were determined using procedures from Section 6.5.1. Significance of canonical correlations (Eqs. 6.3 and 6.4) was reported directly by BMDP6M, as shown in Table 6.7 (page 166). With all four canonical correlations included, $\chi^2(20) = 109.32$, $p < .001$. With the first canonical variate pair removed, χ^2 was still significant: $\chi^2 = 41.07$, $p < .001$. With the first and second canonical correlations removed, χ^2 values were not significant: $\chi^2(6) = 3.30$, $p = .77$. Therefore only the first two pairs of canonical variates are interpreted.

Canonical correlations (r_c) and eigenvalues ($r_c{}^2$) are also shown in Table 6.7. The first canonical correlation is .38, representing 14% overlapping variance between the first pair of canonical variates (see Eq. 6.2). The second canonical correlation is .28, representing 8% overlapping variance between the second pair of canonical variates. Although highly significant, neither of these two canonical correlations represents substantial relationship. Interpretation of the second canonical correlation and its corresponding pair of canonical variates is marginal.

Loading matrices between canonical variates and original variables are

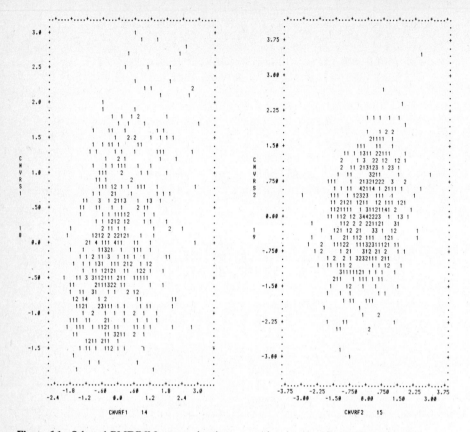

Figure 6.1 Selected BMDP6M output showing scatterplots between first and second pairs of canonical variates for the example.

TABLE 6.7   SELECTED PORTIONS OF BMDP6M OUTPUT SHOWING
CANONICAL CORRELATIONS AND SIGNIFICANCE LEVELS
FOR SETS OF CANONICAL CORRELATIONS

EIGENVALUE	CANONICAL CORRELATION		NUMBER OF EIGENVALUES	BARTLETT"S TEST FOR REMAINING EIGENVALUES		
				CHI-SQUARE	D.F.	SIGNIFICANCE
			0	109.32	20	.00000
.14072	.37512		1	41.07	12	.00005
.08051	.28374		2	3.30	6	.77021
.00633	.07957		3	.44	2	.80143
.00098	.03136					

BARTLETT"S TEST ABOVE INDICATES THE NUMBER OF CANONICAL
VARIABLES NECESSARY TO EXPRESS THE DEPENDENCY BETWEEN THE
TWO SETS OF VARIABLES. THE NECESSARY NUMBER OF CANONICAL
VARIABLES IS THE SMALLEST NUMBER OF EIGENVALUES SUCH THAT
THE TEST OF THE REMAINING EIGENVALUES IS NON-SIGNIFICANT.
FOR EXAMPLE, IF A TEST AT THE .01 LEVEL WERE DESIRED,
THEN   2 VARIABLES WOULD BE CONSIDERED NECESSARY.
HOWEVER, THE NUMBER OF CANONICAL VARIABLES OF PRACTICAL
VALUE IS LIKELY TO BE SMALLER.

TABLE 6.8 SELECTED BMDP6M OUTPUT OF LOADING MATRICES FOR THE TWO SETS OF VARIABLES IN THE EXAMPLE

CANONICAL VARIABLE LOADINGS (CORRELATIONS OF CANONICAL VARIABLES WITH ORIGINAL VARIABLES)

		CNVRF1	CNVRF2	CNVRF3	CNVRF4
		1	2	3	4
SELFESTE	5	.590	.598	-.138	-.525
IESCALE	6	.811	.129	-.325	.469
SHAPSTAT	12	.704	-.322	.561	-.293
WOMENROL	9	-.090	.786	.535	.298

		CNVRS1	CNVRS2	CNVRS3	CNVRS4
		1	2	3	4
LTIMEDHS	10	.122	-.341	-.894	.182
ATTMED	2	.088	.545	-.091	.330
LPHYHEAL	11	.399	-.032	-.648	-.574
MENHEALT	4	.967	-.157	-.192	.038
LDRUGUSE	13	.257	-.561	-.071	-.101

shown in Table 6.8 (Eqs. 6.9 and 6.10). Values in the matrices are converted to *pv* values by application of Eqs. 6.11 and 6.12. For this example:

$$pv_{x1} = \frac{.59^2 + .81^2 + .70^2 + (-.09)^2}{4} = .38$$

$$pv_{y1} = \frac{.12^2 + .09^2 + .40^2 + .97^2 + .26^2}{5} = .24$$

$$pv_{x2} = \frac{.60^2 + .13^2 + (-.32)^2 + .79^2}{4} = .28$$

$$pv_{y2} = \frac{(-.34)^2 + .54^2 + (-.03)^2 + (-.16)^2 + (-.56)^2}{5} = .15$$

Thus the first canonical variate extracts 38% of variance from its own set of variables, while the second canonical variate extracts 28% of variance. Together the two canonical variates account for 66% of variance in the attitudinal set. The first canonical variate extracts 24% and the second variate 15% of the variance from the second set of variables. Together the two canonical variates account for 39% of variance in the health variables.

Redundancies for the canonical variates are found by using Eq. 6.13 and the values in Tables 6.6 and 6.7.

$$rd_{x1} = (.38)(.14) = .05$$

$$rd_{y1} = (.24)(.14) = .03$$

$$rd_{x2} = (.28)(.08) = .02$$

$$rd_{y2} = (.15)(.08) = .01$$

That is, the first attitudinal variate reduces 5% of the uncertainty in the health variables, while the second attitudinal variate reduces 2% of the uncertainty. Together, attitudinal variates "explain" 7% of the variance in health variables.

The first health variate reduces 3% and the second 1% of the variance in the attitudinal set. Together the two health variates overlap the variance in the attitudinal set 4%.

Interpretation of the two significant pairs of canonical variates follows

TABLE 6.9   SELECTED BMDP6M OUTPUT OF UNSTANDARDIZED
AND STANDARDIZED CANONICAL VARIATE COEFFICIENTS

COEFFICIENTS FOR CANONICAL VARIABLES FOR FIRST SET OF VARIABLES

		CNVRF1	CNVRF2	CNVRF3	CNVRF4
		1	2	3	4
SELFESTE	5	.602169D-01	.155752D+00	-.987259D-01	-.215170D+00
IESCALE	6	.489816D+00	.728560D-02	-.280265D+00	.632505D+00
SHAPSTAT	12	.582521D+00	-.520115D+00	.969779D+00	-.192857D+00
WOMENROL	9	-.118055D-01	.931412D-01	.102691D+00	.659171D-01

STANDARDIZED COEFFICIENTS FOR CANONICAL VARIABLES FOR FIRST SET OF VARIABLES
(THESE ARE THE COEFFICIENTS FOR THE STANDARDIZED VARIABLES - MEAN ZERO, STANDARD DEVIATION ONE.)

		CNVRF1	CNVRF2	CNVRF3	CNVRF4
		1	2	3	4
SELFESTE	5	.236	.610	-.387	-.843
IESCALE	6	.623	.009	-.357	.805
SHAPSTAT	12	.494	-.442	.823	-.164
WOMENROL	9	-.079	.626	.690	.443

COEFFICIENTS FOR CANONICAL VARIABLES FOR SECOND SET OF VARIABLES

		CNVRS1	CNVRS2	CNVRS3	CNVRS4
		1	2	3	4
LTIMEDHS	10	-.624800D+00	-.864192D+00	-.216631D+01	.186485D+01
ATTMED	2	.543097D-01	.663215D+00	-.704753D-01	.341852D+00
LPHYHEAL	11	.237093D+00	.286799D+01	-.238039D+01	-.735533D+01
MENHEALT	4	.258280D+00	.820049D-03	.402649D-01	.981477D-01
LDRUGUSE	13	-.170684D+00	-.174347D+01	.904933D+00	-.260858D+00

STANDARDIZED COEFFICIENTS FOR CANONICAL VARIABLES FOR SECOND SET OF VARIABLES
(THESE ARE THE COEFFICIENTS FOR THE STANDARDIZED VARIABLES - MEAN ZERO, STANDARD DEVIATION ONE.)

		CNVRS1	CNVRS2	CNVRS3	CNVRS4
		1	2	3	4
LTIMEDHS	10	-.258	-.357	-.894	.769
ATTMED	2	.063	.767	-.082	.396
LPHYHEAL	11	.039	.477	-.396	-1.224
MENHEALT	4	1.067	.003	.166	.405
LDRUGUSE	13	-.083	-.849	.441	-.127

procedures mentioned in Section 6.5.2. Critical to interpretation are the loading matrices shown in Table 6.8. Correlations (loadings) between variables and variates in excess of .3 are eligible for interpretation. Both the direction of correlations in the loading matrices and the meaning of big numbers on measures have to be considered for a correct interpretation of the pattern of the canonical variates.

Had a goal of the current analysis been production of scores on canonical variates, they would have been readily available. Table 6.9 shows both standardized and unstandardized coefficients for production of canonical variates as well as scores on the variates themselves produced by BMDP6M. Equations 6.5 and 6.6 show one method by which canonical coefficients may be produced, and Eqs. 6.7 and 6.8 show use of coefficients to produce scores on canonical variates should they be desired. (BMDP6M also produces scores on canonical variates if CANV is requested in the /PRINT paragraph.)

A checklist for canonical correlation appears in Table 6.10. An example of a Results section in journal format follows for the complete analysis described in Section 6.7.

TABLE 6.10 CHECKLIST FOR CANONICAL CORRELATION

1. Issues
   a. Normality
   b. Missing data
   c. Outliers
   d. Multicollinearity and singularity
   e. Linearity between variables and between variates
2. Major analyses
   a. Significance of canonical correlations
   b. Correlations of variables and variates
   c. Variance accounted for
      (1) By canonical correlations
      (2) By same-set canonical variates
   d. Redundancy
3. Additional analyses
   a. Canonical coefficients
   b. Canonical variates scores

RESULTS

A canonical correlation analysis was performed between a set of attitudinal variables and a set of health variables with the use of BMDP6M (Dixon, 1981). The attitudinal set included attitudes toward the role of women (WOMENROL), toward locus of control (IESCALE), toward marital status (SHAPSTAT), and toward self (SELFESTE). The health set measured mental health (MENHEALT), physical health (LPHYHEAL), visits to health professionals (LTIMEDHS), attitude toward use of medication (ATTMED), and use of psychotropic drugs (LDRUGUSE). Increasingly large numbers reflect more conservative attitudes toward women's role, external locus of control, dissatisfaction with marital status, low self-esteem, poor mental health, poor physical health, more numerous health visits, favorable attitudes toward drug use, and more drug use.

To improve linearity of relationship between variables and normality of their distributions, a square root transformation was applied to SHAPSTAT. Logarithmic transformations were applied to LTIMEDHS, LPHYHEAL, and LDRUGUSE. No missing data were discovered, although seven cases from the attitudinal set and three from the health set of variables, nine different cases in all (one of the total being an outlier on both sets), were identified as multivariate outliers at $p < .01$ and excluded from analysis.[5] Assumptions regarding within-set multicollinearity were met.

The first canonical correlation was .38 (14% of variance); the second was .28 (8% of variance). The remaining two canonical correlations were effectively zero. With all four canonical correlations included, $\chi^2(20) = 109.32$, $p < .001$, and with the first canonical correlation removed, $\chi^2(12) = 41.07$, $p < .001$. Subsequent χ^2 tests were not statistically significant. The first two canonical correlations therefore account for the significant linkages between the two sets of variables.

Analyses of the two pairs of canonical variates that accompany the first two canonical correlations appear in Table 6.11. Shown in the table are correlations between the variables and the canonical variates, standardized canonical variate coefficients, within-set variance accounted for by the canonical variates (percent of variance), redundancies, and canonical correlations. Total percent of variance and total redundancy indicate that the canonical analysis is more efficient for the first set of variables, and the size of canonical correlations indicate that interpretation of the second pair of canonical variates should proceed cautiously.

Insert Table 6.11 about here

With a cutoff correlation of .3 for interpretation, the variables relevant to the first canonical variate in the attitudinal set were, in order of magnitude, IESCALE, SHAPSTAT, and SELFESTE. Among the health variables, MENHEALT and LPHYHEAL were relevant to the canonical variate. Taken as a pair, the first canonical variates indicate that those with external locus of control (.81), square root transform of feelings of dissatisfaction toward marital status (.70), and lower self-esteem (.59) also tend to have more numerous mental health symptoms (.97) and more numerous physical health symptoms, as logarithmically transformed (.40).

The second canonical variate in the attitudinal set was composed of WOMENROL, SELFESTE, and NOT SHAPSTAT, while the corresponding canonical variate from the health set was composed of NOT LDRUGUSE, ATTMED, and NOT LTIMEDHS. Taken as a pair, these variates suggest that a combination of more conservative attitudes toward the role of women (.79), lower self-esteem (.60), but relative satisfaction with marital status with square root transform applied (−.32)

[5] Of the three outliers with respect to the health set, all scored low on **LDRUGUSE** while scoring high on such other variables as **MENHEALT, LPHYHEAL,** and/or **LTIMEDHS.** The case that proved to be an outlier on both sets also scored extremely low on **SHAPSTAT** ($z = -4.76$), as did four of the other attitudinal set outliers. Another outlying case with respect to attitudes had an extremely low **IESCALE** score ($z = -5.17$). The final attitudinal set outliers had high scores on both **SELFESTE** and **SHAPSTAT.**

corresponds to a combination of lower psychotropic drug use as logarithmically transformed (−.56), more favorable attitudes toward use of drugs (.54), but fewer visits to health professionals, with application of log transform (−.34).

TABLE 6.11   CORRELATIONS, STANDARDIZED CANONICAL COEFFICIENTS, CANONICAL CORRELATIONS, PERCENTS OF VARIANCE, AND REDUNDANCIES BETWEEN ATTITUDINAL AND HEALTH VARIABLES AND THEIR CORRESPONDING CANONICAL VARIATES

| | First canonical variate | | Second canonical variate | | |
	Correlation	Coefficient	Correlation	Coefficient	
Attitudinal set					
WOMENROL	−.09	−.08	.79	.63	
IESCALE	.81	.62	.13	.01	
SHAPSTAT	.70	.49	−.32	−.44	
SELFESTE	.59	.24	.60	.61	
Percent of variance	.38		.28		Total = .66
Redundancy	.05		.02		Total = .07
Health set					
MENHEALT	.97	1.07	−.16	.00	
LPHYHEAL	.40	.04	−.03	.48	
LTIMEDHS	.12	−.26	−.34	−.36	
ATTMED	.09	.06	.54	.77	
LDRUGUSE	.26	−.08	−.56	−.85	
Percent of variance	.24		.15		Total = .39
Redundancy	.03		.01		Total = .04
Canonical correlation	.38		.28		

6.8 SOME EXAMPLES FROM THE LITERATURE

Wingard, Huba, and Bentler (1979) report a canonical analysis of relationships between personality variables and use of various licit and illicit drugs among junior high school students in a metropolitan area. The sample was sufficiently large ($N = 1634$) to be randomly divided into groups for assessment of the stability of the canonical solution. (When a large enough sample is available, cross-validation between randomly selected halves of the sample is highly desirable.) At least two reliable pairs of canonical variates were discovered for both samples with the first pair, but not the second, similar for the samples. The first canonical variate in the drug use set seemed to reflect early patterns of experimentation with the relatively less dangerous, more readily available legal and illegal drugs. The corresponding variate from the personality set reflected "non-abidance with the law, liberalism, leadership, extraversion, lack of diligence, and lack of deliberateness" (p. 139). An attempt to rotate the two dimensions to facilitate interpretation of the second pair of variates failed and also degraded the correspondence on the first dimension for the subsamples. Problems with variable skewness were noted as potentially responsible for difficulties in comparing the second pair of canonical variates for the two subsamples. Although canonical correlations and significance levels were high, redundancies were not.

Cohen, Gaughran, and Cohen (1979) examined the relationships between patterns of fertility across 6 different age groups (as DVs) and 5 different sets of demographic characteristics as IVs (education and occupation; income–labor force; ethnicity; marriage–life cycle; and housing and occupancy) in 5 separate canonical analyses. The units of analysis were 338 New York City health areas. At least 3 of a possible 6 pairs of canonical variates were interpreted for each set of IVs, with the first pair showing substantial r_c values in all cases.

On the DV side, the first canonical variates represented substantial teenage and early 20s childbearing. On the IV side, the first canonical variates represented poorly educated, poverty-level, minority, unmarried persons living in over-crowded housing. Redundancy levels for predicting birthrates from each of the five sets of IVs were about 30%.

On the DV side, the second canonical variates were associated with low rates of childbearing in middle years (20–39). Canonical correlations were around .65 and redundancies around 14% for second pairs of variates. On the IV side, low middle-age childbearing was associated with constellations of demographic characteristics associated with an unmarried, affluent-singles life-style and working, educated women. The third pairs of canonical variates had modest correlations and redundancies. Childbearing among the oldest group of women (over 40) was associated with certain ethnic and religious characteristics. Cohen and colleagues direct those who seek to understand and control population expansion to the relationships found in the first pairs of canonical variates, where the absolute level of childbearing was highest.

In a paper comparing canonical analysis with a method known as external single-set components analysis (ESSCA), Fornell (1979) describes relationships between a set of variables measuring characteristics of 128 consumer affairs departments and a set measuring the ability of the departments to influence management decision making. Two significant pairs of canonical variates were discovered, but the second pair had distinctly marginal redundancy (7%) in the direction of interest (predicting influence on decision making from character-istics of departments). Neither pair was deemed interpretable without rotation.

The rotated solution provided by ESSCA did, however, permit interpreta-tion. The first component predicted impact on consumer service and information, and the second predicted impact on marketing decisions. Fornell recommends ESSCA, which maximizes the sum of squared loadings between IV variates and DV variables, over canonical analysis, which maximizes correlation between pairs of canonical variates, when the distinction between IV and DV is clear, so that only a set of variates from the IVs is required. ESSCA may also prove more interpretable when values in $\mathbf{R}_{xy}$ differ greatly in magnitude.

Chapter 7

Analysis of Covariance

7.1 GENERAL PURPOSE AND DESCRIPTION

Analysis of covariance is an extension of analysis of variance in which the effect of the IV(s) on the DV is assessed after the effects of one or more covariates are partialled out.[1] The major question for ANCOVA is the same as for ANOVA: Are mean differences among groups likely to have occurred by chance?

Analysis of covariance is used for three major purposes. The first, and most common, purpose is to increase the power of the test of the IV(s) by removing predictable variance from the error term. That is, ANCOVA can be used as a noise-reducing device. Although subjects may be randomly assigned to IV groups, this random assignment in no way assures equality among groups on all relevant attributes—it only guarantees, within probability limits, that there are no *systematic* differences between groups. Random individual differences, however, will have the effect of spreading out scores among subjects within groups, that is, among subjects who have been exposed to the same experimental treatment. This makes it hard to show differences among groups who have been exposed to different experimental treatments. One way to diminish the effect of those individual differences is to statistically adjust for them. In this situation, covariates can be seen to be acting as suppressor variables (cf. Chapter 5). The effect of the adjustment is to produce a more powerful test of differences among groups.

In this sense, ANCOVA is similar to a within-subjects (repeated measures)

---

[1] Strictly speaking, ANCOVA, like multiple regression, is not a multivariate technique, since it involves a single DV. For the purposes of this book, however, it is convenient to treat it along with multivariate analyses.

ANOVA (cf. Chapter 2), in which the predictable variance removed from the error term is that associated with consistencies within subjects over treatments. In ANCOVA, the predictable variance removed is that associated with some covariate, usually a continuous variable, that is linearly related to the DV. In the classic experimental situation, subjects are randomly assigned to levels of one or more IVs, and prior information about the subjects is used to form one or more covariates in order to reduce within-cell or error variance. A common situation would be the use of a pretest on some attribute as a covariate. Manipulation of the IV(s) would then be applied. Finally, a posttest would be administered on the same attribute. The posttest acts as the DV.

Suppose an experiment was designed to investigate methods for reducing test anxiety in statistics courses. Volunteers enrolled in statistics classes would be randomly assigned to one of three treatment groups: desensitization, relaxation training, or a control group for which anxiety reduction would be offered after the experiment. Treatment group, then, would be the IV. Before treatment, students in all three groups would be given a standardized measure of test anxiety. This pretest measure would serve as the covariate. Then treatment, or "waiting-list" control, would continue for some specified time. After the treatment period, students in all three groups would be retested on the measure of test anxiety (preferably an alternate form so as to avoid subject memory of the pretest). The posttest measure would then serve as the DV. The goal of statistical analysis would be to test the null hypothesis that differences in treatment had no effect on test anxiety scores, after adjusting for preexisting differences in test anxiety.

A second, nonexperimental application of ANCOVA commonly occurs in situations where subjects cannot be randomly assigned to treatments. Here, the ANCOVA is used as a statistical matching procedure, although interpretation is fraught with difficulty, as will be discussed in Sections 7.3.1 and 7.3.2. The analysis is performed on DV scores that would have occurred had all subjects, within all groups, performed identically on the covariate(s). Here, the variance removed can be seen as primarily affecting the differences between group means. That is, prior differences between subjects in different groups are removed, so that presumably the only remaining differences will be related to the effects of the grouping IV(s). (Differences could also, of course, be due to attributes that have *not* been used as covariates.)

This second application of ANCOVA can be seen as a procedure for descriptive model building. That is, the covariate enhances prediction of the DV, but there is no implication of causality. If the research question to be answered involves causality, ANCOVA is no substitute for running an experiment.

As an example, suppose we were to look at regional differences in political attitudes. Let's say the DV is some composite measure of liberalism-conservatism. Regions of the United States form the IV, say, Northeast, South, Midwest, and West. Two variables that could be expected to vary with political attitude and with geographical region are socioeconomic status and age. These two variables could serve as covariates. The statistical analysis would then test the null hypothesis that political attitudes did not differ with geographical region after adjusting for individual and regional differences in socioeconomic status and age. However, age and socioeconomic differences may be inextricably tied to geography, so the adjustment might not be realistic. And, of course, there is no implication that political attitudes are caused in any way by geographic region. Further, measurement properties of the covariate and the DV-covariate relationship may lead to over- or underadjustment and, therefore, to misleading results. These issues are discussed in greater detail throughout the chapter.

In the third major application of ANCOVA, discussed more fully in Chapter 8, ANCOVA is used to isolate the source of IV differences when several DVs are used. After a multivariate analysis of variance, it is frequently desirable to assess the contribution of the various DVs to significant differences among IV groups. This can be done by testing DVs, in turn, with the effects of other DVs removed. Removal of the effects of other DVs is accomplished by treating them as covariates. This procedure is called a stepdown analysis.

In all three major applications of ANCOVA the statistical operations are identical. Furthermore, ANCOVA can be applied in all ANOVA designs—factorial between-subjects, within-subjects,[2] mixed within-between, nonorthogonal, and so on. Only in a few programs, however, are analyses of these more complex designs possible. Similarly, specific comparisons and trend analysis are possible, although detailed procedures for these analyses are less likely to be readily available.

ANCOVA follows the model of analysis of variance, in that variance among scores is partitioned into variance attributable to differences in scores within groups and variance attributable to differences among groups (on one or more dimensions, or factors). Squared differences between scores and various means are summed (see Chapter 2), and these sums of squares, when divided by appropriate degrees of freedom, provide estimates of variance attributable to different sources (main effects of IVs, interactions between IVs, and error). Ratios of variances then provide tests of hypotheses about the effects of IVs on the DV. In ANCOVA, the regression of the DV on one or more covariates is estimated. Then DV scores and means are adjusted to remove the linear effects of the covariate(s). Finally, analysis of variance is performed on these adjusted values.

Lee (1975) presents an intuitively appealing illustration of the manner in which ANCOVA reduces error variance in a one-way between-subjects design with three levels of the IV (Figure 7.1). Note that the right-hand side of the figure illustrates scores and group means in an ANOVA. The error term is based on the sum of squared deviations of DV scores around their associated group means. In ANCOVA, a regression line is found relating the DV to the covariate. Here, the error term is based on the sum of squared deviations of the DV scores around the DV-covariate regression lines associated with their

---

[2] Note that typically the covariate would provide little help in testing within-subjects effects, since subjects at all levels of the IV (being the same) would already have the same value on all covariates. In some cases, however, a separate covariate measure may be available for each level of the within-subjects IV.

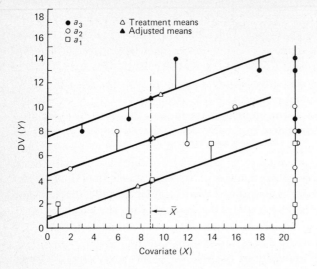

Figure 7.1 Plot of hypothetical data. The straight lines with common slope are those that best fit the data for the three treatments. The data points are also plotted along the single vertical line on the right. (From *Experimental Design and Analysis* by Wayne Lee. W. H. Freeman and Company. Copyright © 1975. Reprinted with permission.)

means. As long as the slope of the regression line is not zero, ANCOVA will result in a smaller sum of squares for error than ANOVA. If the slope is zero, error sum of squares will be the same as if it were calculated around group means. However, error mean square will be larger with zero slope since covariates will use up degrees of freedom.

7.2 KINDS OF RESEARCH QUESTIONS

As with analysis of variance, the goal in using analysis of covariance is the assessment of whether differences in the DV can be attributed to differential levels of the IV rather than to chance. What is added in the ANCOVA is a more precise look at the IV-DV relationship with the effect of covariates partialled out.

7.2.1 Main Effects of IVs

Holding all else constant, does assignment to different levels of a particular IV change behavior? Or, to put it another way: Are changes in behavior associated with different levels of an IV the result of the IV rather than random fluctuations occurring by chance or unexplained individual differences? The procedures described in Section 7.4 are designed to answer this question by testing the null hypothesis that the IV has no systematic effect on the DV.

Several devices are used to "hold all else constant." First, as with ANOVA, any other IVs in the design are held constant by crossing over them. Each IV is analyzed as if it were the only one in the design, with any other factors ignored. Second, controls are built into the research to assure that for differing levels of the IV, *only* the IV changes. Other potentially influential variables either are allowed to randomly vary (so that no systematic differences will be associated with IV levels) or are set at a fixed value for all levels of the IV (such as running all subjects in the same room). Third, the unique contribution of ANCOVA in holding potentially influential variables constant is to statistically adjust for any systematic differences in scores that may be associated with those variables (i.e., covariates).

The examples given in Section 7.1 are designed to answer these questions directly: Is test anxiety affected by treatment, after adjusting for prior individual differences in test anxiety? Does political attitude vary with geographical region, after holding constant differences in socioeconomic status and age?

With more than one IV, separate statistical tests are made for each one. If all groups have equal sample sizes, these separate tests are independent of one another except for a common error term, so that finding a statistically significant effect of one IV gives no clue as to whether an effect will be found for any other IV.

Suppose there were a second IV in the political attitude example. As a second IV we might choose religious affiliation, with four groups: Protestant, Catholic, Jewish, and None-or-other. In addition to testing for differences associated with geographic region, we could also see whether there are differences in liberalism-conservatism associated with religious affiliation, again adjusting for differences in socioeconomic status and age.

7.2.2 Interactions Among IVs

Holding all else constant, does change in behavior over levels of one IV depend on another IV? That is, do IVs interact in their effect on behavior? (See Chapter 2 for a discussion of interactions.) Tests of interactions, while interpreted quite differently from those of main effects, are statistically similar, and will be demonstrated in Section 7.7.

For the political attitude example in Section 7.1, the interaction would test whether differences in liberalism-conservatism over geographic region were the same for all religions, after adjusting for socioeconomic status and age.

With more than two IVs, multiple interactions are generated to be statistically tested. Each interaction is tested separately, and as with main effects, these tests are independent when sample sizes in all groups are equal. Furthermore, tests of interactions are independent from tests of main effects when designs are balanced and have equal sample sizes in all groups.

7.2.3 Specific Comparisons and Trend Analysis

When statistically significant effects are found in a design with more than two levels of a single IV, it is sometimes desirable to evaluate the nature of the differences found. Which groups differ significantly from each other? Is there a simple trend over sequential levels of an IV? Some of these questions may have emerged during the design stage of the research and been set up as specific a priori hypotheses.

Procedures have been developed for testing specific comparisons that are developed a priori as well as those that are generated post hoc—that is, after examining the data. These procedures are discussed in Section 7.5.2.3.

For the test anxiety example in Section 7.1, we might hypothesize prior to data collection that (1) the two treatment groups are more effective in reducing test anxiety than the waiting-list control, after adjusting for individual differences in test anxiety; and (2) that among the two treatment groups, desensitization is more effective than relaxation training in reducing test anxiety, again after adjusting for preexisting differences in test anxiety. These two hypotheses could be tested *instead* of answering the omnibus question regarding the main effect over the three IV groups. Or, with some loss in sensitivity, these two hypotheses could be tested post hoc after finding a main effect of the IV.

7.2.4 Effects of Covariates

Analysis of covariance is based on the possibility of a linear correlation between covariate(s) and the DV. This relationship can be evaluated by statistically testing the effect of covariate(s) as a source of variance in DV scores. This is discussed in Section 7.5.3.

In the previous example, the relationship between pre- and posttest anxiety could be tested. That is, to what extent is it possible to predict posttest anxiety from pretest anxiety, ignoring effects of differential treatment?

7.2.5 Strength of Association

If a main effect or interaction of IVs significantly affects behavior, the next logical question is, How much? What proportion of variance in the adjusted DV scores—adjusted for the covariate(s)—is attributable to variation in the IV(s)? Simple tests for strength of association are demonstrated in Sections 7.4 and 7.5.2.4.

In the test anxiety example, if a main effect of the IV (three groups) is found, the strength of association question is, What proportion of variance in the adjusted test anxiety scores (posttest scores adjusted for pretest) is attributable to differential treatment?

7.2.6 Adjusted Marginal and Cell Means

Again, if any main effects or interactions are statistically significant, what are the estimates of parameters (adjusted means) for groups or levels of the IV? How do these groups score differently, on the average, on the DV, after adjustment for covariates? The reporting of parameter estimates is demonstrated in Section 7.7.

For the test anxiety example, if there is a main effect of treatment, what is the average adjusted posttest anxiety score associated with each of the three groups?

7.3 LIMITATIONS TO ANALYSIS OF COVARIANCE

7.3.1 Theoretical Issues

As with ANOVA, the statistical test in no way assures that changes in the DV were caused by the IV. This is a logical rather than a statistical problem, and depends on the manner in which subjects were assigned to levels of the IV(s) and the controls used in the research. The statistical test is available to legitimately test hypotheses for nonexperimental as well as experimental research, but only in the latter case can attribution of causality be justified.

Choice of covariates is a logical exercise as well, although the number of covariates chosen can affect the power of the test. Calculation of the slope of the regression of the DV on the covariate(s) results in the loss of degrees of freedom for error (1 df per covariate). This means that *the gain in power from decreased sum of squares for error may be offset by the loss in degrees of freedom.* In cases where there is a statistically significant relationship between the DV and a covariate, the gain in reduced error variance should offset the loss of a degree of freedom. With multiple covariates, however, a point of diminishing returns can quickly be reached, especially if the covariates are highly correlated with one another (see Section 7.5.4).

A frequent caution is that the covariates must be independent of the treatment variables, or IVs. It is suggested that this be ensured by administering or gathering data on covariates before treatment is administered. Violation of this precept results in removal of some portion of the effect of the IV on the DV (i.e., that portion of the effect that is associated with the covariate). In this situation, adjusted group means may be closer together than unadjusted means. Further, the adjusted means may be difficult to interpret. This caution, however, really is appropriate only for the first major use of ANCOVA mentioned in Section 7.1. Since the effect of violation will be to reduce power (by reducing the sum of squares for main effects and interactions), it will obviously be inappropriate to use covariates affected by the IV if the ANCOVA is being used as a noise-reducing device.

If ANCOVA is being used as a statistical matching procedure, then adjustment for prior differences in means associated with covariates *should* be made. If the adjustment reduces mean differences on the DV, so be it—unadjusted differences reflect unwanted influences (other than the IV) on the DV. In other words, mean differences on the covariate associated with the IV are quite legitimately corrected for, as long as the covariate differences are not caused by the IV (Overall & Woodward, 1977).

For the third major application, covariance analysis is simply a device for evaluating a series of DVs individually in MANOVA stepdown analysis. Since covariates are actually DVs, it would be expected that they *not* be independent of the IV.

In all uses of ANCOVA, however, the adjusted means must be interpreted with some caution. The mean DV scores after adjustment is made for the covariates may not correspond to any situation in the real world. These are the means that would have occurred *if* all subjects had the same scores on the covariates. Especially in the second use of ANCOVA, such a situation may be so unrealistic as to make the adjusted values meaningless.

As seen in Section 7.3.2, sources of bias in ANCOVA are many and subtle, and can produce under- or overadjustment of the DV. At best, the second form of ANCOVA allows you to look at IV-DV relationships (noncausal) with the effects of covariates, as measured, adjusted for. If causal inference regarding treatment or other manipulated effects is desirable, there is no substitute for random assignment of subjects. Don't expect ANCOVA to permit causal inference of treatment effects with nonequivalent groups. If random assignment is absolutely impossible, or if it breaks down because of nonrandom subject dropout, be sure to thoroughly ground yourself in the literature regarding use of ANCOVA in such cases, starting with Cooke and Campbell (1979).

Limitations to generalizability apply to ANCOVA as they do to ANOVA, or any other statistical test. One can generalize only to those populations from which a random sample has been taken. ANCOVA may, in some limited sense, sometimes adjust for a failure to randomly assign the sample to groups, but it does not affect the relationship between the sample and the population to which one can generalize.

7.3.2 Practical Issues

The ANCOVA model assumes linearity, reliability of covariates, and homogeneity of regression in addition to the usual ANOVA assumptions of normality and homogeneity of variance.

7.3.2.1 Unequal Sample Sizes and Missing Data If scores on the DV are missing in a between-subjects ANCOVA, this is reflected as the problem of unequal n. That is, combinations of IV levels will not contain equal numbers of cases. Consult Section 7.5.2.2 for strategies for dealing with unequal sample sizes.

If some subjects are missing scores on covariate(s), or if, in repeated measures ANOVA, some DV scores are missing for some subjects, this is more clearly a missing-data problem. Consult Chapter 4 for methods of dealing with missing data.

7.3.2.2 Outliers Univariate outliers can occur for any one of the covariates or for the DV. Multivariate outliers can occur among covariates if multiple covariates are used. Outliers among covariates can produce heterogeneity of regression (Section 7.3.2.7). Consult Chapter 4 for methods of dealing with univariate outliers in the DV or covariate(s) and multivariate outliers among multiple covariates. Tests for univariate and multivariate outliers are demonstrated in Section 7.7.1. Since outliers are to be tested for separately within each group, BMDP7D and BMDP7M or BMDPAM are especially handy for checking univariate and multivariate outliers, respectively.

7.3.2.3 Multicollinearity and Singularity If there are multiple covariates, they must not be so redundant that they cause difficulties in computer analysis (see Chapter 4). Programs for ANCOVA in the BMD series automatically guard against this condition, but the SPSS programs do not. Consult Chapter 4 for methods of testing for multicollinearity and singularity among multiple covariates. Covariates producing multicollinearity and singularity should be eliminated, not only because of computational difficulties but also because they add no adjustment to the DV over that of other covariates. A test for multicollinearity and singularity is demonstrated in Section 7.7.

7.3.2.4 Normality As in all ANOVA, it is assumed that the treatment populations (for each level of the IV) from which the samples are taken are normally distributed. (Note that this does *not* mean that the sample scores within each cell need be normally distributed. Instead, it is the sampling distributions of means, as described in Chapter 2, that are assumed to be normal.) Without knowledge of population values, or production of actual sampling distributions of means, there is of course no way to directly test this assumption. However, it has been shown that ANOVA is robust to violations of this assumption, particularly since the central limit theorem suggests that with large samples normal sampling distributions will result even from nonnormal population distributions. *With relatively equal sample sizes, no outliers, and two-tailed tests, robustness should be assured with* 20 *degrees of freedom for error.* (See Chapter 2 for calculation of error degrees of freedom.) Larger samples are necessary for one-tailed tests. With small, unequal samples, or with outliers present, it may be possible to achieve normality with data transformation (cf. Chapter 4).

7.3.2.5 Homogeneity of Variance Just as in ANOVA, it is assumed in ANCOVA that the variance of DV scores within each cell of the design is a separate estimate of the same population variance. Because of the robustness of the analysis to violation of this assumption (as long as there are no outliers inflating variance in some cells), it is typically unnecessary to formally test for homogeneity of variance unless samples are small and the following checks are violated.

In order to assure robustness, Harris (1975) suggests the following checks. *For two-tailed tests, sample sizes should preferably be equal, but in no event should the ratio between largest and smallest sample sizes (within groups) be greater than* 4:1. (See Section 7.5.2.2 for additional problems associated with unequal sample sizes.) *Examine the variances within each cell (standard deviations squared) to assure that the ratio between largest and smallest variance is no greater than approximately* 20:1. If one-tailed tests are used, criteria should be more conservative, since the tests is less robust.

If any of these conditions is violated, a formal test for homogeneity of variance is available in any standard ANOVA text (e.g., Winer, 1971; Keppel, 1973) and in some canned programs (see Table 7.11). Most of these tests are highly sensitive to nonnormality, leading to an overly conservative rejection of the use of ANCOVA. Keppel (1973), however, offers a test that is typically not sensitive to departures from normality. This test is also available in BMDP7D.

In any event, gross violations of homogeneity can be corrected by data transformation, such as a log or square root transform of the DV scores. Interpretation, however, is then limited to the transformed scores. Add to this the difficulty in interpreting adjusted means, and interpretation becomes increasingly speculative. Major heterogeneity should, of course, always be reported by providing within-cell standard deviations.

7.3.2.6 Linearity   The ANCOVA model is based on the assumption that all relationships among covariates and the DV are linear. As with multiple regression (Chapter 5), violation of this assumption reduces the power of the statistical test by minimizing the ability of covariates to reduce error; so violation produces an error in the conservative direction. In this case, results can simply be interpreted as showing less than full adjustment for the covariate. For ANCOVA used in the statistical matching sense, lack of linearity will mean that optimum matching is not achieved, making interpretation of results more difficult.

As an initial screening for linearity, residuals plots may be observed through BMDP1V, SPSS MANOVA, or SPSS REGRESSION (using all covariates, dummy-coded IVs, and dummy-coded IV interactions, as demonstrated in Section 7.5.1.2). Interpretation of residuals plots is described in Section 5.3.2.4. *If there is indication of serious curvilinearity, examine within-cell scatterplots of the DV with each covariate and all covariates with one another. Where curvilinearity is indicated,*[3] *it may be corrected by transforming some of the variables.* Any of the available correlation programs that produce scatterplots (e.g., SPSS SCATTERGRAM and BMDP6D) can be used to evaluate relationships among covariates. Relationships between the DV and each covariate can be produced as one of the sets of plots in BMDP1V.

Because of the difficulties noted in interpreting transformed variables, one may consider eliminating a covariate producing nonlinearity. Or, higher-order powers of the nonlinear covariate may be used to produce additional covariates incorporating nonlinear influences.

7.3.2.7 Homogeneity of Regression   Adjustment of scores in ANCOVA is made on the basis of an average within-cell regression coefficient. This assumes

---

[3] Tests for deviation from linearity are available (cf. Keppel, 1973), but there seems to be little agreement as to the appropriate test or the seriousness of significant deviation in the case of ANCOVA. Therefore formal tests are not recommended.

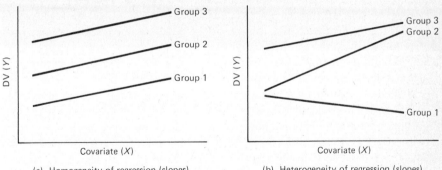

Figure 7.2   DV-covariate regression lines for three groups plotted on the same coordinates for conditions of (a) homogeneity and (b) heterogeneity of regression.

that the slope of the regression of the DV on the covariate(s) within each cell is an estimate of the same population regression coefficient (i.e., the slopes are equal across cells). This is analogous to assumptions behind averaging within-cell variances, as discussed in Section 7.3.2.5. Lack of homogeneity of regression implies an interaction between IV(s) and the covariate(s). This interaction means that the effect of covariates in adjusting the DV differs for various cells. Figure 7.2 illustrates, for three groups, perfect equality of slopes (as in Figure 7.1) and extreme heterogeneity of slopes.

If a between-subjects design is used, test the assumption of homogeneity of regression according to procedures described in Section 7.5.1. If any other design is used, and interaction between IVs and covariates is suspected, ANCOVA is inappropriate.

Violation of the assumption makes ANCOVA not only statistically, but logically, suspect. Useful information regarding the nature of the IV-covariate interaction is unavailable in most ANCOVA programs, although BMDP1V gives separate regression coefficients for each cell. A suggested alternative strategy when the assumption is clearly violated, or an IV-covariate interaction is suspected, is to use the covariate to set up an additional IV—a blocking variable. In this procedure, the continuous covariate is broken into categories (e.g., high, medium, and low scores on the covariate forming three categories). Instead of a k-way ANCOVA, the data are now analyzed as a $(k + 1)$–way ANOVA. For example, a two-way ANCOVA becomes a three-way ANOVA. Not only does ANOVA avoid the problems of linearity and homogeneity of regression, but the IV-covariate interaction is now directly observable. Relative advantages and disadvantages of ANCOVA versus blocking are discussed in Section 7.5.5.

If there is no reason to suspect an IV-covariate interaction with complex designs, it is probably safe to proceed with ANCOVA on the basis of the robustness of the model.

7.3.2.8 Reliability of Covariates   It is assumed in ANCOVA that all covariates are measured without error. That is, covariates are expected to be perfectly reliable. In the case of such variables as sex and age, the assumption can usually be justified. With self-report of demographic variables, and with variables measured psychometrically, such assumptions are not so easily made. Another source of unreliability is a scale that is reliable at any measurement point but fluctuates over short periods. Fluctuation in weight over the course of a day among humans is a prime example. In experimental research, unreliable covariates will only lead to loss of power; that is, they will underadjust. In nonexperimental applications, however, unreliable covariates can lead to errors of either type. Group means may either be spread too far apart (Type I error) or be compressed too closely together (Type II error). The degree of error, of course, depends on how unreliable the covariates are. *In nonexperimental research, attempt to limit use of covariates to those that can be measured reliably* ($r_{yy} > .8$). *If fallible covariates are absolutely unavoidable, make appropriate adjustment.*

There is no universally accepted procedure for adjustment of covariates for unreliability. Nor is there even agreement about which procedure is most appropriate for which application. Because of this disagreement, and because procedures for correction require use of sophisticated programs, they will not be covered in detail in this book. The interested reader is referred to Cohen and Cohen (1975), who recommend a strategy in which the analysis is done both with and without correction for unreliability. Interpretation is then based on the outcome of both analyses. Other procedures, along with an application, are covered by St. Pierre (1978).

7.4 FUNDAMENTAL EQUATIONS FOR ANALYSIS OF COVARIANCE

In the simplest application of analysis of covariance there is, in addition to a single DV score for each experimental unit (subject), a single classification variable (IV) and a single covariate. An example of a small-sample, hypothetical data set is presented in Table 7.1. Here the IV is the kind of treatment given to a sample of nine learning-disabled children. Three children are randomly assigned to each of two treatment groups and a control group, so that sample

TABLE 7.1   SMALL SAMPLE DATA FOR ILLUSTRATION OF ANALYSIS OF COVARIANCE

| | Groups | | | | | |
| | Treatment 1 | | Treatment 2 | | Control | |
	Pre	Post	Pre	Post	Pre	Post
	85	100	86	92	90	95
	80	98	82	99	87	80
	92	105	95	108	78	82
Sums	257	303	263	299	255	257

size within groups is three. For each of the nine children, two scores are given: covariate and DV. The covariate is a pretest score on the reading subtest of the Wide Range Achievement Test (WRAT-R), given before the experiment is begun. The DV is a posttest score on that same measure (WRAT-R), given at the completion of the experiment.

The research question to be asked is, Does differential treatment of learning-disabled children affect reading scores, holding constant differences in the children's prior reading ability? The sample size is, of course, inadequate for a realistic test of this research question but provides convenient data for illustration of the techniques involved in ANCOVA. The reader is encouraged to follow this example with both hand calculations and computer analysis using an available computer program, such as BMD12V, BMDP1V, or SPSS ANOVA.

The ANCOVA equation can be developed by extention of the ANOVA equation. As can be recalled from Chapter 2, differences among scores (or variances) can be partitioned into variance associated with different levels of the IV (between-groups variance) and variance associated with differences in scores within groups (unaccounted for or error variance). Variance is partitioned by first summing and squaring differences between scores and various means.

$$\sum_i \sum_j (Y_{ij} - GM)^2 = n \sum_j (\overline{Y}_j - GM)^2 + \sum_i \sum_j (Y_{ij} - GM)^2 \qquad (7.1)$$

or

$$\text{SS}_{\text{total}} = \text{SS}_{bg} + \text{SS}_{wg}$$

The total sum of squared differences between scores on Y (the DV) and the grand mean (GM) can be partitioned into two components: sum of squared differences between group means ($\overline{Y}_j$) and the grand mean (i.e., systematic or between-groups variability); and sum of squared differences between individual scores (Y_{ij}) and their respective group means.

In ANCOVA, there are two additional partitions involved. First, the differences in covariate scores can be partitioned into between- and within-groups sums of squares:

$$\text{SS}_{\text{total}(X)} = \text{SS}_{bg(X)} + \text{SS}_{wg(X)} \qquad (7.2)$$

The total sum of squared differences on the covariate (X) can be partitioned into differences between groups and differences within groups.

Similarly, the covariance or linear relationship between the DV and the covariate can be partitioned both into sums of products associated with covariance between groups and into sums of products associated with covariance within groups.

$$\text{SP}_{\text{total}} = \text{SP}_{bg} + \text{SP}_{wg} \qquad (7.3)$$

The total sum of products of the DV and the covariate can be partitioned both into sums of products between groups and into sums of products within groups.

While a *sum of squares* involves taking a value (X or Y), squaring it, and then summing, a *sum of products* involves taking two values (both X and Y), multiplying them instead of squaring, and then summing. As discussed in Chapter 2, the order of summing and squaring (or multiplying) is varied in order to assess all sources of variance in the research design.

The two partitions for differences on the covariate (Eq. 7.2) and the two partitions for differences due to the association between the covariate and DV (Eq. 7.3) are used to adjust the sums of squares associated with the DV according to the following equations.

$$SS'_{bg} = SS_{bg} - \left[\frac{(SP_{bg} - SP_{wg})^2}{SS_{bg(X)} - SS_{wg(X)}} - \frac{(SP_{wg})^2}{SS_{wg(X)}} \right] \qquad (7.4)$$

The adjusted between-groups sums of squares (SS'_{bg}) is found by subtracting from the unadjusted between-groups sums of squares a term based on sums of squares associated with the covariate, X, and sums of products associated with the linear relationship between the DV and the covariate.

$$SS'_{wg} = SS_{wg} - \frac{(SP_{wg})^2}{SS_{wg(X)}} \qquad (7.5)$$

The adjusted within-groups sum of squares (SS'_{wg}) is found by subtracting from the unadjusted within-groups sum of squares a term based on within-groups sums of squares and products associated with the covariate and with the linear relationship between the DV and the covariate.

It should be noted that in each case, the adjustment consists of subtracting a value based on the sum of squares associated with the covariate weighted by the regression coefficient representing the regression of the DV (Y) on the covariate (X). Symbolically, for an individual score:

$$(Y - Y') = (Y - GM_Y) - B_{Y.X}(X - GM_X) \qquad (7.6)$$

The adjusted deviation for any subject's Y score ($Y - Y'$) is obtained by subtracting from the unadjusted deviation score ($Y - GM_Y$) the individual's deviation on the X score ($X - GM_X$) weighted by the regression coefficient, $B_{Y.X}$.

Once the adjusted sums of squares are found, mean squares are found as usual by dividing by appropriate degrees of freedom. The only difference in degrees of freedom between ANOVA and ANCOVA is that in analysis of covariance the error degrees of freedom are reduced by one for each covariate (i.e., a degree of freedom is used up in estimating each regression coefficient).

To provide a worked example, calculational equations for the terms in Eqs. 7.4 and 7.5 are necessary. These are presented in Table 7.2. (Note that these equations apply only to equal-n designs.)

With the use of the data from Table 7.1, the six sums of squares and products are as follows:

$$SS_{bg} = \frac{(303)^2 + (299)^2 + (257)^2}{3} - \frac{(859)^2}{(3)(3)} = 432.889$$

$$SS_{wg} = (100)^2 + (98)^2 + (105)^2 + (92)^2 + (99)^2 + (108)^2 + (95)^2 + (80)^2$$
$$+ (82)^2 - \left(\frac{(303)^2 + (299)^2 + (257)^2}{3}\right) = 287.333$$

$$SS_{bg(X)} = \frac{(257)^2 + (263)^2 + (255)^2}{3} - \frac{(775)^2}{(3)(3)} = 11.556$$

$$SS_{wg(X)} = (85)^2 + (80)^2 + (92)^2 + (86)^2 + (82)^2 + (95)^2 + (90)^2 + (87)^2 + (78)^2$$
$$- \left(\frac{(257)^2 + (263)^2 + (255)^2}{3}\right) = 239.333$$

$$SP_{bg} = \frac{(257)(303) + (263)(299) + (255)(257)}{3} - \frac{(775)(859)}{(3)(3)} = 44.889$$

$$SP_{wg} = (85)(100) + (80)(98) + (92)(105) + (86)(92) + (82)(99)$$
$$+ (95)(108) + (90)(95) + (87)(80) + (78)(82)$$
$$- \left(\frac{(257)(303) + (263)(299) + (255)(257)}{3}\right) = 181.667$$

These values can be conveniently summarized in a sum-of-squares and cross-products matrix (cf. Chapter 1). For the between-groups sums of squares and cross products,

$$S_{bg} = \begin{bmatrix} 11.556 & 44.889 \\ 44.889 & 432.889 \end{bmatrix}$$

The first entry (first row and first column) is for the covariate and the second (second row, second column) is for the DV. For the within-groups sum of squares and cross products,

$$S_{wg} = \begin{bmatrix} 239.333 & 181.667 \\ 181.667 & 287.333 \end{bmatrix}$$

Again, the covariate appears as the first entry and the DV as the second. (Recall that entries in the major diagonal represent sums of squares, while off-diagonal entries are sums of products.) From these values, the adjusted sums of squares can be found as per Eqs. 7.4 and 7.5.

TABLE 7.2 COMPUTATIONAL EQUATIONS FOR SUMS OF SQUARES AND CROSS PRODUCTS IN ONE-WAY BETWEEN-SUBJECTS ANALYSIS OF COVARIANCE

Source	Sum of squares for Y (DV)	Sum of squares for X (covariate)	Sum of products
Between groups	$SS_{bg} = \dfrac{\sum\limits^{k}\left(\sum\limits^{n} Y\right)^2}{n} - \dfrac{\left(\sum\limits^{k}\sum\limits^{n} Y\right)^2}{kn}$	$SS_{bg(X)} = \dfrac{\sum\limits^{k}\left(\sum\limits^{n} X\right)^2}{n} - \dfrac{\left(\sum\limits^{k}\sum\limits^{n} X\right)^2}{kn}$	$SP_{bg} = \dfrac{\sum\limits^{k}\left(\sum\limits^{n} Y\right)\left(\sum\limits^{n} X\right)}{n} - \dfrac{\left(\sum\limits^{k}\sum\limits^{n} Y\right)\left(\sum\limits^{k}\sum\limits^{n} X\right)}{kn}$
Within groups	$SS_{wg} = \sum\limits^{k}\sum\limits^{n} Y^2 - \dfrac{\sum\limits^{k}\left(\sum\limits^{n} Y\right)^2}{n}$	$SS_{wg(X)} = \sum\limits^{k}\sum\limits^{n} X^2 - \dfrac{\sum\limits^{k}\left(\sum\limits^{n} X\right)^2}{n}$	$SP_{wg} = \sum\limits^{k}\sum\limits^{n}(XY) - \dfrac{\sum\limits^{k}\left(\sum\limits^{n} Y\right)\left(\sum\limits^{n} X\right)}{n}$

Note: k = number of groups; n = number of subjects per group.

$$SS'_{bg} = 432.889 - \left[\frac{(44.889 + 181.667)^2}{11.556 + 239.333} - \frac{(181.667)^2}{239.333} \right] = 366.202$$

$$SS'_{wg} = 287.334 - \frac{(181.667)^2}{239.333} = 149.439$$

These values can now be entered into a source table. Such a table is illustrated in Table 7.3. Degrees of freedom for between-groups variance are $k - 1$, and for within-groups variance $N - k - c$. (N = total sample size, k = number of levels of the IV, and c = number of covariates.)

As usual, mean squares are found by dividing sums of squares by appropriate degrees of freedom. The hypothesis that there are no differences among groups is tested by the F ratio formed by dividing the mean square between groups by the mean square within groups.

$$F = \frac{183.101}{29.888} = 6.13$$

Using a standard F table, we find that the obtained F of 6.13 exceeds the critical F of 5.79 at $\alpha = .05$ with 2 and 5 df. We can therefore reject the null hypothesis of no change in WRAT reading scores associated with the three treatment levels, after adjusting for pretest reading scores.

Given a statistically significant effect of the IV on the adjusted DV scores, the magnitude can be assessed using η^2, the squared nonlinear association between the IV as fixed for this research and the adjusted DV.

$$\eta^2 = \frac{SS'_{bg}}{SS'_{bg} + SS'_{wg}} \tag{7.7}$$

For the sample data,

$$\eta^2 = \frac{366.202}{366.202 + 149.439} = .7102$$

This indicates that 71% of the variance in the adjusted DV scores (WRAT-R) is explained by variation in the IV (treatment of children).

TABLE 7.3   ANALYSIS OF COVARIANCE FOR SMALL SAMPLE DATA

Source of Variance	Adjusted SS	df	MS	F
Between groups	366.202	2	183.101	6.13*
Within groups	149.439	5	29.888	

* $p < .05$.

TABLE 7.4   ANALYSIS OF VARIANCE FOR SMALL
SAMPLE DATA

Source of Variance	SS	df	MS	F
Between groups	432.889	2	216.444	4.52
Within groups	287.333	6	47.889	

The analysis of variance, omitting the covariate, appears in Table 7.4. Note the increase in sums of squares for both sources of variance, although the increase is much greater for the error term. There is also the gain of a degree of freedom for error (owing to elimination of one covariate). Note especially that the difference among means for treatment groups does not reach statistical significance without inclusion of the covariate.

The model extends to factorial and repeated-measures designs (Section 7.5.2.1), unequal n (Section 7.5.2.2), and multiple covariates (Section 7.5.3). In all cases, the generalization from ANOVA is that the analysis is done on adjusted, rather than raw, DV scores.

7.5 SOME IMPORTANT ISSUES

7.5.1 Test for Homogeneity of Regression

It has been suggested (e.g., Winer, 1971) that with relatively equal n there is evidence to indicate robustness in ANCOVA with respect to assumptions of homogeneity of regression, just as ANOVA is robust with respect to homogeneity of variance. Considering the tedious calculations necessary to verify the regression assumption for complex designs, this is just as well. It turns out, however, that for designs consisting solely of between-subjects variables, computer programs (cf. Table 7.11) are available for testing this assumption, and their use is recommended.

The programs test the hypothesis of no differences among cells in the slope of the regression of the DV on the covariate(s). Stated another way, equality of regression coefficients (B weights as described in Chapter 5) is tested for all cells within the design. Since it is the average of the slopes that is used to adjust the error term, it is assumed that these weights do not differ significantly either from one another or from a single estimate of the population value. Rejection of the null hypothesis of equality among slopes suggests that the analysis of covariance is inappropriate and that a blocking strategy be used (as described in Sections 7.3.2.7 and 7.5.5).

In order to use some BMD and SPSS programs for testing equality of slope, it is necessary to set up the data into a one-way array. That is, a factorial design is analyzed as if it were a simple one-way between-subjects ANCOVA. (Programs specifically designed for factorial ANCOVA do not typically provide tests for homogeneity of regression.) Since the error term in between-subjects designs is based on within-cell variance, irrespective of whether the cells are cross-classified, this is a perfectly legitimate transformation.

7.5.1.1 BMD Series The most straightforward programs are those in the BMD series, in which BMDP1V automatically provides tests for *equality of slope.*

BMDP1V, however, is a one-way ANCOVA program. For factorial designs it is necessary to specify a *grouping* variable, whereby each case is coded as to the cell in which it falls, in the context of a one-way design. For example, in a 3 × 3 factorial design, cells can be coded 1 through 9, and analyzed as a one-way array with 9 levels of the IV. (This type of recoding is demonstrated in Section 7.7.1, Table 7.13.) The output includes the test for slope equality in each cell as part of the overall analysis of covariance. For example, for the small sample data of Table 7.1, selected output appears in Table 7.5. As can be seen, the obtained F value of .0337, with 2 and 3 df, falls far short of the critical $F = 9.55$, at the .95 level of confidence. Therefore the assumption of homogeneity of regression is tenable. The slopes (standardized regression coefficients) for the three groups are also shown in Table 7.5. Since these sample

TABLE 7.5 INPUT AND SELECTED OUTPUT FROM BMDP1V FOR SMALL
SAMPLE DATA

```
PROGRAM CONTROL INFORMATION
        /PROBLEM    TITLE IS "HOMOGENEITY OF REGRESSION, SMALL SAMPLE ANCOVA".
        /INPUT      VARIABLES ARE 3.
                    FORMAT IS "(6X,F1.0,2F4.0)". UNIT=70.
        /VARIABLE   NAMES ARE TREAT, PRE, POST.
                    GROUPING IS TREAT.
        /GROUP      CODES(1) ARE 1,2,3.
                    NAMES(1) ARE TREAT1, TREAT2, CONTROL.
        /DESIGN     DEPENDENT IS POST.
                    INDEPENDENT IS PRE.
                CONTRAST = 1,0,-1.
                CONTRAST = 1,-2,1.
                CONTRAST = 1,1,-2.
        /END
```

ANALYSIS OF VARIANCE

SOURCE OF VARIANCE	D.F.	SUM OF SQ.	MEAN SQ.	F-VALUE	TAIL AREA PROBABILITY
EQUALITY OF ADJ. CELL MEANS	2	366.2012	183.1006	6.1263	.0452
ZERO SLOPE	1	137.8946	137.8946	4.6138	.0845
ERROR	5	149.4387	29.8877		
EQUALITY OF SLOPES	2	3.2868	1.6434	.0337	.9672
ERROR	3	146.1519	48.7173		

SLOPE WITHIN EACH GROUP

		TREAT1	TREAT2	CONTROL
		1	2	3
PRE	2	.5917	.8759	.7821

data represent a one-way analysis, the test for EQUALITY OF ADJ. CELL MEANS is a test of the major ANCOVA hypothesis. (It will be noted that the results are identical to those presented in Table 7.3.) For between-subjects factorial designs, however, BMDP1V can be used only for the preliminary test of slope equality, and the tests of major hypotheses (main effects and interactions) require additional runs on programs appropriate for factorial ANCOVA (BMD12V, BMDP2V, BMDP4V, SPSS ANOVA, and SPSS MANOVA).

7.5.1.2 SPSS Package   A somewhat more complicated procedure for testing homogeneity of regression is available within the SPSS REGRESSION program. It is useful for those without access to BMDP1V, SPSS MANOVA, or SPSS NEW REGRESSION, or for data sets in which cells are coded factorially and codes representing a grouping variable cannot be conveniently developed. (This procedure could also be used with BMDP2R.) SPSS REGRESSION tests the assumption of equality of slopes within the context of a test of the IV by covariate interaction. If such an interaction exists, it means that the relationship between the DV and the covariate(s) differs as a function of the IV. In other words, for differing levels of the IV (i.e., different cells) the regression of the DV on the covariate(s) differs.

First, the IV must be recoded so as to form a one-way array. This can be done within the IF procedure described in the SPSS manual (Nie et al., 1975). The one-way IV must then be additionally coded by creating dummy variables. Next, the IV-covariate dummy variables are created by multiplying the IV dummy variable value for each case by the value of the covariate, using the COMPUTE procedure. With multiple covariates, a separate set of interaction dummy variables is necessary for each covariate. Finally, a hierarchical SPSS REGRESSION run provides the data needed for a test of the IV-covariate interaction. The procedures for creating the dummy variables and setting up the run are fully described in the SPSS manual in the chapter on special topics in general linear models. Again, if the design actually is one-way, the output also tests the major hypothesis of equality of adjusted cell means.

Selected output for an application of this procedure for the small sample data in Table 7.1 is presented in Table 7.6. In this application the IV has been labeled TREAT, with the three groups coded: Treatment 1 = 1, Treatment 2 = 2, and Control = 3. The covariate is labeled PRE, and the DV is POST. The F test for homogeneity of regression is provided by Eq. 7.8.

$$F = \frac{[R^2(\text{step 3}) - R^2(\text{step 2})]/k_1 k_2}{[1 - R^2(\text{step 3})]/(N - K - 1)} \tag{7.8}$$

The F ratio for testing homogeneity of regression is based on six values: (1) $R^2(\text{step 2})$ is the squared multiple correlation based on inclusion of dummy variables for the IV and the covariate(s). (2) $R^2(\text{step 3})$ is the squared multiple correlation based on information in step 2 plus dummy variables used to represent the IV-covariate interaction. (3) k_1 is the number of covariates [VARIABLE(S) ENTERED ON STEP NUMBER 1].

TABLE 7.6 SELECTED OUTPUT FROM SPSS REGRESSION FOR TEST
OF ASSUMPTION OF HOMOGENEITY OF REGRESSION, SMALL
SAMPLE DATA (WITH COVARIATE ENTERED ON STEP 1, IV
ON STEP 2, AND IV-COVARIATE INTERACTION ON STEP 3)

```
VALUE LABELS   TREAT (1) TREATMENT 1 (2)TREATMENT 2 (3)CONTROL
IF             (TREAT EQ 1) T1=1
IF             (TREAT EQ 2) T2=1
COMPUTE        T1COV=T1*PRE
COMPUTE        T2COV=T2*PRE
REGRESSION     VARIABLES=PRE,POST,T1,T2,T1COV,T2COV/
               REGRESSION=POST WITH PRE(8), T1,T2(6), T1COV,T2COV(4)
STATISTICS     ALL
```

DEPENDENT VARIABLE.. POST

VARIABLE(S) ENTERED ON STEP NUMBER   1..   PRE

MULTIPLE R   .53297

R SQUARE   .28405

ADJUSTED R SQUARE   .18178

VARIABLE(S) ENTERED ON STEP NUMBER   2..   T1
                                            T2

MULTIPLE R   .89023

R SQUARE   .79251

ADJUSTED R SQUARE   .66802

VARIABLE(S) ENTERED ON STEP NUMBER   3..   T2COV
                                            T1COV

MULTIPLE R   .89279

R SQUARE   .79707

ADJUSTED R SQUARE   .45886

(4) k_2 is the number of dummy variables used to represent the IV [VARIA-BLE(S) ENTERED ON STEP NUMBER 2]. (5) N is the total sample size. (6) K is the total number of coded variables [VARIABLE(S) EN-TERED ON STEP NUMBER 1 plus step 2 plus step 3].

This is tested with numerator df $= k_1 k_2$ and denominator df $= N - K - 1$. For the sample data:

$$F = \frac{(.79707 - .79251)/(1)(2)}{(1 - .79707)/(9 - 5 - 1)} = 0.0337 \qquad \text{with df} = 2, 3$$

With multiple covariates, all the interaction dummy variables are entered together on the last step.

For a factorial design, SPSS MANOVA would provide a more convenient program within this package (see Section 8.7.1 for demonstration).

7.5.2 Design Complexity

Extension of ANCOVA to factorial between-subjects designs is straightforward as long as sample sizes within cells are equal. Partitioning of sources of variance follows ANOVA (cf. Chapter 2) with "subjects nested within cells" as the single error term. Sums of squares for the DV are adjusted for differences in the covariate and differences due to the association between the covariate and the DV, just as they are for the one-way design demonstrated in Section 7.4.

There are, however, two major design complexities that can arise: inclusion of within-subjects IVs, and unequal sample sizes in the cells of the factorial designs.

7.5.2.1 Within-Subjects and Mixed Within-Between Designs If each subject has a single value on each covariate, there is little value to ANCOVA in a simple one-way within-subjects design. Since DV scores on all levels are from the same subjects, each score from a given subject would have the same adjustment at all levels of the IV, which would be the same as no adjustment at all. With separate measures of the covariate(s) for each level of the IV, however, the design might prove fruitful. Further, ANCOVA could well be useful in a mixed design, even with single measures of each covariate. Here the adjustment could increase the power of the statistical test of the between-subjects IV(s).

As can be recalled from Chapter 2, one problem in ANOVA with the inclusion of within-subjects variables is the requirement for additional (or different) error terms. Another complication is the assumption of homogeneity of covariance. (See Chapter 2 for an explanation of this homogeneity).

For the user with access to BMDP, BMDP2V handles all between, within, and mixed within-between designs, and provides a test of homogeneity of covariance.[4] Both types of covariates can be entered—those that are constant over levels of the within-subjects IV(s) and those that change over levels.

BMDP4V, described in Section 8.6, also can be used for complex ANCOVA. Of the SPSS procedures, only MANOVA can be used for complex designs.

BMD12V can be used for complex designs with equal sample sizes, but the way in which the source table is set up is very deceptive. When nesting arrangements are specified, all appropriate error terms for complex designs will appear in the output table. However, these will *not* be the error terms used in testing the various main effects and interactions. Instead, all sources use the last term in the source table as the error term.[5] But the program can be used for the more complex designs as long as the user can specify which error term is appropriate for each test of a main effect or interaction. That is, the user must be able to specify the design in terms of associated effects and error terms (cf. Chapter 2). Then each appropriate F ratio can be formed by simply dividing the adjusted mean square for effect by the adjusted mean square for error. The mean squares listed in the output table are correct, as well as the degrees of freedom. Changing or constant measures of each covariate can be used. (If constant, the covariate is reentered for each level of the within-subjects IVs.)

No test for homogeneity of covariance is available in BMD12V. If there is reason to doubt the assumption of homogeneity of covariance, such as suspected strong carry-over effects from one level of the within-subjects IV to another, a conservative adjustment of the within-subjects IV(s) can be made. For the purposes of finding critical F (at a given level of alpha), the degrees of freedom associated with a within-subjects variable are considered to be 1, instead of $k - 1$. Whenever the degrees of freedom for the within-subjects variable appear in a main effect, interaction, or error term as $k - 1$, a value of 1 is substituted. (This is only for finding critical F, not for calculation of mean squares.) This is an extremely conservative adjustment, resulting in great loss of power, and should be used cautiously.

In BMDP4V, a program designed for MANOVA but that can be used for ANCOVA, a correction to degrees of freedom is applied that is more precise and therefore less conservative. The more accurate correction is also available in the most recent releases of BMDP2V. Other alternatives that avoid the requirement for homogeneity of covariance are to substitute MANCOVA for repeated-measures ANCOVA (cf. Section 8.2.6) or to do a trend analysis on the within-subjects IV.

7.5.2.2 Unequal Sample Sizes When cells in a factorial design have unequal numbers of scores per cell, the hypotheses tested for main effects and

---

[4] Be sure to note the reservations mentioned in the manual (Dixon, 1981) with respect to the compound symmetry test. The test actually involves only that portion of compound symmetry that is relevant to the underlying assumptions (Frane, personal communication). Corrective procedures for failure of the homogeneity of covariance test are discussed at the end of this section.

[5] Results for a one-way within-subjects design are correct as printed out.

interactions are no longer independent; the design has become nonorthogonal (cf. Section 2.2.5.3). The problem generalizes directly to ANCOVA.

If artificially equalizing cell sizes is inappropriate, there are a number of strategies for dealing with unequal n, depending on the type of research. Three major methods are described by Overall and Spiegel (1969). There are a number of ways of viewing these methods, and the terminology associated with these viewpoints is quite different and, sometimes, seemingly contradictory. Table 7.7 delineates research situations calling for different methods and notes some of the jargon styles associated with these distinctions.

Differences in these methods can be conceptualized within the context of multiple regression (Chapter 5). In Method 1, all significance tests are given equal weight, with each main effect and interaction assessed after adjustment is made for all other main effects and interactions, as well as for covariates. That is, it is like a standard multiple regression. In SPSS this is called the *regression* approach.[6] It tests the same hypotheses as the unweighted-means approach in which each cell mean is given equal weight regardless of its sample size. This is the approach to be recommended in experimental research unless there is a reason for doing otherwise. Reasons would include a desire to give heavier weighting to some cells than others, due to unequal importance or unequal population sizes for treatments when occurring naturally. If cells, initially designed to be equal-n, end up grossly unequal, the problem is not one of adjustment but, more seriously, of differential dropout.

Method 2 imposes a hierarchy of testing of effects, whereby main effects are adjusted only for one another and for covariates, while interactions are adjusted for one another, for covariates, and for main effects and lower-order interactions, and so on. That is, there is an implicit order of priority that places more emphasis on main effects than interactions, and more emphasis on lower-order interactions than higher-order interactions. This is called the *classic experimental* approach by SPSS and is sometimes referred to elsewhere as the *least-squares* approach. It is recommended when there is a desire to weight cell means by sample sizes by assigning higher weighting to cells with larger sample sizes when evaluating marginal means and lower-order interactions.

Finally, Method 3 allows the researcher to set up the hierarchy of testing of covariates, main effects, and interactions. In addition, BMDP4V allows cells to be weighted by importance or by population size, regardless of sample size.

Three simple programs in the BMD and SPSS packages allow analysis of covariance with unequal sample sizes. BMDP2V provides for unequal n in addition to complex designs, but only one method of adjustment is available, Method 1 as described earlier. To achieve other varieties of adjustment, multiple runs are necessary, in which the effect to be adjusted for is entered as a covariate. SPSS ANOVA allows Methods 1 and 2, and many varieties of Method 3, but

---

[6] Note that the terms *regression* and *classic experimental* as used by SPSS do *not* imply that these approaches are most appropriately used for nonexperimental and experimental designs, respectively. If anything, the opposite is true.

TABLE 7.7 TERMINOLOGY FOR STRATEGIES FOR ADJUSTMENT FOR UNEQUAL CELL SIZES

Research type	Overall and Speigel	SPSS	BMD	Other common terminology
1. Experiments designed to be equal-n, with random dropout. All cells equally important	Method 1	*Regression* approach. Option 9 in SPSS ANOVA program. In SPSS MANOVA, METHOD-SSTYPE (UNIQUE)	Equal cell weights. Default for BMDP2V	Unweighted means
2. Nonexperimental research in which sample sizes reflect importance of cells. Main effects have equal priority	Method 2	*Classic experimental* approach. Default in SPSS ANOVA program	*Cell size weights*[a]	Least squares
3. Like number 2 above, except main effects have unequal priority	Method 3	*Hierarchical* (sequential) approach. Option 10 in SPSS ANOVA. Default in SPSS MANOVA	*Cell size weights*[a]	
4. Research in which cells are given unequal weight on basis of prior knowledge	N.A.	N.A.	User-defined cell weights	

[a] Output table in BMDP4V gives information for interpretation as either research type 2 or research type 3.

requires a between-subjects design. Only SPSS MANOVA and BMDP4V provide for design complexity as well as flexibility in adjustment for unequal n.

The more complicated general linear model programs can be used for unequal n, but they require far more sophistication on the part of the user, and there is no significant advantage over the simpler programs for most applications.

7.5.2.3 Specific Comparisons and Trend Analysis

As for ANOVA, the finding of a significant main effect or interaction is sometimes insufficient for full interpretation of the effects of IV(s) on the DV. If there are more than two levels of an IV, the omnibus F test gives no information as to which levels are significantly different from which other levels. For an interaction, the omnibus test gives no information as to the cells that differ significantly. With a quantitative IV, interpretation can frequently be enhanced by the discovery of a simple trend of behavior change over sequential levels of the IV.

As with ANOVA (Chapter 2), these specific comparisons or trends can be hypothesized a priori as part of the research design, or can be tested as part of a data-snooping procedure after omnibus analyses have been completed (see Section 2.2.6.6 regarding Type IV errors). In either case, comparisons are achieved by specifying contrast coefficients and running analyses based on these coefficients. For post hoc analysis, the probability of Type I error increases with the number of comparisons made. Therefore some adjustment must be made for inflated α error. For a priori contrasts, protection against inflated Type I error is achieved by running a number of contrasts instead of omnibus F, where the number does not exceed the available degrees of freedom, and by working with a set of contrasts that is mutually orthogonal. (See Chapter 2 for a review of a priori orthogonal contrasts.)

Specific comparisons can easily be achieved through use of BMDP1V if all IVs are between subjects and cell sample sizes are equal.[7] First, it is necessary to convert data to a one-way design (as discussed in Section 7.5.1.1). A program run then automatically produces a T-TEST MATRIX FOR ADJUSTED GROUP MEANS, in which differences among all pairs of cells are evaluated. For the hypothetical data of Table 7.1, the matrix of t tests is shown in Table 7.8. The significance of a comparison is tested with df $= N - K - c$, where N = total sample size (9 in the example), k = number of cells (3 in the example), and c = number of covariates (1 in the example). (Note that BMDP1V prints out both the degrees of freedom and a matrix of probability levels.)

If one of these comparisons had been run a priori, say TREAT1 versus CONTROL, the obtained value of t for that comparison (3.3171) would be considered statistically significant for a two-tailed test, $\alpha = .05$, since $p = .0211$ (i.e., $p < .05$).

If this comparison is to be evaluated post hoc, a Scheffé-type adjust-

---

[7] If cell sizes are unequal, BMDP4V can be used for a test of appropriately adjusted (weighted) means (see Section 8.6). At the present time, however, adjusted means are not printed out. See Chapter 8 for specific comparisons with SPSS MANOVA.

TABLE 7.8 COMPARISONS OF ALL PAIRS OF CELL MEANS AS
PRODUCED BY BMDP1V FOR SMALL SAMPLE DATA

T-TEST MATRIX FOR ADJUSTED GROUP MEANS ON   5 DEGREES OF FREEDOM

		TREAT1	TREAT2	CONTROL
		1	2	3
TREAT1	1	0.0000		
TREAT2	2	-.6309	0.0000	
CONTROL	3	-3.3171	-2.6250	0.0000

PROBABILITIES FOR THE T-VALUES ABOVE

		TREAT1	TREAT2	CONTROL
		1	2	3
TREAT1	1	1.0000		
TREAT2	2	.5558	1.0000	
CONTROL	3	.0211	.0468	1.0000

Note: Program input appears in Table 7.5.

ment (or some other adjusting procedure) must be made to counteract the
inflation in Type I error rate. First, the obtained t value is squared to produce
"obtained F." For example, the F obtained for TREAT1 versus CONTROL
is $(-3.3171)^2 = 11.00$. This is then compared with an adjusted value of critical
F, whereby the tabled value of F (in this case 6.61 for 1 and 5 df, $\alpha = .05$)
is multiplied by the degrees of freedom associated with the number of cells,
or $k - 1$. For the example, then, the adjusted critical F value is $2(6.61) =$
13.22, and the difference between TREAT1 and CONTROL now fails to reach
statistical significance.

For comparisons other than simple cell pairs, a set of contrast coefficients
is entered in the deck set-up for each comparison. The program then prints
out the coefficients entered, the t-test value associated with each contrast, and
the probability value for each test, P(T), as illustrated in Table 7.9 for the
sample data. The first two contrasts represent a trend analysis, with linear and
quadratic coefficients given in the first and second rows, respectively.[8] These

[8] Trend analysis is shown for illustration only; it would have no theoretical relevance for
this hypothetical data.

TABLE 7.9   CONTRASTS AS SPECIFIED FOR SMALL SAMPLE DATA;
OUTPUT PRODUCED BY BMDP1V

```
T-VALUES FOR CONTRASTS IN ADJUSTED GROUP MEANS
```

CONTRAST NUMBER	T	P(T)	GROUP TREAT1	GROUP TREAT2	GROUP CONTROL
1	3.3171	.0211	1.0000	0.0000	-1.0000
2	-1.1542	.3006	1.0000	-2.0000	1.0000
3	3.4272	.0187	1.0000	1.0000	-2.0000

Note: Program input appears in Table 7.5.

coefficients for orthogonal polynomials are available in all standard ANOVA texts, such as Keppel (1973) and Winer (1971). The third contrast represents a test of the average of both treatment groups against the control group, testing the null hypothesis of no difference between treatment and control. Degrees of freedom for error remain $N - k - c$. With df $= 5$ in the sample data, producing critical $t = 2.57$, a priori analysis would indicate a significant linear trend and rejection of the null hypothesis of no effect of treatment.

For specified contrasts with more than one IV, adjustment for inflated α error depends on whether the contrasts are associated with marginal or cell means (main effects or interactions, respectively). For a two-way design, for example, a test of marginal means on the IV labeled A would require an adjustment of $(a - 1)$. That is, the tabled value of F would be multiplied by $(a - 1)$ to produce critical F. Comparisons among marginal means of the IV labeled B would require tabled F to be multiplied by $(b - 1)$. Cell comparisons call for a multiplication of tabled F by $(a - 1)(b - 1)$.

If appropriate programs are unavailable, hand calculations for specific comparisons are not particularly difficult, as long as sample sizes are equal for each cell. The general equation for evaluating contrasts is

$$F = \frac{n_c (\Sigma w_j \overline{Y}_j)^2 / \Sigma w_j{}^2}{MS_{error}} \tag{7.9}$$

An F ratio is formed to test the significance of a comparison. This is based on (1) the number of scores, n_c, in each of the means to be compared; (2) a squared sum of weighted means, $(\Sigma w_j \overline{Y}_j)^2$, based on the multiplication of each adjusted mean, $\overline{Y}_j$, by its contrast coefficient, w_j; (3) the sum of the squared contrast coefficients, $\Sigma w_j{}^2$; and (4) the mean square for error in the ANCOVA design.

The error mean square is readily available from the SPSS ANOVA, SPSS MANOVA, or BMD12V program on which omnibus ANCOVA can be run for multifactor designs. In the case of BMD12V it is important to choose the appropriate error term for the comparison (i.e., the one that would be used to test the main effect or interaction for which the specific comparison is a part). Adjusted marginal and cell means are directly available in BMD12V and SPSS MANOVA. In SPSS ANOVA, adjusted marginal means for main effects can be found from the "multiple classification analysis" feature of the program, but adjusted cell means for interactions in multifactor designs are not readily available.

For the small sample data of Table 7.1, the omnibus ANCOVA and multiple classification analysis as produced by SPSS ANOVA are illustrated in Table 7.10. The three adjusted group means are derived algebraically by adding the adjusted deviations for each group to the grand mean. These are

TREATMENT 1: $95.44 + 5.89 = 101.33$

TREATMENT 2: $95.44 + 3.04 = 98.48$

CONTROL: $95.44 + (-8.93) = 86.51$

The error mean square (residual) $= 29.887$, as printed out in the source table. For the test of treatment average versus control (third contrast of Table 7.9), the contrast coefficients are 1, 1, -2. Applying Eq. 7.9, then, we obtain

$$F = \frac{3[(1)(101.33) + (1)(98.48) + (-2)(86.51)]^2/[(1)^2 + (1)^2 + (-2)^2]}{29.888}$$

$$= 12.01$$

This is an approximation (within rather substantial rounding error) to the 11.75 that would be produced by squaring the $t = 3.43$ from Table 7.9.

7.5.2.4 Strength of Association For factorial designs η^2 as a test of strength of association can be found as an extension of Eq. 7.7 The numerator consists of the adjusted sum of squares for the main effect or interaction being evaluated, and the denominator consists of the total adjusted sum of squares. The latter total adjusted sum of squares includes components for all individual main effects and interactions and all error terms. It does *not* include components for covariates or the mean, which are typically printed out by SPSS and BMD programs. Nor does it include such summary terms as *main effects, two-way interactions, explained* variance, and the like, which are printed out by SPSS ANOVA (cf. Table 7.10).[9] To find the strength of association between an effect and the adjusted DV scores, then,

---

[9] The total sum of squares printed out in SPSS ANOVA is inappropriate for this purpose, since it includes covariates. That is, it is the sum of squares for the original DV scores rather than for the adjusted DV scores.

TABLE 7.10   ANALYSIS OF COVARIANCE AND MULTIPLE
CLASSIFICATION ANALYSIS PRODUCED BY SPSS ANOVA
FOR SMALL SAMPLE DATA (SELECTED OUTPUT)

```
            VALUE LABELS    TREAT (1) TREATMENT 1 (2)TREATMENT 2 (3)CONTROL
            ANOVA           POST BY TREAT(1,3) WITH PRE
            STATISTICS      ALL



* * * * * * * * A N A L Y S I S   O F   V A R I A N C E * * * * * * * *
            POST
         BY TREAT
         WITH PRE
* * * * * * * * * * * * * * * * * * * * * * * * * * * * * * * * * * * * *


                              SUM OF             MEAN           SIGNIF
SOURCE OF VARIATION           SQUARES    DF     SQUARE     F     OF F

COVARIATES                    204.582    1     204.582   6.845   .047
    PRE                       204.582    1     204.582   6.845   .047

MAIN EFFECTS                  366.201    2     183.101   6.126   .045
    TREAT                     366.201    2     183.101   6.126   .045

EXPLAINED                     570.784    3     190.261   6.366   .037

RESIDUAL                      149.439    5      29.888

TOTAL                         720.222    8      90.028



COVARIATE       REGRESSION COEFFICIENT ADJUSTED FOR
                ALL OTHER COVARIATES


PRE                 .903



* * M U L T I P L E   C L A S S I F I C A T I O N   A N A L Y S I S * *
            POST
         BY   TREAT
         WITH PRE
* * * * * * * * * * * * * * * * * * * * * * * * * * * * * * * * * * * * *

GRAND MEAN =   95.44
```

| | | | | ADJUSTED FOR | | ADJUSTED FOR INDEPENDENTS | |
| | | | UNADJUSTED | INDEPENDENTS | | + COVARIATES | |
VARIABLE + CATEGORY	N	DEV"N	ETA	DEV"N	BETA	DEV"N	BETA
TREAT							
1 TREATMENT 1	3	5.56				5.89	
2 TREATMENT 2	3	4.22				3.04	
3 CONTROL	3	-9.78				-8.93	
			.78				.72

$$\eta^2 = \frac{SS'_{effect}}{SS'_{total}} \tag{7.10}$$

In multifactorial designs, the size of η^2 for a particular effect will depend on the strength of other effects within the design. That is, in a design in which several main effects and interactions are significant, the test of a particular effect will be diminished because of the increasing denominator. Keppel (1973) suggests an alternative method of viewing strength of association. Applied to η^2, the test for each effect includes in the denominator only the adjusted sum of squares for the effect being tested and the adjusted sum of squares for the appropriate error term for that effect (see Chapter 2 for appropriate error terms).

$$\eta^2_{alt} = \frac{SS'_{effect}}{SS'_{effect} + SS'_{error}} \tag{7.11}$$

Since this alternative form is not standard, an explanatory footnote is appropriate whenever the technique is used in published results.

7.5.3 Evaluation of Covariates

Covariates in analysis of covariance can be evaluated as sources of prediction of the DV. In the multiple regression sense (Chapter 5), each covariate can be considered a continuous IV with the remaining IVs (main effects and interactions) evaluated as if entering the regression equation after the covariate.

Significance tests for covariates can provide information as to their utility in adjusting the DV. If a single covariate is used, and it is statistically significant, it is providing adjustment of the DV scores. For the small sample data, reference to Table 7.10 shows that the covariate, PRE, is significantly related to the DV, POST, with $F(1, 5) = 6.845$, $p < .05$.

With multiple covariates, all covariates enter the multiple regression equation at once and, as a set, are treated in a standard multiple regression format (Chapter 5). That is, within a set of covariates, each covariate is assessed as if it entered the equation after all other covariates. Therefore, although each covariate may be significantly correlated with the DV when considered individually, when considered last, each may add no significant predictability (or adjustment) of the DV once the other covariates have entered. So it is helpful to consult both tables of correlations among covariates and the DV and significance levels for each covariate as reported in ANCOVA source tables when assessing the contribution of each covariate to adjustment.[10] Evaluation of covariates in exploratory research is useful in developing a maximally effective set of covariates (see the next section). Section 7.7 demonstrates evaluation of covariates.

Beta estimates (standardized regression coefficients), provided by most

---

[10] If covariates enter the analysis before any other effects (e.g., SPSS ANOVA, default or Option 10), simple bivariate correlations are appropriate. Otherwise, pooled-within-cell correlations should be used (cf. Section 8.7.2.1).

canned computer programs (see Table 7.10), have the same meaning as beta weights described in Chapter 5. However, interpretation of these weights depends on the method used for adjustment of unequal n. In the unweighted-means approach (equal cell weights or Method 1—see Table 7.7), covariates are assessed as if entering the regression equation after all the main effects and interactions. Beta weights, then, are like those of standard multiple regression. In other methods, however, covariates may enter the equation first, or after main effects but before interactions. Here, the beta weights are evaluated at the point they enter the equation.

7.5.4 Choosing Covariates

If several covariates are potentially available, there is the desirability of an optimum set. A point of diminishing returns in adjustment of the DV is quickly reached with a large set of intercorrelated covariates. Additional covariates only reduce power by using up degrees of freedom. A preliminary analysis of the set of covariates, however, can avoid this problem.

First, it may be possible to pick a small set on an a priori basis. If questions remain, you can look at correlations among covariates and eliminate obvious redundancies. Or, if even fewer covariates are desired, a look at all squared multiple correlations may be useful, with each covariate, in turn, acting as the DV and all other covariates as the IVs.[11] Programs designed for principal components analysis (see Chapter 10) do this automatically, if all covariates are entered as a set of variables. Or, if N is very large and power is not a problem, it still may be worthwhile to find a reduced set of covariates for the sake of parsimony. Useless covariates can be identified from the first ANCOVA run. Then further ANCOVAs can be run with nonworking covariates sequentially eliminated, until a set of covariates is found that, in future research, could be expected to be optimal.

For purposes of reporting, the simpler analysis—with useless covariates eliminated—can be given in detail. However, mention should be made of the discarded covariate(s) and the fact that the pattern of results did not change.

7.5.5 Alternatives to ANCOVA

Because of the stringent limitations of ANCOVA, and the frequent ambiguity in interpreting results, alternative analysis strategies are often sought. For example, with pretest scores available, one alternative is to base an ANOVA on difference scores (posttest minus pretest) as the DV rather than using the pretest as a covariate. If the research question is phrased in terms of "change," then difference scores will provide the answer. For example, does one year of belly dance training change self-esteem scores more than participation in aerobic

---

[11] This information is also used to check for multicollinearity among covariates. However, the criterion in choosing a reduced set would be a much lower R^2 than that which indicates multicollinearity.

dance classes? This is subtly different from the covariance question, Does belly dance training produce greater self-esteem over that of aerobic dance classes, using pretreatment self-esteem scores as a baseline? Problems with difference scores involve ceiling and floor effects (or, in general, problems with regression toward the mean). In some research situations, either approach can provide valid answers. If so, ANCOVA is usually the better approach since, in effect, the approach involving difference scores assumes that the correlation between the DV and the covariate is perfect.

Another alternative would be to treat the pre- and posttest scores as two levels of a within-subjects IV. The problem with this strategy is that the interpretation of the effect of the IV of interest now becomes difficult, as it is reflected in the interaction between the IV of interest and the pre- versus posttest IV. It is also a less powerful test of the IV of interest than would be available under other strategies. These first two approaches, of course, are possible only with potential covariates that are measured on the same scale as the DV.

Additional approaches are available with potential covariates measured on any continuous scale. One strategy is to use a randomized block design, in which subjects are matched into blocks on the basis of the pretest or other potential covariate(s). Each block has as many members as the number of levels of the IV (or number of cells—cf. Section 2.2.3), and subjects are assigned to levels randomly within each block. Members of each block are then treated as if they were the same person, in a repeated-measures design. The disadvantages of this approach are the loss of degrees of freedom and strong assumptions associated with all within-subjects designs. In addition, it requires the added step of random assignment between administration of the potential covariate and the DV, which may be inconvenient, if not impossible, in some applications.

A final alternative requires the added step of random assignment between collection of the potential covariate and DV (or at least some type of assignment if randomness is not possible) but none of the assumptions of ANCOVA or within-subjects ANOVA. This is the strategy of blocking in which the potential covariate is simply converted into another full-scale IV. Scores on the potential covariate are cut into two or more levels. This new IV is then crossed with the IV of interest in a factorial design. The design can be expanded to multiple covariates. Interpretation of the main effect of the IV of interest is now straightforward, with any variation due to the potential covariate(s) removed from the estimate of error variance. Furthermore, if there is interaction between a potential covariate and the IV of interest (i.e., if the assumption of homogeneity of regression is violated), it will be directly testable statistically. This test may provide highly valuable information. Another advantage of blocking is that the relation between the potential covariate and the DV need not be linear.

In many situations, then, blocking will be preferable to ANCOVA—particularly in experimental, rather than correlational, research. In some applications, however, ANCOVA will be the preferable approach. With a linear correlation between DV and covariate, ANCOVA is more powerful than blocking. With multiple potential covariates, designs based on blocking can become horrendously large and complex. And, if the assumptions for covariance analysis are met, a conversion from a continuous covariate to a discrete IV can result in loss of information. Finally, practical limitations may prevent collection of potential covariate information sufficiently in advance of the administration of the IV to accomplish random assignment to the original IV within each level of the IV developed from the covariate. With post hoc blocking, sample sizes within cells are likely to be highly discrepant, leading to all the problems associated with nonorthogonality. In some applications, a combination of blocking and ANCOVA may turn out to be the optimal solution. Some potential covariates can be used to create blocking variables, while others are analyzed as covariates.

7.6 COMPARISON OF PROGRAMS

For the novice, there is a bewildering array of canned computer programs in the BMD and SPSS packages for ANCOVA: SPSS REGRESSION, ANOVA, and MANOVA; BMD03V, 04V, 05V, 06V, 09V, 11V, 12V, P1V, P2V, and P4V.

On the basis of complexity of use, those programs based on the general linear model can be set aside, since they offer little advantage over programs more easily used. This eliminates BMD05V, 06V, 11V, and SPSS REGRESSION.

7.6.1 BMD Series

A second reduction can be made within the BMD series. BMDP1V incorporates features of BMD03V, BMD04V, and BMD09V, making the earlier programs obsolete. Features of the remaining programs are summarized in Table 7.11.

BMDP1V is designed for one-way, between-subjects ANCOVA, but offers features unavailable in factorial programs. Most notably, it provides for tests of homogeneity of regression (equality of slopes) and for user-designated contrasts. As discussed in Sections 7.5.1 and 7.5.2.3, data from factorial between-subjects designs can be recoded in order to take advantage of these features, in addition to standard analysis by factorial programs.

For factorial ANCOVA, the best all-around program in the BMD series designed for ANCOVA is BMDP2V. It allows for both unequal n and complex designs, in terms of within-subjects IVs and unbalanced layouts. The recommended test for homogeneity of covariance, labeled *compound symmetry,* in within-subjects designs is available within the program. (Only the portion of the compound-symmetry test relevant to homogeneity of covariance is evaluated.) In recent releases, adjustment is made for violation of homogeneity of covariance. There is also a full orthogonal decomposition (trend analysis) of within-subjects variables. Spacing between levels of the within-subjects variable need not even be equal. In combination with P1V to test for equality of slope and user-specified contrasts in between-subjects designs, the user with access to BMD P-series programs has almost as much flexibility of analysis of covariance as is available in the most complicated general linear model programs (such

TABLE 7.11 COMPARISON OF SELECTED PROGRAMS FOR ANALYSIS OF COVARIANCE

Feature	BMDP4V	BMDP1V	BMD12V	BMDP2V	SPSS ANOVA	SPSS MANOVA[a]
Input						
Maximum number of IVs	8 B-S + 8 W-S	1	9	9 B-S + 9 W-S	5	10
Allows unequal n	Yes	Yes	No	Yes	Yes	Yes
Choice of equal-n adjustment	No	N.A.	N.A.	No	Yes	Yes
Data must be ordered	No	No	Yes	No	No	No
Within-subjects IVs	Yes	No	Yes[b]	Yes	No	Yes
Unbalanced designs	Yes	N.A.	No	Yes	No	Yes
Output						
Source table	Yes	Yes	Yes	Yes	Yes	Yes
Unadjusted cell means	Yes	Yes	Yes	Yes	Yes	Yes
Unadjusted marginal means	Yes	Yes	Yes	No	Yes	Yes
Cell standard deviations	No	No	No	Yes	No	Yes
Adjusted cell means	No	Yes	Yes	Yes	No	Yes
Adjusted marginal means	No	Yes	Yes	No	No	Yes
Grand mean and adjusted marginal deviation	No	No	No	No	Yes	No
Test for equality of slope (homogeneity of regression)	No	Yes	No	No	No[c]	Yes
t test for all pairs of cell means	No	Yes	No	No	No[c]	Yes
Trend analysis (including unequal spacing)	Yes	No	No	Within-subjects IVs only	No[c]	Yes
User-specified contrasts (including equal-space trend analysis)	Yes	Yes	No	No	No[c]	Yes
Variance-covariance matrices (effects)	No	Yes	No	No	No	Yes[d]
Sum-of-squares and cross-products matrices (effects)	Yes	No	Yes	No	No	Yes[d]
Regression coefficients for each covariate	Yes	Yes	Yes	No	No	Yes[d]
Regression coefficient for each main effect	No	No	No	No	Yes	No

TABLE 7.11 (Continued)

Feature	BMDP4V	BMDP1V	BMD12V	BMDP2V	SPSS ANOVA	SPSS MANOVA[a]
Multiple R and R^2	No	No	No	No	Yes	Yes
Pooling specified interactions with error	Yes	N.A.	No	Yes	Yes	Yes
Correlation matrix for regression coefficients	No	Yes	No	No	No	No
Estimates of slopes within groups (regression coefficients)	No	Yes	No	No	No	Yes
Observed maximums and minimums	Yes	Yes	No	No	No	No
Correlation matrix for adjusted group means	No	Yes	No	No	No	No
Number of cases per group	Yes	Yes	N.A.	Yes	No	Yes
Correlation matrix for all covariates and DV (by groups)	No	Yes	No	No	No	Yes
Pooled-within-cell correlation matrix	No	No	No	No	No	Yes
Covariance matrix for all covariates and DV (by groups)	No	Yes	No	No	No	Yes
Test homogeneity of variance	No	No	No	No	Yes[e]	Yes
Test for homogeneity of covariance	Sphericity indices	N.A.	No	No	No	Yes
Predicted values and residuals	No	No	No	Yes	No	No
User-specified tolerance	No	No	No	Yes	No	Yes
Greenhouse-Geissen-Imhof statistic	Yes	No	No	Yes[f]	No	No

[a] Additional features described in Chapter 8 (MANOVA). Based on Release 9.0 (Hull & Nie, 1981).
[b] F must be recalculated. See Section 7.5.2.1.
[c] Can be done through SPSS REGRESSION.
[d] Pooled-within-cells matrix only.
[e] Through subprogram ONEWAY only.
[f] Most recent releases.

as BMD11V). BMDP4V, described more fully in Section 8.6, is more comprehensive but is also more difficult to use. It allows a variety of methods for dealing with unequal n. In addition, specific contrasts can be made among means.

The remaining BMD program, BMD12V, is extremely easy to use but far less flexible. Factorial designs are acceptable, including those with within-subjects variables, but output is misleading with respect to error terms. F ratios are inappropriately calculated for factorial designs that include within-subjects IVs and must be recalculated by hand, as discussed in Section 7.5.2.1. The program requires equal sample sizes in all cells of the design, and does not provide tests for assumptions. Nor does it provide specific comparisons, but information is available to simplify hand calculation of these tests, as described in Section 7.5.2.3.

7.6.2 SPSS Package

The SPSS program designed directly for ANCOVA is ANOVA—that is, the analysis of variance program includes analysis of covariance. Features of the program are illustrated in Table 7.11.

This program has nowhere near the flexibility of BMDP2V, but with between-subjects designs and an assist from SPSS REGRESSION, a fairly large subset of ANCOVA problems can be analyzed. The one major restriction is the limitation to between-subjects IVs. Beyond that, the program handles unequal sample sizes well, allowing the user to specify the method for adjustment in nonorthogonal designs (see Section 7.5.2.2). Tests for assumptions and specific comparisons, including trend analysis, cannot be accomplished directly through SPSS ANOVA, but with some effort on the part of the user they can be set up through SPSS REGRESSION. Procedures for testing homogeneity of regression (equality of slopes or IV by covariate interaction) with this program are discussed in Section 7.5.1.2. Specific comparisons require knowledge of contrast coding beyond a level described in the manual (cf. Cohen & Cohen, 1975). Another major limitation is lack of provision for adjusted cell means, although marginal means are available from the marginal deviations (see Section 7.5.2.3).

SPSS MANOVA, though designed for the more general case of multivariate analysis of variance and covariance, can also be used for univariate ANCOVA. Because the program is so flexible, it may take more time to master than SPSS ANOVA, but is well worth the effort. SPSS MANOVA can be used for nonorthogonal and complex designs, and can provide tests for homogeneity of regression as well as adjusted cell and marginal means. Specific comparisons and trend analysis are readily available through the program.

7.7 COMPLETE EXAMPLE OF ANALYSIS OF COVARIANCE

For the large sample illustration of the use of analysis of covariance, the research described in Appendix B again provides the data. The question to be addressed in this example is whether attitudes toward medication among women are affected by current employment status and/or religious affiliation.

In examining other data available for this sample of women, three variables stand out that could be expected to relate to attitudes toward medication and might obscure effects of employment status and religion. These variables are general state of physical health, mental health, and the use of psychotropic drugs. In order to control for the effects of these three variables on attitudes toward medication, they are treated as covariates. Attitude toward medication (ATTMED) then serves as the DV, with increasingly high scores reflecting more favorable attitudes. The two IVs, factorially combined, are current employment status (CUEMPST) with two levels, employed and unemployed; and religious affiliation (RELIGION) with four levels. The categories of RELIGION are (1) None-or-other, (2) Catholic, (3) Protestant, and (4) Jewish. Covariates are physical health (PHYHEALT), mental health (MENHEALT), and sum of all psychotropic drug uses, prescription and over-the-counter (ALLSUM). For PHYHEALT and MENHEALT larger scores reflect increasingly poor health. The 2 × 4 analysis of covariance, then, provides a test of the effects of employment status and religion on attitudes toward medication, holding constant physical health, mental health, and use of psychotropic drugs. Note that this is a form of ANCOVA in which no causal inference can be made.

7.7.1 Evaluation of Assumptions

With respect to these variables, practical limitations of the analysis of covariance technique can be assessed, as described in Section 7.3.2.

7.7.1.1 Unequal *n* and Missing Data
Three women of the original sample of 465 failed to provide information on religious affiliation. Since this formed one of the IVs, for which cell sizes were unequal in any event, these three cases were dropped from analysis. The cell-size approach (Method 2 of Section 7.5.2.2) to dealing with unequal *n* was chosen for this study. This method, it will be recalled, weights cells by their sample size, which in this study represents relative population sizes for the groups as well.

7.7.1.2 Outliers
As an initial screening for univariate outliers, BMDP7D was used to look at histograms as well as minimum and maximum values within each cell for the DV and covariates. Although no outliers were evident for the DV, several cases were clearly outliers for two of the covariates, PHYHEALT and ALLSUM. Histograms and descriptive statistics for ALLSUM within each group, as printed out by BMDP7D, are shown in Table 7.12. On the basis of this output, as well as indication of significant skewness through SPSS CON-DESCRIPTIVE (overall skewness = 2.07 for ALLSUM and 1.03 for PHY-HEALT), transformations were performed. LALLSUM was created as the logarithm of ALLSUM (incremented by 1 since many of the values were at zero)

TABLE 7.12   INPUT AND PARTIAL OUTPUT OF SCREENING RUNS FOR
UNIVARIATE OUTLIERS ON BMDP7D

```
PROGRAM CONTROL INFORMATION

        /PROBLEM      TITLE IS "LARGE SAMPLE ANCOVA, OUTLIERS".
        /INPUT        VARIABLES ARE 6.
                      FORMAT IS "(6F4.0)".
                      UNIT=75.
        /VARIABLE     NAMES ARE ATTMED,PHYHEALT,MENHEALT,ALLSUM,CUEMPST,RELIGION,
                      LALLSUM,SPHYHEAL.  ADD=2.
        /TRANSFORM    IF (RELIGION EQ 9) THEN USE=0.
                      LALLSUM=LOG (ALLSUM+1).
                      SPHYHEAL=SQRT (PHYHEALT).
        /GROUP        CODES(5) ARE 1, 2.
                      NAMES(5) ARE EMPLOYED, UNEMPLOYED.
                      CODES(6) ARE 1 TO 4.
                      NAMES(6) ARE NONOTHER, CATHOLIC, PROTESTANT,JEWISH.
        /HISTOGRAM    GROUPING IS CUEMPST,RELIGION.
        /END
```

```
              ************                                          ************
HISTOGRAM OF * ALLSUM  * (VARIABLE    4). CASES DIVIDED INTO GROUPS BASED ON VALUES OF * CUEMPST  * (VARIABLE   5)
              ************                                          * RELIGION * (VARIABLE   6)
                                                                    ************
                                                                              CASES WITH
                                                                               UNUSED
                                                                              VALUES FOR
                                                                              CUEMPST
        EMPLOYED    EMPLOYED    EMPLOYED    EMPLOYED    UNEMPLOY    UNEMPLOY    UNEMPLOY    UNEMPLOY

        NONOTHER    CATHOLIC    PROTESTA    JEWISH      NONOTHER    CATHOLIC    PROTESTA    JEWISH      RELIGION
        ..........+..........+..........+..........+..........+..........+..........+..........+..........+..........+
MIDPOINTS
   48.000)
   46.000)
   44.000)
   42.000)                                                                     *
   40.000)
   38.000)
   36.000)
   34.000)                                                                     *
   32.000)*
   30.000)                                                                     *
   28.000)
   26.000)                             **                                     **
   24.000)*                                                 *                          **
   22.000)*              *                                  *                  *
   20.000)               *                                  *
   18.000)                                         *                          **
   16.000)**             ***         ***           *                 **       ****
   14.000)***            *****       *                       **       *        *
   12.000)*              *           *             *         **       *        **
   10.000)*     ***      ********    **            *         ***      *****    ***
    8.000)****  **       ****        ***           **        ****     *******  *
    6.000)M**   ******   *******     ****          *         ***      M*****   M*
    4.000)**    ****     M****       M*            *         M        ***      ***
    2.000)****  M********* *******   **            M***      ***      *****    ******
    0.000)*********23 *********36 *********50 ********24 *********20 ********37 ********47 *******21
   -2.000)
GROUP MEANS ARE DENOTED BY M"S IF THEY COINCIDE WITH *"S, N"S OTHERWISE

MEAN         5.348      2.349      4.185      4.932      2.067      3.232      5.096      6.083
STD.DEV.     7.657      3.413      5.736      7.212      3.629      5.579      8.641      7.582
S. E. M.     1.129       .430       .598      1.087       .663       .746       .949      1.094
MAXIMUM     32.000     14.000     23.000     27.000     13.000     25.000     43.000     25.000
MINIMUM      0.000      0.000      0.000      0.000      0.000      0.000      0.000      0.000
SAMPLE SIZE     46         63         92         44         30         56         83         48
```

and SPHYHEAL was created as the square root of PHYHEALT.[12] After
transformation, all cases were within about three standard deviations of their
group means.

The three covariates (two of them transformed) were then tested in
BMDPAM for multivariate outliers. In this program, outliers are defined as

---

[12] Note that this transformation, made on the basis of within-cell distributions, is not as
strong as the log transform of PHYHEALT required for the large sample regression and canonical
analyses of Chapters 5 and 6.

TABLE 7.13   INPUT AND PARTIAL OUTPUT OF TEST FOR
                 MULTIVARIATE OUTLIERS ON BMDPAM

```
PROGRAM CONTROL INFORMATION

    /PROBLEM   TITLE IS "LARGE SAMPLE ANCOVA, MULTIVARIATE OUTLIERS, BMDPAM".
    /INPUT           VARIABLES ARE 6.
                     FORMAT IS "(6F4.0)".
                     UNIT=75.
    /VARIABLE        NAMES ARE ATTMED,PHYHEALT,MENHEALT,ALLSUM,CUEMPST,RELIGION,
                     GROUP,LALLSUM,SPHYHEAL.   ADD=3.
                     GROUPING IS GROUP.    USE=3,7,8,9.
    /TRANSFORM       IF (CUEMPST EQ 1 AND RELIGION EQ 1) THEN GROUP=1.
                     IF (CUEMPST EQ 2 AND RELIGION EQ 1) THEN GROUP=2.
                     IF (CUEMPST EQ 1 AND RELIGION EQ 2) THEN GROUP=3.
                     IF (CUEMPST EQ 2 AND RELIGION EQ 2) THEN GROUP=4.
                     IF (CUEMPST EQ 1 AND RELIGION EQ 3) THEN GROUP=5.
                     IF (CUEMPST EQ 2 AND RELIGION EQ 3) THEN GROUP=6.
                     IF (CUEMPST EQ 1 AND RELIGION EQ 4) THEN GROUP=7.
                     IF (CUEMPST EQ 2 AND RELIGION EQ 4) THEN GROUP=8.
                     IF (RELIGION EQ 9) THEN USE=0.
                     LALLSUM=LOG(ALLSUM+1).
                     SPHYHEAL=SQRT(PHYHEALT).
    /GROUP           CODES(7) ARE 1 TO 8.
    /EST             METHOD=REGR.
    /PRINT           MATR=DIS.
    /END
```

ESTIMATES OF MISSING DATA, MAHALANOBIS D-SQUARED (CHI-SQUARED)
AND SQUARED MULTIPLE CORRELATIONS WITH AVAILABLE VARIABLES

CASE LABEL	CASE NUMBER	MISSING VARIABLE	ESTIMATE	R-SQUARED	GROUP	CHI-SQ	CHISQ/DF	D.F.	SIGNIFICANCE
	51				* 5.0000	2.350	.783	3	.5030
	52				* 4.0000	8.190	2.730	3	.0422
	53				* 5.0000	3.598	1.199	3	.3083
	54				* 8.0000	2.779	.926	3	.4270
	55				* 6.0000	2.682	.894	3	.4433
	56				* 3.0000	1.165	.388	3	.7613
	57				* 3.0000	3.186	1.062	3	.3638
	58				* 5.0000	2.398	.799	3	.4940
	59				* 3.0000	2.573	.858	3	.4622
	60				* 5.0000	2.839	.946	3	.4171
	61				* 3.0000	1.515	.505	3	.6788
	62				* 1.0000	2.467	.822	3	.4813
	63				* 3.0000	.532	.177	3	.9119
	64				* 5.0000	12.656	4.219	3	.0054
	65				* 1.0000	7.501	2.500	3	.0575
	66				* 5.0000	2.526	.842	3	.4706
	67				* 3.0000	.546	.182	3	.9088

cases with extreme Mahalanobis distance from their groups. A portion of the
output is displayed in Table 7.13. (Notice that groups have been transformed
to create a one-way design, necessary for use of BMDPAM.) The Mahalanobis
distance is evaluated as χ^2 with degrees of freedom equal to the number of
variables (in this case the three covariates). BMDPAM prints significance values
for each case.

At $\alpha = .01$, the first outlier to be encountered was case 64 in group 5,
with $\chi^2 = 12.66$ for her group, as seen in Table 7.13. Only one additional
outlier was found, in group 8. The use of BMDP9R and BMDP1D, as described
in Section 5.8.1, revealed characteristics of the two outlying cases. Elimination
of the two outliers brings the overall $N = 460$.

7.7.1.3 Normality   With data transformation and deletion of outliers, as well as the large sample size and use of two-tailed tests, there was every reason to expect normality of sampling distributions of means.

7.7.1.4 Multicollinearity and Singularity   To test for multicollinearity and singularity among covariates, SPSS REGRESSION was run with each covariate as DV, and the remaining covariates (transformed as needed) as IVs. Since the largest $R^2 = .29$, no danger of multicollinearity or singularity was evident.

7.7.1.5 Homogeneity of Variance   Examination of the sample variances revealed little difference among the eight groups. The ratio of largest to smallest variance is 1.78 to 1 after transformation and deletion of outliers. With the use of two-tailed tests, with ratio of largest to smallest sample size less than 4 to 1, and with absence of outliers, robustness was assured and no formal test of homogeneity seemed necessary.

7.7.1.6 Linearity   Plots of residuals against predicted scores on the DV, ATTMED, were examined for each group, as produced by BMDP1V in the same run used for the test of homogeneity of regression. As an example, the residuals plots for group 6 are shown in Figure 7.3, along with input showing the setup for BMDP1V with groups recoded into a one-way array. There was no notable curvilinearity in any of the residuals plots, so individual scatterplots were not examined.

7.7.1.7 Homogeneity of Regression   Since a between-subjects design was being used, a test of homogeneity of regression was readily available through canned computer programs, as discussed in Section 7.5.1. For these data, the eight groups were recoded as shown in Figure 7.3, and analyzed through BMDP1V as a one-way analysis of covariance. Output is in the format of Table 7.5. The obtained $F(21, 430) = 0.984$ for equality of slopes indicates that regression coefficients for the eight groups are homogeneous at the .05 level of significance.

7.7.1.8 Reliability of Covariates   The three covariates, MENHEALT, PHYHEALT, and ALLSUM, were simply measured on the basis of counts of symptoms or drug use—"have you ever . . . ?" It was assumed that the nature of these variables was such that high reliability could be expected. Therefore no adjustment in ANCOVA was made for unreliability of covariates.

7.7.2 Analysis of Covariance

The program chosen for the major two-way analysis of covariance was SPSS ANOVA. The approach used by the program as default, Method 2, is appropriate for this data set (see Table 7.7). Ease of use, then, makes it the most convenient one available for unequal-n data.

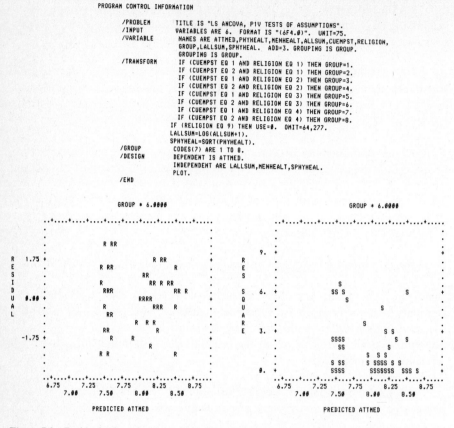

Figure 7.3   Residuals plots showing relationship between predicted scores in the DV (ATT-MED) and errors of prediction. Input and partial output from BMDP1V.

Selected output from application of SPSS ANOVA to these data appears in Table 7.14. Included are printouts of unadjusted cell means and sizes, the source table, and adjusted marginal deviations from which marginal means can be computed (cf. Section 7.5.2.3). Sources of variance are the three covariates, individually and combined; the main effects for CUEMPST and RELIGION, individually and combined; the interaction; and the error term shown as RESI-DUAL. For each computed F, significance level is provided so that it is unnecessary to consult a table of critical F. Any SIGNIF OF F less than .05 is statistically significant with $\alpha = .05$, so that for our example the main effect of RELIGION reaches statistical significance.

Entries in the sum of squares column of the source table can be used to calculate η^2 as a measure of strength of association for significant main effects and interactions (Eq. 7.10 or 7.11). Since there is only one significant effect, the more standard equation, Eq. 7.10, is preferable.

TABLE 7.14   SELECTED INPUT AND OUTPUT OF APPLICATION OF SPSS
ANALYSIS OF COVARIANCE TO LARGE SAMPLE DATA

```
SELECT IF      (NOT(SEQNUM EQ 64 OR 277))
SELECT IF      (RELIGION NE 9)
COMPUTE        LALLSUM=LG1Ø(ALLSUM+1)
COMPUTE        SPHYHEAL=SQRT(PHYHEALT)
ANOVA          ATTMED BY CUEMPST(1,2), RELIGION(1,4) WITH
               LALLSUM, MENHEALT, SPHYHEAL
STATISTICS     ALL
```

```
* * * * * * * * * * * *    C E L L   M E A N S   * * * * * * * * * * * *
              ATTMED
           BY CUEMPST
              RELIGION
* * * * * * * * * * * * * * * * * * * * * * * * * * * * * * * * * * * * *


TOTAL POPULATION

    7.68
 (    46Ø)


CUEMPST
        1          2

     7.6Ø      7.78
 (   244) (    216)


RELIGION
        1          2          3          4

     7.45      7.84      7.68      7.69
 (    76) (   119) (   174) (    91)



          RELIGION
             1          2          3          4
CUEMPST
        1    7.67      7.67      7.53      7.59
         (   46) (    63) (    91) (    44)

        2    7.1Ø      8.Ø4      7.84      7.79
         (   3Ø) (    56) (    83) (    47)
```

TABLE 7.14 (*Continued*)

```
* * * * * * * *  A N A L Y S I S   O F   V A R I A N C E  * * * * * * * *
                ATTMED
             BY CUEMPST
                RELIGION
           WITH LALLSUM
                MENHEALT
                SPHYHEAL
* * * * * * * * * * * * * * * * * * * * * * * * * * * * * * * * * * * * * *
```

SOURCE OF VARIATION	SUM OF SQUARES	DF	MEAN SQUARE	F	SIGNIF OF F
COVARIATES	55.865	3	18.622	15.626	.001
LALLSUM	45.199	1	45.199	37.928	.001
MENHEALT	1.002	1	1.002	.841	.360
SPHYHEAL	.647	1	.647	.543	.462
MAIN EFFECTS	13.559	4	3.390	2.844	.024
CUEMPST	3.291	1	3.291	2.762	.097
RELIGION	9.914	3	3.305	2.773	.041
2-WAY INTERACTIONS	8.796	3	2.932	2.460	.062
CUEMPST RELIGION	8.796	3	2.932	2.460	.062
EXPLAINED	78.221	10	7.822	6.564	.001
RESIDUAL	535.073	449	1.192		
TOTAL	613.293	459	1.336		

COVARIATE	REGRESSION COEFFICIENT ADJUSTED FOR ALL OTHER COVARIATES
LALLSUM	.691
MENHEALT	-.013
SPHYHEAL	.090

TABLE 7.14 (*Continued*)

```
* *  M U L T I P L E   C L A S S I F I C A T I O N   A N A L Y S I S  * *
            ATTMED
      BY    CUEMPST
            RELIGION
     WITH LALLSUM
            MENHEALT
            SPHYHEAL
* * * * * * * * * * * * * * * * * * * * * * * * * * * * * * * * * * * *
```

				ADJUSTED FOR		ADJUSTED FOR INDEPENDENTS	
GRAND MEAN =	7.68		UNADJUSTED	INDEPENDENTS		+ COVARIATES	
VARIABLE + CATEGORY		N	DEV"N ETA	DEV"N BETA		DEV"N BETA	

```
CUEMPST
   1                        244    -.08                     -.08
   2                        216     .09                      .09
                                         .08                      .07


RELIGION
   1                         76    -.24                     -.22
   2                        119     .16                      .22
   3                        174    -.01                     -.03
   4                         91     .01                     -.06
                                         .11                      .13


MULTIPLE R SQUARED                                          .113
MULTIPLE R                                                  .336
```

$$\eta^2 = \frac{9.914}{(3.291 + 9.914 + 8.796) + 535.073} = .0178$$

Significance tests of covariates, along with a table of intercorrelations among covariates and the DV, ATTMED, can be used to assess the utility of the covariates, as described in Section 7.5.3. Since covariates enter the equation first in Method 2 as reported by SPSS, covariates are adjusted only for each other in predicting the DV.

Following the source table in SPSS ANOVA, adjusted marginal deviations for the DV, ATTMED, are given. Since only a main effect of RELIGION was found, a journal report would provide adjusted marginal means for the four religion groups. (Had an interaction been found, it would have been necessary to enlist one of the BMD programs or SPSS MANOVA to obtain adjusted cell means.)

To find the adjusted marginal mean for the first religion group, None-or-other, the deviation (−.22) is algebraically added to the grand mean (7.68):

$$7.68 + (-.22) = 7.46$$

TABLE 7.15 ADJUSTED AND UNADJUSTED MEAN
ATTITUDE TOWARD MEDICATION
FOR FOUR LEVELS OF RELIGION

Religion	Adjusted mean	Unadjusted mean
None-or-other	7.46	7.44
Catholic	7.90	7.84
Protestant	7.65	7.67
Jewish	7.62	7.69

Application of this procedure to the other three religion groups results in the adjusted marginal means shown in Table 7.15. The table also includes the unadjusted means for comparison. Note a slight difference in unadjusted means depending on whether they are copied from CELL MEANS output or computed from DEV-N output of Table 7.14.

Since no a priori hypotheses about differences among religion groups were generated, planned comparisons are not appropriate. A glance at the four adjusted means in Table 7.15 suggests a straightforward interpretation, so in the absence of specific questions about differences between means, there seems to be no compelling reason to take a post hoc look at all possible contrasts.

A checklist for analysis of covariance appears as Table 7.16.

TABLE 7.16 CHECKLIST FOR ANALYSIS OF COVARIANCE

1. Issues
   a. Unequal sample size and missing data
   b. Within-cell outliers
   c. Normality
   d. Homogeneity of variance
   e. Within-cell linearity
   f. Homogeneity of regression
   g. Reliability of covariates
2. Major analyses
   a. Main effect(s) or planned comparison. If significant:
      (1) Adjusted marginal means
      (2) Strength of association
   b. Interactions or planned comparisons. If significant:
      (1) Adjusted cell means (in table or interaction graph)
      (2) Strength of association
3. Additional analyses
   a. Evaluation of covariate effects
   b. Evaluation of intercorrelations
   c. Post hoc comparisons (if appropriate)
   d. Unadjusted marginal and/or cell means (if significant main effect and/or interaction) if nonexperimental application

An example of a Results section, in journal format, follows for the analysis described above.

RESULTS

A 2 × 4 between-groups analysis of covariance was performed on attitude toward medication. Independent variables consisted of two levels of current employment status (employed and unemployed) and four levels of religious identification (None-or-other, Catholic, Protestant, and Jewish), factorially combined. Covariates were physical health, mental health, and the sum of psychotropic drug uses. Analyses were performed by SPSS ANOVA, using default strategy, that is, weighting cells by their sample sizes in order to deal with unequal-cell sample sizes.

Results of the evaluations of the assumptions of linearity, homogeneity of variance, homogeneity of regression, and reliability of covariates were satisfactory. Presence of outliers and skewness led to transformation of two of the covariates. A logarithmic transform was made of the sum of psychotropic drug uses, and the square root of physical health was used. After transformation, two cases remained outliers within their group and were eliminated from subsequent analysis, resulting in total $N = 460$. Both cases were characterized by highest scores for their groups on physical health (after square root transform) and lowest possible scores on the sum of psychotropic drug use (logarithmically transformed).

After adjustment by linear components of transformed covariates, attitude toward medication varied significantly with religious identification, as demonstrated in Table 7.17, with $F(3, 449) = 2.77$, $p < .05$. Examination of adjusted marginal means, as displayed in Table 7.15, reveals that most favorable attitudes toward medication were associated with Catholic women, and least favorable attitudes with women who either were unaffiliated with a religion or identified with some religion other than the major three. Attitudes among Protestant and Jewish women were almost identical on the average for this sample, and fell between those of the two other groups. The strength of the relationship between adjusted attitudes toward medication and religion is weak, however, with $\eta^2 = .02$.

Insert Tables 7.15 and 7.17 about here

No statistically significant main effect of current employment status was found. Nor was there a significant interaction between employment status and religion in affecting attitude toward medication among women, after adjustment for covariates.

Intercorrelations among covariates and attitude toward medication are shown in Table 7.18. Two of the covariates, square root of physical health and logarithm of drug use, are significantly associated with the dependent variable. However, only the latter covariate significantly accounted for adjustment of the attitude scores, $F(1, 449) = 37.928$, $p < .01$. The remaining two covariates, mental health and square root of physical health, provided no additional adjustment. Neither did they add to predictability of attitude scores beyond that of drug uses. [Correlations obtained through SPSS PEARSON CORR; output not shown.]

Insert Table 7.18 about here

TABLE 7.17 ANALYSIS OF COVARIANCE OF ATTITUDE TOWARD
           MEDICATION

Source of Variance	Adjusted SS	df	MS	F
Current employment status	3.291	1	3.291	2.762
Religion	9.914	3	3.305	2.773*
Interaction	8.796	3	2.932	2.460
Covariates	55.865	3	18.622	15.626
Physical health (SQRT)	0.647	1	0.647	0.543
Mental health	1.002	1	1.002	0.841
Drug uses (LOG)	45.199	1	45.199	37.928**
Error	535.073	449	1.192	

* $p < .05$.

** $p < .01$.

TABLE 7.18 INTERCORRELATIONS AMONG THREE COVARIATES
           AND THE DEPENDENT VARIABLE, ATTITUDE TOWARD
           MEDICATION

	Physical health (SQRT)	Mental health	Drug uses (LOG)
Attitude toward medication	.1317*	.0736	.2987*
Physical health (SQRT)		.5180*	.3913*
Mental health			.3380*

* $p < .01$.

7.8 SOME EXAMPLES FROM THE LITERATURE

A straightforward, classic experimental application of ANCOVA is demonstrated in a study by Hall and colleagues (1977). Obese members of a weight
reduction club were randomly assigned to five experimental treatments: self-
management, external management, self-management plus external management,
psychotherapy, and no-treatment control. Using pretreatment weight as a covariate, a one-way ANCOVA was done on posttreatment weight. After a 10-week
treatment period, the groups differed significantly in weight after adjustment
for the covariate. A Tukey HSD test showed that the three behavioral conditions
did not differ significantly from one another, but did differ from the two remaining
conditions. The remaining conditions, psychotherapy and no-treatment control,
did not differ from each other. ANCOVAs performed on weight at 3- and
6-month follow-ups showed that differences among groups disappeared.

Merrill and Towle (1976) investigated the effects of the availability of
objectives in a graduate course in programmed instruction. Analyses of covariance were used to evaluate the effect of presence versus absence of course objec-

tives on student performance and anxiety level. The course consisted of 12 cognitive units. For the first half of the course, consisting of six units, half of the 32 students received objectives and half did not.

Using pretest scores on the material as the covariate, no significant effect of course objectives was found on posttest performance (evaluated as the scores on tests for the first six units and the review test). For evaluation of anxiety level, the two covariates were A-Trait and A-State scales of the State-Trait Anxiety Inventory administered to graduate students during the first class session. The DV was the average of the short-form A-State scales administered after each unit test for the first six units. Availability of objectives significantly decreased A-State scores after adjustment for covariates. Separate analyses of covariance were then done on the A-State scores for each of the first six units.[13] These analyses showed that anxiety was significantly reduced only for the first three units in which objectives were given.

O'Kane and co-workers (1977) studied the relationship between anticipated social mobility and resultant social and political ideology. A total of 307 male Catholic adolescents were divided into four groups. They were first separated into either middle or working class on the basis of the Duncan Index score of the heads of households. These two groups were then further subdivided on the basis of their own predicted future class, either same as class of origin or different—upward or downward depending on class of origin. The four groups, then, were middle-middle, working-middle, working-working, and middle-working. IQ served as the covariate for analyses of covariance on three measures of ideology: economic liberalism, noneconomic liberalism, and ethnocentrism.[14]

For economic liberalism, no significant effects were found for class of origin, class of destination, or the interaction between origin and destination classes. For both noneconomic liberalism and ethnocentrism, significant main effects were found for class of destination. Those who expect to join (or remain in) the working class show more conservative scores on the scale of noneconomic liberalism, and score higher in ethnocentrism, than those who expect to join (or remain in) the middle class. For these two measures, no significant effects of class of origin were found, nor was there a significant interaction between class of origin and class of destination. Additional *t* tests, presumably on the basis of planned comparisons, showed that the two stable groups differed from each other, in the expected direction, on measures of noneconomic liberalism and ethnocentrism, but not on economic liberalism.

[13] This is a questionable practice, considering the inflation of Type I error inherent in multiple testing, especially when the DVs are correlated. The stepdown analysis described in Chapter 8 would be more appropriate. Or, if assumptions are met, "unit" could act as a six-level within-subjects IV in a mixed within-between analysis of variance.

[14] Since the three measures are probably correlated, a better strategy would have been the use of multivariate analysis of covariance, followed by stepdown analysis, as discussed in Chapter 8.

Multivariate Analysis of Variance and Covariance

8.1 GENERAL PURPOSE AND DESCRIPTION

Multivariate analysis of variance is a generalization of analysis of variance to a situation in which there are several DVs. By measuring several DVs instead of only one, a researcher may improve the chance of discovering changes produced by different treatments and interactions—at the expense, however, of increased complexity of analysis. An advantage of MANOVA over a series of ANOVAs, one for each DV, is in protection against Type I error (see Chapter 2). But this advantage is seen only when a two-tailed significance test is appropriate. If a one-tailed test is desired, use of MANOVA may result in an unacceptable loss of power.

Another advantage to MANOVA is that, under certain conditions, it may reveal differences not shown in separate ANOVAs. Such a situation is shown in Figure 8.1 for a two-group design. In this figure, the axes represent frequency distributions for each of two DVs, Y_1 and Y_2. Notice that from the point of view of either axis, the distributions are sufficiently overlapping that a mean difference might not be found in ANOVA. The ellipses in the quadrant, however, represent the *joint* distribution of Y_1 and Y_2 for each group separately. When responses to two DVs are considered in combination, group differences become apparent. Thus MANOVA, which considers DVs in combination, may sometimes be more powerful than separate ANOVAs.

It can be recalled from Chapter 2 that the purpose of analysis of variance is to test whether mean differences among groups are likely to have occurred by chance. That is, do changes in the IVs (groups) produce reliable changes in the DV means? In MANOVA, the question is whether the IVs (group assignments) significantly affect an optimal linear combination of DV means (optimal

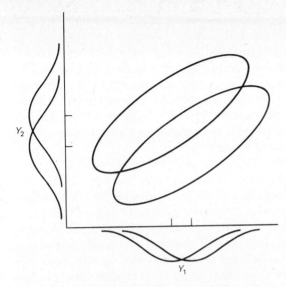

Figure 8.1 Advantage of MANOVA, which combines DVs, over ANOVA. Each axis represents a DV; frequency distributions projected to axes show considerable overlap, while ellipses, showing DVs in combination, do not.

in the sense that group differences are maximized). From one view, in MANOVA a new DV is formed, based on the best linear combination of all DVs. An ANOVA is then performed on the newly created DV. As in ANOVA, hypotheses about means are tested by comparing variances—hence multivariate analysis of *variance*.

In factorial or more complicated MANOVA, a different "best linear combination" of DVs is formed for each main effect and interaction. The combination of DVs that best separates the groups of the first main effect may be different from the combination that best separates the groups of the second main effect or the cells from an interaction. Further, the optimal combination of DVs for a single main effect (or interaction) may be further subdivided into separate dimensions, as many dimensions as there are degrees of freedom in the main effect (or interaction) or DVs, whichever is the smaller.

As an example of a one-way MANOVA, let's say that we are interested in the effect of different treatment procedures on a variety of measures of anxiety. Using the example of the preceding chapter, suppose we were interested in reducing not only test anxiety but also anxiety in reaction to minor life stresses as well as so-called free-floating anxiety. The IV would still consist of three groups: desensitization, relaxation training, and a waiting-list control. After a specified period of treatment time, all three groups would be given the three dependent measures: test anxiety, stress anxiety, and free-floating anxiety. The statistical goal of MANOVA would be to test whether anxiety, composed of the three test scores, varied as a function of differential treatment.

Similarly, multivariate analysis of covariance (MANCOVA) is the multivariate extension of analysis of covariance. That is, after statistically adjusting for one or more covariates, are there significant differences among groups in the best linear combination of DV means?

To provide a more sensitive measure of the effect of treatment on anxiety, three covariates could be added to the preceding example. Before the treatment period, volunteers in all three treatment groups could be pretested on the three measures of anxiety. The statistical test would now investigate whether anxiety (as a composite) differed with differential treatment, after adjusting for preexisting differences in the three types of anxiety.

Although computing procedures and programs for MANOVA and MANCOVA are not so well developed as for ANOVA and ANCOVA, there is in theory no limit to the generalization of the model, despite the complications that do arise. There is no reason why all types of analysis of variance designs—one-way, factorial, mixed within-between, nonorthogonal, and so on—cannot be extended to research with several DVs. All questions of specific comparisons and trend analysis are equally interesting with MANOVA. In addition, there is the question of importance of DVs—that is, which DVs most clearly reflect the behavior or behaviors affected by the IVs.

MANOVA techniques have been developed within the context of experimental research, in the tradition of analysis of variance. But it should be noted that there is no mathematical distinction between MANOVA and discriminant function analysis, to be covered in Chapter 9. Indeed, one of the BMDP programs to be discussed in detail (in both chapters) applies equally to MANOVA and discriminant function analysis. Traditionally, however, discriminant function analysis was developed in the context of nonexperimental research, and the question of differences between groups is phrased somewhat differently: How can the DVs be linearly combined in a way that best discriminates between groups? Typically, MANOVA procedures are applied to experimental situations in which all, or at least some, IVs are manipulated as part of the research design. Subjects are randomly assigned to groups, usually in such a way as to achieve equal sample sizes within groups. Discriminant function analysis, on the other hand, is typically used in situations in which groups are formed naturally, and they would not necessarily be expected to be of equal size.

Because computer programs for discriminant analysis are more elaborate and informative and, for the most part, are limited to one-way designs, the analysis on one-way MANOVA will be deferred to Chapter 9. The present chapter will be limited to factorial MANOVA and MANCOVA.

Multivariate analysis of covariance is useful in the three ways discussed for ANCOVA in Chapter 7. First, it can be seen as a noise-reducing device in an experiment, where variance associated with the covariate(s) is removed from error variance. This reduced error variance, provides a more powerful test of differences among IV groups.

The second use of MANCOVA involves statistical matching between IV groups when random assignment to these groups has not been possible. In this case, prior differences among IV groups are diminished by equating all subjects on covariate scores. (But review Chapter 7 for a discussion of the logical difficulties of using covariates this way.)

The third use of MANCOVA has a quite different purpose. It is to test whether, after adjusting for the effects of some of the DVs, there are any significant group differences in the remaining DVs. That is, do changes on the remaining DVs depend on prior DVs rather than IVs? In this sense, MANCOVA, like ANCOVA, can be used as a tool in interpreting MANOVA results.

MANOVA follows the model of analysis of variance, in that variance among scores is partitioned into variance attributable to differences in scores within groups and into variance attributable to differences among groups (on one or more dimensions, or factors). To do this in ANOVA, squared differences between scores and various means are summed (see Chapter 2), and these sums of squares, when divided by appropriate degrees of freedom, provide estimates of variance attributable to different sources (main effects of IVs, interactions among IVs, and error). Ratios of variances then provide tests of hypotheses about the effects of IVs on the DV.

In MANOVA, however, there is a set of DVs. That is, each subject has a score on each of several DVs. When different subjects as well as different DVs are considered, a *matrix* of scores is formed, rather than a simple set of scores, within each group. Matrices of difference scores are formed by subtracting from each score an appropriate mean, squaring by multiplying the matrix by its transpose, and then summing. The analog of a sum of squares, then, is a sum-of-squares and cross-products or cross-products matrix (**S** matrix). Determinants[1] of the various **S** matrices are found and ratios between them provide tests of hypotheses about the effects of IVs on the linear combination of DVs.

In MANCOVA, the sums of squares and cross products in the **S** matrix are adjusted for intercorrelations with the covariates, just as sums of squares are adjusted in ANCOVA (Chapter 7).

8.2 KINDS OF RESEARCH QUESTIONS

The goal of research using MANOVA is to discover whether behavior, as reflected by the DVs, is changed by manipulation (or other action) of the IVs. Making this discovery, of course, will depend on the choice of DVs. To the extent that DVs are uncorrelated, it is possible to discover a variety of behaviors that may be affected by the IVs. To the extent that DVs are correlated, the various DVs can be seen as measuring the same or similar facets of behavior in slightly different ways.

Choice of number and composition of DVs should be a practical, research decision, rather than a statistical one. Now that MANOVA techniques are accessible for more elaborate research designs through recent programs by both

---

[1] A determinant, as described in Appendix A, can be viewed as a measure of generalized variance of a matrix.

SPSS (MANOVA) and BMDP4V, it is likely that designs with large numbers of interrelated DVs will become more common.

Statistical techniques are currently available for answering the types of questions posed in Sections 8.2.1 through 8.2.8.

8.2.1 Main Effects of IVs

Holding all else constant, does differing assignment to various levels of a particular IV change behavior? In other words, are changes in behavior associated with different levels of an IV due to something other than random fluctuations or individual differences occurring by chance? The statistical procedures described in Sections 8.4.1 and 8.4.2 are designed to answer this question, by testing the null hypothesis that the IV has no systematic effect on the optimal linear combination of DVs.

As in ANOVA "holding all else constant" can refer to a variety of devices: (1) All other IVs are ignored by "crossing over" them. That is, data are analyzed as if they were produced by a one-way design, with any other factors ignored or adjusted for. (2) Control procedures in designing the research serve to hold extraneous variables constant either by setting them at a given level (e.g., running females only as subjects) or by allowing them to vary randomly. (3) Covariates can be used to produce an "as if constant" state by statistically adjusting for differences that may exist.

In the anxiety-reduction example of Section 8.1, the omnibus test answers this major question: Are there differences in anxiety—measured by test anxiety, stress anxiety, and free-floating anxiety—associated with differences in treatment? With the addition of covariates, the question becomes: Are there differences in anxiety associated with treatment, after adjusting for individual differences in anxiety prior to treatment?

When there are two or more IVs, separate tests are made for each IV. Further, when there are equal sample sizes in all groups, these separate tests are independent of one another except for use of a common error term, so that finding an effect of one IV in no way predicts whether an effect will be found for another IV.

Another IV could be added to the anxiety-reduction example. Suppose we were also interested in differences in anxiety associated with the setting in which treatments were offered. Volunteers in all three treatment groups would report either to a university counseling center for the experiment or to a lab in the psychology building. The main effect of setting on the three anxiety measures could be tested in MANOVA. Or the main effect of setting on anxiety measures could be tested after adjusting for pretreatment anxiety levels.

8.2.2 Interactions Among IVs

Holding all else constant, does change in behavior over levels of one IV depend on another IV? That is, do IVs interact in their effect on behavior? These tests of interactions, though interpreted quite differently from those of main effects, are statistically similar, and likewise follow procedures covered in Sections 8.4.1 and 8.4.2. (The concept of interaction is more fully discussed in Chapter 2.)

For the preceding two-way anxiety example (treatment by setting), the interaction could also be tested. In MANOVA, we would test whether differential treatment effects on the anxiety measures varied with setting. In MANCOVA, the interaction question would be asked after adjustment for pretest anxiety.

With more than two IVs, there will be multiple interactions to test. Each of the interactions is tested separately, and as with main effects, these tests, but for a common error term in the F ratio, are independent when sample sizes in all groups are equal. Furthermore, hypotheses about interactions are independent of tests of main effects with equal sample sizes and balanced designs.

8.2.3 Importance of DVs

If there are significant differences among groups (reflected in either main effects and/or interactions), you may want to know which of the DVs are being changed and which are basically unaffected by action of the IVs. Procedures for assessing importance appear in Section 8.5.3.

Suppose a main effect of treatment on anxiety is found in MANOVA. It may be that only test anxiety is amenable to treatment. Stress anxiety and free-floating anxiety may not differ with the different treatment groups.

8.2.4 Specific Comparisons and Trend Analysis

Given that you have a significant main effect for an IV with more than two levels, you may want to assess the nature of the difference to discover which levels are significantly different from which other levels. Indeed, you may have had some specific prior hypotheses about the pattern of expected differences so that planned comparisons were possible. Similarly, with significant interactions, you may want to further assess the nature of the differences—that is, just where within the design do cell differences diverge significantly.

For example, after a significant treatment-by-setting interaction on anxiety, it might be interesting to see if for the two waiting-list control groups there is a significant difference in anxiety associated with setting.

In ANOVA, a variety of procedures have been developed to test specific comparisons, both those that were planned and those that became interesting after examination of results of the omnibus analysis of variance. For MANOVA, and especially for MANCOVA, fewer procedures are readily available. These tests are discussed in Section 8.5.2.

8.2.5 Strength of Association

Given that a main effect or interaction of IVs is shown to affect behavior, the logical next question is, How much? That is, what proportion of variance in the linear combination of DV scores can be attributable to action of the IV(s)?

A simple test of strength of association between IVs and DVs is described in Section 8.4.1.

If a main effect of treatment is found on the three measures of anxiety (as a composite), we can find the proportion of variance in the composite DV associated with differences in treatment.

Procedures are also available for finding the strength of association between the IV and an individually significant DV. These are demonstrated in Section 8.7.

8.2.6 Repeated-Measures Analysis of Variance

In repeated-measures analysis of variance, the levels of the within-subjects variable can be viewed as separate DVs. For example, separate measures on each subject for different trials can be treated as separate "tests" of that subject, where each DV is being measured on the same scale. One can then do a MANOVA on the between-subjects IV(s), with the within-subjects variable producing the several DVs.

A major assumption in repeated-measures analysis of variance is that correlations among levels of the within-subjects variable are constant over all combinations of levels. [Tests of this assumption of homogeneity of covariance or of the stronger assumption of compound-symmetry of the variance-covariance matrix are available in such comprehensive analysis of variance texts as Winer (1971) and in BMDP2V.] In cases where the assumption of homogeneity of covariance is violated, significance tests in repeated-measures ANOVA may become too liberal.[2] It may be worthwhile to repeat the analysis, using MANOVA to verify the pattern of significant differences. If MANOVA and ANOVA produce the same results, they can be believed.

If MANOVA and repeated-measures ANOVA produce different results, however, the researcher must decide between them by paying careful attention to the details of the experimental design. If the design was "clean" (e.g., counterbalancing was done to reduce carry-over effects), and repeated measures were collected with approximately equal amounts of time between them, then the repeated-measures ANOVA may be more appropriate. If the experiment was not clean, then the results of MANOVA may be more appropriate.

The problem is circumvented if, instead of looking at omnibus main effects and interactions, a trend analysis or orthogonal polynomial breakdown (cf. Chapter 2) is done. For these 1 df tests, there is no possibility of heterogeneity of covariance. This is most likely to be useful in just those situations that can be expected to violate homogeneity of covariance—that is, where the repeated-measures IV is, in itself, confounded with passage of time. Examples are sequential effects in memory and practice effects over learning trials.

---

[2] An additional assumption of repeated-measures ANOVA is homogeneity of variance over within-subjects IVs, tested in SPSS MANOVA as an FMAX test of "homogeneity of variance over measures." If the test fails (i.e., is statistically significant), MANOVA is preferred over repeated-measures ANOVA.

As an example of alternative strategies for conceptualizing the same experiment, suppose we want to include follow-up testing in the one-way ANCOVA of Section 7.1. That is, volunteers from a statistics class are assigned to the three treatment groups. They are pretested on a measure of test anxiety to provide a covariate, then are treated in one of two ways (sensitization or relaxation training) or are put on a waiting list, and then are retested on test anxiety immediately after treatment. Six months later they are again retested on test anxiety, and finally they are retested one year later. The three posttesting periods could be viewed as a second, within-subjects IV with three levels: immediate, six months, and one year. We would then have a 3×3 mixed within-between ANCOVA. Alternatively, this experiment could be analyzed as a one-way MANCOVA with three groups and three DVs: test anxiety immediately after treatment, test anxiety six months later, and test anxiety after a year. We could test differences among groups in a measure of test anxiety combined over the three posttests. Individual periods could also be assessed by the procedures in Section 8.5.3, or through profile analysis (cf. Section 8.5.2), or a priori trend analysis could be used, replacing the omnibus tests for the within-subjects variable, time, and the time by group interaction.

8.2.7 Adjusted Marginal and Cell Means

If there is a significant effect of the IV(s) or interaction(s) on the DVs, how do scores on the various DVs vary in the different groups? The population parameters in a stepdown analysis (cf. Section 8.5.3.2) are best estimated by the adjusted cell means for interactions and the adjusted marginal means for main effects. In stepdown analysis, DVs are tested individually by adjusting for higher-priority DVs. For instance, in the one-way MANOVA example of test anxiety of Section 8.1, let's say the three DVs had been ordered in the following manner: Free-floating anxiety is given highest priority, stress anxiety is given second highest priority, and test anxiety is given lowest priority. Now let's suppose that a stepdown analysis shows that only test anxiety is affected by differential treatment. In reporting differences among the three groups in test anxiety, it is the adjusted average test anxiety score for each group that would be reported—adjusted for differences in free-floating anxiety and stress anxiety. These adjusted means are the ones for which differences are being tested statistically. A look at the unadjusted means, too, may enhance interpretation. If covariates were to be used (e.g., pretest scores on the three anxiety measures), test anxiety would be adjusted for the covariates as well as for the two higher-priority DVs. Section 8.7 demonstrates procedures for finding and reporting adjusted marginal means.

8.2.8 Effects of Covariates

As with univariate ANCOVA (Chapter 7), it is frequently useful to evaluate the covariates chosen in MANCOVA. What is the nature of the DV-covariate relationship? Analysis of covariates is demonstrated in Section 8.7.3.1.

For example, with the three covariates mentioned in the one-way anxiety-reduction MANCOVA, it is possible to assess the degree to which each covariate (pretest free-floating anxiety, pretest stress anxiety, and pretest test anxiety) contributes to the adjustment of the composite DV.

8.3 LIMITATIONS TO MULTIVARIATE ANALYSIS OF VARIANCE AND COVARIANCE

8.3.1 Theoretical Issues

Because of the increase in complexity and ambiguity of results with MANOVA, one of the best overall recommendations is: Avoid it if you can. Ask yourself *why* you are including correlated DVs in the same analysis. Might there be some way of combining them or deleting some of them so that ANOVA can be performed? If, however, the logic of your research question requires MANOVA, there are other theoretical issues to be considered.

As with all other procedures, attribution of causality to IVs is in no way assured by the statistical test. This caution is especially relevant for MANOVA since the statistical test, as an extension of ANOVA, stems from experimental research in which IVs are typically manipulated by the experimenter, and desire for causal inference provides the reason behind elaborate controls. But the statistical test is available whether or not the IVs have been manipulated and whether or not proper controls have been instituted. Therefore the inference that significant changes in the DVs were caused by concomitant changes in the IVs is a logical exercise, not a statistical one. Care must be taken in interpreting the results of the analysis if an IV-DV relationship is found.

Choice of variables is also a question of logic and research design rather than of statistics. Whether or not appropriate behaviors are being sampled depends on theory and the skill of the researcher in choosing those DVs that have some chance of showing effects of the IVs.

In addition to choice of *which* DVs to measure is choice of the *order* in which these DVs enter analysis. If stepdown analysis (see Section 8.5.3.2) is used in interpretation of DVs, higher priority for entry is usually given to more important DVs or to DVs that are theoretically causally prior to the others. The choice is not trivial if stepdown F is to be interpreted. The significance of a DV may well depend on how high a priority it has been given, just as in hierarchical multiple regression the significance of an IV may depend on its position in the hierarchy.

Some further logical problems surround MANCOVA as they do univariate ANCOVA. If the purpose of the covariate is to reduce experimental error, then covariance analysis is difficult to interpret if covariates are affected by the behavior being measured by the DVs or are affected by the IVs. For example, it may or may not make any sense to adjust for within-group differences on covariates that are caused by the DVs, since that would reduce rather than increase the power of the test. Another problem with any covariance analysis is that the "adjusted" means for the DVs—the values on which MANOVA is done after the DVs have been adjusted for differences on the covariates—may not correspond to any realistic situation. This caution is especially applicable to situations where MANCOVA is being used to adjust for differences in intact groups in an attempt to reveal relationships between IVs and DVs. As discussed in Chapter 7, inference of causality may be impossible.

Sometimes the purpose of the covariance analysis, however, is to test the effects of some linear combinations of DVs in the context of other DVs that are, for analytic purposes, treated as covariates. In that case, possible causal relationships between covariates and IVs or DVs are not so critical.

Finally, the usual limits to generalizability apply. The results of MANOVA and MANCOVA can be validly generalized only to those populations from which the researcher has randomly sampled. Although covariance analysis may, in some very limited sense, adjust for a failure of random assignment to groups within the design, it does not adjust for failure to sample from portions of the population to which one wishes to generalize.

8.3.2 Practical Issues

In addition to the theoretical and logical issues, the statistical test itself demands consideration of some practical matters.

8.3.2.1 Unequal Sample Sizes and Missing Data   All of the issues raised in Chapter 7 apply here. Consult Section 7.3.2.1 and Chapter 4 for a review of problems caused by and potential solutions to incomplete data sets. Unfortunately, as experiments become more complicated with addition of DVs and, perhaps, covariates, the probability of missing data increases.

When planning research to be analyzed through MANOVA, it is important for two reasons to include more cases than DVs in every cell. First, the power of the analysis may be lowered unless there are more cases than DVs because of reduced degrees of freedom for error. One likely outcome of reduced power is a nonsignificant multivariate F, but one or more significant univariate F's (and a very unhappy researcher).

The second reason for having more cases than DVs in every cell has to do with the assumption of homogeneous variance-covariance matrices (Section 8.3.2.4). Recall that error is estimated from pooled within-cell variance-covariance matrices and that pooling them is ill-advised if they are very different. If one or more cells has more DVs than cases, the cell becomes singular, leading to rejection (actually untestability) of the assumption of homogeneity of variance-covariance matrices. If one or more cells has only one or two more cases than DVs, the likelihood of rejection of the assumption of homogeneity of variance-covariance matrices is increased. Thus MANOVA as an analytic strategy may be discarded because of heterogeneity of variance-covariance matrices when the heterogeneity was really created as an artifact of a ratio of cases to DVs that is too low.

8.3.2.2 Multivariate Normality   The mathematical model that underlies MANOVA, MANCOVA, and other multivariate techniques is based on the multivariate normal distribution. This means that the sampling distributions of means of the various DVs in each cell are normally distributed as are linear combinations of them. With univariate F, the central limit theorem suggests that, for large samples, the sampling distribution of means can be expected to approach normality. Mardia (1971) shows that MANOVA is also robust to modest violation of normality if the violation is created by skewness rather than by outliers. *A sample size that would produce* 20 *degrees of freedom*[3] *for error in the univariate case should ensure robustness of the test, as long as sample sizes are equal and two-tailed tests are used.* Even with unequal *n*, a sample size of about 20 in the smallest group should ensure robustness with a few DVs. With small, unequal samples, normality can really be assessed only by some reliance on judgment. Would the DVs individually be expected to be fairly normally distributed in the population? If not, would some transformation be expected to produce normality? (Cf. Chapter 4.)

8.3.2.3 Outliers   One of the more serious limitations of MANOVA is its sensitivity to outliers; inclusion of an outlier in MANOVA may well make the statistical test uninterpretable. Especially worrisome is that an outlier can produce either a Type I or a Type II error, with no clue in the analysis as to which is occurring. Therefore it is highly recommended by Frane (1977) that a test for outliers accompany any use of MANOVA.

Several programs are available for checking for univariate and multivariate outliers (cf. Chapter 4). *Run tests for univariate and multivariate outliers separately for each cell of the design and transform or eliminate any significant outliers from MANOVA or MANCOVA analysis.* Whenever transforming or eliminating outliers, it is necessary to report this fact. Tests for within-cell univariate and multivariate outliers are demonstrated in Sections 7.7.1 and 8.7.1.

8.3.2.4 Homogeneity of Variance-Covariance Matrices   The multivariate generalization of homogeneity of variance[4] for individual DVs is homogeneity of variance-covariance matrices. The assumption is that variance-covariance matrices within each cell in the design are sampled from the same population variance-covariance matrix. If they are not homogeneous, the pooled matrix is misleading as an estimate of error variance.[5]

The following guidelines for testing this assumption in MANOVA are

---

[3] See Chapter 2 for review of calculation of degrees of freedom for univariate F.

[4] Homogeneity of variance for each of the DVs is also assumed. See Section 7.3.2.5 for discussion and recommendations.

[5] This requirement is to be distinguished from the assumption of homogeneity of covariance required by repeated-measures univariate analysis of variance. The latter assumption, *not* required for MANOVA, is that all covariances in the *pooled* matrix are equivalent. Violation of homogeneity of covariances is justification for deciding in favor of MANOVA over repeated-measures ANOVA. A test for homogeneity of covariance is available in BMDP2V. (But of course the assumption of homogeneity of covariance is necessary for repeated-measures MANOVA in the matrices representing repeated measures on a single DV.)

based on a generalization of Monte Carlo tests of robustness in T^2 (Hakstian, Roed & Lind, 1979). *If sample sizes are equal, regardless of the outcome of Box's M test* (a notoriously sensitive test of homogeneity of variance-covariance matrices available through SPSS MANOVA), *robustness of significance tests is expected.* If sample sizes are unequal *and* Box's M test leads to rejection, at $p < .001$, of the assumption of homogeneity of variance-covariance matrices, then robustness is *not* guaranteed. Whether the direction of likely Type I error is conservative or liberal depends on whether or not the cells with the larger samples also generated the larger variances and covariances. If larger samples produced larger covariances, a conservative α level is produced so that a test of mean differences, if rejected, may be rejected with confidence. If, however, smaller samples produced larger variances and covariances, the significance test is too liberal. As the number of DVs increases and discrepancy between sample sizes increases, greater distortion of α levels may be produced. In this case, the null hypothesis may be retained with confidence but indications of mean differences are suspect. *Use of Pillai's criterion instead of Wilks' Lambda* (see Section 8.5.1) *for evaluating significance* (Olson, 1979) *may improve the robustness of the test.* If sample sizes can be equalized by randomly deleting cases without too great a loss in power, robustness is once again likely.

8.3.2.5 Linearity The MANOVA and MANCOVA models assume that the interrelationships among all DVs and covariates are linear within each cell. These include all pairs of DVs, all pairs of covariates, and all DV-covariate pairs. Deviation from linearity will reduce power of the statistical tests in that (1) linear combinations of DVs will not show maximum relationship with the IVs (i.e., group differences will be smaller), and (2) the effects of covariates in reducing error will be minimized.

With a small number of DVs and/or covariates, examine all within-cell scatterplots between pairs of DVs, pairs of covariates, and pairs of DV-covariate combinations. These plots are available through SPSS SCATTERGRAM or BMDP6D. As the number of DVs and/or covariates increases, however, the procedure becomes cumbersome. A spot check of those bivariate relationships for which nonlinearity is suspect would be in order.

An alternative screening procedure for linearity with large numbers of variables is available through BMDP1V, as described in Section 7.3.2.6. One DV, preferably the one with the lowest entry priority, can be designated DEPEN-DENT, and the remainder designated INDEPENDENT (covariates). Residuals plots may suggest the presence of nonlinearity, as described in Section 5.3.2.4.

If serious curvilinearity is found, transformation of variables can correct the problem (cf. Section 4.5) but at the risk of complicating interpretation.

8.3.2.6 Homogeneity of Regression An additional assumption in MANCOVA (as well as in univariate ANCOVA) is that regression coefficients within each group are homogeneous across groups. That is, it is assumed that the relationship between covariates and DVs in any one group is the same as the relationship in any other group, because the adjustment for the error term is based on an average of regression planes over all groups. Homogeneity of regression is relevant not only to MANCOVA but also to stepdown analysis of individual DVs in MANOVA. In this analysis each DV is assessed, in turn, with higher-priority DVs acting as covariates. *For MANCOVA, test for overall and stepdown homogeneity of regression using a program such as SPSS MANOVA. For MANOVA, test for stepdown homogeneity of regression.* (See Section 8.7.1 for examples of the test for MANOVA and MANCOVA.)

Lack of homogeneity of regression suggests there is some interaction between the IV(s) and the covariates that implies that the effect of covariates in adjusting the DVs would be different for different groups. If interaction between IVs and covariates is suspected, MANCOVA is an inappropriate analytic strategy, both statistically and logically. As an alternative, one could form new IVs by blocking on the covariates (cf. Section 7.3.2.7).

Similarly, the analysis of a subset of DVs after adjusting for other DVs is uninterpretable if this assumption is violated. If violation occurs within the stepdown analysis, then the IV-"covariate" interaction must itself be interpreted and the DV causing violation must be eliminated from further steps.

8.3.2.7 Reliability of Covariates

As for univariate ANCOVA, the F test for mean differences is more powerful if covariates have been reliably measured in experimental research. Errors of either type can occur with unreliable covariates in nonexperimental research. The covariate issue is discussed in Section 7.3.2.8. For stepdown analysis, in which most DVs act as covariates in assessing other DVs, the issue of unreliability becomes even more muddy. If any of the DVs are unreliable (say $r_{yy} < .8$), the results of the stepdown analysis may be meaningless. In such cases, other procedures for assessing individual DVs are recommended in place of, or in addition to, stepdown analysis (cf. Section 8.5.3). In any event, known or suspected unreliability of covariates and high-priority DVs should be discussed in reporting of research results.

8.3.2.8 Multicollinearity and Singularity

When correlations among DVs are too high, and almost all of the variance in the matrices is covariance, a condition of multicollinearity or singularity may occur (cf. Chapter 4). When multicollinearity or singularity is present, one DV is a near-linear combination of other DVs; that is, one DV has information that is redundant to the information available in one or more of the other DVs. It is both statistically and logically suspect, therefore, to include all the DVs in analysis. *Identify and deal with the source of multicollinearity or singularity.* Most commonly, this will involve deletion of the offending DV. If there is some need to retain all DVs for theoretical reasons, a principal components analysis (cf. Chapter 10) may be done on the pooled within-cell correlation matrix, and the component scores entered as an alternative set of DVs.

BMDP4V and BMDP7M protect against multicollinearity and singularity through calculation of pooled within-cell tolerance $(1 - SMC)$ on each DV.

In SPSS MANOVA, singularity or multicollinearity may be present when the determinant of the within-cell correlation matrix is near zero (say, less than .0001). The source of the problem may be identified by using the within-cell correlation matrix (from MANOVA) as input to the REGRESSION program and then performing several regressions, with each DV in turn serving as DV for all the other DVs serving as IVs in analysis. If any R^2 approaches .99, the DV is causing multicollinearity or singularity. Alternatively, the within-cell correlation matrix can be input to the SPSS FACTOR program where SMCs (called EST COMMUNALITY) are given as output in a default or PA2 run. BMDP4M, a factor analysis program, routinely prints SMCs.

8.4 FUNDAMENTAL EQUATIONS FOR MULTIVARIATE ANALYSIS OF VARIANCE AND COVARIANCE

8.4.1 Multivariate Analysis of Variance

A data set appropriate for multivariate analysis of variance consists of one or more classification (or independent) variables and two or more measures (or DVs) on each sampling unit (subject) within each combination of IVs, or group. There may be two or more levels of each classification variable. For example, a fictitious small sample with two DVs and two IVs is illustrated in Table 8.1.

The example in Table 8.1 has two IVs: degree of disability and treatment. Degree of disability has three levels—mild, moderate, and severe—while the treatment variable has two levels—treatment and control (no treatment). Three children are assigned to each of the six groups, or combinations of IVs, so that sample size within groups is three. For each of these 3×6 or 18 children, two dependent measures are given: score on the reading subtest of the Wide Range Achievement Test (WRAT-R) and score on the arithmetic subtest (WRAT-A). In addition, an IQ score is given in parenthesis for each child. This score is treated as a covariate in Section 8.4.2.

The major questions to be answered by the MANOVA are: (1) Disregarding degree of disability, does treatment affect performance on the two subtests of the WRAT? and (2) Does the effect of treatment on the two subtests differ as a function of degree of disability? The first question is answered by testing the main effect of treatment, while the second question requires a test of the interaction between treatment and degree of disability. A third question will automatically be answered by any computer program providing omnibus F tests: Are scores on the WRAT affected by degree of disability? But in the context of this experiment, that is a trivial question. We would assume that degree of disability would at least partially be defined by difficulty in reading and/or arithmetic, and a significant effect would provide no useful information. On the other hand, absence of such an effect might lead us to question the adequacy

TABLE 8.1 SMALL SAMPLE DATA FOR ILLUSTRATION OF MULTIVARIATE ANALYSIS OF VARIANCE

	Mild			Moderate			Severe		
	WRAT-R	WRAT-A	(IQ)	WRAT-R	WRAT-A	(IQ)	WRAT-R	WRAT-A	(IQ)
Treatment	115	108	(110)	100	105	(115)	89	78	(99)
	98	105	(102)	105	95	(98)	100	85	(102)
	107	98	(100)	95	98	(100)	90	95	(100)
Control	90	92	(108)	70	80	(100)	65	62	(101)
	85	95	(115)	85	68	(99)	80	70	(95)
	80	81	(95)	78	82	(105)	72	73	(102)

of classification. (Although if we were to test the effect for adequacy of classification, the test should be limited to the control group, since treatment might tend to diminish disability differences.)

The sample size of three children used in this example would probably be inadequate for a realistic test, but it should serve to illustrate the techniques of MANOVA. Additionally, if causal inference were to be made in this situation, assignment of children to treatment and control groups should be random and independent. The reader is encouraged to work problems involving these data by hand as well as by computer programs, such as BMDP4V, BMD12V, BMDP7M, or SPSS MANOVA.

The MANOVA equation for equal n can most easily be developed through extension from analysis of variance. It can be recalled (from Chapter 2) that analysis of variance involves a partitioning of total variance, or differences among scores, into variance associated with sources that can be identified within the research design. The simplest partition apportions variance to systematic sources (variance attributable to differences between groups) and to unknown sources or error (variance attributable to differences in scores within groups). To do this partitioning, differences between scores and various means are squared and summed.

$$\sum_i \sum_j (Y_{ij} - GM)^2 = n\sum_j (\overline{Y}_j - GM)^2 + \sum_i \sum_j (Y_{ij} - \overline{Y}_j)^2 \tag{8.1}$$

The total sum of squared differences between scores on Y (the DV) and the grand mean (GM) can be partitioned into two components: sum of squared differences between group means $(\overline{Y}_j)$ and the grand mean (i.e., systematic or between-groups variability), and sum of squared differences between individual scores (Y_{ij}) and their respective group means.

or

$$\text{SS}_{\text{total}} = \text{SS}_{bg} + \text{SS}_{wg}$$

For factorial designs, or those with more than one IV, the variance associated with systematic sources (between groups) can be further partitioned into variance associated with the first IV (in our illustration, degree of disability, abbreviated D), variance associated with the second IV (in our illustration treatment, or T), and variance associated with the interaction between degree of disability and treatment (or DT).

$$n_{km}\sum_k \sum_m (DT_{km} - GM)^2 = n_k\sum_k (D_k - GM)^2 + n_m\sum_m (T_m - GM)^2$$
$$+ \left[n_{km}\sum_k \sum_m (DT_{km} - GM)^2 - n_k\sum_k (D_k - GM)^2 \right.$$
$$\left. - n_m\sum_m (T_m - GM)^2 \right] \tag{8.2}$$

The sum of squared differences between group or cell (DT_{km}) means and the grand mean can be partitioned into (1) sum of squared differences between means associated with differing degree of disability (D_k) and the grand mean; (2) sum of squared differences between means associated with treatment and control (T_m) and the grand mean; and (3) sum of squared differences associated with combinations of treatment and degree of disability (DT_{km}) and the grand mean, from which differences associated with D_k and T_m have been subtracted. Each n is the number of scores composing the relevant marginal or cell mean.

or

$$SS_{bg} = SS_D + SS_T + SS_{DT}$$

So that, in the full partition for a factorial between-subjects design, we have

$$\sum_i \sum_k \sum_m (Y_{ikm} - GM)^2 = n_k \sum_k (D_k - GM)^2 + n_m \sum_m (T_m - GM)^2$$
$$+ \left[n_{km} \sum_k \sum_m (DT_{km} - GM)^2 - n_k \sum_k (D_k - GM)^2 \right.$$
$$\left. - n_m \sum_m (T_m - GM)^2 \right] + \sum_i \sum_k \sum_m (Y_{ikm} - DT_{km})^2$$

$$(8.3)$$

For MANOVA, there is no single DV, Y. Instead, there is a column matrix (or vector) of Y_{ikm} values, containing a score on each DV. For example, in the data illustrated in Table 8.1, column matrices of Y scores for the first three children (in the first cell of the design, mild disability with treatment) are

$$\mathbf{Y}_{i\,11} = \begin{bmatrix} 115 \\ 108 \end{bmatrix} \quad \begin{bmatrix} 98 \\ 105 \end{bmatrix} \quad \begin{bmatrix} 107 \\ 98 \end{bmatrix}$$

Similarly, there is a column matrix of D_k means for each level of D, with one entry in each matrix for each DV.

$$\mathbf{D}_1 = \begin{bmatrix} 95.83 \\ 96.50 \end{bmatrix} \quad \mathbf{D}_2 = \begin{bmatrix} 88.83 \\ 88.00 \end{bmatrix} \quad \mathbf{D}_3 = \begin{bmatrix} 82.67 \\ 77.17 \end{bmatrix}$$

where 95.83 is the mean on WRAT-R for mild disability, averaged over children in treatment and control groups, and 96.50 is the mean score on WRAT-A for those same children.

Matrices for the T_m means, averaged over children of all levels of disability are

$$\mathbf{T}_1 = \begin{bmatrix} 99.89 \\ 96.33 \end{bmatrix} \quad \mathbf{T}_2 = \begin{bmatrix} 78.33 \\ 78.11 \end{bmatrix}$$

Similarly, there would be six matrices for the cell means (DT_{km}) averaged over the three children in each group. Finally, there is a single matrix of grand means (GM), averaged over all children in the experiment, with one value for each DV:

$$\mathbf{GM} = \begin{bmatrix} 89.11 \\ 87.22 \end{bmatrix}$$

Matrices can simply be subtracted from one another, to produce difference matrices. For example, for the error term, the matrix of grand means **(GM)** can be subtracted from each of the matrices of individual scores $(\mathbf{Y}_{ikm})$ by the methods of matrix subtraction, reviewed in Appendix A. The matrix counterpart of a difference score, then, is a difference matrix. Thus for the first child in the example:

$$(\mathbf{Y}_{i11} - \mathbf{GM}) = \begin{bmatrix} 115 \\ 108 \end{bmatrix} - \begin{bmatrix} 89.11 \\ 87.22 \end{bmatrix} = \begin{bmatrix} 25.89 \\ 20.78 \end{bmatrix}$$

In ANOVA, difference scores must be squared. The matrix counterpart of the squaring operation is multiplication by a transpose. That is, each column matrix is multiplied by its corresponding row matrix (see Appendix A for matrix transposition and multiplication) to produce a squares and cross-products matrix. For example, for the first child in the first group of the design:

$$(\mathbf{Y}_{111} - \mathbf{GM})(\mathbf{Y}_{111} - \mathbf{GM})' = \begin{bmatrix} 25.89 \\ 20.78 \end{bmatrix} [25.89 \quad 20.78]$$

$$= \begin{bmatrix} 670.29 & 537.99 \\ 537.99 & 431.81 \end{bmatrix}$$

These matrices are then added over subjects and over groups, just as squared differences are summed in univariate ANOVA.[6] The resulting matrix **(S)** is called by various names: sum-of-squares and cross-products, cross-products, or sum-of-products. The MANOVA partitioning of sums of squares and cross products for our factorial example can be represented as a matrix form of Eq. 8.3:

$$\sum_i \sum_k \sum_m (\mathbf{Y}_{ikm} - \mathbf{GM})(\mathbf{Y}_{ikm} - \mathbf{GM})' = n_k \sum_k (\mathbf{D}_k - \mathbf{GM})(\mathbf{D}_k - \mathbf{GM})'$$

$$+ n_m \sum_m (\mathbf{T}_m - \mathbf{GM})(\mathbf{T}_m - \mathbf{GM})'$$

$$+ \left[n_{km} \sum_k \sum_m (\mathbf{DT}_{km} - \mathbf{GM})(\mathbf{DT}_{km} - \mathbf{GM})' \right.$$

$$- n_k \sum_k (\mathbf{D}_k - \mathbf{GM})(\mathbf{D}_k - \mathbf{GM})'$$

$$\left. - n_m \sum_m (\mathbf{T}_m - \mathbf{GM})(\mathbf{T}_m - \mathbf{GM})' \right]$$

$$+ \sum_i \sum_k \sum_m (\mathbf{Y}_{ikm} - \mathbf{DT})(\mathbf{Y}_{ikm} - \mathbf{DT})'$$

or

$$\mathbf{S}_{total} = \mathbf{S}_D + \mathbf{S}_T + \mathbf{S}_{DT} + \mathbf{S}_{S(DT)}$$

---

[6] Order of summing and squaring follows the ANOVA order for a comparable design. That is, for a between-groups term, matrices are summed over subjects within groups before "squaring," then summed over groups.

The total cross-products matrix (S_{total}) is partitioned into cross-products matrices for differences associated with degree of disability, with treatment, with the interaction between disability and treatment, and for subjects within groups ($S_{S(DT)}$).

For the example in Table 8.1, the four resulting cross-products matrices are

$$S_D = \begin{bmatrix} 520.78 & 761.72 \\ 761.72 & 1126.78 \end{bmatrix} \qquad S_T = \begin{bmatrix} 2090.89 & 1767.56 \\ 1767.56 & 1494.22 \end{bmatrix}$$

$$S_{DT} = \begin{bmatrix} 2.11 & 5.28 \\ 5.28 & 52.78 \end{bmatrix} \qquad S_{S(DT)} = \begin{bmatrix} 544.00 & 31.00 \\ 31.00 & 539.33 \end{bmatrix}$$

(Numbers producing these matrices were carried to 8 digits before rounding.) Notice that all these matrices are symmetric, with the major diagonal elements (top left to bottom right) representing sums of squares that, when divided by degrees of freedom, would produce variances, and with the off-diagonals representing sums of cross products that, when divided by degrees of freedom, would produce covariances. That is, the first element in the major diagonal is the sum of squares for the first DV, WRAT-R, and the second diagonal element is the sum of squares for the second DV, WRAT-A. The off-diagonal elements represent the covariance between WRAT-R and WRAT-A.

In analysis of variance, sums of squares are divided by degrees of freedom to produce variances, or mean squares. In MANOVA, the variance components used are based on the cross-products matrices. The matrix analog for a variance is a determinant (see Appendix A). The determinant is found for each of the cross-products matrices.

To test hypotheses about main effects and interactions, using the Wilks' Lambda criterion (see Section 8.5.1 for alternative criteria), it is necessary to form ratios of determinants. These ratios follow the general form:

$$\Lambda = \frac{|S_{error}|}{|S_{effect} + S_{error}|} \tag{8.4}$$

Wilks' Lambda (Λ) is the ratio of the determinant of the error cross-products matrix to the determinant of the sum of the error and effect cross-products matrices.

To find Wilks' Λ for various tests, the matrices corresponding to main effects and interactions must be added to the within-groups matrix before the determinants are found. For example, the matrix produced by adding the S_{DT} matrix for the interacting effects to the $S_{S(DT)}$ matrix for subjects within groups (error) is

$$[S_{DT} + S_{S(DT)}] = \begin{bmatrix} 2.11 & 5.28 \\ 5.28 & 52.78 \end{bmatrix} + \begin{bmatrix} 544.00 & 31.00 \\ 31.00 & 539.33 \end{bmatrix} = \begin{bmatrix} 546.11 & 36.28 \\ 36.28 & 592.11 \end{bmatrix}$$

For the four matrices needed to test all three hypotheses—main effects of degree of disability, main effects of treatment, and the treatment-disability interaction—the determinants are

$$|S_{S(DT)}| = 292436.34$$

$$|S_D + S_{S(DT)}| = 1145629.58$$

$$|S_T + S_{S(DT)}| = 2123390.88$$

$$|S_{DT} + S_{S(DT)}| = 322042.38$$

At this point a source table, similar to those used to report results in ANOVA, becomes useful. Table 8.2 was produced by running the sample data through BMD12V. The first column, as in the univariate case, lists sources of variance, in this case the two main effects, the interaction, and the error term. Next comes the determinant or, since it is usually such a large number, its natural logarithm, typically reported as log (generalized variance). The third column contains the value of Wilks' Lambda or, as it is sometimes reported, the U statistic.

The U statistic is calculated by forming an appropriate ratio of determinants, as described in Eq. 8.4. For example, for the interaction between disability and treatment, Wilks' Lambda is

$$\Lambda = \frac{|S_{S(DT)}|}{|S_{DT} + S_{S(DT)}|} = \frac{292436.34}{322042.38} = .908068$$

A column for three degrees-of-freedom parameters for a direct test of Wilks' Lambda is then shown. These represent (1) the number of DVs, (2) the degrees of freedom associated with the effect being tested, and (3) the degrees of freedom associated with the error term. (Degrees of freedom for effect and for error are calculated in the same manner as for univariate ANOVA.) Tables for evaluating Wilks' Lambda are not common, though they do appear in Harris (1975). However, an F approximation has been derived that closely fits Lambda. The last two columns of Table 8.2, then, represent the approximate F values and their associated degrees of freedom.

The following procedure for calculating approximate F is based on Wilks' Lambda and the various degrees of freedom associated with it.[7]

$$\text{approximate } F(df_1, df_2) = \left(\frac{1-y}{y}\right)\left(\frac{df_2}{df_1}\right) \tag{8.5}$$

[7] Different formulas are required when $df_{\text{effect}} = 1$. (Of course, when $p = 1$, we have univariate ANOVA.) For 1 df for effect:

$$F = \left(\frac{1-\Lambda}{\Lambda}\right)\left(\frac{N-p-1}{p}\right)$$

with df $= p$, $N - p - 1$, where $N =$ total number of scores.

TABLE 8.2 MULTIVARIATE ANALYSIS OF VARIANCE OF WRAT-R AND WRAT-A SCORES

SOURCE	LOG (GENERALIZED VARIANCE)	U-STATISTIC	DEGREES OF FREEDOM		APPROXIMATE F-STATISTIC	DEGREES OF FREEDOM	
D	13.95146	0.255263	2	12	5.3860*	4	22.00
T	14.56852	0.137721	1	12	34.4357**	2	11.00
DT	12.68244	0.908068	2	12	0.2717	4	22.00
FULL MODEL	12.58600						

* $p < .01$.
** $p < .001$.

where df_1 and df_2 are the degrees of freedom for testing the F ratio and are defined below, and y is

$$y = \Lambda^{1/s}$$

Λ is defined as in Eq. 8.4, and s is

$$s = \sqrt{\frac{p^2(df_{effect})^2 - 4}{p^2 + (df_{effect})^2 - 5}}$$

where p is the number of DVs, and df_{effect} is the degrees of freedom associated with the effect being tested. Thus

$$df_1 = p(df_{effect})$$

and

$$df_2 = s\left[(df_{error}) - \frac{p - df_{effect} + 1}{2}\right] - \left[\frac{p(df_{effect}) - 2}{2}\right]$$

where df_{error} are the degrees of freedom associated with the error term.

In the sample problem, looking at the effect of the interaction, we see that

$p = 2$ the number of DVs

$df_{effect} = 2$ the number of treatment levels minus 1 times the number of disability levels minus 1 or $(t - 1)(d - 1)$

$df_{error} = 12$ the number of treatment levels, times the number of disability levels, times the quantity $n - 1$ (where n is the number of scores per cell for each DV)—that is, $df_{error} = dt(n - 1)$

Thus

$$s = \sqrt{\frac{(2)^2(2)^2 - 4}{(2)^2 + (2)^2 - 5}} = 2$$

$$y = .908068^{1/2} = .952926$$

$$df_1 = 2(2) = 4$$

$$df_2 = 2\left[12 - \frac{2 - 2 + 1}{2}\right] - \left[\frac{2(2) - 2}{2}\right] = 22$$

approximate $F(4, 22) = \left(\frac{.047074}{.952926}\right)\left(\frac{22}{4}\right) = 0.2717$

Once these approximate F values are available, significance is tested by using the usual tables of F for the preset value of alpha. In our example, the test of the interaction between disability and treatment is not statistically significant with 4 and 22 df, with the observed value of 0.2717 obviously not exceeding the critical value of 2.82 with $\alpha = .05$. The effect of treatment is statistically significant with the observed value of 34.44 exceeding the critical value of 3.98 with 2 and 11 df, $\alpha = .05$. The effect of degree of disability is also statistically significant, with the critical F for 4 and 22 df, at the .05 level of significance, equal to 2.82—less than the observed value of 5.39. (As noted previously, this main effect would probably not be of research interest.) In Table 8.2, significance is indicated for the highest level of alpha reached, following standard practice.

In addition, a measure of strength of association is readily available from Wilks' Lambda.[8] For MANOVA:

$$\eta^2 = 1 - \Lambda \qquad (8.6)$$

The reason that this equation represents the variance accounted for by the best linear combination of DVs can be seen from Eq. 8.4. In a one-way analysis, Eq. 8.4 indicates that Wilks' Lambda is a ratio of the determinant (variance) of the error matrix and the determinant (variance) of the effect plus error (or total sum-of-squares and cross-products) matrix. If Λ, then, is the proportion of variance *not* accounted for in the linear combination of DVs, then $1 - \Lambda$ is the proportion that is accounted for. Thus for each statistically significant effect, the proportion of variance accounted for by that effect can easily be calculated using Eq. 8.6. For example, for the main effect of treatment:

$$\eta_T{}^2 = 1 - \Lambda_T = 1 - .137721 = .862279$$

This equation indicates that 86% of the variance in the linear combination of WRAT-R and WRAT-A scores is accounted for by assignment to treatment versus control.[9] Taking the square root, we get $\eta = .93$. That is, the correlation between WRAT scores and assignment to treatment is .93.

8.4.2 Multivariate Analysis of Covariance

In MANCOVA, the linear combination of DVs is statistically adjusted for differences in the covariates. The new "adjusted" linear combination of DVs represents the combination that would have been obtained if all participants had started out with the same scores on all covariates.

---

[8] An alternative measure of strength of association is the canonical correlation, printed out by some computer programs. It is the correlation between the linear combination of IV levels and the best linear combination of DVs. Canonical correlation as a general procedure is discussed in Chapter 6, and the relation between canonical correlation and MANOVA is described briefly in Chapter 11.

[9] It should be noted that, unlike η^2 in the analogous ANOVA design, the sum of MANOVA η^2 for all effects may be greater than 1.0. This lessens the appeal of an interpretation in terms of proportion of variance accounted for.

For our example, assume that in addition to postexperimental WRAT-R and WRAT-A scores for each child, we had recorded preexperimental IQ scores (say, on the WISC-R, full scale). These hypothetical IQ scores are listed in parentheses in Table 8.1.

In developing MANCOVA from MANOVA, the basic partition of sources of variance is the same. Score matrices in MANCOVA, however, include entries for covariates as well as DVs. All of the column matrices, then—Y_{ikm}, D_k, DT_{km}, and **GM**—would have three entries in our example. The first entry represents the covariate (IQ score), while the second two entries represent the two DV scores (WRAT-R and WRAT-A).

For example, for the first child in the sample (mild disability with treatment), the column matrix of covariate and DV scores is

$$Y_{111} = \begin{bmatrix} 110 \\ 115 \\ 108 \end{bmatrix} \begin{matrix} \text{(IQ)} \\ \text{(WRAT-R)} \\ \text{(WRAT-A)} \end{matrix}$$

As in MANOVA, the difference matrices are found by subtraction, and the squares and cross-products matrices are formed by multiplying each difference matrix by its transpose. After appropriate summing and "squaring" (multiplication by transpose), the **S** (or cross-products) matrices are formed.

It is at this point that another departure from MANOVA occurs. The **S** matrices are partitioned into sections corresponding to the covariates, the DVs, and the cross products between covariates and DVs. For example, the cross-products matrix for the main effect of treatment for our small sample data is

$$S_T = \begin{bmatrix} [2.00] & [64.67 & 54.67] \\ \begin{bmatrix} 64.67 \\ 54.67 \end{bmatrix} & \begin{bmatrix} 2090.89 & 1767.56 \\ 1767.56 & 1494.22 \end{bmatrix} \end{bmatrix}$$

The lower right-hand partition represents the S_T matrix for the DVs (or $S_T^{(Y)}$), and is identical to the S_T matrix developed in Section 8.4.1. The upper left matrix represents the sum of squares for the covariate (or $S_T^{(X)}$). (With additional covariates, this would become a full sum-of-squares and cross-products matrix.) Finally, the two off-diagonal matrices represent cross products between covariates and DVs: $S_T^{(YX)}$ in the lower left and $S_T^{(XY)}$ in the upper right.

From these partitioned matrices, *reduced* or **S*** matrices are formed. An **S*** matrix is formed for the sums of squares and the cross products for DVs adjusted for the effects of the covariates. In effect, each sum of squares and each cross product is adjusted by a value that reflects variance due to differences in the covariate.

In matrix terms, the adjustment takes the following form:

$$S^* = S^{(Y)} - S^{(YX)}(S^{(X)})^{-1} S^{(XY)} \tag{8.7}$$

The cross-products matrix of DVs ($S^{(Y)}$) is adjusted to produce a new DV cross-products matrix **S***. The adjustment involves subtraction of a product based on the cross-products matrix for covariate(s) ($S^{(X)}$) and cross-products matrices for the relation between the covariates and the DVs ($S^{(YX)}$ and $S^{(XY)}$).

A glance at this adjustment makes it clear that the adjustment is being made for the regression of the DVs (Y) on the covariates (X). Since $S^{(XY)}$ is the transpose of $S^{(YX)}$, their multiplication is analogous to a squaring operation. Multiplying by the inverse of $S^{(X)}$ is analogous to division. It can be recalled from Chapter 2 that with simple scalar numbers, the regression coefficient is a function of the sum of cross products between X and Y, divided by the sum of squares for X.

An adjustment is made for each S matrix for the partition of sources of variances (Eq. 8.3). In our example, then, each of the S^* matrices will now be a 2×2 matrix, but entries can be expected to be smaller than in the original MANOVA S matrices. These reduced S^* matrices are

$$S_D^* = \begin{bmatrix} 388.18 & 500.49 \\ 500.49 & 654.57 \end{bmatrix} \qquad S_T^* = \begin{bmatrix} 2059.50 & 1708.24 \\ 1708.24 & 1416.88 \end{bmatrix}$$

$$S_{DT}^* = \begin{bmatrix} 2.06 & 0.87 \\ 0.87 & 19.61 \end{bmatrix} \qquad S_{S(DT)}^* = \begin{bmatrix} 528.41 & -26.62 \\ -26.62 & 324.95 \end{bmatrix}$$

Note that, as in the lower right-hand partition, cross-products matrices *may* have negative values for entries other than the major diagonal (which represents sums of squares).

Any of the tests appropriate for MANOVA can now be applied to the reduced S^* matrices. For example, ratios of determinants can be formed to test hypotheses about main effects and interactions by using Wilk's Lambda criterion (Eq. 8.4). For our example, the determinants of the four matrices needed to test the three hypotheses (two main effects and the interaction) are

$$|S_{S(DT)}^*| = 171032.69$$
$$|S_D^* + S_{S(DT)}^*| = 673383.31$$
$$|S_T^* + S_{S(DT)}^*| = 1680076.69$$
$$|S_{DT}^* + S_{S(DT)}^*| = 182152.59$$

The source table as produced by BMD12V, analogous to that produced for MANOVA, for our sample data is illustrated in Table 8.3.

One new item appears in this source table that was missing from the MANOVA table of Section 8.4.1. One source of variance in the DVs is the covariate. (With more than one covariate, there would be one row representing combined covariates, and one row for each of the individual covariates.) In addition, degrees of freedom for testing the approximate F ratio are used up, as in any covariance analysis (see Chapter 7). This means that Eq. 8.5 must be modified. The modification affects only the df_2 term (i.e., the denominator

TABLE 8.3 MULTIVARIATE ANALYSIS OF COVARIANCE OF WRAT-R AND WRAT-A SCORES

SOURCE	LOG(GENERALIZED VARIANCE)	U-STATISTIC	DEGREES OF FREEDOM		APPROXIMATE F-STATISTIC	DEGREES OF FREEDOM	
D	13.42007	0.253991	2	11	4.9211*	4	20.00
T	14.33435	0.101801	2	11	44.1155**	2	10.00
DT	12.11260	0.938956	2	11	0.1600	4	20.00
COVARIATE	20.60274	0.584855	2	11	3.5491	2	10.00
FULL MODEL	12.04961						

* $p < .01$.
** $p < .001$.

for testing F) and the s term (used to modify df_2 and y in Eq. 8.5). For multivariate analysis of covariance, then,

$$s = \sqrt{\frac{(p + q)^2(df_{effect})^2 - 4}{(p + q)^2 + (df_{effect})^2 - 5}} \qquad (8.8)$$

where q represents the number of covariates, and all other terms are defined as in Eq. 8.5. And

$$df_2 = s\left[(df_{error}) - \frac{(p + q) - df_{effect} + 1}{2}\right] - \left[\frac{(p + q)(df_{effect}) - 2}{2}\right]$$

The approximate F ratio can be used to test the significance of the covariate-DV relationship, in addition to the main effects and interactions. That is, is there a statistically significant relationship between DVs and covariate(s)? If so, the Wilks' Lambda can be used to find the degree of relationship. As previously shown (Eq. 8.6), a measure of strength of association is η^2.

8.5 SOME IMPORTANT ISSUES

8.5.1 Criteria for Statistical Inference

Different criteria for inferring population differences on the basis of sample data have been proposed: Wilks' Lambda, Roy's *gcr* criterion, Hotelling's trace criterion, and Pillai's criterion. For many applications, all four test statistics for assessing mean differences provide the same answer. However, occasionally one or more of them may give indications of significant differences while the others do not, and the researcher is left wondering which result to believe.

Wilks' Lambda (or U statistic) is a criterion used by all canned computer programs covered in this chapter for MANOVA, and is the *only* criterion available in most programs (except SPSS MANOVA and BMDP4V). Wilks' Lambda, as described in Section 8.4, is a likelihood ratio criterion, testing the likelihood of the data under the assumption of equal population mean vectors for all groups against the likelihood under the assumption that population mean vectors are identical to those of the sample mean vectors for the different groups.

Another criterion, preferred by some researchers (e.g., Harris, 1975) but recommended against by others (e.g., Olson, 1976), is Roy's greatest characteristic root (*gcr*) criterion. With more than one DV, groups may differ from one another along a number of independent dimensions, or principal components or factors. That is, the differences between groups may not be fully described in terms of points along a single continuum. Analogous to procedures for principal component extraction (see Chapter 10), the first component (or, in this case, discriminant function) creates the maximum possible discrimination among groups (i.e., the largest possible group differences). Succeeding discriminant functions are all mutually orthogonal, and account for ever-decreasing discriminating power. The characteristic root (eigenvalue) is a single value for each discriminant function based on the relative proportion of variance shared between the IV and the DVs. (See Appendix A for a definition of characteristic roots.) Roy's criterion tests the statistical significance of group differences based on the characteristic root of only the first discriminating function, while Wilks' Lambda, Hotelling's trace criterion, and Pillai's criterion, are based on all of the characteristic roots. Equations describing all four criteria are available in the SPSS Update 7-9 manual (Hull & Nie, 1981) and the SPSS MANOVA manual (Cohen & Burns, 1977). Thus Roy's *gcr* criterion, with only one dimension to consider, may be the most easily interpreted of the criteria. It is not, however, widely accepted as the test statistic of choice.

Among the other three, with respect to availability, Wilks' Lambda is superior, with only SPSS MANOVA automatically providing results in terms of all four criteria, while BMDP4V provides all except Pillai's criterion. Wilks' Lambda has already been defined in this section and in Eq. 8.4. It, like Hotelling's trace and Pillai's criterion, is an evaluation of the null hypothesis when group differences based on more than one dimension (discriminant function) have been pooled. Hotelling's trace is the sum of the eigenvalues (see Appendix A) for each dimension. Stated differently, Hotelling's trace is a sum of the ratio of variance due to effect and variance due to error for each dimension. Pillai's criterion, on the other hand, is simply the sum of variances due to effect for each dimension. Roy's *gcr* criterion differs from Pillai's in that, as already stated, only the variance associated with the most important dimension, the one with the largest eigenvalue, is considered. Wilks' Lambda, Hotelling's trace, and Roy's *gcr* criteria may be more powerful than Pillai's criterion when one dimension (the first characteristic root) provides most discrimination and less powerful when discrimination is distributed among several dimensions. But Pillai's criterion is said to be more robust than the other three (Olson, 1979). As sample size decreases, unequal *n*'s appear, and the assumption of homogeneity of variance-covariance matrices is violated (Section 8.3.2.4), then any advantage of Pillai's criterion in terms of robustness becomes more important. Under less than ideal design conditions, therefore, Pillai's criterion may be the criterion of choice.

In addition to ambiguity regarding choice among potentially conflicting significance tests for multivariate F is the irritation of a nonsignificant multivariate F but a significant univariate F for one (or more, although this seems unlikely) of the DVs. Had the researcher measured only the one DV, a significant finding was available, but since more DVs were measured, it is not. And why, one might ask, does not MANOVA calculate multivariate F by forming a linear combination of DVs in which the significant one is assigned a weighting of 1 and the nonsignificant DVs weights of zero? In fact, MANOVA comes close to doing just that, but the calculation of multivariate F is simply not always as powerful as separate univariate F's (or even stepdown F's; cf. Section 8.5.3.2) and significance may be lost. In this case, about the best one can do is report the nonsignificant multivariate F but offer the univariate and/or stepdown result as a guide to future research, with tentative interpretation.

8.5.2 Specific Comparisons and Trend Analysis

Sometimes an overall omnibus F does not provide sufficient information about the nature of specific group differences. To answer fine-grained questions, tests of specific comparisons are necessary. These may be based on a priori hypotheses, or on questions that may arise after examining the results of the omnibus MANOVA. (See Chapter 2 regarding Type IV error.)

A priori contrasts in MANOVA, like those in univariate ANOVA, can be done by specifying the appropriate contrast coefficients and running MANOVA based on those coefficients. As in all a priori tests, protection against inflated Type I error is achieved by performing the contrasts *instead of* omnibus F, and making sure that all contrasts for a given main effect or interaction are orthogonal (refer to Chapter 2 for a review of a priori orthogonal contrasts).

For MANOVA, the most convenient program for omnibus F (BMD12V) has no provision for specifying contrasts. Instead, it is necessary to use one of the more sophisticated programs, BMD11V, BMDP2V, BMDP4V, BMDP7M, or SPSS MANOVA. Desired contrasts may easily be specified, but if the design is factorial, indexing of groups in BMDP7M may be inconvenient. In any event, setting up contrasts requires knowledge of IV coding, which, incidentally, is the same for MANOVA as for ANOVA. Comprehensive coverage of such coding is provided by Cohen and Cohen (1975), while the basics are provided in most texts on ANOVA (e.g., Keppel, 1973; Winer, 1971).

As an example of an a priori test for the sample data of Section 8.4, it was pointed out that the omnibus main effect of degree of disability was trivial since it was simply related to the adequacy of classifying children as progressively more disabled. To test directly the adequacy of classification, however, only the control group should be examined, because the effect of disability may be overshadowed for those who had been treated. We can test the classification adequacy, then, by testing the linear component of the disability effect for control groups. (It is the linear component of the effect that tests whether the disability is *progressively* more severe over the three control groups.) The contrast coefficients are presented in Table 8.4.

TABLE 8.4 CONTRAST COEF-FICIENTS FOR TESTING LINEAR COMPONENT OF DISABILITY EFFECT FOR CONTROL GROUPS, SMALL SAMPLE DATA	Degree of disability		
	Mild	Moderate	Severe
Treatment	0	0	0
Control	1	0	−1

Note that zeros have been assigned to all treated groups. The coefficients for the untreated control groups are those for testing linear trend when the number of groups $= 3$, and can be found in any table of orthogonal polynomial coefficients (e.g., as presented in Keppel, 1973; Winer, 1971).

Results of the contrast, as printed out by BMDP7M, are presented in Table 8.5. The approximate F statistic for the contrast, with 2 and 11 df, is 8.74, exceeding both the .05 and .01 significance levels at 3.98 and 7.21, respectively.

Table 8.5 calls for a bit of further discussion. The program control information for the sample problem appears in the table, but much of the output relevant to discriminant function analysis has been omitted. Note, for example, that the statistical information is printed only for step 2. Since BMDP7M is a *stepwise* program, analogous to stepwise regression as discussed in Chapter 5, DVs are analyzed cumulatively. For the purposes of MANOVA, we are typically interested in the analysis of all of the DVs, which will be contained in the last step of the progression. So in using this program for MANOVA, it is essential to ensure that all DVs enter the equation. This can be achieved by the "forcing" procedure. Extension to MANCOVA simply involves forcing the covariate(s) in before the DVs.

TABLE 8.5 TEST OF LINEAR COMPONENT OF DISABILITY EFFECT FOR CONTROL GROUPS, SMALL SAMPLE DATA (SELECTED OUTPUT FROM BMDP7M)

```
PROGRAM CONTROL INFORMATION

        /PROBLEM      TITLE IS "SMALL SAMPLE MANOVA, CONTRASTS".
        /INPUT        VARIABLES ARE 3.  FORMAT IS "(12X,3F4.0,5X,F1.0)".  UNIT=81.
        /VARIABLE     NAMES ARE WRATR,WRATA,GROUPNO.
                      GROUPING IS GROUPNO.
        /GROUP        CODES(3) ARE 1,2,3,4,5,6.
                      NAMES(3) ARE TMILD,CMILD,TMOD,CMOD,TSEVERE,CSEVERE.
        /DISCRIM      LEVEL=1,2,0.  FORCE=2.  CONTRAST=0,1,0,0,0,-1.
        /END
```

```
STEP NUMBER    2
VARIABLE ENTERED    2 WRATA
    VARIABLE          F TO  FORCE  TOLERANCE  *      VARIABLE          F TO  FORCE    TOLERANCE
                      REMOVE LEVEL            *                        ENTER LEVEL
                      DF=  1   11             *                        DF=  1   10
     1 WRATR          1.795   1     .996725   *
     2 WRATA          8.750   2     .996725   *
  U-STATISTIC OR WILKS" LAMBDA      .3861290      DEGREES OF FREEDOM   2   1        12
  APPROXIMATE F-STATISTIC              8.744      DEGREES OF FREEDOM   2.00    11.00
  F - MATRIX          DEGREES OF FREEDOM =   2   11
                 THILD    CHILD    TMOD     CMOD    TSEVERE
  CHILD          9.75
  TMOD            .91     4.70
  CMOD          22.61     3.12    14.59
  TSEVERE        7.21     1.19     3.31     4.66
  CSEVERE       34.97     8.74    24.89     1.42    10.65
```

Finally, note the F matrix presented at the bottom of Table 8.5. This is automatically computed for any BMDP7M run, no matter what the specified contrast, and actually contains F statistics for all pairwise comparisons of the cells in the design. So, for example, the obtained F for the contrast between treatment and control for mild disability is 9.75.

For post hoc comparisons, the most straightforward procedure for protection against Type I error is a direct extension of the Scheffé test. (Just as in ANOVA, this is a *very conservative* test, but places no limit on the number of contrasts to be tested.) The procedure calls for an adjustment of the test value by multiplying the critical value of F by the degrees of freedom for the effect being tested. If marginal means for an effect are being contrasted, the degrees of freedom will be those associated with that main effect. If cell means are being contrasted, the degrees of freedom will be those associated with the interaction. (See Chapter 2 for a review of degrees of freedom for effects.) For example, in our sample if the contrast had been between the marginal means of mild versus moderate disability, averaged over treatment groups, the critical F value of 3.98 (2 and 11 df, $\alpha = .05$) would be multiplied by 2—the number of levels of disability minus 1.

Had our sample test for the linear component of the disability effect for control groups (a cell rather than marginal contrast) been tested post hoc, the critical F value would again be multiplied by 2. Degrees of freedom associated with an interaction equal the degrees of freedom for one effect (e.g., disability df = 2) times degrees of freedom for the other effect (e.g., treatment df = 1). After multiplying the critical F value of 3.98 by 2, the adjusted critical F of 7.96 is still exceeded by the obtained value of 8.74. By either planned or post hoc contrasts, then, the classification of children into disability groups was consistent with their performance on the two subtests of the WRAT.

The Scheffé procedure as just outlined holds only for the situation where contrasts among means are based on the "best" linear combination of all DVs, as determined by the omnibus tests for MANOVA. Procedures are also available for adjusting critical F in the case where both the linear combination of outcome measures and the contrast to be tested are determined post hoc. The procedures involve use of Roy's *gcr* criterion (see Section 8.5.1) and are beyond the scope of this book. A full discussion is provided by Harris (1975).

Data snooping may in addition be concerned with profile analysis, or relative effect of the IV on the various DVs. For example, does one level of the IV elevate all DVs, or does it elevate some DVs and depress others? Since profile analysis is limited to situations where all DVs are measured on the same scale, it will not be discussed in this book. The interested reader is referred to Harris (1975).

8.5.3 Assessing DVs

The problem of assessing DVs in MANOVA is much the same as the problem of assigning importance to IVs in multiple regression (Chapter 5).

8.5.3.1 Univariate *F* If the pooled within-group correlations among the DVs were zero, a series of univariate ANOVAs would give most of the information about their importance. Those DVs that produce significant univariate *F*'s would be the important ones. DVs could be ranked in importance by magnitude of univariate *F*. This is analogous to assessing importance of IVs in multiple regression by the magnitude of their individual correlations with the DV.

In the MANOVA case where DVs are correlated, they can be seen as measuring overlapping aspects of the same behavior. To say that two correlated DVs are both "significant" mistakenly suggests that the IV is affecting two different behaviors. For example, suppose two DVs were Stanford-Binet IQ and WISC IQ. Any IV that affected one measure would surely affect the other, since they are so highly correlated that they basically measure the same thing.

A further problem with reporting univariate *F*'s is that of inflation of Type I error rate. With correlated DVs the univariate *F*'s are not independent, and no straightforward adjustment of the error rate is possible. Nevertheless, some (Cooley & Lohnes, 1971) recommend reporting only univariate *F*'s as an aid in assessing DVs following a significant multivariate *F*. Though this is certainly a simple tactic, it should be accompanied by a table of pooled within-group correlations among DVs (such as might be provided by SPSS MANOVA) so that the reader might make the necessary interpretive adjustments. (BMDP7M provides the pooled within-group variance-covariance matrix that can be converted to correlations as per Chapter 1.)

For the small sample data, univariate ANOVAs (run as subproblems on BMD12V) of the two DVs, WRAT Reading and WRAT Arithmetic, are shown in Tables 8.6 and 8.7, respectively. The pooled within-group correlation between WRAT-R and WRAT-A is .057 with 12 df.

TABLE 8.6   UNIVARIATE ANALYSIS OF VARIANCE OF WRAT-R SCORES

SOURCE	SUM OF SQUARES	DEGREES OF FREEDOM	MEAN SQUARE	F
MEAN	142934.2222	1	142934.2222	3152.9608
D	520.7778	2	260.3889	5.7439
T	2090.8889	1	2090.8889	46.1225
DT	2.1111	2	1.0556	0.0233
S(DT)	544.0000	12	45.3333	

TABLE 8.7   UNIVARIATE ANALYSIS OF VARIANCE OF WRAT-A SCORES

SOURCE	SUM OF SQUARES	DEGREES OF FREEDOM	MEAN SQUARE	F
MEAN	136938.8889	1	136938.8889	3046.8480
D	1126.7778	2	563.3889	12.5352
T	1494.2222	1	1494.2222	33.2460
DT	52.7778	2	26.3889	0.5871
FULL MODEL	539.3333	12	44.9444	

8.5.3.2 Stepdown Analysis.[10]   A resolution to the problems of inflated Type I error rate and the nonindependence of univariate F tests is provided by the procedures of stepdown analysis, as described by Bock (1966) and Bock and Haggard (1968). This procedure calls for a determination of priority of DVs in terms of theoretical or practical interest.[11] The DVs are then tested in a series of ANCOVAs. The "most interesting" DV is tested first, with appropriate adjustment of alpha, in a univariate ANOVA. Each successive DV is then tested with the higher-priority DVs as covariates, to see if the new DV significantly adds to the combination of DVs already tested. In effect, this is analogous to testing the importance of IVs in multiple regression by running a hierarchical analysis.

Succeeding tests are statistically independent, so that the adjustment for inflated Type I error for the series of tests is relatively straightforward. The researcher assigns alpha for each of the tests in such a way that alpha for the series of tests does not exceed some critical value.

$$\alpha = 1 - (1 - \alpha_1)(1 - \alpha_2) \cdots (1 - \alpha_p)$$ (8.9)

The Type I error rate (α) for a stepdown analysis is based on the error rate for testing the first DV (α_1), the second DV (α_2), and all other DVs to the pth, or last, DV (α_p).

This procedure makes it possible to set all of the individual alphas at the same level, or to set alpha for the higher-priority DVs at a higher (less stringent) level than that for the less interesting DVs. For example, suppose there were four DVs. Setting the individual $\alpha = .01$, the overall α level according to Eq. 8.9 would be .039, acceptably below a critical level of .05. Or, the alpha for the two highest-priority DVs could be set at .02, and the two lower-priority DVs set at .001. The overall alpha would then be .041, again below a critical level of .05.

Using our sample data, let's say that the WRAT Reading score might be assigned a higher priority, since reading problems represent the most common presenting symptoms for learning disabled children. As there are only two DVs, it is possible to set the individual alpha as high as .025 for each, while keeping the overall $\alpha = .049$. The WRAT-R scores are then run as univariate ANOVA, the results of which have already been displayed in Table 8.6. The only effect of interest is that of treatment, since the main effect of disability produced no information and the interaction was not statistically significant in the MANOVA

---

[10] Frane (personal communication) recommends use of the term *sequential* rather than *stepdown* here to clearly distinguish between this analysis and stepwise analyses. We agree that the terms are confusing, but have chosen to retain the more widely used term while warning readers that stepdown is *not* stepwise.

[11] If one did not have any reason for assigning priority to the DVs, it would be possible to order the variables based on a statistical criterion. However, this tactic would suffer all the problems inherent in stepwise regression, discussed in Chapter 5.

TABLE 8.8 ANALYSIS OF COVARIANCE OF WRAT-A SCORES, WITH WRAT-R SCORES AS THE COVARIATE

SOURCE	SUM OF SQUARES	DEGREES OF FREEDOM	MEAN SQUARE	F
MEAN	460.5102	1	460.5102	9.4232
D	538.3662	2	269.1831	5.5082
T	268.3081	1	268.3081	5.4903
DT	52.1344	2	26.0672	0.5334
COVARIATES	1.7665	1	1.7665	0.0361
COVARIATE 1	1.7665	1	1.7665	0.0361
FULL MODEL	537.5668	11	48.8697	

(Table 8.2).[12] The critical value for testing the treatment effect, with 1 and 12 df at $\alpha = .025$ is 6.55. This is clearly exceeded by the obtained $F = 46.12$.

For the analysis of WRAT Arithmetic as the DV, an ANCOVA is run with WRAT Reading score as the covariate. The results of this analysis, using BMD12V, appear in Table 8.8.[13] Again, looking only at the treatment effect, the critical value with 1 and 11 df at $\alpha = .025$ is 6.72. This exceeds the obtained $F = 5.49$. Thus the stepdown analysis suggests that all the significant effect of treatment is represented in the increased WRAT-R scores, with no additional change uniquely represented by WRAT-A scores.

Note that WRAT-A scores show significant univariate but not stepdown F. Because WRAT-A scores are not significant in stepdown analysis does not mean they are unaffected by treatment, but simply that no unique variability is shared with treatment after adjusting for WRAT-R. This is analogous to the potential difference between bivariate and semipartial correlations between the DV and an IV in hierarchical regression. The significant bivariate correlation shows the IV to be a predictor. The nonsignificant semipartial r shows that it adds no additional predictability to the DV after other IVs are considered. Univariate F can be thought of as similar to the bivariate r between IV and DV; stepdown F is analogous to the semipartial r for a variable entering the equation after the first step in the hierarchy.

A stepdown analysis using a statistical criterion, such as might be considered in the absence of meaningful a priori ordering of variables, could be run through a stepwise discriminant function program (e.g., BMDP7M). Tests of main effects and interactions would be specified by appropriate contrast coefficients, and DVs could then be tested on the step in which they enter the equation. Further discussion of stepwise analysis will be provided in Chapter 9, on discriminant function analysis. Also discussed in that chapter (Section 9.6.4) is a proce-

---

[12] The entire stepdown analysis can be run *instead* of MANOVA (Bock & Haggard, 1968), where the finding of at least one significant stepdown F is interpreted as a significant multivariate effect of that IV (or interaction).

[13] A full stepdown analysis can be automatically produced by SPSS MANOVA by simple specification. For illustrative purposes, however, it is instructional to show how the analysis develops.

dure for evaluating predictor variables (DVs in MANOVA) by contrasting each group with all other groups.

The analysis of covariance procedure can be extended to tests of hypotheses about effects of IVs on a lower-priority *set* of DVs, after the effects of higher-order DVs have been eliminated by covariance. This involves multivariate analysis of covariance, with a set of high-priority DVs acting as covariates and the remaining DVs analyzed in combination. This procedure might be especially useful in situations where the DVs form categories of interest, such as scholastic variables and attitudinal variables. One could then test whether there is any change in attitudinal variables as a result of the IV, after the effect on scholastic variables has been partialled out. This is demonstrated in Chapter 9.

8.5.3.3 Standardized Discriminant Function Coefficients

One suggested way to assess the importance of DVs in relation to an IV is in terms of their relative weights in an equation set up to predict the levels of the IV from knowledge of DV scores. A discriminant function can be seen as this kind of equation, as it assigns coefficients to each of the DVs. The standardized discriminant function coefficients are analogous to beta weights in a multiple regression equation. Though these weights can provide some information, they suffer from some of the same problems of interpretation as beta weights—instability and, in the case of suppressor variables, distortion. Since these equations are derived in the context of discriminant function analysis, further discussion will be reserved for Chapter 9.

8.5.3.4 Loading Matrices in MANOVA

An alternative to interpreting discriminant function coefficients is interpreting a loading matrix. The elements in the loading matrix are correlations between the linear combination of DVs that maximizes treatment differences and the DVs themselves. DVs that correlate highly with the combination are more important to discrimination among groups. Interpretation of loading matrices in this context is fully described in Section 9.6.3.2.

8.5.3.5 Choosing Among Strategies for Assessing DVs

Although several procedures are available for assessing the importance of DVs and interpreting the pattern of results, there is a certain ambiguity associated with the use of each. Procedures that are statistically honest (e.g., stepdown analysis) are frequently difficult to interpret. Procedures that best describe the pattern of results (e.g., loading matrices, univariate F) do not realistically reflect Type I error rate.

We have found the following strategy to be useful. If the priority ordering of DVs is compelling on theoretical grounds, a stepdown analysis is clearly called for, with univariate F's and pooled within-cell correlations reported simply as supplemental information. Appropriate adjusted marginal and/or cell means are reported and interpreted (i.e., means on DVs adjusted for higher-priority DVs).

If the ordering of DVs is more arbitrary, an initial decision in favor of stepdown analysis may still be desirable on the grounds of statistical purity. If the pattern of results from the stepdown analysis makes sense logically in the light of the pattern from the univariate analysis, interpretation can take into account both patterns, with emphasis placed on those DVs that are significant in the stepdown analysis. For example, a DV may show a significant univariate effect, but a nonsignificant stepdown effect. In this case interpretation is straightforward: The variance the DV shares with the IV has already been accounted for through overlapping variance with a higher-priority DV. For those significant stepdown DVs, adjusted marginal and/or cell means can be reported. This is the strategy followed in the large sample example in Section 8.7.

Frequently, however, the pattern of results from the stepdown analysis will be incompatible with that of the univariate analyses. For example, a DV may show no significant univariate effect but nevertheless show a significant stepdown effect. That is, in the presence of higher-order DVs as covariates, the DV suddenly takes on "importance." In this case, interpretation becomes extremely complex and is tied to the context in which the DVs entered the stepdown analysis.

It may be worthwhile at this point, especially if there is only a weak basis for ordering DVs, to forgo evaluation of statistical significance of DVs and resort to simple description. Loading matrices (detailed in Section 9.6.3.2) provide one such description, although they are not without ambiguity. First, they do not remove overlapping variance among DVs (i.e., they are raw, unadjusted correlations of DVs with discriminant functions). Second, plots may be necessary to aid interpretation of the discriminant functions if there is more than one significant discriminant function associated with any main effect or interaction. For high-loading DVs, appropriate marginal and/or cell means (unadjusted) can be reported.

Or, for greatest simplicity, unadjusted marginal and/or cell means might be reported for those DVs with high univariate F's (after finding a significant multivariate effect), although it would be inappropriate to report significance levels for those univariate F's. That is, stepdown results could be ignored if uninterpretable.

8.5.4 Design Complexity

Extension of MANOVA to more than two between-subjects IVs is straightforward as long as sample sizes are equal within each cell of the design. Partitioning of sources of variance follows that of ANOVA, with a variance component associated with each main effect and interaction. Variance due to differences between subjects within cells serves as the single error term. The simple BMD12V program can handle up to 10 IVs. Specific comparisons and assessment of DVs can proceed as described in Sections 8.5.2 and 8.5.3. Two major design complexities that can arise, however, are inclusion of within-subjects IVs and unequal sample sizes in the various cells of the design.

8.5.4.1 Within-Subjects and Mixed Between-Within Designs As can be recalled from Chapter 2, multiple error terms are required whenever IVs are included in which subjects are measured at all levels of the IV. The simplest such design is one-way within-subjects, with all subjects measured in all conditions. With several IVs, all can be within-subjects, or there can be any combination of within- and between-subjects IVs. For MANOVA, analysis becomes increasingly complex because separate matrices, with subjects treated as an IV, are required for each error term.

Of available MANOVA programs, the easiest to interpret for repeated-measures designs are BMDP4V and SPSS MANOVA. But because of the algorithm used for repeated measures in SPSS MANOVA, computer memory requirements can become enormous with moderate size n. You may find that your computer either cannot handle the problem or charges you outrageously for the runs.

BMDP12V, the simplest MANOVA program to use, is quite deceptive in regard to repeated-measures designs. When nesting arrangements are specified, all appropriate error terms will appear in the output source table. However, these will *not* be the error terms used in testing the various main effects and interactions. Instead, a single error term labeled FULL MODEL is used to test *all* effects. The analysis proceeds as if a between-subjects design had been specified. The source of variance containing indices for all IVs (including subjects as an index) is considered the error term in all tests of main effects and interactions, even though components corresponding to appropriate error terms are listed in the source table.

With some extra effort, however, the user can apply the program for the more complex design. First, it is necessary to specify which error term is appropriate for each test of a main effect or interaction. The program (BMD12V) can be requested to produce cross-products matrices (**S** matrices) for each source of variance (i.e., variance component) specified by the design. The second step is to add associated effect and error matrices, as discussed in Section 8.4.

For example, for a two-way mixed design, suppose you wanted to test the main effect of the between-subjects variable, designated A. The appropriate error term would be $S(A)$, or subjects nested within levels of A. You would then add the $S_{S(A)}$ matrix to the S_A matrix to form the $[S_A + S_{S(A)}]$ matrix.

The third step would be to find the determinants of the two appropriate matrices. Wilks' Lambda could then be found as per Eq. 8.4, for example.

$$\Lambda = \frac{|S_{S(A)}|}{|S_A + S_{S(A)}|}$$

Finally, approximate F can be found as illustrated in Eq. 8.5, and the significance of the multivariate effect can be tested against the critical value of F, with the two degree-of-freedom parameters shown in that equation. Each main effect and interaction of interest can be tested in this way. This procedure, however, is limited to MANOVA. In the case of MANCOVA, there is no provision for printing out the *reduced* cross-products matrices in BMD12V (see Section 8.4.2).

8.5.4.2 Unequal Sample Sizes When cells in a factorial ANOVA have an unequal number of scores, the tests of main effects and interactions are no longer independent and the design becomes nonorthogonal. This problem generalizes to MANOVA.

There are a number of alternative methods for adjusting sums of squares to test ANOVA hypotheses, and these can all be generalized to cross-product or variance-covariance matrices for MANOVA. Of the simpler programs, only SPSS MANOVA and BMDP4V are set up to adjust for unequal n. In SPSS MANOVA, the default option is Method 3 (cf. Section 7.5.2.2), not the most popular of choices. Other alternatives are available, however, with a simple procedure for Method 1: METHOD = SSTYPE(UNIQUE). BMDP4V has no default option, but specification of method is very simple through choice of cell-weighting procedure.

A thorough discussion of adjustment methods for nonorthogonal designs is provided by Woodward and Overall (1975) in the context of applying multiple regression procedures to MANOVA.

8.6 COMPARISON OF PROGRAMS

8.6.1 SPSS Package

Within the original SPSS series of programs (Nie et al., 1975), there is none specifically designed for MANOVA. Applications of REGRESSION to the general linear model are restricted to the univariate model. One-way MANOVA can be run on the discriminant function analysis program, DISCRIMINANT, and, as suggested earlier, this simple type of MANOVA frequently is best run through that type of program. Discussion of the program will be reserved for Chapter 9, in which discriminant function analysis is covered.

A newer program, however, was added after publication of the original package, and it vastly expands capabilities of SPSS for both MANOVA and discriminant function analyses. SPSS MANOVA is a comprehensive and flexible program. Features of the program are summarized in Table 8.9.

The program is the only one available that deals automatically with nonorthogonality produced by unequal sample sizes and offers alternative methods of adjustment. (Note, however, that the default option, hierarchical, typically is not the most popular choice.) It is also the only program that can automatically perform a stepdown analysis, as described in Section 8.5.3.2. Further, should some statistical criterion other than Wilks' Lambda be desired (cf. Section 8.5.1), SPSS MANOVA provides alternatives.

Tests are available for evaluation of two MANOVA assumptions—multicollinearity and homogeneity of variance-covariance matrices—but not for outliers. Another feature, test for sphericity, allows evaluation of the efficacy of

TABLE 8.9 COMPARISON OF PROGRAMS FOR MULTIVARIATE ANALYSIS OF VARIANCE AND COVARIANCE

Feature	BMDP4V	BMD12V	BMDP7M	SPSS MANOVA[a]
Input				
Contrast coding required	No	No	If factorial	No
Data must be ordered	No	Yes	No	No
Equal n required	No	Yes	No	No
Variety of strategies for unequal n	Yes	N.A.	No	Yes
Alternative languages or procedures to specify design	Yes	No	No	Yes
Output				
Source table	Yes	Yes	No	PRINT = SIGNIF(BRIEF)
Stepwise solution	No	No	Yes	No
Omnibus F tests	Yes	Yes	Yes	Yes
Specific comparisons (user-specified)	Yes	No	Yes	Yes
Tests of simple effects	Yes	No	No	Yes
F matrix of cell comparisons	No	No	Yes	No
Design matrix	Yes	No	No	Yes
Covariance matrices	No	No	No	Yes
Covariance matrix determinants	No	(Logs)	No	Yes
Correlation matrices	No	No	No	Yes
Cross-products matrices	Yes	No	No	Yes
Cross-products matrix determinants	No	Partial (Logs)	No	Yes
Marginal standard deviation	Yes	No	No	No
Marginal means	Yes	No	No	Yes
Weighted marginal means	Yes	No	No	Yes
Cell means	Yes	Yes	Yes	Yes
Cell standard deviations	Yes	No	Yes	Yes
Cell and marginal variance	Yes	No	No	No
Confidence interval around cell means	No	No	No	Yes
Cell and marginal minimum and maximum values	Yes	No	No	No
Adjusted cell and marginal means	No	Yes	No	Yes

Wilks' Lambda (U statistic)	LRATIO	Yes	Yes
Approximate F statistic	Yes	Yes	Yes
Greenhouse-Geisser-Imhof statistic	Yes	No	No
Test for outliers	No	Yes	No
Classification of cases	No	Yes	No
Canonical (discriminant function) statistics[b]	Eigenvalues and eigenvectors	Yes	Yes
Univariate F tests	Yes	STEP 0 F TO ENTER	Yes
Averaged univariate F tests	No	No	Yes
Stepdown F tests (DVs)	No[c]	No	Yes
Criteria other than Wilks'	No	No	Yes
Bartlett test of sphericity	No	No	Yes
Tests for univariate homogeneity of variance	No	No	Yes
Test for homogeneity of covariance matrices	No	No	BOX'S M
Principal component analysis of residuals	No	No	Yes
Pooled within-cell correlations	No	Yes	Yes
Pooled within-cell covariances	No	No	Yes
Determinant of pooled within-cell correlation matrix	Yes	No	Yes
Profile analysis	No	No	Yes[d]
Homogeneity of regression	No	No	Yes
ANCOVA with separate regression estimates	No	No	Yes
Sphericity indices for homogeneity of covariance	Yes	No	No
Tukey's test for nonadditivity	No	No	Yes
Normal plots for each DV and covariate	No	No	Yes
Predicted values and residuals for each case	No	No	Yes
Residuals plots	No	No	Yes

a Based on Release 9.0 (Hull & Nie, 1981).

b Discussed more fully in Chapter 9.

c Can be done through successive ANCOVAs as subproblems.

d Procedures for doing this are described in a preliminary SPSS-MANOVA manual (Cohen & Burns, 1977).

univariate versus multivariate analysis and is described in the SPSS Update 7–9 manual. An FMAX test for homogeneity of variance among DVs facilitates the choice between repeated-measures univariate ANOVA and MANOVA.

Homogeneity of regression is readily tested. If the assumption is violated, the manual describes procedures for ANCOVA with separate regression estimates. Adjusted marginal as well as cell means can be obtained. However, there is no test for homogeneity of covariance.

Finally, a principal components analysis can be performed, as described in the manual. In the case of multicollinearity or singularity among DVs (see Chapter 4) principal components analysis can be used to produce composite variables that are orthogonal to one another. (Note, however, that the PCA is just descriptive. The MANOVA analysis is still done on the raw DV scores, not the component scores.)

8.6.2 BMD Series

There are several programs available in the BMD series, some specifically designed for MANOVA and others basically discriminant function analysis programs.

The earliest of the programs, BMD04M (Discriminant Analysis for Two Groups), BMD05M (Discriminant Analysis for Several Groups), and BMD07M (Stepwise Discriminant Analysis) are, like the SPSS DISCRIMINANT program mentioned earlier, limited to one-way MANOVA. Furthermore, features of all three programs have been consolidated into the newest P-series program, BMDP7M (Dixon, 1981). For users who do not have access to the newer package, the three original BMD discriminant analysis programs will be discussed in Chapter 9. BMD11V is a highly flexible program based on the multivariate general linear model. But it is very difficult to use, and most of its features have been incorporated into the newer BMDP4V program. Features of BMD12V, BMDP4V, and BMDP7M are summarized in Table 8.9.

One P-series program, BMDP7M, is actually designed for MANOVA as well as discriminant function analysis as titled, and handles factorial as well as one-way designs. Tests of hypotheses (main effects, interactions, specific comparisons, and trends) are set up by specifying contrasts. For each contrast a complete stepwise analysis is done, similar to the procedure described in Chapter 5 on multiple regression. This is especially useful in identifying a subset of DVs "working" for each main effect and interaction. By manipulating order of entry of variables (as in hierarchical regression analysis), one can run a form of multivariate analyses of covariance. Covariates are forced to enter the equation early, and the effect of the DVs is assessed after adjusting for the covariates. The test criterion is Wilks' Lambda (or U statistic) with associated approximate F. A useful procedure in this program, unique among factorial MANOVA programs, is a test for outliers.

There are, however, some serious complications in using BMDP7M for factorial MANOVA and MANCOVA. First, specification of the contrasts may pose some difficulty for the novice user. Little help is given in the manual

(Dixon, 1981) and the user unfamiliar with contrast coding in analysis of variance would need to consult other sources, such as Cohen and Cohen (1975). Further, in order to produce meaningful contrasts for a factorial design, a *new* indexing variable must be produced that forms a single set of groups from the original groups within factors. For example, if one has a 2 × 3 design (i.e., two groups within the first factor and three groups within the second factor, crossed to produce six groups), cases will have to be coded in such a way that they fall into groups 1 through 6 by use of variable transformation procedures to combine two indices into one, or by rearranging cases into groups and creating a "group" variable corresponding to the six groups.

A second complication is produced by unequal sample sizes. Although the program does not require equal *n,* the results of factorial designs cannot be taken at face value because no adjustment is made for nonorthogonality (cf. Section 8.5.4.2). This takes the use of the program for factorial designs with unequal sample sizes beyond the reach of all but the most sophisticated user. As a third complication, the program is limited to simple between-subjects IVs, with no provision for within-subjects variables. A final disadvantage of the program is that no source table is produced.

The simplest, most straightforward program to use for MANOVA and MANCOVA is BMD12V (Dixon, 1974).[14] This program is well adapted to the novice user, and sets up in a fashion similar to the most popular analysis of variance program in the series, BMD08V. IVs need not be coded, but it is necessary to order data in a matter consistent with the design specified. Tests of main effects and interactions are automatically produced using the Wilks' Lambda criterion (*U*-statistic), and associated approximate *F*'s are given in a source table. Marginal and cell means for each DV can be requested, and, in the case of covariance, adjusted as well as unadjusted means can be obtained.

The program does, however, have some serious limitations. First, it requires equal sample sizes. Second, there is no procedure for specifying contrasts to test specific comparisons—only omnibus tests can be done. Specific comparisons would require runs of the data on some other program, such as BMDP4V, BMDP7M, or SPSS MANOVA. Finally, although setup procedures call for specification of between- and within-subjects variables, the analysis proceeds as if all factors were between-subjects factors. Procedures for circumventing this problem, by use of program-produced cross-products matrices, are discussed in Section 8.5.4.1. However, in multivariate analysis of covariance the reduced matrix is not produced, so the procedures outlined in Section 8.5.4.1 would become even more complex. In its most direct form, then, BMD12V is limited to a between-subjects, equal-*n* design in which only omnibus tests are of interest.

The newest, most flexible, and most comprehensive MANOVA program in the BMD series is P4V. The program is also known as URWAS (University of Rochester Weighted Analysis of Variance System). It is unique among the programs reviewed here in that it uses a cell-weighting system to specify hypothe-

---

[14] In the earlier BMD X-series publication (Dixon, 1972) this program was known as BMDX69.

ses to be tested, particularly with respect to adjustment for unequal n. And strategies for unequal n include weighting for situations in which population sizes are unequal even though sample sizes are equal.

Information is given regarding the tenability of the assumption of homogeneity of covariance, and a Greenhouse-Geisser-Imhof correction for degrees of freedom is made in case the assumption is not met.

Profile analysis statistics (in which levels of a within-subjects IV are treated as separate DVs) are automatically provided, even when multiple DVs are specified as well. Weighted marginal means are given that are useful if weights other than sample sizes are used (see Section 7.5.2.2). Adjusted (for covariates) cell and marginal means are planned but had not been implemented at the time of this writing. Three multivariate test criteria, as well as univariate F tests, are automatically provided. Specific comparisons include a simple effects procedure. For example, in a three-way design, factorial analysis of two of the IVs can be done at each level of the third IV.

This is the only program that provides alternative language systems for specifying complex and nonorthogonal designs. For a factorial design in which tests of all main effects and interactions are desired, a simple statement so indicates. For less routine analysis, a Model Description Language is available, in which the user specifies the design matrix of contrast coefficients and/or other design goodies. Finally, a structural-equation procedure allows the user to write out an equation that specifies the design and desired tests.

No test for homogeneity of regression is available for MANCOVA, although tests for MANOVA stepdown analysis could be done through BMDP1V, as described in Chapter 7. No direct evidence is available to evaluate multicollinearity, but the program does deal with the problem. If a sum-of-squares and cross-products matrix is multicollinear, it is conditioned and a pseudoinverse calculated. And if there is any problem with the accuracy of a statistic, warning messages regarding precision are printed out.

8.7 COMPLETE EXAMPLES OF MULTIVARIATE ANALYSIS OF VARIANCE AND COVARIANCE

In the research described in Appendix B, there was interest in the manner in which a variety of variables differed as a function of sex role identification. What are the differences among women who vary in the way they identify with traits characteristic of the sexes?

Sex role identification was defined as performance on the two scales of the Bem Sex Role Inventory (Bem, 1974). Performance can be distinguished into four groups, defined by the factorial combination of high and low femininity with high and low masculinity. These four groups are formed by dividing the masculinity and femininity scales each at their median. The groups, then, are (1) Undifferentiated, with low scores on both the femininity and masculinity scales, (2) Feminine, with high femininity scores and low masculinity scores, (3) Masculine, with high masculine and low feminine scores, and (4) Androgynous, with high scores on both scales.

Given this breakdown of scores on the two scales, it is possible to assess the effect of masculinity, femininity, and the interaction between masculinity and femininity on any DVs of interest. Chosen as DVs for this analysis were self-esteem (SELFESTE), internal versus external locus of control (IESCALE), attitudes toward women's role (WOMENROL), socioeconomic level (DUN), an introversion-extraversion scale (INTEXT), and neuroticism (NEUROTIC).

An omnibus MANOVA shows us whether these DVs are associated with the two IVs (femininity and masculinity) or their interaction. Then a stepdown analysis, in conjunction with the univariate F values, allows us to examine the pattern of relationships between DVs and each IV.

In a second MANCOVA analysis, DUN, IESCALE, and WOMENROL are considered to be covariates, with SELFESTE, INTEXT, and NEUROTIC serving as DVs. Here the question is whether the three personality DVs vary as a function of sex role identification (the two IVs and their interaction) after adjusting for socioeconomic status, attitudes toward women's role, and beliefs regarding locus of control of reinforcements.

8.7.1 Evaluation of Assumptions

Before preceeding with the multivariate analyses of variance and covariance, our variables will be assessed with respect to practical limitations of the technique.

8.7.1.1 Unequal Sample Sizes and Missing Data   No datum was missing on any of the DVs chosen for the 369 women who were administered the Bem Sex Role Inventory. The hierarchical approach to adjusting for nonorthogonality produced by unequal n was used, with order of effects: F (femininity), M (masculinity), and F by M (interaction between femininity and masculinity). SPSS MANOVA, used for this analysis, provides this approach as default. F was chosen over M as the highest-order effect for this female sample.

8.7.1.2 Outliers   BMDP7D was run as a check for univariate outliers. Distributions of variables are shown separately for each group. As can be seen in Table 8.10, one of the IESCALE values in the feminine group diverges sharply from the rest of the cases, with $z = -4.94$. After elimination of this, the only univariate outlier, multivariate outliers were tested through BMDPAM in the manner illustrated in Section 7.7.1.

With $p < .01$ as a cutoff criterion, six additional outlying cases were found; two in the feminine group, one in the masculine group, and three who were androgynous. This forced a reduction in sample size to 362, with 35 cases in the smallest group. Outlying cases were examined through BMDP9R and BMDP1D, as described in Section 5.8.1.

8.7.1.3 Multivariate Normality   The reduced sample size of 362 includes over 30 cases for each cell of the 2 × 2 between-subjects design (2 levels each of masculinity and femininity) after elimination of outliers. This produces far

TABLE 8.10 DECK SETUP AND SELECTED OUTPUT FROM BMDP7D RUN
TO CHECK DISTRIBUTIONS AND UNIVARIATE OUTLIERS

```
PROGRAM CONTROL INFORMATION

        /PROBLEM      TITLE IS "LARGE SAMPLE MANOVA, UNIVARIATE OUTLIERS".
        /INPUT        VARIABLES ARE 10.     UNIT = 80.
                      FORMAT IS "(F5.0,9F8.5)".
        /VARIABLE     NAMES ARE CASENO,ANDRM,FEM,MASC,SELFESTE,IESCALE,
                      WOMENROL,DUN,INTEXT,NEUROTIC.
                      LABEL=CASENO.
        /GROUP        CODES(2) ARE 1,2,3,4.
                      NAMES(2) ARE UNDIFFERENTIATED,FEMININE,MASCULINE,ANDROGYNOUS.
        /HISTOGRAM    GROUPING IS ANDRM.
        /END


                 ************                                                ************
HISTOGRAM OF * IESCALE  * (VARIABLE    6). CASES DIVIDED INTO GROUPS BASED ON VALUES OF * ANDRM   * (VARIABLE    2)
                 ************                                                ************

                                                                                       CASES WITH
                                                                                       UNUSED
                                                                                       VALUES FOR
                UNDIFFER            FEMININE            MASCULIN         ANDROGYN        ANDRM

            ...................+..................+...................+...................+................
MIDPOINTS
  17.000)
  16.000)
  15.000)
  14.000)
  13.000)
  12.000)
  11.000)
  10.000)***                 ***                                  **
   9.000)********            *******************  **              ***
   8.000)****************    ******************* *****            ************
   7.000)M*************      M****************51 **********        ************************
   6.000)*****************24 ***********************54 M*********  M*******************26
   5.000)********            **********************26 *********    ***********************
   4.000)
   3.000)
   2.000)
   1.000)
   0.000)
  -1.000)
  -2.000)
  -3.000)
  -4.000)
  -5.000)
  -6.000)
  -7.000)
GROUP MEANS ARE DENOTED BY M"S IF THEY COINCIDE WITH *"S, N"S OTHERWISE

MEAN        7.042               6.734               6.500            6.461
STD.DEV.    1.346               1.363               1.183            1.235
S. E. M.     .160                .104                .197             .131
MAXIMUM    10.000              10.000               9.000           10.000
MINIMUM     5.000               0.000               5.000            5.000
SAMPLE SIZE   71                 173                  36               89
```

more than the 20 df for error suggested to assure multivariate normality of
the sampling distribution of means, even with unequal sample sizes in the 4
cells. Further, no glaring skewness was observed in the distributions produced
by BMDP7D (e.g., Table 8.10).

The provision for two-tailed tests is handled by the computer programs
used. That is, the F test provides for a test of differences between means in
either direction.

8.7.1.4 Homogeneity of Variance-Covariance Matrices

As a preliminary
check for robustness, sample variances for each of the 6 DVs in each of the 4
IV groups were examined. For no DV did the ratio of the largest to smallest
variance approach $20:1$. As a matter of fact, the largest ratio was about $1.5:1$
for the Undifferentiated versus Androgynous groups on IESCALE.

Sample sizes are widely discrepant, with a ratio of almost $5:1$ for Feminine to Masculine groups. However, with the minimal differences in variance and the use of two-tailed tests, the discrepancy in sample sizes does not invalidate use of MANOVA. A test for homogeneity of covariance matrices performed through SPSS MANOVA produced $F(63, 63020) = 1.07$, $p > .05$ for Box's M, showing no statistically significant deviation from homogeneity of covariance matrices.

8.7.1.5 Linearity   A few of the 60 within-cell scatterplots were examined for linearity through SPSS SCATTERGRAM. As an example, Figure 8.2 illustrates the plot of scores on NEUROTIC against scores on IESCALE for women defined as androgynous ($r = .39$). Since none of the scatterplots shows gross

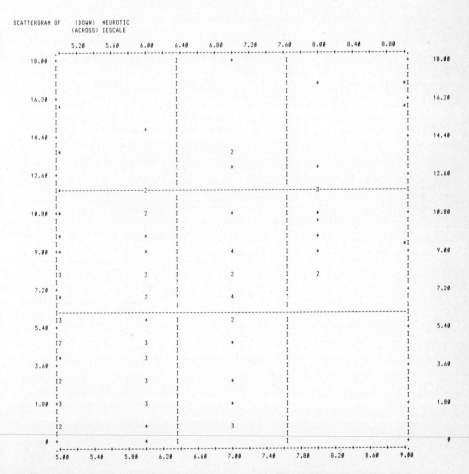

Figure 8.2 Scatterplot of relationship between scores on NEUROTIC and IESCALE for androgynous women. Selected input and output from SPSS SCATTERGRAM.

deviation from linearity, and since minor violations of the assumption of linearity should do nothing except reduce the power of the MANOVA, the decision was made to retain all variables in their original form.

8.7.1.6 Homogeneity of Regression   In the omnibus MANOVA no covariates were used. In the stepdown tests, however, some DVs act as covariates in some analyses. Tests for homogeneity of regression were therefore run for all steps of the stepdown analysis. Homogeneity of regression was established for all steps. Table 8.11 illustrates the SPSS MANOVA deck setup for tests of homogeneity of regression, and output for the two last steps. At each step, the relevant effect is the one appearing last, so that for DUN, for example,

TABLE 8.11   TEST FOR HOMOGENEITY OF REGRESSION FOR LARGE
SAMPLE MANOVA STEPDOWN ANALYSIS (DECK SETUP AND
PARTIAL OUTPUT FOR LAST TWO TESTS FROM SPSS
MANOVA)

```
          MANOVA        SELFESTE,WOMENROL,NEUROTIC,INTEXT,IESCALE,DUN BY
                        F(1,2) M(1,2)/
                        PRINT=SIGNIF(BRIEF)/
                        NOPRINT=PARAMETERS(ESTIM)/
                         ANALYSIS=WOMENROL/
                         DESIGN=SELFESTE,F,M,F BY M, SELFESTE BY F + SELFESTE BY M +
                         SELFESTE BY F BY M/
                         ANALYSIS=NEUROTIC/
                         DESIGN=SELFESTE,WOMENROL,F,M,F BY M,CONTIN(SELFESTE,
                         WOMENROL) BY F + CONTIN(SELFESTE,WOMENROL) BY M + CONTIN
                         (SELFESTE,WOMENROL) BY F BY M/
                         ANALYSIS=INTEXT/
                         DESIGN=SELFESTE,WOMENROL,NEUROTIC,F,M,F BY M,CONTIN
                         (SELFESTE,WOMENROL,NEUROTIC) BY F + CONTIN(SELFESTE,WOMENROL,
                         NEUROTIC) BY M + CONTIN(SELFESTE,WOMENROL,NEUROTIC) BY F BY M/
                         ANALYSIS=IESCALE/
                         DESIGN=SELFESTE,WOMENROL,NEUROTIC,INTEXT,F,M,F BY M,
                         CONTIN(SELFESTE,WOMENROL,NEUROTIC,INTEXT) BY F + CONTIN
                         (SELFESTE,WOMENROL,NEUROTIC,INTEXT) BY M + CONTIN(SELFESTE,
                         WOMENROL,NEUROTIC,INTEXT) BY F BY M/
                         ANALYSIS=DUN/
                         DESIGN=SELFESTE,WOMENROL,NEUROTIC,INTEXT,IESCALE,F,M,
                         F BY M,CONTIN(SELFESTE,WOMENROL,NEUROTIC,INTEXT,IESCALE) BY F +
                         CONTIN(SELFESTE,WOMENROL,NEUROTIC,INTEXT,IESCALE) BY M +
                         CONTIN(SELFESTE,WOMENROL,NEUROTIC,INTEXT,IESCALE) BY F BY M/
```

TESTS OF SIGNIFICANCE FOR IESCALE USING SEQUENTIAL SUMS OF SQUARES

SOURCE OF VARIATION	SUM OF SQUARES	DF	MEAN SQUARE	F	SIG. OF F
WITHIN+RESIDUAL	416.56500	342	1.21803		
CONSTANT	16258.22099	1	16258.22099	13348.00468	0
SELFESTE	91.74519	1	91.74519	75.32283	0
WOMENROL	3.51935	1	3.51935	2.88938	.090
NEUROTIC	48.71343	1	48.71343	39.99374	0
INTEXT	.50525	1	.50525	.41481	.520
F	.02343	1	.02343	.01924	.890
M	.23588	1	.23588	.19366	.660
F BY M	.00387	1	.00387	.00318	.955
CONTIN(SELFESTE WOMENROL NEUROTIC INTEXT) BY F + CONTIN(SELFESTE WOMENROL NEUROTIC INTEXT) BY M + CONTIN(SELFESTE WOMENROL NEUROTIC INTEXT) BY F BY M	12.46761	12	1.03897	.85299	.596

SOURCE OF VARIATION	SUM OF SQUARES	DF	MEAN SQUARE	F	SIG. OF F
WITHIN+RESIDUAL	217304.06573	338	642.91144		
CONSTANT	611707.76443	1	611707.76443	951.46505	0
SELFESTE	797.52660	1	797.52660	1.24049	.266
WOMENROL	4.30139	1	4.30139	.00669	.935
NEUROTIC	3.00406	1	3.00406	.00467	.946
INTEXT	707.57127	1	707.57127	1.10057	.295
IESCALE	1123.90168	1	1123.90168	1.74814	.187
F	.91767	1	.91767	.00143	.970
M	355.23229	1	355.23229	.55254	.458
F BY M	198.70914	1	198.70914	.30988	.579
CONTIN(SELFESTE WOMENROL NEUROTIC INTEXT IESCALE) BY F + CONTIN(SELFESTE WOMENROL NEUROTIC INTEXT IESCALE) BY M + CONTIN(SELFESTE WOMENROL NEUROTIC INTEXT IESCALE) BY F BY M	13800.36614	15	920.02441	1.43103	.130

$F(15, 338) = 1.43$, $p > .01$. (An appropriate cutoff here would be .01, since robustness might be expected.)

For the MANCOVA, an overall test of homogeneity of regression is required, in addition to stepdown tests. All runs showed sufficient homogeneity of regression for this analysis. Deck setup and output for the overall test and the last stepdown test are shown in Table 8.12. Note that multivariate output is printed for the overall test since there are three DVs rather than a single one, as for stepdown tests.

8.7.1.7 Reliability of Covariates   For the stepdown analysis in MANOVA, all DVs except SELFESTE act as covariates, and are therefore required to be reliable. Based on the nature of scale development and data collection procedures, there is no reason to expect unreliability of a magnitude harmful to covariance analysis for WOMENROL, NEUROTIC, INTEXT, IESCALE, and DUN.

TABLE 8.12   TESTS FOR HOMOGENEITY OF REGRESSION FOR LARGE SAMPLE MANCOVA AND STEPDOWN ANALYSIS (DECK SETUP AND PARTIAL OUTPUT FOR OVERALL TEST AND LAST STEPDOWN TEST FROM SPSS MANOVA)

```
MANOVA       SELFESTE,WOMENROL,NEUROTIC,INTEXT,IESCALE,DUN BY
             F(1,2) M(1,2)/
             PRINT=SIGNIF(BRIEF)/
             NOPRINT=PARAMETERS(ESTIM)/
             ANALYSIS=SELFESTE,INTEXT,NEUROTIC/
             DESIGN=IESCALE,WOMENROL,DUN,F,M,F BY M,
             CONTIN(IESCALE,WOMENROL,DUN) BY F +
             CONTIN(IESCALE,WOMENROL,DUN) BY M +
             CONTIN(IESCALE,WOMENROL,DUN) BY F BY M/
             ANALYSIS=SELFESTE/
             DESIGN=IESCALE,WOMENROL,DUN,F,M,F BY M,
             CONTIN(IESCALE,WOMENROL,DUN) BY F +
             CONTIN(IESCALE,WOMENROL,DUN) BY M +
             CONTIN(IESCALE,WOMENROL,DUN) BY F BY M/
             ANALYSIS=INTEXT/
             DESIGN=IESCALE,WOMENROL,DUN,SELFESTE,F,M,F BY M,
             CONTIN(IESCALE,WOMENROL,DUN,SELFESTE) BY F +
             CONTIN(IESCALE,WOMENROL,DUN,SELFESTE) BY M +
             CONTIN(IESCALE,WOMENROL,DUN,SELFESTE) BY F BY M/
             ANALYSIS=NEUROTIC/
             DESIGN=IESCALE,WOMENROL,DUN,SELFESTE,INTEXT,F,M,F BY M,
             CONTIN(IESCALE,WOMENROL,DUN,SELFESTE,INTEXT) BY F +
             CONTIN(IESCALE,WOMENROL,DUN,SELFESTE,INTEXT) BY M +
             CONTIN(IESCALE,WOMENROL,DUN,SELFESTE,INTEXT) BY F BY M/
```

TESTS OF SIGNIFICANCE FOR WITHIN+RESIDUAL USING SEQUENTIAL SUMS OF SQUARES

SOURCE OF VARIATION	WILKS LAMBDA	MULT. F	HYPOTH. DF	ERROR DF	SIG. OF F
CONSTANT	.02277	4921.05419	3.00	344.00	0
IESCALE	.71731	45.18900	3.00	344.00	0
WOMENROL	.92311	9.55094	3.00	344.00	0
DUN	.99572	.49300	3.00	344.00	.687
F	.94658	6.47065	3.00	344.00	2.852E-004
M	.83917	21.97644	3.00	344.00	0
F BY M	.99448	.63647	3.00	344.00	.592
CONTIN(IESCALE WOMENROL DUN) BY F + CONTIN(IESCALE WOMENROL DUN) BY M + CONTIN(IESCALE WOMENROL DUN) BY F BY M	.93605	.85209	27.00	1005.30	.683

TESTS OF SIGNIFICANCE FOR NEUROTIC USING SEQUENTIAL SUMS OF SQUARES

SOURCE OF VARIATION	SUM OF SQUARES	DF	MEAN SQUARE	F	SIG. OF F
WITHIN+RESIDUAL	6361.84441	338	18.82202		
CONSTANT	27054.61395	1	27054.61395	1437.39125	0
IESCALE	1512.77043	1	1512.77043	80.37235	0
WOMENROL	58.02196	1	58.02196	3.08266	.080
DUN	1.32880	1	1.32880	.07060	.791
SELFESTE	362.81197	1	362.81197	19.27593	1.515E-005
INTEXT	25.40394	1	25.40394	1.34969	.246
F	3.15822	1	3.15822	.16779	.682
M	.57084	1	.57084	.03033	.862
F BY M	.29518	1	.29518	.01568	.900
CONTIN(IESCALE WOMENROL DUN SELFESTE INTEXT) BY F + CONTIN(IESCALE WOMENROL DUN SELFESTE INTEXT) BY M + CONTIN(IESCALE WOMENROL DUN SELFESTE INTEXT) BY F BY M	347.93030	15	23.19535	1.23235	.245

These same variables act as true or stepdown covariates in the MANCOVA analysis.

8.7.1.8 Multicollinearity and Singularity   The determinant of the pooled within-cells correlation matrix was found, through SPSS MANOVA, to be 0.81. This is sufficiently different from zero that neither multicollinearity nor singularity is judged to be a problem.

8.7.2 Multivariate Analysis of Variance

Deck setup and partial output of the omnibus MANOVA as produced by SPSS MANOVA appears in Table 8.13. Four parallel tests of significance are given for each of the three effects in turn: F by M (the interaction between femininity and masculinity), M (low vs. high masculinity), and F (low vs. high femininity). Because there are only two levels of each of the IVs, three of the tests—Pillai's, Hotelling's, and Wilks'—produce the same F.[15] (With only 1 df, F is not approximate.) The tests show highly significant main effects, with no statistically significant interaction. Following Eq. 8.5, the value of Wilks' Lambda can be used to find a measure of the strength of association between the IV effect and the combination of DVs.

Since the omnibus MANOVA shows significant multivariate effects, it is appropriate to investigate further the nature of the relationships among the IVs and DVs. Three kinds of information can help clarify these relationships.

First, the degree to which DVs are intercorrelated provides information as to the independence of behaviors being assessed. Pooled within-cell correlations, as produced by SPSS MANOVA [PRINT=ERROR(COR)], appear in Table 8.14 (page 272).[16] (Diagonal elements are pooled standard deviations.)

Second, it is usually interesting to investigate the univariate F's for each of the DVs. Although the statistical significance of these F values is misleading, investigators frequently are interested in the ANOVA that would have been produced if each DV had been investigated in isolation. These univariate analyses are produced automatically by SPSS MANOVA. These are shown in Table 8.15 (page 272) for the three effects in turn: F by M, M, and F.

Finally, stepdown analysis allows a look at the significance of DVs in context, with Type I error rate controlled. For the purpose of this study, the following priority order of DVs was developed, from most to least important: SELFESTE, WOMENROL, NEUROTIC, INTEXT, IESCALE, DUN. Following the procedures for stepdown analysis (see Section 8.5.3.2), the highest-priority DV, SELFESTE, is tested in a univariate ANOVA. The second most important DV, WOMENROL, is assessed in an analysis of covariance with SELFESTE

---

[15] For more complex designs, it might be worthwhile to obtain a single source table containing all effects through PRINT=SIGNIF(BRIEF). However, that table displays only Wilks' criterion.

[16] For a MANCOVA problem, SPSS MANOVA prints pooled within-cell correlations among DVs (called criteria), adjusted for covariates. To get a full within-cell correlation matrix (covariates as well as DVs), you need a run in which covariates are included in the set of DVs.

TABLE 8.13 MULTIVARIATE ANALYSIS OF VARIANCE OF SELFESTE,
IESCALE, WOMENROL, DUN, INTEXT, AND NEUROTIC, AS A
FUNCTION OF (TOP TO BOTTOM) F BY M INTERACTION,
MASCULINITY, AND FEMININITY (DECK SETUP AND
SELECTED OUTPUT FROM SPSS MANOVA)

```
    MANOVA          SELFESTE,WOMENROL,NEUROTIC,INTEXT,IESCALE,DUN BY
                    F(1,2) M(1,2)/
                    ANALYSIS=SELFESTE,INTEXT,NEUROTIC WITH IESCALE,WOMENROL,DUN/
                    PRINT=SIGNIF(STEPDOWN), ERROR(COR),
                    HOMOGENEITY(BARTLETT,COCHRAN,BOXM)/
                    NOPRINT=PARAMETERS(ESTIM)/
```

EFFECT .. F BY M

MULTIVARIATE TESTS OF SIGNIFICANCE (S = 1, M = 2, N = 175 1/2)

TEST NAME	VALUE	APPROX. F	HYPOTHESIS D. F.	ERROR D. F.	SIGNIF. OF F
PILLAIS	.00963	.57188	6.00000	353.00000	.75273
HOTELLINGS	.00972	.57188	6.00000	353.00000	.75273
WILKS	.99037	.57188	6.00000	353.00000	.75273
ROYS	.00963				

EFFECT .. M

MULTIVARIATE TESTS OF SIGNIFICANCE (S = 1, M = 2, N = 175 1/2)

TEST NAME	VALUE	APPROX. F	HYPOTHESIS D. F.	ERROR D. F.	SIGNIF. OF F
PILLAIS	.26192	20.87838	6.00000	353.00000	.00001
HOTELLINGS	.35487	20.87838	6.00000	353.00000	.00001
WILKS	.73808	20.87838	6.00000	353.00000	.00001
ROYS	.26192				

EFFECT .. F

MULTIVARIATE TESTS OF SIGNIFICANCE (S = 1, M = 2, N = 175 1/2)

TEST NAME	VALUE	APPROX. F	HYPOTHESIS D. F.	ERROR D. F.	SIGNIF. OF F
PILLAIS	.08795	5.67368	6.00000	353.00000	.00001
HOTELLINGS	.09644	5.67368	6.00000	353.00000	.00001
WILKS	.91205	5.67368	6.00000	353.00000	.00001
ROYS	.08795				

as the covariate. The third most important DV, NEUROTIC, is tested with
SELFESTE and WOMENROL as covariates, and so on, until all DVs are
analyzed. Stepdown analyses for the two main effects and the interaction are
presented in Table 8.16 (page 273).

For purposes of journal reporting, critical information from Tables 8.15
and 8.16 can be consolidated into a single table representing the univariate
and stepdown analyses, as shown in Table 8.17 (page 274).

For those DVs that are significantly associated with the IVs, SELFESTE,

TABLE 8.14   POOLED WITHIN-CELL CORRELATIONS AMONG SIX
VARIABLES (SELECTED OUTPUT FROM SPSS MANOVA)

WITHIN CELLS CORRELATIONS OF CRITERIA (STD. DEVS. ON DIAGONAL)

	SELFESTE	WOMENROL	NEUROTIC	INTEXT	IESCALE	DUN
SELFESTE	3.41739					
WOMENROL	.16198	6.19715				
NEUROTIC	.33436	.06234	4.85764			
INTEXT	-.12618	.02341	.00910	3.50359		
IESCALE	.35736	-.02131	.39904	-.05820	1.24211	
DUN	-.02654	.00165	-.01295	.04818	-.07278	25.50332

TABLE 8.15   UNIVARIATE ANALYSES OF VARIANCE OF SIX DVs FOR
EFFECTS OF (TOP TO BOTTOM) F BY M INTERACTION,
MASCULINITY, AND FEMININITY (SELECTED OUTPUT FROM
SPSS MANOVA)

UNIVARIATE F-TESTS WITH (1,358) D. F.

VARIATE	HYPOTHESIS SUM OF SQ.	ERROR SUM OF SQ.	HYPOTHESIS MEAN SQ.	ERROR MEAN SQ.	F	SIGNIF. OF F
SELFESTE	30.75423	4180.92359	30.75423	11.67856	2.63339	.10552
WOMENROL	35.02338	13748.87765	35.02338	38.40469	.91196	.34024
NEUROTIC	4.58769	8447.61180	4.58769	23.59668	.19442	.65953
INTEXT	.00274	4394.48894	.00274	12.27511	.00022	.98809
IESCALE	.44280	552.33159	.44280	1.54283	.28701	.59248
DUN	226.31987	232850.12954	226.31987	650.41936	.34796	.55564

UNIVARIATE F-TESTS WITH (1,358) D. F.

VARIATE	HYPOTHESIS SUM OF SQ.	ERROR SUM OF SQ.	HYPOTHESIS MEAN SQ.	ERROR MEAN SQ.	F	SIGNIF. OF F
SELFESTE	982.91202	4180.92359	982.91202	11.67856	84.16382	.00001
WOMENROL	1394.80740	13748.87765	1394.80740	38.40469	36.31868	.00001
NEUROTIC	187.06555	8447.61180	187.06555	23.59668	7.92762	.00514
INTEXT	365.43689	4394.48894	365.43689	12.27511	29.77056	.00001
IESCALE	17.74801	552.33159	17.74801	1.54283	11.50357	.00077
DUN	1169.10475	232850.12954	1169.10475	650.41936	1.79746	.18087

UNIVARIATE F-TESTS WITH (1,358) D. F.

VARIATE	HYPOTHESIS SUM OF SQ.	ERROR SUM OF SQ.	HYPOTHESIS MEAN SQ.	ERROR MEAN SQ.	F	SIGNIF. OF F
SELFESTE	103.63115	4180.92359	103.63115	11.67856	8.87363	.00309
WOMENROL	592.39654	13748.87765	592.39654	38.40469	15.42511	.00010
NEUROTIC	34.87101	8447.61180	34.87101	23.59668	1.47779	.22492
INTEXT	103.07972	4394.48894	103.07972	12.27511	8.39746	.00399
IESCALE	3.25660	552.33159	3.25660	1.54283	2.11080	.14714
DUN	50.04180	232850.12954	50.04180	650.41936	.07694	.78165

TABLE 8.16   STEPDOWN ANALYSES OF SIX ORDERED DVs FOR (TOP TO BOTTOM) F BY M INTERACTION, MASCULINITY, AND FEMININITY (SELECTED OUTPUT FROM SPSS MANOVA)

ROY-BARGMAN STEPDOWN F - TESTS

VARIATE	HYPOTHESIS MEAN SQ.	ERROR MEAN SQ.	STEP-DOWN F	HYPOTHESIS D. F.	ERROR D. F.	SIGNIF. OF F
SELFESTE	30.75423	11.67856	2.63339	1.00000	358.00000	.10552
WOMENROL	18.26190	37.50178	.48696	1.00000	357.00000	.48574
NEUROTIC	.27022	21.07471	.01282	1.00000	356.00000	.90991
INTEXT	.33059	12.12115	.02727	1.00000	355.00000	.86892
IESCALE	.00387	1.21196	.00320	1.00000	354.00000	.95496
DUN	198.70914	654.68678	.30352	1.00000	353.00000	.58203

ROY-BARGMAN STEPDOWN F - TESTS

VARIATE	HYPOTHESIS MEAN SQ.	ERROR MEAN SQ.	STEP-DOWN F	HYPOTHESIS D. F.	ERROR D. F.	SIGNIF. OF F
SELFESTE	982.91202	11.67856	84.16382	1.00000	358.00000	.00001
WOMENROL	641.04114	37.50178	17.09362	1.00000	357.00000	.00004
NEUROTIC	1.53358	21.07471	.07277	1.00000	356.00000	.78750
INTEXT	190.86630	12.12115	15.74655	1.00000	355.00000	.00009
IESCALE	.23799	1.21196	.19637	1.00000	354.00000	.65794
DUN	377.08010	654.68678	.57597	1.00000	353.00000	.44840

ROY-BARGMAN STEPDOWN F - TESTS

VARIATE	HYPOTHESIS MEAN SQ.	ERROR MEAN SQ.	STEP-DOWN F	HYPOTHESIS D. F.	ERROR D. F.	SIGNIF. OF F
SELFESTE	103.63115	11.67856	8.87363	1.00000	358.00000	.00309
WOMENROL	728.83301	37.50178	19.43462	1.00000	357.00000	.00001
NEUROTIC	1.43864	21.07471	.06826	1.00000	356.00000	.79403
INTEXT	62.08791	12.12115	5.12228	1.00000	355.00000	.02422
IESCALE	.01780	1.21196	.01469	1.00000	354.00000	.90361
DUN	3.06865	654.68678	.00469	1.00000	353.00000	.94546

WOMENROL, and INTEXT, interpretation requires the relevant marginal means for main effects (and cell means, had any interactions been statistically significant). Table 8.18 (page 275) contains deck setup and marginal means for SELFESTE, WOMENROL, and INTEXT (adjusted for appropriate effects and covariates) as produced through SPSS MANOVA. Marginal means for effects with univariate but not stepdown differences are shown in Table 8.19 (page 276).

As in all analyses of variance and covariance, strength of association between an IV and the DV for which it shows a significant relationship can be evaluated as η^2: the ratio between sum of squares for effect (hypothesis) and total sum of squares (see Eq. 7.10).[17] Required information is available from SPSS MANOVA stepdown tables (see Table 8.16) but not in a form most convenient for calculation of η^2. Thus, appropriate information is summarized in Table 8.20 (page 276). Each sum of squares has been found by multiplying the appropriate mean square by its associated degrees of freedom. The total sum of squares is then calculated by adding the sums of squares over all effects

---

[17] Note also Eq. 7.11 as an alternative for multifactorial designs.

TABLE 8.17   TESTS OF FEMININITY, MASCULINITY, AND THEIR
                 INTERACTION

IV	DV	Univariate F	df	Stepdown F	df	α
Femininity	SELFESTE	8.87[a]	1/358	8.87**	1/358	.01
	WOMENROL	15.43[a]	1/358	19.43**	1/357	.01
	NEUROTIC	1.48	1/358	0.07	1/356	.01
	INTEXT	8.40[a]	1/358	5.12	1/355	.01
	IESCALE	2.11	1/358	0.01	1/354	.01
	DUN	0.08	1/358	0.00	1/353	.001
Masculinity	SELFESTE	84.16[a]	1/358	84.16**	1/358	.01
	WOMENROL	36.32[a]	1/358	17.09**	1/357	.01
	NEUROTIC	7.93[a]	1/358	0.07	1/356	.01
	INTEXT	29.77[a]	1/358	15.75**	1/355	.01
	IESCALE	11.50[a]	1/358	0.20	1/354	.01
	DUN	1.80	1/358	0.58	1/353	.001
Femininity by	SELFESTE	2.63	1/358	2.63	1/358	.01
masculinity	WOMENROL	0.91	1/358	0.49	1/357	.01
interaction	NEUROTIC	0.19	1/358	0.01	1/356	.01
	INTEXT	0.00	1/358	0.03	1/355	.01
	IESCALE	0.29	1/358	0.00	1/354	.01
	DUN	0.35	1/358	0.30	1/353	.001

[a] Significance level cannot be evaluated but would reach $p < .01$ in univariate context.
** $p < .01$.

and error for that DV. For example, WOMENROL is significantly related to femininity (F). So for WOMENROL the adjusted sum of squares for femininity (hypothesis) is the mean square for hypothesis times the degrees of freedom from the bottom section of Table 8.16.

$$SS'_F = (728.83301)(1) = 728.83$$

For the denominator of η^2, adjusted sums of squares for the remaining effects and error are necessary and are also available from Table 8.16. The adjusted sum of squares for masculinity (middle part of Table 8.16) is

$$SS'_M = (641.04114)(1) = 641.04$$

and for interaction, from the top part of Table 8.16:

$$SS'_{F \text{ by } M} = (18.26190)(1) = 18.26$$

Error mean square and degrees of freedom can be found from any of the three portions of Table 8.16. Adjusted sum of squares for error, then, is

$$SS'_{error} = (37.50178)(357) = 13388.14$$

TABLE 8.18   ADJUSTED MARGINAL MEANS FOR SELFESTE; WOMENROL
WITH SELFESTE AS A COVARIATE; AND INTEXT WITH
SELFESTE, WOMENROL, AND NEUROTIC AS COVARIATES
(DECK SETUP AND PARTIAL OUTPUT FROM SPSS MANOVA)

```
MANOVA        SELFESTE,WOMENROL,NEUROTIC,INTEXT,IESCALE,DUN BY
              F(1,2) M(1,2)/
              ANALYSIS=SELFESTE/DESIGN=CONSPLUS F/
                          DESIGN=F,CONSPLUS M/
              ANALYSIS=WOMENROL WITH SELFESTE/DESIGN=CONSPLUS F/
                          DESIGN=F,CONSPLUS M/
              ANALYSIS=INTEXT WITH SELFESTE,WOMENROL,NEUROTIC/
              DESIGN=F,CONSPLUS M/
              NOPRINT=SIGNIF(MULTIV,EIGEN,DIMENR,UNIV)/
```

ESTIMATES FOR SELFESTE

CONSPLUS F

D. F.	COEFF.
1	16.50000
2	15.32422

ESTIMATES FOR SELFESTE

CONSPLUS M

D. F.	COEFF.
1	17.07555
2	13.58240

ESTIMATES FOR WOMENROL
ADJUSTED FOR 1 COVARIATE

CONSPLUS F

D. F.	COEFF.
1	32.71802
2	35.87457

ESTIMATES FOR WOMENROL
ADJUSTED FOR 1 COVARIATE

CONSPLUS M

D. F.	COEFF.
1	35.34049
2	32.20538

ESTIMATES FOR INTEXT
ADJUSTED FOR 3 COVARIATES

CONSPLUS M

D. F.	COEFF.
1	10.98862
2	12.73998

Note: COEFF. = adjusted marginal mean; 1 = low; 2 = high.

TABLE 8.19 UNADJUSTED MARGINAL MEANS FOR INTEXT, NEUROTIC, AND IESCALE (DECK SETUP AND PARTIAL OUTPUT FROM SPSS MANOVA)

```
MANOVA          SELFESTE,WOMENROL,NEUROTIC,INTEXT,IESCALE,DUN BY
                F(1,2) M(1,2)/
                ANALYSIS=INTEXT/DESIGN=CONSPLUS F/
                ANALYSIS=NEUROTIC/DESIGN=F,CONSPLUS M/
                ANALYSIS=IESCALE/DESIGN=F,CONSPLUS M/
                NOPRINT=SIGNIF(MULTIV,EIGEN,DIMENR,UNIV)/




                ESTIMATES FOR INTEXT

                CONSPLUS F

                    PARAMETER              COEFF.

                         1         10.9386792453
                         2         12.1113281250
```

```
CONSPLUS M                                    CONSPLUS M

PARAMETER          COEFF.                     PARAMETER              COEFF.

    2        9.2938905612                          2        6.9011775578
    3        7.7699898897                          3        6.4317867404
```

Note: COEFF. = marginal means; 1 = low; 2 = high.

TABLE 8.20 SUMMARY OF ADJUSTED SUMS OF SQUARES AND η^2 FOR EFFECTS OF FEMININITY, MASCULINITY, AND THEIR INTERACTION ON SELFESTE; WOMENROL WITH SELFESTE AS A COVARIATE; AND INTEXT WITH SELFESTE, WOMENROL, AND NEUROTIC AS COVARIATES

Source of variance	SELFESTE		WOMENROL		INTEXT	
	SS'	η^2	SS'	η^2	SS'	η^2
F	103.63	.02	728.83	.05	62.09	–
M	982.91	.19	641.04	.04	190.87	.04
F by M	30.75	–	18.26	–	0.33	–
Error	4180.92		13388.14		4303.01	
Total	5298.21		14776.27		4556.30	

Note: See Table 8.16 for MEAN SQ. values. SS = (MEAN SQ.) (D.F.)

TABLE 8.21 CHECKLIST FOR MULTIVARIATE ANALYSIS OF VARIANCE

1. Issues
   a. Unequal sample sizes and missing data
   b. Normality
   c. Outliers
   d. Homogeneity of variance-covariance matrices
   e. Linearity
   f. In stepdown, when DVs act as covariates
      (1) Homogeneity of regression
      (2) Reliability of DVs
   g. Multicollinearity and singularity
2. Major analyses
   a. Planned comparisons or omnibus main effects and interactions
      If significant:
      (1) Multivariate Strength of Association
      (2) Stepdown analysis
         (a) Stepdown and univariate F
         (b) Strength of association for significant stepdown F
         (c) Adjusted marginal and/or cell means for significant stepdown F
3. Additional analyses
   a. Post hoc comparisons
   b. Interpretation of IV-covariates interaction (if homogeneity of regression violated)
   c. Pooled within-cell correlations among DVs
   d. Interpretation of loading matrices

Finally, the strength of association is found by writing

$$\eta^2 = \frac{SS'_F}{SS'_{total}} = \frac{SS'_F}{SS'_F + SS'_M + SS'_{F \text{ by } M} + SS'_{error}}$$

$$= \frac{728.83}{728.83 + 641.04 + 18.26 + 13388.14} = \frac{728.83}{14776.27} = .05$$

A checklist for MANOVA appears in Table 8.21. An example of a Results section, in journal format, follows for the study just described.

RESULTS

A 2 × 2 between-subjects multivariate analysis of variance was performed on the six dependent variables: SELFESTE, WOMENROL, NEUROTIC, INTEXT, IESCALE, and DUN. Independent variables were masculinity (low and high) and femininity (low and high).

SPSS MANOVA was used for the analyses with the hierarchical (default) adjustment for nonorthogonality. Order of entry of IVs was femininity, then masculinity. Total $N = 369$ was reduced to 362 with deletion of all within-cell outliers with $p <$

.01, as determined through BMDPAM.[18] Results of evaluation of assumptions of normality, homogeneity of variance-covariance matrices, linearity, and multicollinearity were satisfactory after deletion of outliers.

With the use of Wilks' criterion, the combined DVs were significantly affected by both masculinity, $F(6, 353) = 20.88$, $p < .001$, and femininity, $F(6,353) = 5.67$, $p < .001$, but not by their interaction, $F(6, 353) = 0.57$, $p > .05$. The results reflected a moderate association between masculinity scores (low vs. high) and the combined DVs, $\eta^2 = .26$. The association was less substantial between femininity and the DVs, $\eta^2 = .09$. [F and lambda are from Table 8.13; η^2 is calculated according to Eq. 8.6.]

To investigate the effects of each main effect and interaction on the individual DVs, a stepdown analysis was performed, on the basis of an a priori ordering of the importance of the DVs. Thus each DV was analyzed, in turn, with higher-priority DVs treated as covariates and with the highest-priority DV tested in a univariate ANOVA. Homogeneity of regression was achieved for all components of the stepdown analysis. All DVs were judged to be sufficiently reliable to warrant stepdown analysis. Results of this analysis are summarized in Table 8.17. An experimentwise error rate of 5% was achieved by the apportionment of alpha as shown in the last column of Table 8.17 for each of the DVs.

Insert Table 8.17 about here

A unique contribution to predicting differences between those low and high on femininity was made by SELFESTE, stepdown $F(1, 358) = 8.87$, $p < .01$, $\eta^2 = .02$. Noting that self-esteem was scored inversely, women with higher femininity scores showed greater self-esteem (mean self-esteem = 15.32) than those with lower femininity (mean self-esteem = 16.50). After the pattern of differences measured by SELFESTE was entered, a difference was also found on WOMENROL, stepdown $F(1, 357) = 19.43$, $p < .01$, $\eta^2 = .05$. Women with higher femininity scores had more conservative attitudes toward women's role (adjusted mean attitude = 35.87) than those lower in femininity (adjusted mean attitude = 32.72). Although a univariate comparison revealed that those higher in femininity also were more extraverted on INTEXT, univariate $F(1, 358) = 8.40$, this difference was already represented in the stepdown analysis by higher-priority DVs.

Three DVs—SELFESTE, WOMENROL, and INTEXT—made unique contributions to the composite DV that best distinguished between those high and low in masculinity.

---

[18] Two of the three outliers in the high femininity–low masculinity group were extremely low in self-esteem for their group while scoring high either on neuroticism or on introversion. The third outlier of the group had an extremely low score on locus of control ($z = -4.94$), indicating belief in internal control of her life. The outlying case in the high masculinity–low femininity group had the highest neuroticism score for her group and was highly introverted. Two of the three androgynous outliers scored at the top of the locus-of-control scale for their group, indicating belief in external control over their lives. Of these, one scored relatively low on neuroticism while the other scored high for her group. The final outlier was highly conservative in her attitude toward women's role and was the most introverted in her group.

The greatest contribution was made by SELFESTE, the highest-priority DV, stepdown $F(1, 358) = 84.16$, $p < .01$, $\eta^2 = 19$. Women scoring high in masculinity had higher self-esteem (mean self-esteem = 13.58) than those scoring low (mean self-esteem = 17.08). With differences due to self-esteem already entered, WOMENROL made a unique contribution, stepdown $F(1, 357) = 17.09$, $p < .01$, $\eta^2 = .04$. Women scoring lower in masculinity had more conservative attitudes toward the proper role of women (adjusted mean attitude = 35.34) than those scoring higher (adjusted mean attitude = 32.21). Introversion-extraversion, adjusted by self-esteem, attitudes toward women's role, and neuroticism also made a unique contribution to the composite DV, stepdown $F(1, 355) = 15.75$, $p < .01$, $\eta^2 = .04$. Women with higher masculinity were more extraverted (mean adjusted introversion-extraversion score = 12.74) than lower masculinity women (mean adjusted introversion-extraversion score = 10.99). Univariate analyses revealed that women with higher masculinity scores were also less neurotic, univariate $F(1, 350) = 7.93$, and had a more internal locus of control (IESCALE), univariate $F(1, 358) = 11.50$, differences that were already accounted for in the composite DV by higher-priority DVs. [Adjusted means are from Table 8.18, η^2 values are from Table 8.20. Unadjusted means for univariate interpretation are in Table 8.19.]

High-masculinity women, then, have greater self-esteem, have less conservative attitudes toward the role of women, and are more extraverted than women scoring low on masculinity. High femininity is associated with greater self-esteem and more conservative attitudes toward women's role than is low femininity. Of the five effects, however, only the association between masculinity and self-esteem shows even a moderate proportion of shared variance.

Pooled within-cell correlations among DVs are shown in Table 8.14.

Insert Table 8.14 about here

8.7.3 Multivariate Analysis of Covariance

For this example, the same six variables are used as for MANOVA. The three personality variables—SELFESTE, INTEXT, and NEUROTIC—are treated as DVs for this analysis. The remaining three variables—IESCALE, WOMENROL, and DUN—reflecting attitudes and demography are treated as covariates.

Deck setup and partial output of omnibus MANCOVA as produced by SPSS MANOVA appear in Table 8.22. As in the MANOVA example (Section 8.7.2), the default (hierarchical) approach was used for adjusting IV and interaction effects for nonorthogonality. Again, the multivariate results show highly significant main effects with no statistically significant interaction.

8.7.3.1 Assessing Covariates The omnibus MANOVA shows a significant relationship between the set of DVs (SELFESTE, INTEXT, and NEUROTIC) and the set of covariates (IESCALE, WOMENROL, and DUN). These relationships can be profitably analyzed by looking at the multiple regression

TABLE 8.22 MULTIVARIATE ANALYSIS OF COVARIANCE OF SELFESTE, INTEXT, AND NEUROTIC AS A FUNCTION OF (TOP TO BOTTOM) F BY M INTERACTION, MASCULINITY, AND FEMININITY; COVARIATES ARE WOMENROL, IESCALE, AND DUN (DECK SETUP AND SELECTED OUTPUT FROM SPSS MANOVA)

```
MANOVA        SELFESTE,WOMENROL,NEUROTIC,INTEXT,IESCALE,DUN BY
              F(1,2) M(1,2)/
              PRINT=SIGNIF(STEPDOWN), ERROR(COR),
              HOMOGENEITY(BARTLETT,COCHRAN,BOXM)/
              NOPRINT=PARAMETERS(ESTIM)/
```

EFFECT .. WITHIN CELLS REGRESSION

MULTIVARIATE TESTS OF SIGNIFICANCE (S = 3, M = - 1/2, N = 175 1/2)

TEST NAME	VALUE	APPROX. F	HYPOTHESIS D. F.	ERROR D. F.	SIGNIF. OF F
PILLAIS	.24962	10.73986	9.00000	1065.00000	.00001
HOTELLINGS	.32297	12.61979	9.00000	1055.00000	.00001
WILKS	.75351	11.77362	9.00000	859.25985	.00001
ROYS	.23645				

EFFECT .. F BY M

MULTIVARIATE TESTS OF SIGNIFICANCE (S = 1, M = 1/2, N = 175 1/2)
 (CONT.)

TEST NAME	VALUE	APPROX. F	HYPOTHESIS D. F.	ERROR D. F.	SIGNIF. OF F
HOTELLINGS	.00543	.63929	3.00000	353.00000	.59015
WILKS	.99460	.63929	3.00000	353.00000	.59015
ROYS	.00540				

EFFECT .. M

MULTIVARIATE TESTS OF SIGNIFICANCE (S = 1, M = 1/2, N = 175 1/2)

TEST NAME	VALUE	APPROX. F	HYPOTHESIS D. F.	ERROR D. F.	SIGNIF. OF F
PILLAIS	.15901	22.24709	3.00000	353.00000	.00001
HOTELLINGS	.18907	22.24709	3.00000	353.00000	.00001
WILKS	.84099	22.24709	3.00000	353.00000	.00001
ROYS	.15901				

EFFECT .. F

MULTIVARIATE TESTS OF SIGNIFICANCE (S = 1, M = 1/2, N = 175 1/2)

TEST NAME	VALUE	APPROX. F	HYPOTHESIS D. F.	ERROR D. F.	SIGNIF. OF F
PILLAIS	.04380	5.39042	3.00000	353.00000	.00123
HOTELLINGS	.04581	5.39042	3.00000	353.00000	.00123
WILKS	.95620	5.39042	3.00000	353.00000	.00123
ROYS	.04380				

analyses of each DV in turn, with covariates acting as multiple continuous IVs (see Chapter 5). These analyses are automatically produced by SPSS MANOVA with the deck setup depicted in Table 8.22. The analyses are done on the pooled within-cell correlation matrix, so that effects of the IVs (masculinity and femininity) and their interaction are eliminated. The results of the DV-covariate multiple regressions are shown in Table 8.23. Note that for SELFESTE, two of the covariates, IESCALE and WOMENROL, are significantly related but DUN is not. None of the three covariates is related to the second DV, INTEXT. Finally, for NEUROTIC, only one of the covariates, IESCALE, is significantly related. A look at all three analyses shows that the covariate representing socioeconomic level, DUN, is providing no adjustment for any of the DVs and could be omitted from future analyses. Also shown in Table 8.23

TABLE 8.23   MULTIPLE REGRESSION, UNIVARIATE, AND STEPDOWN
ANALYSIS FOR THREE DVS WITH THREE COVARIATES
(PARTIAL OUTPUT FROM SPSS MANOVA)

```
REGRESSION ANALYSIS FOR WITHIN CELLS ERROR TERM

DEPENDENT VARIABLE .. SELFESTE
```

COVARIATE	B	BETA	STD. ERROR	T-VALUE	SIGNIF. OF T	LOWER .95 CONF. LIM.	UPPER .95 CONF. LIM.
IESCALE	.99305	.36094	.13450	7.38331	.00000	.72853	1.25756
WOMENROL	.09357	.16967	.02689	3.48003	.00056	.04069	.14644
DUN	-.00007	-.00055	.00655	-.01132	.99097	-.01295	.01281

```
DEPENDENT VARIABLE .. INTEXT
```

COVARIATE	B	BETA	STD. ERROR	T-VALUE	SIGNIF. OF T	LOWER .95 CONF. LIM.	UPPER .95 CONF. LIM.
IESCALE	-.15376	-.05451	.14970	-1.02709	.30508	-.44817	.14066
WOMENROL	.01254	.02218	.02993	.41894	.67551	-.04632	.07139
DUN	.00607	.04418	.00729	.83256	.40565	-.00827	.02040

```
DEPENDENT VARIABLE .. NEUROTIC
```

COVARIATE	B	BETA	STD. ERROR	T-VALUE	SIGNIF. OF T	LOWER .95 CONF. LIM.	UPPER .95 CONF. LIM.
IESCALE	1.57110	.40173	.19027	8.25715	.00000	1.19690	1.94530
WOMENROL	.05556	.07088	.03804	1.46064	.14500	-.01925	.13036
DUN	.00308	.01618	.00926	.33253	.73968	-.01514	.02130

```
UNIVARIATE F-TESTS WITH (3,355) D. F.
```

VARIATE	SQUARED MULTIPLE R	MULTIPLE R	ADJUSTED R-SQUARED	HYPOTHESIS MEAN SQ.	ERROR MEAN SQ.	F	SIGNIF. OF F
SELFESTE	.15649	.39558	.14223	218.08469	9.93428	21.95274	.00001
INTEXT	.00582	.07629	0	8.52519	12.30680	.69272	.55698
NEUROTIC	.16452	.40561	.15040	463.26063	19.88121	23.30143	.00001

```
ROY-BARGMAN STEPDOWN F - TESTS
```

VARIATE	HYPOTHESIS MEAN SQ.	ERROR MEAN SQ.	STEP-DOWN F	HYPOTHESIS D. F.	ERROR D. F.	SIGNIF. OF F
SELFESTE	218.08469	9.93428	21.95274	3.00000	355.00000	.00001
INTEXT	5.88061	12.16632	.48335	3.00000	354.00000	.69406
NEUROTIC	256.93878	19.00786	13.51750	3.00000	353.00000	.00001

are the results of the univariate and stepdown analysis, summarizing the results of multiple regressions for the three DVs independently and in priority order (see Section 8.7.3.2).

8.7.3.2 Assessing DVs   Procedures for evaluating DVs, now adjusted for covariates, follow those specified in Section 8.7.2 for MANOVA. Correlations among all DVs are informative, as are correlations among covariates and between DVs and covariates. That is, all of the correlations in Table 8.14 are still relevant.

Univariate F's are now reported as adjusted for covariates. The univariate ANCOVAs produced by the SPSS MANOVA run depicted in Table 8.22 are shown in Table 8.24.

For interpretation of effects of IVs on DVs, adjusted for covariates, a comparison of the stepdown analysis with univariate F's again provides the best information. The priority order of DVs developed for this analysis was, in descending order, SELFESTE, INTEXT, and NEUROTIC. This means that SELFESTE is evaluated in terms of its relationships with the IVs and interaction after adjustment only for the three covariates. INTEXT is adjusted for effects on SELFESTE as well as covariates, and NEUROTIC is adjusted for SELFESTE and INTEXT in addition to the three covariates. In effect, then, INTEXT is adjusted for four covariates and NEUROTIC is adjusted for five.

Stepdown analysis for the interaction and two main effects is illustrated

TABLE 8.24   UNIVARIATE ANALYSES OF COVARIANCE OF THREE DVS ADJUSTED FOR THREE COVARIATES FOR (TOP TO BOTTOM) F BY M INTERACTION, MASCULINITY, AND FEMININITY (SELECTED OUTPUT FROM SPSS MANOVA)

UNIVARIATE F-TESTS WITH (1,355) D. F.

VARIATE	HYPOTHESIS SUM OF SQ.	ERROR SUM OF SQ.	HYPOTHESIS MEAN SQ.	ERROR MEAN SQ.	F	SIGNIF. OF F
SELFESTE	18.66929	3526.66954	18.66929	9.93428	1.87928	.17128
INTEXT	.00448	4368.91336	.00448	12.30680	.00036	.98479
NEUROTIC	.65977	7057.82991	.65977	19.88121	.03319	.85555

UNIVARIATE F-TESTS WITH (1,355) D. F.

VARIATE	HYPOTHESIS SUM OF SQ.	ERROR SUM OF SQ.	HYPOTHESIS MEAN SQ.	ERROR MEAN SQ.	F	SIGNIF. OF F
SELFESTE	491.74387	3526.66954	491.74387	9.93428	49.49970	.00001
INTEXT	307.86072	4368.91336	307.86072	12.30680	25.01550	.00001
NEUROTIC	22.72778	7057.82991	22.72778	19.88121	1.14318	.28571

UNIVARIATE F-TESTS WITH (1,355) D. F.

VARIATE	HYPOTHESIS SUM OF SQ.	ERROR SUM OF SQ.	HYPOTHESIS MEAN SQ.	ERROR MEAN SQ.	F	SIGNIF. OF F
SELFESTE	108.48356	3526.66954	108.48356	9.93428	10.92012	.00105
INTEXT	86.57804	4368.91336	86.57804	12.30680	7.03498	.00835
NEUROTIC	18.83671	7057.82991	18.83671	19.88121	.94746	.33103

in Table 8.25. Consolidation of information from Tables 8.24 and 8.25, as well as some information from Table 8.23, appears in Table 8.26, along with apportionment of the 5% α error to the various tests.

TABLE 8.25   STEPDOWN ANALYSES OF THREE ORDERED DVS ADJUSTED
FOR THREE COVARIATES FOR (TOP TO BOTTOM) F BY M
INTERACTION, MASCULINITY, AND FEMININITY (SELECTED
OUTPUT FROM SPSS MANOVA)

```
ROY-BARGMAN STEPDOWN F - TESTS

                HYPOTHESIS      ERROR       STEP-DOWN   HYPOTHESIS      ERROR       SIGNIF.
    VARIATE     MEAN SQ.        MEAN SQ.        F       D. F.           D. F.       OF F

    SELFESTE    18.66929        9.93428     1.87928     1.00000         355.00000   .17128
    INTEXT        .40746       12.16632      .03349     1.00000         354.00000   .85490
    NEUROTIC      .29518       19.00786      .01553     1.00000         353.00000   .90090


ROY-BARGMAN STEPDOWN F - TESTS

                HYPOTHESIS      ERROR       STEP-DOWN   HYPOTHESIS      ERROR       SIGNIF.
    VARIATE     MEAN SQ.        MEAN SQ.        F       D. F.           D. F.       OF F

    SELFESTE   491.74387        9.93428    49.49970     1.00000         355.00000   .00001
    INTEXT     187.19889       12.16632    15.38665     1.00000         354.00000   .00011
    NEUROTIC      .60538       19.00786      .03185     1.00000         353.00000   .85846


ROY-BARGMAN STEPDOWN F - TESTS

                HYPOTHESIS      ERROR       STEP-DOWN   HYPOTHESIS      ERROR       SIGNIF.
    VARIATE     MEAN SQ.        MEAN SQ.        F       D. F.           D. F.       OF F

    SELFESTE   108.48356        9.93428    10.92012     1.00000         355.00000   .00105
    INTEXT      60.90533       12.16632     5.00606     1.00000         354.00000   .02588
    NEUROTIC     3.03999       19.00786      .15993     1.00000         353.00000   .68946
```

TABLE 8.26   TEST OF COVARIATES, FEMININITY, MASCULINITY, AND
INTERACTION

Effect	DV	Univariate F	df	Stepdown F	df	α
Covariates	SELFESTE	21.95[a]	3/355	21.95**	3/355	.02
	INTEXT	0.69	3/355	0.48	3/354	.02
	NEUROTIC	23.30[a]	3/355	13.52**	3/353	.01
Femininity	SELFESTE	10.92[a]	1/355	10.92**	1/355	.02
	INTEXT	7.03[a]	1/355	5.01	1/354	.02
	NEUROTIC	0.95	1/355	0.16	1/353	.01
Masculinity	SELFESTE	49.50[a]	1/355	49.50**	1/355	.02
	INTEXT	25.02[a]	1/355	15.39**	1/354	.02
	NEUROTIC	1.14	1/355	0.03	1/353	.01
Femininity by masculinity interaction	SELFESTE	1.88	1/355	1.88	1/355	.02
	INTEXT	0.00	1/355	0.03	1/354	.02
	NEUROTIC	0.03	1/355	0.02	1/353	.01

[a] Significance level cannot be evaluated but would reach $p < .01$ in univariate context.
** $p < .01$.

For those DVs associated with significant main effects, interpretation requires associated marginal means. Table 8.27 contains deck setup and adjusted marginal means for SELFESTE and for INTEXT (which is adjusted for SELFESTE as well as covariates). Marginal means for the main effect of femininity on INTEXT (univariate but not stepdown effect) appear in Table 8.28. A checklist for MANCOVA appears in Table 8.29.

TABLE 8.27   ADJUSTED MARGINAL MEANS FOR SELFESTE ADJUSTED FOR THREE COVARIATES AND INTEXT ADJUSTED FOR SELFESTE PLUS THREE COVARIATES (DECK SETUP AND PARTIAL OUTPUT FROM SPSS MANOVA)

```
MANOVA          SELFESTE,WOMENROL,NEUROTIC,INTEXT,IESCALE,DUN BY
                F(1,2) M(1,2)/
                ANALYSIS=SELFESTE WITH IESCALE,WOMENROL,DUN/
                DESIGN=CONSPLUS F/DESIGN=F,CONSPLUS M/
                ANALYSIS=INTEXT WITH IESCALE,WOMENROL,DUN,SELFESTE/
                DESIGN=F,CONSPLUS M/
                NOPRINT=SIGNIF(MULTIV,EIGEN,DIMENR,UNIV)/
```

ESTIMATES FOR SELFESTE ADJUSTED FOR 3 COVARIATES		ESTIMATES FOR SELFESTE ADJUSTED FOR 3 COVARIATES	
CONSPLUS F		CONSPLUS M	
D. F.	COEFF.	D. F.	COEFF.
1	16.53959	1	16.80213
2	15.30783	2	14.16473

ESTIMATES FOR INTEXT
ADJUSTED FOR 4 COVARIATES

CONSPLUS M

D. F.	COEFF.
1	10.99529
2	12.73230

Note: COEFF. = adjusted marginal mean; 1 = low; 2 = high.

TABLE 8.28   MARGINAL MEANS FOR INTEXT ADJUSTED FOR THREE
COVARIATES ONLY (DECK SETUP AND PARTIAL OUTPUT
FROM SPSS MANOVA)

```
MANOVA          SELFESTE,WOMENROL,NEUROTIC,INTEXT,IESCALE,DUN BY
                F(1,2) M(1,2)/
                ANALYSIS=INTEXT WITH IESCALE,WOMENROL,DUN/
                DESIGN=CONSPLUS F/
                NOPRINT=SIGNIF(MULTIV,EIGEN,DIMENR,UNIV)/




        ESTIMATES FOR INTEXT ADJUSTED FOR 3 COVARIATES

        CONSPLUS F

           PARAMETER              COEFF.

                  1        10.9897727011
                  2        12.0901722409
```

Note: COEFF. = adjusted marginal mean; 1 = low; 2 = high.

TABLE 8.29   CHECKLIST FOR MULTIVARIATE ANALYSIS OF COVARIANCE

1. Issues
    a. Unequal sample sizes and missing data
    b. Normality
    c. Outliers
    d. Homogeneity of variance-covariance matrices
    e. Linearity
    f. Homogeneity of regression
        (1) Covariates
        (2) DVs for stepdown analysis
    g. Reliability of covariates (and DVs for stepdown)
    h. Multicollinearity and singularity
2. Major analyses
    a. Planned comparisons or omnibus main effects and interactions. If significant:
        (1) Multivariate strength of association
        (2) Stepdown analysis. For significant effects:
            (a) Stepdown and univariate F
            (b) Strength of association for significant stepdown F
            (c) Adjusted marginal and/or cell means for significant stepdown F
3. Additional analyses
    a. Assessment of covariates
    b. Interpretation of IV-covariates interaction (if homogeneity of regression violated for stepdown analysis)
    c. Post hoc comparisons
    d. Pooled within-cell correlations among DVs and covariates
    e. Interpretation of loading matrices

An example of a Results section, as might be appropriate for journal presentation, follows.

RESULTS

A 2×2 between-subjects multivariate analysis of covariance was performed on three dependent variables associated with personality of respondents: SELFESTE, INTEXT, and NEUROTIC. Adjustment was made for three covariates. Two covariates, WOMENROL and IESCALE, reflected attitudes toward role of women and locus of control, respectively, and the third, DUN, is a measure of socioeconomic status. Independent variables were masculinity (low and high) and femininity (low and high). Analysis was done through SPSS MANOVA, with hierarchical (default) ordering of effects to adjust for nonorthogonality. Order of entry of IVs was femininity, then masculinity. Total $N = 369$ was reduced to 362 with deletion of within-cell outliers identified through BMD10M with $p < .01$.[19] Results of evaluation of assumptions of normality, homogeneity of covariance matrices, linearity, homogeneity of regression, and multicollinearity were satisfactory after deletion of outliers. Covariates were judged to be adequately reliable for covariance analysis.

With the use of Wilks' criterion, the combined DVs were significantly related to the combined covariates, approximate $F(9, 859) = 11.77$, $p < .001$, to femininity, $F(3, 353) = 5.39$, $p < .01$, and to masculinity, $F(3, 353) = 22.25$, $p < .001$. The multivariate test for the masculinity by femininity interaction failed to reach statistical significance, $F(3, 353) = 0.64$, $p > .05$. The results reflected a moderate association between DVs and covariates, with $\eta^2 = .25$. A somewhat smaller association was found between combined DVs and the main effect of masculinity, $\eta^2 = .16$, and the association between the main effect of femininity and the combined DVs was smaller yet, $\eta^2 = .04$. [F and lambda are from Table 8.22; η^2 is calculated from Eq. 8.6.]

To investigate more specifically the power of the covariates to adjust dependent variables, multiple regressions were run for each DV in turn, with covariates acting as multiple predictors. Two of the three covariates, locus of control and attitudes toward women's role, provided significant prediction of self-esteem. The β value of .36 for locus of control was significantly different from zero, $t(355) = 7.38$, $p < .001$, as was that for attitudes toward women's role at .17, $t(355) = 3.48$, $p < .001$. None of the covariates showed statistically significant relationships with the introversion-extraversion scale. For neuroticism, only locus of control reached statistical significance, with $\beta = .40$, $t(355) = 8.26$, $p < .001$. For none of the DVs did socioeconomic status provide significant adjustment.

[19] Two of the three outliers in the high femininity–low masculinity group were extremely low in self-esteem for their group while scoring high on either neuroticism or on introversion. The third outlier of the group had an extremely low score on locus of control ($z = -4.94$), indicating belief in internal control of her life. The outlying case in the high masculinity–low femininity group had the highest neuroticism score for her group and was highly introverted. Two of the three androgynous outliers scored at the top of the locus-of-control scale for their group, indicating belief in external control over their lives. Of these, one scored relatively low on neuroticism while the other scored high for her group. The final outlier was highly conservative in her attitude toward women's role and was the most introverted in her group.

Effects of masculinity and femininity on the DVs after adjustment for covariates were investigated in univariate and stepdown analysis, in which self-esteem was given the highest priority in an a priori hierarchy of importance among the DVs. Second highest priority was given to introversion-extraversion score so that adjustment was made for self-esteem as well as for the three covariates. Last in the hierarchy was neuroticism, adjusted for self-esteem and introversion-extraversion, in addition to the three covariates. Homogeneity of regression was satisfactory for this analysis, and DVs were judged to be sufficiently reliable to act as covariates. Results of this analysis are summarized in Table 8.26. An experimentwise error rate of 5% for each effect was achieved by apportioning alpha according to the values shown in the last column in Table 8.26.

Insert Table 8.26 about here

After statistically adjusting for differences in attitudes toward women's role, locus of control, and socioeconomic level, one DV, SELFESTE, made a significant contribution to the composite DV that best distinguished between women who were high or low in femininity, stepdown $F(1, 355) = 10.92$, $p < .01$, $\eta^2 = .03$. With self-esteem scored inversely, women with higher femininity scores showed greater self-esteem after adjustment for covariates (adjusted mean self-esteem $= 15.31$) than those scoring lower on femininity (adjusted mean self-esteem $= 16.54$). Univariate analysis revealed that a reliable difference was also present on the introversion-extraversion (INTEXT) measure, with higher femininity women more extraverted, univariate $F(1, 355) = 7.03$, a difference already accounted for by covariates and the higher-priority DV. [Adjusted means are from Tables 8.27 and 8.28; η^2 is calculated as in Section 8.7.2.]

Lower- versus higher-masculinity women differed in self-esteem, the highest-priority DV, after adjustment for covariates, stepdown $F(1, 355) = 49.50$, $p < .01$, $\eta^2 = .12$. Greater self-esteem was found among higher-masculinity women (adjusted mean $= 14.16$) than among lower-masculinity women (adjusted mean $= 16.80$). The measure of introversion and extraversion, adjusted for covariates and self-esteem, was also related to differences in masculinity, stepdown $F(1, 354) = 15.39$, $p < .01$, $\eta^2 = .04$. Women scoring higher on the masculinity scale were more extraverted (adjusted mean extraversion $= 12.73$) than those showing lower masculinity (adjusted mean extraversion $= 11.00$).

High-masculinity women, then, are characterized by greater self-esteem and extraversion than low-masculinity women when adjustments are made for differences in socioeconomic status as well as attitudes toward women's role and locus of control. High-femininity women show greater self-esteem than low-femininity women with adjustment for those covariates.

Pooled within-cell correlations among dependent variables and covariates are shown in Table 8.14.

Insert Table 8.14 about here

8.8 SOME EXAMPLES FROM THE LITERATURE

8.8.1 Examples of MANOVA

Wade and Baker (1977) surveyed clinical psychologists on their reasons for decisions to use tests. Psychologists were divided into low and high test users, as the two levels of the IV. Six "reasons," each rated on a 5-point scale of importance, served as the DVs. Separate MANOVAs were performed on users of objective and projective tests. After finding significant multivariate effects—that is, that high and low test users rated importance of reasons differently—for both objective and projective users, the researchers performed univariate ANOVAs on each of the reasons. They found that for objective test users, the distinguishing reasons were reliability and validity, and agency requirements, for which high users gave higher ratings of importance than did low users. For the projective test users, high users rated the following reasons as more important than low users: graduate training experience, previous experience with tests, and reliability and validity. A MANOVA testing reasons for use as a function of 8 major therapeutic orientations was not statistically significant.

In a study of romantic attraction, Giarrusso (1977) investigated the effects of physical attractiveness and similarity of attitudes of a target date in a 2 × 2 between-groups MANOVA. Male students rated the target date on degree of liking, desire to date, comparability with previous dates, goodness of personality, and consideration of whether the target date liked them. These five ratings served as DVs. A significant multivariate effect of physical attractiveness was found, but not of similarity of attitudes nor of the interaction between attractiveness and attitudes. After an a priori ordering of DVs, it was found that only the highest-priority variable, degree of liking, was significantly affected by physical attractiveness. That is, the more attractive the target date, the more she was liked. The remaining variables showed no statistically significant differences once the effect on degree of liking was partialled out.

Junior high school students were evaluated in terms of their behavioral changes after being exposed in class for one week to materials on "responsibility," in a study by Singh, Greer, and Hammond (1977). IVs were (1) whether or not materials were presented; (2) grade level: 7, 8, and 9; and (3) ability levels, three levels based on the School and College Ability Tests. The three DVs consisted of an attitude test designed to accompany the "responsibility" materials, and experimenter-designed essay and objective tests, which tapped comprehension of various aspects of responsibility. With the use of a 2 × 3 × 3 between-groups multivariate analysis of variance on gain scores, only the main effect of program was found to be statistically significant. Bonferroni-type confidence intervals constructed for the three DVs indicated that the attitude test alone accounted for the significant multivariate effect,[20] but the authors felt that the magnitude of change was insufficient to support the program as implemented in the study.

---

[20] This procedure represents an alternative to a stepdown analysis.

Concerned that policy and decision makers seldom use evaluation information, Brown, Braskamp, and Newman (1978) investigated effects on readers' reactions to use of jargon and to objective versus subjective statements in evaluation reports. After reading one of four statements (with presence vs. absence of jargon, and subjective vs. objective statements factorially combined), readers judged the reports on two dependent measures, difficulty and technicality, and judged the writer of the evaluation on nine measures (including such variables as thoroughness and believability). Separate MANOVAs were performed on the two types of DVs—judgment of report and judgment of writer. Significant multivariate effects were interpreted in terms of cell means for DVs with significant univariate F's, rather than with the combination of univariate and stepdown analysis recommended here.

The only significant multivariate effect was that of jargon on ratings of the report. Both report DVs, technicality and difficulty, produced significant univariate F's, indicating that reports high in jargon were considered to be more technical and more difficult. Ratings of the report writer were not affected by jargon, objectivity, or the interaction. Further, a univariate ANOVA revealed that extent of agreement with the recommendations of the evaluation report was independent of use of jargon, objectivity, or the interaction.

Jakubczak (1977) was interested in age differences in regulation of calorie intake in rats. For the first experiment, he used as IVs three age levels, three levels of dilution of food with cellulose, two levels of previous food deprivation experience, and six daysets. The last was a within-subjects factor. DVs were calorie intake and body weight. Significant multivariate main effects were found on all factors, in addition to a number of two- and three-way interactions. Univariate ANOVAs, rather than stepdown and univariate analyses, were then used for interpreting effects on individual DVs. An age-related decrement in calorie intake and body weight was found in response to dilution with cellulose. Previous deprivation experience did not affect this relationship.

In a second experiment, two age levels, two levels of dilution by water, and nine daysets were used as factors. As an additional IV, rats were or were not given quinine in their diets. Older rats behaved like younger rats in response to dilution of the diet by water, maintaining calorie intake and weight. With addition of quinine, however, older rats decreased calorie intake and body weight to a greater degree than younger rats.

For both experiments, MANCOVAs run with initial levels of calorie intake and body weight as covariates revealed no difference in decisions with respect to within-subjects effects, as could be expected (cf. Section 7.5.2.1). Jakubczak made no mention of changes in between-subjects effects with the use of covariates.

Willis and Wortman (1976) studied public acceptance of randomized control-group designs in medical experimentation. Factors were sex of subject, three levels of science emphasis in describing the medical experiment, and three levels of treatment scarcity situations. For the last IV, descriptions of the medical experiment explicitly mentioned nonscarcity (i.e., the proposed treatment was available to all), mentioned scarcity (the treatment was expensive and scarce), or failed to mention anything about scarcity. Dependent measures were nine opinion questions, answered on bipolar scales. MANOVA revealed significant multivariate effects for two of the three IVs, scarcity of treatment and sex of subject. None of the interactions was statistically significant. DVs were analyzed in terms of cell means for scarcity and sex associated with those DVs with significant univariate F's. For post hoc comparisons among the three scarcity levels, multivariate F's were appropriately adjusted for inflated α error. It was concluded that the most favorable reactions occurred in situations in which scarcity was not mentioned at all, and that females tended to be less accepting than males of placebos and withholding of treatment.

8.8.2 Examples of MANCOVA

Cornbleth (1977) studied the effect of a protected hospital ward on geriatric patients. IVs were ward assignment as protected or unprotected and identification of the patient as a wanderer or nonwanderer. These two IVs were factorially combined, forming a 2×2 between-subjects design. Separate analyses were run for two periods in time (rather than including time period as a within-subjects factor). DVs were two physical measures, five cognitive measures, and six psychosocial measures. Separate analyses were done for each of these three categories of functioning. In each analysis, preassessment measures on the appropriate DVs were used as covariates. Therefore there were six separate MANCOVAs. Contribution of the DVs to multivariate effects were evaluated through examination of standardized discriminant function coefficients (cf. Chapter 9). In addition, adjusted (for covariates) cell means were reported for all DVs.

On physical variables, a significant multivariate interaction was found (marginal at time 1, $p < .05$ at time 2). The protected ward produced greater range of motion for wanderers, while the regular nonprotected ward facilitated performance for nonwanderers. In additon, independent of ward assignment, wanderers showed a lower level of psychosocial functioning than nonwanderers at both time points.

Modification of perceived locus of control in high school students was studied by Bradley and Gaa (1977). Subjects were blocked on sex and on two levels of prior achievement, and then randomly assigned to one of three treatment groups: goal-setting, conference (serving as a placebo), and control. A total of five variables measuring aspects of perceived locus of control was used. The covariate was a teacher-developed achievement pretest. In the report of results, no mention was made of main effects of the blocking IVs nor any interactions between them or with the manipulated IV.

Three separate analyses were reported, reflecting specific comparisons among groups rather than a single omnibus analysis of the three groups. These three comparisons were not mutually orthogonal, and they required more degrees of freedom than the omnibus test. However, no mention was made of adjusting (by Scheffé procedures or otherwise) for inflated α error produced by multiple testing. Further, although it was reported that three MANCOVAs were performed, no multivariate statistics were presented. Instead, univariate F's for each DV were reported for each of the three analyses. It was concluded that goal-setting treatment was effective in promoting more internal orientation among students, but only in a limited sense. That is, only measures related to locus of control in academic situations were found to vary with treatment.

Chapter 9

Discriminant Function Analysis

9.1 GENERAL PURPOSE AND DESCRIPTION

The major purpose of discriminant function analysis is to predict group membership on the basis of a variety of predictor variables. What is the best combination of predictor variables to maximize differences among groups? For example, can a differential diagnosis between learning disability and emotional disorder be made reliably on the basis of a set of psychological test scores? The two classification groups are learning disability and emotional disorder. Predictor (independent) variables are the psychological test scores (e.g., Illinois Test of Psycholinguistic Ability, subtests of the Wide Range Achievement Test, Figure Drawing scores, and Wechsler Intelligence Scale for Children).

It can be seen that the question of best group prediction is simply the question of MANOVA turned around. In MANOVA, we ask whether group membership produces significant differences on a combination of variables. If the answer to that question is yes, then that combination of variables can be used to discriminate among groups. Or in univariate terms, a significant difference between groups implies that given a score, you can predict which group it came from. This is not to suggest that statistical significance implies a high rate of prediction—only that the discrimination among groups is better than chance. Statistical significance is necessary, but not sufficient, for meaningful classification or discrimination.

Semantically, however, confusion can arise between MANOVA and discriminant function analysis (DISCRIM). In MANOVA, IVs are defined as in ANOVA and multiple regression—that is, they are the predictors. DVs in MANOVA are the *multivariables,* that is, the behaviors being measured. In

DISCRIM, these DVs or behaviors have now become predictors, which are traditionally termed IVs. We can avoid this semantic silliness by simply referring to IVs as *predictors* and to outcome variables (the variables being predicted) as *groups,* or grouping variables.[1]

Mathematically, MANOVA and DISCRIM are the same. The emphases, however, differ and these differences are reflected in many of the available canned computer programs. With the emphasis on predicting group membership, there is less attention to factorial arrangements of grouping variables, although factorial DISCRIM programs are available. Similarly, DISCRIM programs make no direct provision for within-subjects variables. But if these kinds of analyses are desired, the question can usually be rephrased in terms of MANOVA. For this reason the emphasis in this chapter will be on one-way between-subjects DISCRIM.

The DISCRIM analog to MANCOVA is available, since discriminant function analysis can be set up in a hierarchical manner. Actually, the covariate can be viewed as just another predictor that is given top priority. For example, what happens to discrimination between learning disability and emotional disorder among children on the basis of the Wide Range Achievement Test, Illinois Test of Psycholinguistic Ability, and Figure Drawings after differences in IQ are evaluated?

The emphasis on predicting group membership also leads to greater attention to the variety of dimensions on which group members differ. With more than two groups, the best discrimination between groups may require that more than a single dimension of discrimination be considered. For example, suppose there are three groups: learning disability, emotional disorder, and normal (no known disorder). It may be that one linear combination of psychological test scores separates the two disorder groups from the normal. To discriminate best between the learning disability and emotional disorder groups, however, a different linear combination of test scores may be required.

Note that there are two facets of discriminant analysis, one or both of which may be emphasized for any given research application. The researcher may simply be interested in a decision rule for classifying new cases, in which case the number of dimensions and their meaning may be irrelevant. Or, the emphasis may be on interpreting the discrimination space in terms of the variables contributing most heavily to separation of the groups in that space.

Finally, programs for DISCRIM typically provide a test for adequacy of classification. If groups can be reliably (i.e., significantly) discriminated or classified, just how well does the classification procedure do? How many cases in the original sample, or some cross-validation sample, are classified correctly? Classification is done on the basis of the discriminant function (prediction

---

[1] Many texts also refer to IVs, or predictor variables, as *discriminating variables* and to DVs, or grouping variables, as *classification variables.* However, there are also discriminant functions and classification functions to contend with, so that the terminology becomes quite confusing. We have tried to simplify it by use of only the terms *predictor variables* and *grouping variables.*

equation)[2] if there are only two groups or on the basis of classification functions if there are more than two groups.

The analytic model follows that of MANOVA, described in Chapter 8. A column matrix of scores on predictor variables is formed within each group, one for each case. By subtracting an appropriate mean from each score, matrices of difference scores are formed. These matrices are squared, by multiplying each matrix by its transpose, and summed. In this way, cross-products matrices (S matrices) are formed, analogous to sums of squares in ANOVA. Determinants[3] of the various matrices are found and ratios between them provide tests of hypotheses about the ability of the predictor variables to differentiate among groups. Since differentiation may be possible from a series of discriminant functions (one fewer function than the number of groups or equal to the number of predictor variables, whichever is the smaller), a hypothesis is tested about each function.[4] The first discriminant function is always the most powerful, with succeeding, orthogonal functions becoming successively less powerful. Significance tests indicate which discriminant functions (or dimensions) reliably discriminate among groups and which provide no additional information. When there are only two groups, the coefficients of the significant discriminant function can be used to predict group membership. When there are more than two groups, prediction of group membership involves the development of classification functions for each group, on the basis of the within-groups variance-covariance matrix and the group means. Cases are then classified into the group for which they have the highest classification score.

Discriminant functions are found in the same manner as canonical variates in problems in canonical correlation. DISCRIM, then, can also be thought of as a *multivariate multiple regression* (canonical correlation) in which grouping variables are discrete, and predictor variables are either continuous or discrete.

9.2 KINDS OF RESEARCH QUESTIONS

The primary goals of DISCRIM are to find the dimension or dimensions along which groups are maximally different and to predict group membership on the basis of those predictor variables used to create the dimensions. The degree to which this can be successfully done depends, of course, on the choice of predictors. Typically, this choice will be made either on the basis of theory—that is, which variables should provide information about group membership—

---

[2] The term *discriminant function* has been differently defined by various authors. The programs described in this text use the term synonymously with canonical variables (as defined in Chapter 6).

[3] A determinant, as described in Appendix A, can be viewed as a measure of generalized variance of a matrix.

[4] In MANOVA, hypotheses are usually tested on the basis of all discriminant functions at once (except for Roy's *gcr* test). For DISCRIM, it is frequently useful also to look at discriminant functions individually.

or on the basis of convenience. Sometimes one may simply seek out those pre-dictors that are easiest to obtain unobtrusively.

If groups can reasonably be cross-classified into a factorial design, it is frequently convenient to rephrase the research questions so that they can be answered within the framework of MANOVA. However, DISCRIM can in some circumstances be directly applied to factorial designs (cf. Section 9.6.5). It should be emphasized that the same data can be profitably analyzed through either MANOVA or DISCRIM programs, and frequently both, depending on the kinds of questions you want answered. In any event, statistical procedures are readily available within canned computer programs for answering the follow-ing types of questions generally associated with DISCRIM.

9.2.1 Significance of Prediction

Can group membership be reliably predicted from the set of predictor variables? For example, can we do better than chance in predicting whether children are learning disabled, emotionally disordered, or normal on the basis of the set of psychological test scores? This is the major question of DISCRIM, which the statistical procedures described in Section 9.6.1 are designed to answer. This question is identical to the question about "main effects of IVs" for a one-way MANOVA.

9.2.2 Number of Significant Discriminant Functions

Along how many dimensions do groups reliably differ? The DISCRIM model is set up so that the first discriminant function maximally separates groups. Then a second dimension, orthogonal to the first, is found that best separates groups on the basis of information not accounted for by the first discriminant function. This procedure of finding successive orthogonal discriminant functions continues until all possible dimensions are evaluated. The total number of possible dimensions is either one fewer than the number of groups or equal to the number of predictor variables, whichever is the smaller. Typically, only the first one or two dimensions reliably discriminate among groups, so that remaining dimen-sions provide no additional information about group membership and are better ignored. Tests of significance for deciding which discriminant functions are worth keeping are discussed in Section 9.6.2.

For the three groups of children in our example, two discriminant functions are possible. It might be that both are statistically significant and therefore enhance discrimination among the three groups. For example, suppose the first function separates the normal group from the other two, and the second separates those with learning disability from those with emotional disorder. Or it might be the case that a single linear combination of predictors is sufficient to discrimi-nate among the three groups. That is, the group with learning disability might be midway between the normal group and the group with emotional disorder on a single dimension representing the combined psychological test scores (highly unlikely with this example).

9.2.3 Dimensions of Discrimination

How can the dimensions along which groups are separated be interpreted? Where are groups located with respect to the discriminant functions? How do predictor variables correlate with the discriminant functions? These questions are discussed in Section 9.6.3. As in principal components or factor analysis, the discriminant functions (as components or factors) can be rotated to enhance interpretability.[5] Discussion of rotation will be reserved for Chapter 10.

In our example, if the two significant discriminant functions are found as described above, which predictor variables are correlated highly with each function? Which types of test scores discriminate the normal children from the two other groups (first discriminant function)? Which types of scores are most useful in discriminating learning disability from emotional disorder (second discriminant function)?

9.2.4 Classification Functions

After determining the number of reliable discriminant functions, what is the best linear equation for classifying new cases into groups? For example, suppose we had the battery of psychological test scores for a group of new, undiagnosed children. How would we best combine (weight) these scores so as to achieve the most reliable diagnosis? If there are only two groups, or if only the first discriminant function is statistically significant, the single discriminant function provides sufficient information for classification. A case with a discriminant score above zero belongs in one group, while a case with a score below zero on the discriminant function is predicted to belong to the other group. With more than one significant discriminant function, however, information from all reliable functions will be necessary for optimum classification. Such information is provided more conveniently by classification functions, one for each group. A case is classified into the group for which it has the highest classification score. Procedures for deriving and using classification functions are discussed in Sections 9.4.2 and 9.6.6.

9.2.5 Adequacy of Classification

Given a classification scheme (through a single discriminant function or a series of classification functions), what proportion of cases is correctly classified? When errors occur, into which groups do cases tend to be incorrectly classified? If the researcher has knowledge that some groups are more likely to occur, or if some kinds of misclassification are especially undesirable, the classification procedure can be modified by specifying prior probabilities.

When the same cases are classified that were used to set up the original classification functions, the adequacy of classification can be assessed. Alterna-

---

[5] Rotation is available in some programs (cf. Table 9.8) but interpretation of rotated functions is still considered an experimental procedure.

tively, a cross-validation sample can be used to assess adequacy. When new cases are classified for whom group membership is unknown, their pattern of predicted group membership can be compared with the pattern produced by the "standardization sample" that produced the classification functions. Procedures for deriving classification functions are discussed in Section 9.4.2, and procedures for testing them in Section 9.6.6.

With our example, adequacy of classification could be assessed by omitting from the original discriminant function analysis one-third of the available children for whom diagnosis is known. After developing the best weighting equations for psychological test scores from the remaining two-thirds of the sample, we could test out the equations on the one-third of the children withheld. How many of these withheld children are correctly classified (diagnosed) on the basis of the statistical weighting? What kinds of misclassification occur? For example, are normal children more likely to be misclassified as learning disabled or as emotionally disordered?

9.2.6 Strength of Association

What is the degree of relationship between group membership on the one hand and the set of predictor variables on the other? This is basically a question regarding percent of variance accounted for and, as seen in Section 9.4.1., is answered through canonical correlation. For each discriminant function (canonical variable) a canonical correlation is found that, when squared, indicates the proportion of variance shared between grouping variables and predictor variables on that dimension. These correlations will change if the discriminating components or factors are rotated (see Section 9.2.3).

For instance, if the second discriminant function in our example separates the group with learning disability from that with emotional disorder, what proportion of variance in this dichotomy is associated with the best linear combination of psychological test scores? One could think of this as a squared correlation between a dichotomous and a continuous variable. On the one hand there is the dichotomous variable representing two groups, one with learning disability and the other with emotional disorder. On the other hand there is a continuous variable, derived from combining all of the psychological test scores in such a way as to maximize group differences. This point biserial correlation is analogous to the canonical correlation for the second discriminant function. (Unfortunately, discriminant functions rarely separate groups out into such clearcut dichotomies.)

9.2.7 Importance of Predictor Variables

Which predictor variables are most important in predicting group membership? Questions about importance of predictor variables are analogous to those of importance of DVs in MANOVA and to those of IVs in multiple regression. One indication of importance is the absolute magnitude of the standardized discriminant function coefficients (which are similar to the β weights in a multiple regression equation). Magnitude of discriminant function coefficients, however, can be misleading just as β weights can be misleading. As in MANOVA, univariate F's represent the ability of each predictor variable by itself to predict group membership. By themselves, univariate F's can be misleading, too, because they neither take into account correlations among predictor variables nor compensate for increased Type I errors with multiple testing. Another procedure is interpretation of a loading matrix of correlations between the predictor variables and each discriminant function, as discussed in Section 9.6.3.2. A final alternative is to evaluate predictor variables in the context of how well they separate each group from all the others. This strategy, discussed in Section 9.6.4, deals with problems of correlations among predictors and multiple testing.

9.2.8 Significance of Prediction with Covariates

After removing statistically the effects of one or more covariates, can one reliably predict group membership? In DISCRIM, as in MANCOVA, group separation can be assessed after adjustment for some prior variables. With currently available programs, however, the procedure frequently is more conveniently conceptualized in terms of MANCOVA.

In our example, scores on the Wechsler Intelligence Scale for Children (WISC) could be considered the covariate. In such a case, the major question of analysis would be whether discrimination among the three groups is better than chance on the basis of scores on the Illinois Test of Psycholinguistic Ability (ITPA), Figure Drawings, and Wide Range Achievement Test (WRAT), after adjusting for differences measured on the WISC. Rephrased in terms of hierarchical discriminant function analysis, the question becomes, Do scores on the Figure Drawings, WRAT, and ITPA provide significantly better classification among the three groups than that afforded by scores on the WISC? Hierarchical DISCRIM is discussed in Section 9.5.2 and demonstrated in Section 9.8.3.

9.2.9 Group Means

If predictor variables are shown to discriminate among groups, it will be interesting to note just how the groups differ on those variables. The best estimates of population parameters for the groups are, of course, the sample means. Therefore, for example, if the ITPA were found to discriminate between groups with learning disability and emotional disorder, it would be worthwhile to compare the average ITPA score for learning disabled children with the average ITPA score for emotionally disordered children.

9.3 LIMITATIONS TO DISCRIMINANT FUNCTION ANALYSIS

9.3.1 Theoretical Issues

Since DISCRIM is typically used to predict membership in groups that are naturally occurring, rather than groups into which the experimenter has randomly assigned cases, questions of causality will typically not be answered. DISCRIM does not tell us why we can reliably predict group membership, or what causes differential membership. The statistical procedures, however, are the same as for MANOVA, and if it is the case that group membership has occurred by random assignment, inferences of causality are justifiable as long as proper experimental controls have been instituted. The DISCRIM question then becomes, Has treatment following random assignment to groups produced enough difference in the predictor variables that we can now reliably separate groups on the basis of those variables?

As implied, limitations to DISCRIM are identical to limitations to MANOVA. However, problems with covariates are typically less difficult for DISCRIM than those for MANOVA, because covariates are less ambiguous in correlational than in experimental research. In correlational research, we are simply asking about the predictive power of variables after accounting for covariates. We are typically not interested in such things as "adjusted group means" in DISCRIM, so we need not worry about their interpretation.

The usual difficulties of generalizability apply to DISCRIM. But there is a procedure available that gives some indication about the ability of the solution to generalize, at least to a cross-validation sample (see Section 9.6.6).

9.3.2 Practical Issues

Practical issues for DISCRIM are basically the same as those for MANOVA. Therefore they will be discussed here only to the extent of identifying the analogies between MANOVA and DISCRIM and distinguishing situations in which assumptions for MANOVA and DISCRIM might differ. Because meaningful classification makes greater demands on the variables than statistical inference does, it is sometimes possible to relax some of the following requirements (except for outliers). For example, if you could achieve 95% accuracy in classification, you would hardly worry about the shape of the distribution. Nevertheless, DISCRIM is optimal under those conditions under which MANOVA is valid. And if the classification rate is unsatisfactory, it might be because of violation of assumptions or limitations.

9.3.2.1 Unequal Sample Sizes and Missing Data   As DISCRIM is typically a one-way analysis, no special problems are posed by unequal sample sizes. (For effects of unequal sample sizes on factorial analysis, see Chapter 7.) In classification, however, a decision must be made as to the a priori probabilities with which cases are assigned to groups. You may or may not want these probabilities to be influenced by sample sizes. Section 9.4.2 discusses this issue, and use of unequal probabilities is demonstrated in Section 9.8.

Regarding missing data (absence of scores on predictor variables for some cases), consult Section 7.3.2.1 and Chapter 4 for a review of problems and potential solutions.

As discussed in Section 8.3.2.1, the sample size of the smallest group should exceed the number of predictor variables. Although hierarchical and stepwise analyses avoid the problem of multicollinearity and singularity (prevented by the tolerance test at each step), overfitting can occur with stepwise as well as direct discriminant analysis if the number of cases does not notably exceed the number of variables.

9.3.2.2 Multivariate Normality For DISCRIM, multivariate normality assumes that the predictor variable scores are independently and randomly sampled from a population of scores, and that the sampling distribution of any linear combination of predictor variables is normally distributed. DISCRIM, like MANOVA, is robust to failures of normality if violation is caused by skewness rather than by outliers. *A sample size that would produce* 20 *df for error in the univariate case should ensure robustness with respect to multivariate normality, as long as sample sizes are equal and two-tailed tests are used.* (Calculation of error degrees of freedom is discussed in Chapter 2.)

Since tests for DISCRIM typically are not one-sided, this requirement poses no difficulty. Sample sizes, however, are *not* usually equal for applications of DISCRIM. Naturally occurring groups rarely occur or are sampled with equal numbers of members. The greater the difference between groups in sample size, the larger the overall sample size necessary to assure robustness. As a conservative recommendation, robustness could be expected with 20 cases in the smallest group if there are only a few predictors. In many applications of DISCRIM, however, very large samples (in the hundreds) are available from survey research. In these cases, robustness of the statistical procedures need not be worrisome. If samples are both small and unequal in size, assessment of normality is a matter of judgment. Would DVs reasonably be expected to be normally distributed in the population being sampled? If not, transformation of some variable(s) (cf. Chapter 4) might be worthwhile.

9.3.2.3 Outliers DISCRIM, like MANOVA, is highly sensitive to multivariate outliers. Therefore, *for each group separately, run a test for univariate and multivariate outliers, and transform or eliminate any significant outliers before DISCRIM analysis* (see Chapter 4). When transforming or eliminating data from analysis, you should so indicate in your report.

9.3.2.4 Homogeneity of Variance-Covariance Matrices As is true for MANOVA (Section 8.3.2.4), robustness can be expected for DISCRIM with respect to violation of the assumption of equal variance-covariance (dispersion) matrices with equally sized or large samples. When sample sizes are unequal and small, however, results of significance testing can be misleading if there is also heterogeneity of the variance-covariance matrices. In that case, homogeneity of variance-covariance matrices should be assessed (see Section 8.3.2.4 for guidelines). Alternatively, the scatterplots of scores on the first two canonical discriminant functions can be examined separately for each group. If the scatterplots (produced by BMDP7M and SPSS DISCRIMINANT) are roughly equal in size, there is evidence for homogeneity of variance-covariance matrices.

Though the test of discriminant functions is usually robust with respect to heterogeneity of variance and covariance, classification procedures are not. Cases tend to be overclassified into groups with greater dispersions. *If classification is an important goal in the analysis, examine plots or test for homogeneity of variance and covariance.* If covariance matrices are grossly different, transformation of predictor variables may equalize them. (Skewness of individual predictors should be examined to determine form of transformation.) Or, it may be possible with some implementations of SPSS DISCRIMINANT to classify on the basis of separate covariance matrices, although this procedure may lead to overfitting.

9.3.2.5 Linearity   The DISCRIM model assumes a linear relationship among all predictor variables within each group. The assumption is less serious than some others, however, in that violation simply leads to reduced power rather than increase in Type I error. Apply procedures in Section 7.3.2.6 to all relations among predictor variables.

9.3.2.6 Multicollinearity and Singularity   With highly redundant discriminating variables, multicollinearity or singularity may occur, making the inversion of matrices unreliable. Fortunately, *most computer programs for discriminant function analysis protect against this possibility by specifying a tolerance value.* Variables not meeting tolerance are not allowed to participate in the prediction. Guidelines for assessing multicollinearity and singularity for programs that do not include tolerance values, and for dealing with multicollinearity or singularity when it occurs, appear in Section 8.3.2.8. Note that analysis will be done on predictor variables (not "DVs").

9.4 FUNDAMENTAL EQUATIONS FOR DISCRIMINANT FUNCTION ANALYSIS

For demonstration of DISCRIM, hypothetical scores on four predictor variables are given for three groups of learning-disabled children. Scores for the three cases in each of the three groups are illustrated in Table 9.1. The groups, classified on the basis of major learning problem, are (1) MEMORY, including children whose major difficulty seems to be with tasks related to memory; (2) PERCEPTION, including children who show difficulty in visual perception; and (3) COMMUNICATION, characterized by children with language difficulty. The four predictor variables are (1) PERF, or WISC-R (WISC Revised) Performance Scale IQ; (2) INFO, scaled score on the Information subtest of the WISC-R; (3) VERBEXP, scaled score on the Verbal Expression subtest of the ITPA; and (4) AGE, or chronological age in years. The grouping variable, then, is type of learning disability, and the predictor variables include selected scores from psychodiagnostic instruments and age.

TABLE 9.1 HYPOTHETICAL SMALL SAMPLE DATA FOR
ILLUSTRATION OF DISCRIMINANT FUNCTION
ANALYSIS

	Predictors			
Group	PERF	INFO	VERBEXP	AGE
	87	5	31	6.4
MEMORY	97	7	36	8.3
	112	9	42	7.2
	102	16	45	7.0
PERCEPTION	85	10	38	7.6
	76	9	32	6.2
	120	12	30	8.4
COMMUNICATION	85	8	28	6.3
	99	9	27	8.2

Fundamental formulas will be presented for the two major parts of discriminant function analysis: derivation and statistical tests of the discriminant functions; and classification of cases.

9.4.1 Test and Derivation of Discriminant Functions

The fundamental formula for testing the significance of a set of discriminant functions is the same as that for MANOVA, discussed in Chapter 8. That is, variance in the set of predictors is partitioned into two sources: variance attributable to differences between groups; and variance attributable to differences within groups. With the use of procedures illustrated in Eqs. 8.1 to 8.3, cross-products matrices are formed.

$$S_{total} = S_{bg} + S_{wg} \tag{9.1}$$

The total cross-products matrix (S_{total}) is partitioned into cross-products matrices associated with differences between learning-disability groups (S_{bg}) and differences associated with children within groups (S_{wg}).

For the example in Table 9.1, the resulting cross-products matrices are

$$S_{bg} = \begin{bmatrix} 314.89 & -71.56 & -180.00 & 14.41 \\ -71.56 & 32.89 & 8.00 & -2.22 \\ -180.00 & 8.00 & 168.00 & -10.40 \\ 14.49 & -2.22 & -10.40 & 0.74 \end{bmatrix}$$

$$S_{wg} = \begin{bmatrix} 1286.00 & 220.00 & 348.33 & 50.00 \\ 220.00 & 45.33 & 73.67 & 6.37 \\ 348.33 & 73.67 & 150.00 & 9.73 \\ 50.00 & 6.37 & 9.73 & 5.49 \end{bmatrix}$$

Determinants for these matrices are

$$|\mathbf{S}_{wg}| = 4.70034789 \times 10^{13}$$

$$|\mathbf{S}_{bg} + \mathbf{S}_{wg}| = 448.63489 \times 10^{13}$$

Following procedures in Eq. 8.4, we can find Wilks' Lambda,[6] or the U statistic, for these matrices:

$$\Lambda = \frac{|\mathbf{S}_{wg}|}{|\mathbf{S}_{bg} + \mathbf{S}_{wg}|} = .010477$$

To find an associated approximate F ratio, as delineated in Eq. 8.5, the following values are used:

$p = 4$ — the number of predictor variables

$df_{effect} = 2$ — the number of groups minus one, or $k - 1$

$df_{error} = 6$ — the number of groups times the quantity $n - 1$, where n is the number of cases per group. Since n is frequently not equal for all groups in DISCRIM, an alternative formula for df_{error} is $N - k$, where N is the total number of cases in all groups—9 in this case.

Thus we obtain

$$s = \sqrt{\frac{(4)^2(2)^2 - 4}{(4)^2 + (2)^2 - 5}} = 2$$

$$y = (.010477)^{1/2} = .102357$$

$$df_1 = 4(2) = 8$$

$$df_2 = (2)\left[6 - \frac{4 - 2 + 1}{2}\right] - \left[\frac{4(2) - 2}{2}\right] = 6$$

$$\text{approximate } F(8, 6) = \left(\frac{1 - .102357}{.102357}\right)\left(\frac{6}{8}\right) = 6.58$$

For 8 and 6 df at $\alpha = .05$, the critical level of F is 4.15. Since our obtained F of 6.58 exceeds that value, we conclude that the three groups of children *can* be distinguished on the basis of the combination of the four predictor variables.

To solve for a discriminant function score, the basic formula for the *i*th function is

$$D_i = d_{i1}z_1 + d_{i2}z_2 + \cdots + d_{ip}z_p \tag{9.2}$$

---

[6] Alternative statistical criteria are discussed in Section 9.6.1.1.

A child's score on the ith discriminant function (D_i) is found by multiplying the standardized score on each predictor variable (z) by its associated standardized discriminant function coefficient (d_i) and adding the products over all predictor variables.

Finding the d_i is basically a problem in canonical correlation, where we solve for successive canonical variates (here called discriminant functions) according to the procedures in Chapter 6. But here d_i are chosen so as to maximize differences between groups relative to differences within groups. Just as in multiple regression (cf. Chapter 5), the equation can be formulated on the basis of raw scores as well as standardized scores. A discriminant function score, then, can also be produced by multiplying the raw score on each predictor variable by its associated unstandardized weight, adding the products over all predictor variables, and adding a constant to adjust for the means. The score produced in this way is the same D_i as produced in Eq. 9.2.

As can be seen, the mean of each discriminant function, combining data over all cases, will be zero, since the mean of each of the variables, when standardized, is zero. Further, the standard deviation of D_i is equal to 1. If there were only two groups, then, the discriminant function could be used to classify cases into groups. A case would be classified into one group if the D score were above zero, and into the other group if the D score were below zero. As the sample data are based on more than two groups, however, classification is more complicated and discussion of it will be reserved for the following section.

Just as D_i can be calculated for each case, a mean value of D_i can be calculated for each group. That is, the members of each group considered together will have a mean score on D_i that is the distance of the group in standard deviation units from the zero mean of the discriminant function. Group means on D_i are typically called centroids in reduced space, the space being reduced from that of the p variables to a single dimension, or discriminant function.

As discussed in Section 9.2.2, the maximum number of discriminant functions is limited by the number of groups or number of predictor variables. It is the smaller of either (1) the number of predictor variables or (2) the number of groups minus one. Since the smaller value for the sample is two (the number of groups minus one), a maximum of two discriminant functions may be produced for these data.

For each of these discriminant functions (canonical variates), the canonical correlation is found following procedures in Chapter 6. In addition, successive discriminant functions can be evaluated for significance, as discussed in Section 9.6.2. For the sample data, the evaluation of discriminant functions, standardized and unstandardized discriminant coefficients, loading matrix, and the group centroids appear in Table 9.2.

For the sample data, examination of the centroids shows that the first discriminant function distinguishes group 1 (MEMORY) from the other two groups, while the second discriminant function distinguishes group 3 (COMMUNICATION) from the other two groups. Canonical correlations of .965 and

TABLE 9.2   DECK SETUP AND PARTIAL OUTPUT FROM SPSS
DISCRIMINANT FOR SAMPLE DATA OF TABLE 9.1

```
                      VALUE LABELS    GROUP (1)MEMORY (2)PERCEP (3)COMMUN
                      VAR LABELS      PERF, WISCR PERFORMANCE IQ/
                                      INFO, WISCR INFORMATION/
                                      VERBEXP, ITPA VERBAL EXPRESSION
                      DISCRIMINANT    GROUPS=GROUP(1,3)/
                                      VARIABLES=PERF TO AGE/
                                      ANALYSIS=PERF TO AGE
                      OPTIONS         5,6,11,12
                      STATISTICS      10
```

CANONICAL DISCRIMINANT FUNCTIONS

FUNCTION	EIGENVALUE	PERCENT OF VARIANCE	CUMULATIVE PERCENT	CANONICAL CORRELATION	-	AFTER FUNCTION	WILKS LAMBDA	CHI-SQUARED	D.F.	SIGNICANCE
					-	0	.0104766	20.514	8	.0086
1*	13.48590	70.70	70.70	.9648665	-	1	.1517629	8.4845	3	.0370
2*	5.58923	29.30	100.00	.9209979	-					

* MARKS THE   2 FUNCTION(S) TO BE USED IN THE REMAINING ANALYSIS.

STANDARDIZED CANONICAL DISCRIMINANT FUNCTION COEFFICIENTS

	FUNC 1	FUNC 2
PERF	-2.50352	1.47406
INFO	3.48961	.28380
VERBEXP	-1.32466	-1.78881
AGE	.50273	-.23625

POOLED WITHIN-GROUPS CORRELATIONS BETWEEN CANONICAL DISCRIMINANT FUNCTIONS AND DISCRIMINATING VARIABLES
VARIABLES ARE ORDERED BY THE FUNCTION WITH LARGEST CORRELATION AND THE MAGNITUDE OF THAT CORRELATION.

	FUNC 1	FUNC 2
INFO	.22796*	-.06642
VERBEXP	-.02233	-.44630*
PERF	-.07546	.17341*
AGE	-.02786	.14861*

UNSTANDARDIZED CANONICAL DISCRIMINANT FUNCTION COEFFICIENTS

	FUNC 1	FUNC 2
PERF	-.1710041	.1006866
INFO	1.269534	.1032488
VERBEXP	-.2649326	-.3577618
AGE	.5254062	-.2469000
(CONSTANT)	9.673740	3.452930

CANONICAL DISCRIMINANT FUNCTIONS EVALUATED AT GROUP MEANS (GROUP CENTROIDS)

GROUP	FUNC 1	FUNC 2
1	-4.10234	-.69097
2	2.98068	-1.94169
3	1.12166	2.63265

.921 indicate that each discriminant function provides a high degree of association between discriminant function scores and group membership.

9.4.2 Classification

To classify cases into groups, a classification equation is developed for each group. Each case has a classification score for each group. Then a case is assigned to the group for which it has the highest classification score. Note that classifying by these classification equations is simpler than using an alternative technique based on sequential classification using a set of discriminant equations.

In its simplest form, the basic classification equation for the jth group $(j = 1, 2, \ldots, k)$ is

$$C_j = c_{j0} + c_{j1} Y_1 + c_{j2} Y_2 + \cdots + c_{jp} Y_p \tag{9.3}$$

A score on the classification function for group j (C_j) is found by first multiplying the raw score on each predictor variable (Y) by its associated classification function coefficient (c_j). These products are summed over all predictor variables and are added to a constant, c_{j0}.

The coefficients, c_j, are found from the means of the p predictor variables and the pooled within-group variance-covariance matrix, $\mathbf{W}$. The within-group covariance matrix is produced by dividing each element in the cross-products matrix, $\mathbf{S}_{wg}$, by the within-group degrees of freedom, $N - k$. In matrix form,

$$C_j = \mathbf{W}^{-1} \mathbf{M}_j \tag{9.4}$$

The row matrix of classification coefficients for group j ($\mathbf{C}_j = c_{j1}, c_{j2}, \ldots, c_{jp}$) is found by multiplying the inverse of the within-group variance-covariance matrix ($\mathbf{W}^{-1}$) by a column matrix of means for group j on the p variables ($\mathbf{M}_j = Y_{j1}, Y_{j2}, \ldots, Y_{jp}$).

The constant for group j, c_{j0}, is found as follows:

$$c_{j0} = \left(-\frac{1}{2}\right) C_j M_j \tag{9.5}$$

The constant for the classification function for group j (c_{j0}) is formed by multiplying the row matrix of classification coefficients for group j (C_j) by the column matrix of means for group j (M_j).

For the sample data, each element in the $\mathbf{S}_{wg}$ matrix from Section 9.4.1 is divided by $\mathrm{df}_{wg} = \mathrm{df}_{error} = 6$. This produces the following variance-covariance matrix:

$$W = \begin{bmatrix} 214.33 & 36.67 & 58.06 & 8.33 \\ 36.67 & 7.56 & 12.28 & 1.06 \\ 58.06 & 12.28 & 25.00 & 1.62 \\ 8.33 & 1.06 & 1.62 & 0.92 \end{bmatrix}$$

The inverse of the within-groups variance-covariance matrix is

$$W^{-1} = \begin{bmatrix} 0.04362 & -0.20195 & 0.00956 & -0.17990 \\ -0.20195 & 1.62970 & -0.37073 & 0.60623 \\ 0.00956 & -0.37073 & 0.20071 & -0.01299 \\ -0.17990 & 0.60623 & -0.01299 & 2.05006 \end{bmatrix}$$

Multiplying W^{-1} by the column matrix of means for the first group gives the matrix of classification coefficients for that group, as per Eq. 9.4.

$$C_1 = W^{-1} \begin{bmatrix} 98.67 \\ 7.00 \\ 36.33 \\ 7.30 \end{bmatrix} = [1.92, -17.56, 5.55, 0.99]$$

The constant for group 1, then, according to Eq. 9.5, is

$$c_{1,0} = \left(-\frac{1}{2}\right) \left([1.92, -17.56, 5.55, 0.99] \begin{bmatrix} 98.67 \\ 7.00 \\ 36.33 \\ 7.30 \end{bmatrix} \right) = -137.83$$

(Values used in these calculations were carried to several decimal places.) By following these procedures for groups 2 and 3, the full set of classification equations can be produced. These are illustrated in Table 9.3.

In its simplest form, classification can now proceed. For the first case in group 1, the classification scores, applying Eq. 9.3, for the three groups, respectively, are

TABLE 9.3 CLASSIFICATION FUNCTION COEFFICIENTS FOR SAMPLE DATA OF TABLE 9.1

	Group 1: MEMORY	Group 2: PERCEP	Group 3: COMMUN
PERF	1.92420	0.58704	1.36552
INFO	-17.56221	-8.69921	-10.58700
VERBEXP	5.54585	4.11679	2.97278
AGE	0.98723	5.01749	2.91135
(CONSTANT)	-137.82892	-71.28563	-71.24188

$$C_1 = -137.83 + (1.92)(87) + (-17.56)(5) + (5.55)(31) + (0.99)(6.4) = 119.80$$

$$C_2 = -71.29 + (0.59)(87) + (-8.70)(5) + (4.12)(31) + (5.02)(6.4) = 96.39$$

$$C_3 = -71.24 + (1.37)(87) + (-10.59)(5) + (2.97)(31) + (2.91)(6.4) = 105.69$$

Since the group with the highest classification score for this child is group 1, that is the group to which the child would be assigned.

This simplest classification scheme is most appropriate when the expected population sizes of all groups are equal. In the event of unequal population sizes, the classification procedure can be modified to reflect the a priori inequalities. Although a number of highly sophisticated classification schemes have been suggested (e.g., Tatsuoka, 1975), the most straightforward simply involves adding to each classification equation a term that adjusts for group size.[7] The classification equation for group j (C_j) then becomes

$$C_j = c_{j0} + \sum_{i=1}^{p} c_{ji} Y_i + \ln(n_j/N) \tag{9.6}$$

where n_j = size of group j and N = total sample size over all groups.

It should be reemphasized that the classification procedure is highly sensitive to heterogeneity of variance-covariance matrices. Cases are more likely to be classified into the group with the greatest dispersion—that is, the group for which the determinant of the within-group covariance matrix is greatest. Section 9.3.2.4 provides suggestions for dealing with this problem. Uses of classification procedures are discussed more fully in Section 9.6.6.

9.5 TYPES OF DISCRIMINANT FUNCTION ANALYSIS

The three basic types of discriminant function analysis—standard (direct), hierarchical, and stepwise—are analogous to the three types of multiple regression discussed in Section 5.5. Actually, the analogy between multiple regression and DISCRIM is complete in the case of DISCRIM with two groups: the IVs in a regression context are the predictor variables, and the DV in regression is the grouping variable in DISCRIM. This, then, is a case of multiple regression with a dichotomous DV, and it is perfectly legitimate to perform DISCRIM with two groups using multiple regression techniques. With more than two groups, there may be more than one dimension (function) for discriminating groups in DISCRIM, a situation that cannot occur in multiple regression.

Criteria for choosing among the three strategies follow directly from those discussed in Section 5.5.4 for multiple regression.

---

[7] In SPSS DISCRIMINANT (starting with Release 8.0) and BMDP7M the constants in the classification equations are adjusted for prior probabilities. This means that Eq. 9.3 can be used with that output regardless of prior probabilities.

9.5.1 Direct Discriminant Function Analysis

In standard (direct) DISCRIM, the discriminant function equations are solved simultaneously on the basis of all predictor variables. That is, like standard multiple regression, all of the predictor variables enter the equations at once. The standard model is the one used in the multivariate test MANOVA, in which all DVs are considered simultaneously. This is also the model demonstrated in the sample problem in Section 9.4.1.

All of the canned computer programs listed later, in Table 9.8, can be used for standard (direct) discriminant function analysis. The information in Table 9.2 was produced by running the small sample data of Section 9.4 through SPSS DISCRIMINANT, in DIRECT (default) method. Another SPSS program, MANOVA, is available for standard discriminant function analysis, but does not include classification. The BMD programs, BMD07M and BMDP7M, can be used for stepwise and hierarchical DISCRIM as well as standard, and will be illustrated in forthcoming sections.

9.5.2 Hierarchical Discriminant Function Analysis

The hierarchical (or, as some prefer to call it, sequential) mode of DISCRIM can be used to evaluate contributions to group discrimination by predictor variables as they enter the equations in some priority order as determined by the researcher. This means that the researcher can assess the predictive power gained by adding a variable to a set of prior variables. It is possible, then, to determine if classification of cases to groups improves by adding a set of variables (cf. Section 9.6.6.4).

When prior variables are viewed as covariates and the added variable is viewed as a DV, this can be seen as similar to a problem in analysis of covariance. Indeed, hierarchical DISCRIM might be used to perform a stepdown analysis following MANOVA (cf. Section 8.5.3.2).

Another useful application of hierarchical DISCRIM might be when a reduced set of predictors is desired, and there is some basis for setting up predictors in priority order. For example, some predictors might be very easy to obtain, while others might require increased time or expense. A useful, cost-effective set of predictors might then be determined through the hierarchical procedure.

BMD07M and BMDP7M are available for hierarchical analysis. Application of BMDP7M to the small sample data is illustrated in Table 9.4. Age was given the highest priority of entry since it is obviously the easiest piece of information to obtain about a child, and furthermore might be a useful candidate as a covariate. The next level of priority was assigned to the two WISC-R results, allowed to compete with each other in a stepwise fashion for entry (see Section 9.5.3). Third level of priority was assigned to the ITPA subtest score since it is the least likely to be readily available without special testing. At each step, significance of discrimination is assessed, as well as adequacy of classification.

TABLE 9.4   DECK SETUP AND SELECTED BMDP7M OUTPUT FOR
HIERARCHICAL DISCRIMINANT FUNCTION ANALYSIS OF
SAMPLE DATA IN TABLE 9.1

```
                    PROGRAM CONTROL INFORMATION

                    /PROBLEM     TITLE IS "SMALL SAMPLE DISCRIMINANT EXAMPLE".
                    /INPUT       VARIABLES ARE 6.  FORMAT IS "(4X,5F4.0,F4.1)".  UNIT=41.
                    /VARIABLE    NAMES ARE CASENO,GROUP,PERF,INFO,VERBEXP,AGE.
                                 LABEL=CASENO.
                                 GROUPING IS GROUP.
                    /GROUP       CODES(2) ARE 1,2,3.
                                 NAMES(2) ARE MEMORY,PERCEP,COMMUN.
                    /DISCRIM     LEVEL=0,0,3,3,4,2.  FORCE=4.
                    /END

STEP NUMBER   0
   VARIABLE        F TO FORCE  TOLERANCE  *    VARIABLE       F TO  FORCE   TOLERANCE
                   REMOVE LEVEL                *                     ENTER LEVEL
             DF=  2    7                   *                  DF=  2    6
                                           *     3 PERF        .735   3    1.000000
                                           *     4 INFO       2.176   3    1.000000
                                           *     5 VERBEXP    3.360   4    1.000000
                                           *     6 AGE         .402   2    1.000000

STEP NUMBER   1
VARIABLE ENTERED   6 AGE
   VARIABLE        F TO FORCE  TOLERANCE  *    VARIABLE       F TO  FORCE   TOLERANCE
                   REMOVE LEVEL                *                     ENTER LEVEL
             DF=  2    6                   *                  DF=  2    5
   6 AGE           .402   2   1.000000     *     3 PERF        .308   3     .646114
                                           *     4 INFO       2.471   3     .837231
                                           *     5 VERBEXP    3.487   4     .885027
U-STATISTIC OR WILKS" LAMBDA    .8819122   DEGREES OF FREEDOM  1   2       6
APPROXIMATE F-STATISTIC             .402   DEGREES OF FREEDOM    2.00     6.00
  F - MATRIX         DEGREES OF FREEDOM =   1    6
            MEMORY    PERCEP
PERCEP      .22
COMMUN      .18      .80
CLASSIFICATION FUNCTIONS
         GROUP =   MEMORY      PERCEP      COMMUN
VARIABLE
  6 AGE          7.97330     7.57282     8.33738
CONSTANT       -30.20116   -27.35104   -32.91961
******************************************************************************************

STEP NUMBER   2
VARIABLE ENTERED   4 INFO
   VARIABLE        F TO FORCE  TOLERANCE  *    VARIABLE       F TO  FORCE   TOLERANCE
                   REMOVE LEVEL                *                     ENTER LEVEL
             DF=  2    5                   *                  DF=  2    4
   4 INFO         2.471   3    .837231     *     3 PERF       9.303   3     .108093
   6 AGE           .767   2    .837231     *     5 VERBEXP   13.133   4     .201401
U-STATISTIC OR WILKS" LAMBDA    .4435523   DEGREES OF FREEDOM  2   2       6
APPROXIMATE F-STATISTIC            1.254   DEGREES OF FREEDOM    4.00    10.00
  F - MATRIX         DEGREES OF FREEDOM =   2    5
            MEMORY    PERCEP
PERCEP     2.65
COMMUN      .59     1.12

CLASSIFICATION FUNCTIONS
         GROUP =   MEMORY      PERCEP      COMMUN
VARIABLE
  4 INFO          -.23089      .57401      .12959
  6 AGE          8.24090     6.90754     8.18718
CONSTANT       -30.36978   -28.39318   -32.97273
******************************************************************************************
```

TABLE 9.4 *(Continued)*

```
STEP NUMBER   3
VARIABLE ENTERED   3 PERF
    VARIABLE        F TO FORCE  TOLERANCE  *     VARIABLE        F TO FORCE   TOLERANCE
                    REMOVE LEVEL                 *                ENTER LEVEL
                 DF=  2    4                     *                          DF=  2    3
    3 PERF         9.303    3   .108093   *     5 VERBEXP       9.737    4        .199297
    4 INFO        18.010    3   .140067   *
    6 AGE           .445    2   .532999   *
U-STATISTIC OR WILKS" LAMBDA     .0784812    DEGREES OF FREEDOM   3    2      6
APPROXIMATE F-STATISTIC             3.426    DEGREES OF FREEDOM   6.00     8.00
 F - MATRIX         DEGREES OF FREEDOM =    3    4
            MEMORY    PERCEP
PERCEP     13.85
COMMUN      1.78     5.95
CLASSIFICATION FUNCTIONS
        GROUP =   MEMORY      PERCEP      COMMUN
VARIABLE
    3 PERF        1.65998      .39091     1.22389
    4 INFO       -7.31838    -1.09502    -5.09593
    6 AGE         1.34609     5.28388     3.10371
CONSTANT       -62.29002   -30.16334   -50.32441

STEP NUMBER   4
VARIABLE ENTERED   5 VERBEXP
    VARIABLE        F TO  FORCE  TOLERANCE  *     VARIABLE        F TO  FORCE   TOLERANCE
                    REMOVE LEVEL                  *                ENTER LEVEL
                 DF=  2    3                      *                          DF=  2    2
    3 PERF         6.893    3   .106964   *
    4 INFO        18.837    3   .081213   *
    5 VERBEXP      9.737    4   .199297   *
    6 AGE           .266    2   .532781   *
U-STATISTIC OR WILKS" LAMBDA     .0104766    DEGREES OF FREEDOM   4    2      6
APPROXIMATE F-STATISTIC             6.577    DEGREES OF FREEDOM   8.00     6.00
 F - MATRIX         DEGREES OF FREEDOM =    4    3
            MEMORY    PERCEP
PERCEP      9.70
COMMUN      7.19     4.57
CLASSIFICATION FUNCTIONS
        GROUP =   MEMORY      PERCEP      COMMUN
VARIABLE
    3 PERF        1.92420      .58704     1.36552
    4 INFO      -17.56221    -8.69921   -10.58700
    5 VERBEXP     5.54585     4.11679     2.97278
    6 AGE          .98723     5.01749     2.91135
CONSTANT      -138.91108   -72.38436   -72.34031
CLASSIFICATION MATRIX
GROUP     PERCENT   NUMBER OF CASES CLASSIFIED INTO GROUP -
          CORRECT
                    MEMORY   PERCEP    COMMUN
  MEMORY   100.0      3        0         0
  PERCEP   100.0      0        3         0
  COMMUN   100.0      0        0         3
  TOTAL    100.0      3        3         3
```

TABLE 9.4 (*Continued*)

```
SUMMARY TABLE
    STEP           VARIABLE           F VALUE TO        NUMBER OF         U-STATISTIC   APPROXIMATE    DEGREES OF
  NUMBER     ENTERED    REMOVED    ENTER OR REMOVE  VARIABLES INCLUDED                  F-STATISTIC     FREEDOM
     1      6 AGE                        .4017              1                .8819          .402    2.00    6.00
     2      4 INFO                      2.4707              2                .4436         1.254    4.00   10.00
     3      3 PERF                      9.3034              3                .0785         3.426    6.00    8.00
     4      5 VERBEXP                   9.7366              4                .0105         6.577    8.00    6.00

              INCORRECT                 MAHALANOBIS D-SQUARE FROM AND
           CLASSIFICATIONS              POSTERIOR PROBABILITY FOR GROUP -

 GROUP   MEMORY                MEMORY          PERCEP           COMMUN

   CASE
     1                     2.0 1.000     50.0  .000     31.2  .000
     2                     2.8 1.000     41.1  .000     33.7  .000
     3                     3.3 1.000     72.2  .000     58.1  .000
 GROUP   PERCEP                MEMORY          PERCEP           COMMUN

   CASE
     4                    75.5  .000      3.8 1.000     37.8  .000
     5                    38.5  .000      2.9 1.000     27.2  .000
     6                    50.0  .000      2.2  .999     17.1  .001
 GROUP   COMMUN                MEMORY          PERCEP           COMMUN

   CASE
     7                    47.3  .000     40.5  .000      2.9 1.000
     8                    33.3  .000     15.0  .003      3.4  .997
     9                    41.5  .000     24.7  .000       .8 1.000
 EIGENVALUES
                13.48590        5.58923
 CUMULATIVE PROPORTION OF TOTAL DISPERSION
                 .70699        1.00000
 CANONICAL CORRELATIONS
                 .96487         .92100
 VARIABLE     COEFFICIENTS FOR CANONICAL VARIABLES

   3 PERF          .17100        -.10069
   4 INFO        -1.26953        -.10325
   5 VERBEXP       .26493         .35776
   6 AGE          -.52541         .24690
 CONSTANT        -9.67374       -3.45293
 GROUP        CANONICAL VARIABLES EVALUATED AT GROUP MEANS
   MEMORY        4.10234         .69097
   PERCEP       -2.98068        1.94169
   COMMUN       -1.12166       -2.63265
```

Hierarchical DISCRIM is also available through SPSS DISCRIMINANT, to be illustrated in Section 9.5.3. Although classification information is provided only at the last step of the analysis, there is other useful information, unavailable in BMD programs, at each step. Particularly handy for stepdown analysis is the "change in Rao's V" (see Section 9.6.1) at every step.[8]

Hierarchical DISCRIM differs from hierarchical multiple regression in the way variables enter at each step. For DISCRIM, only one variable enters at each step. If several predictors are given the same entry level (priority), they compete with one another in a stepwise fashion for order of entry. This makes DISCRIM slightly less flexible than multiple regression, and in situations in which there are only two groups it might be more profitable to use multiple regression procedures for the analysis. If classification of cases, which is not available with multiple regression, is desired, preliminary multiple regression analysis could be followed by DISCRIM.

[8] In using DISCRIM for stepdown analysis following MANOVA, change in Rao's V can replace stepdown F as the criterion for evaluating significance of each successive DV. Rao's V statistics are optionally available in SPSS DISCRIMINANT (see Section 9.6.1).

9.5.3 Stepwise Discriminant Function Analysis

When the researcher has no a priori reason for ordering entry of variables into the discriminating equations, statistical criteria are available for determining order of entry. For example, if the researcher desires a reduced set of predictors, but has no preferences among them, stepwise DISCRIM can be used to produce the reduced set.

Stepwise DISCRIM carries the same controversial aspects as stepwise multiple regression (see Section 5.5.3). Order of entry can be determined by trivial sample differences in relationships among variables that do not reflect population differences. However, if use is made of classification with a cross-validation sample, a feature readily available in SPSS DISCRIMINANT, this bias can be reduced.

Application of stepwise analysis to our small sample example through SPSS DISCRIMINANT is illustrated in Table 9.5. Notice first of all that AGE has been dropped as a predictor, so that the analysis proceeded for only three steps. On comparing this output with that of Table 9.4, elimination of AGE

TABLE 9.5 DECK SETUP AND SELECTED SPSS OUTPUT FOR STEPWISE DISCRIMINANT FUNCTION ANALYSIS ON SAMPLE DATA IN TABLE 9.1

```
                          VALUE LABELS    GROUP (1)MEMORY (2)PERCEP (3)COMMUN
                          VAR LABELS      PERF, WISCR PERFORMANCE IQ/
                                          INFO, WISCR INFORMATION/
                                          VERBEXP, ITPA VERBAL EXPRESSION
                          DISCRIMINANT    GROUPS=GROUP(1,3)/
                                          VARIABLES=PERF TO AGE/
                                          ANALYSIS=PERF TO AGE/
                                          METHOD=WILKS
                          OPTIONS         5,6,7,8,11,12
                          STATISTICS      1,2,3,4,5,6,9,10




        ANALYSIS NUMBER        1

        STEPWISE VARIABLE SELECTION

           SELECTION RULE-  MINIMIZE WILKS LAMBDA
           MAXIMUM NUMBER OF STEPS..................      8
           MINIMUM TOLERANCE LEVEL..................    .00100
           MINIMUM F TO ENTER......................   1.0000
           MAXIMUM F TO REMOVE.....................   1.0000

        CANONICAL DISCRIMINANT FUNCTIONS

           MAXIMUM NUMBER OF FUNCTIONS.............      2
           MINIMUM CUMULATIVE PERCENT OF VARIANCE...  100.00
           MAXIMUM SIGNIFICANCE OF WILKS  LAMBDA....   1.0000


        PRIOR PROBABILITY FOR EACH GROUP IS   .33333

        -------------- VARIABLES NOT IN THE ANALYSIS AFTER STEP   0 --------------
                                 MINIMUM
           VARIABLE   TOLERANCE  TOLERANCE   F TO ENTER  WILKS  LAMBDA

           PERF       1.0000000  1.0000000      .7346      .80330
           INFO       1.0000000  1.0000000     2.1765      .57955
           VERBEXP    1.0000000  1.0000000     3.3600      .47170
           AGE        1.0000000  1.0000000      .4017      .88191
```

TABLE 9.5 (*Continued*)

```
AT STEP   1, VERBEXP  WAS INCLUDED IN THE ANALYSIS.

                                    DEGREES OF FREEDOM  SIGNIF.  BETWEEN GROUPS
WILKS LAMBDA           .4716981     1    2      6.0
EQUIVALENT F          3.360000           2      6.0    .1050


--------------- VARIABLES IN THE ANALYSIS AFTER STEP   1 ----------------


VARIABLE   TOLERANCE  F TO REMOVE  WILKS  LAMBDA

VERBEXP   1.0000000      3.3600


-------------- VARIABLES NOT IN THE ANALYSIS AFTER STEP   1 ---------------

                         MINIMUM
VARIABLE   TOLERANCE   TOLERANCE    F TO ENTER  WILKS  LAMBDA

PERF       .3709896    .3709896      5.4218      .14886
INFO       .2019444    .2019444     13.1336      .07543
AGE        .8850270    .8850270       .7023      .36825


F STATISTICS AND SIGNIFICANCES BETWEEN PAIRS OF GROUPS AFTER STEP   1
EACH F STATISTIC HAS   1 AND        6.0 DEGREES OF FREEDOM.

                   GROUP       1          2
                          MEMORY       PERCEP
      GROUP

         2  PERCEP        .24000
                         .6416

         3  COMMUN       3.8400      6.0000
                        .0978       .0498




AT STEP   2, INFO    WAS INCLUDED IN THE ANALYSIS.

                                    DEGREES OF FREEDOM  SIGNIF.  BETWEEN GROUPS
WILKS  LAMBDA          .0754301     2    2      6.0
EQUIVALENT F          6.602645           4     10.0    .0072


--------------- VARIABLES IN THE ANALYSIS AFTER STEP   2 ----------------


VARIABLE   TOLERANCE  F TO REMOVE  WILKS  LAMBDA

INFO       .2019444     13.1336      .47170
VERBEXP    .2019444     16.7080      .57955


-------------- VARIABLES NOT IN THE ANALYSIS AFTER STEP   2 ---------------

                         MINIMUM
VARIABLE   TOLERANCE   TOLERANCE    F TO ENTER  WILKS  LAMBDA

PERF       .1676353    .0912506     10.2314      .01233
AGE        .8349769    .1905241       .5735      .05862


F STATISTICS AND SIGNIFICANCES BETWEEN PAIRS OF GROUPS AFTER STEP   2
EACH F STATISTIC HAS   2 AND        5.0 DEGREES OF FREEDOM.

                   GROUP       1          2
                          MEMORY       PERCEP
      GROUP

         2  PERCEP       5.6607
                         .0519

         3  COMMUN      19.419      5.9714
                        .0044       .0473
```

TABLE 9.5 (*Continued*)

```
            AT STEP   3, PERF    WAS INCLUDED IN THE ANALYSIS.

                                    DEGREES OF FREEDOM  SIGNIF.  BETWEEN GROUPS
            WILKS  LAMBDA     .0123339    3    2      6.0
            EQUIVALENT F     10.67242          6      8.0     .0019


            --------------- VARIABLES IN THE ANALYSIS AFTER STEP   3 ----------------


            VARIABLE   TOLERANCE  F TO REMOVE  WILKS  LAMBDA

            PERF       .1676353    10.2314      .07543
            INFO       .0912506    22.1384      .14886
            VERBEXP    .1993789    13.5551      .09593


            -------------- VARIABLES NOT IN THE ANALYSIS AFTER STEP   3 --------------

                                 MINIMUM
            VARIABLE   TOLERANCE  TOLERANCE  F TO ENTER  WILKS  LAMBDA

            AGE        .5327805   .0812130     .2659      .01048



            F STATISTICS AND SIGNIFICANCES BETWEEN PAIRS OF GROUPS AFTER STEP   3
            EACH F STATISTIC HAS   3 AND       4.0 DEGREES OF FREEDOM.

                                 GROUP      1        2
                                       MEMORY     PERCEP
                       GROUP

                       2 PERCEP        14.603
                                        .0128

                       3 COMMUN        12.177     7.4056
                                        .0176      .0414


            F LEVEL OR TOLERANCE OR VIN INSUFFICIENT FOR FURTHER COMPUTATION.
                            SUMMARY TABLE

            ACTION        VARS   WILKS
  STEP ENTERED REMOVED     IN    LAMBDA   SIG.   LABEL

   1  VERBEXP              1    .471698  .1050  ITPA VERBAL EXPRESSION
   2  INFO                 2    .075430  .0072  WISCR INFORMATION
   3  PERF                 3    .012334  .0019  WISCR PERFORMANCE IQ


CLASSIFICATION FUNCTION COEFFICIENTS
(FISHER*S LINEAR DISCRIMINANT FUNCTIONS)

GROUP  =       1           2           3
           MEMORY      PERCEP      COMMUN


PERF         2.010831    1.027340    1.620995
INFO       -17.85415   -10.18294   -11.44792
VERBEXP      5.552108    4.148576    2.991223
(CONSTANT) -138.6734   -66.24425   -70.27306


                        CANONICAL DISCRIMINANT FUNCTIONS


                PERCENT OF  CUMULATIVE   CANONICAL  -  AFTER
FUNCTION EIGENVALUE VARIANCE  PERCENT   CORRELATION - FUNCTION  WILKS  LAMBDA  CHI-SQUARED  D.F.  SIGNICANCE

                                                   -    0       .0123339        21.977      6     .0012
   1*    11.71790    68.55     68.55    .9598805   -    1       .1568607         9.2620      2     .0097
   2*     5.37508    31.45    100.00    .9182262   -

   * MARKS THE   2 FUNCTION(S) TO BE USED IN THE REMAINING ANALYSIS.

STANDARDIZED CANONICAL DISCRIMINANT FUNCTION COEFFICIENTS

           FUNC  1    FUNC  2

PERF        1.88474    1.42704
INFO       -3.30213     .07042
VERBEXP     1.50580   -1.64500
```

can be seen to have enhanced F for discrimination ($F = 6.58$ with AGE included, $F = 10.67$ without AGE). That is, AGE is worthless as a predictor in this sample. If this result is replicated, AGE should be dropped from consideration when classifying future cases.

Progression of the stepwise analysis is summarized at the bottom of Table 9.5. In addition, contents of Tables 9.3 and 9.2 follow—that is the summary table is followed by classification function coefficients, information on the discriminant functions, and group centroids. Moreover, a great deal of classification information is available, as summarized later, in Table 9.8, and discussed in Section 9.6.6. Entry of predictors is determined by user-specified criteria, of which five are available (discussed in Section 9.6.1). Additional statistical criteria for entry are also available.

Stepwise DISCRIM in BMD, like hierarchical DISCRIM, is produced by BMD07M and BMDP7M. Format of output for stepwise analysis is identical to that of hierarchical analysis, as illustrated in Table 9.4. Available are four methods of entry and, as for SPSS, additional statistical entry criteria.

For BMD07M, information available at each step is similar to that of BMDP7M. However, no summary table is produced.

9.6 SOME IMPORTANT ISSUES

9.6.1 Statistical Inference

The section begins with a discussion of criteria for evaluating the overall statistical significance of the set of predictors for predicting group membership. Then, methods for directing the progress of stepwise DISCRIM are noted, along with statistical criteria for entry of predictors.

9.6.1.1 Criteria for Statistical Inference   The basic criteria for evaluating statistical significance of discriminant functions are identical to those for MANOVA. The choice between Wilks' Lambda, Roy's *gcr*, Hotelling's trace, and Pillai's criterion is based on the same considerations as discussed in Section 8.5.1.

Two additional statistical criteria, Mahalanobis' D^2 and Rao's V, are especially relevant to stepwise DISCRIM and classification of cases. Mahalanobis' D^2 is based on distance between *pairs* of group centroids (which can then be generalized to distances over multiple pairs of groups). Rao's V is a generalized distance measure, and attains its largest value when there is greatest overall separation among groups. These two criteria are available for use both in directing the progression of stepwise DISCRIM and in evaluating the discriminating power of a set of predictors. They are, like Wilks' Lambda, based on a combination of discriminant functions (characteristic roots) rather than a single one. Note that these three criteria—D^2, V, and Lambda—are descriptive statistics. They are not, themselves, inferential statistics, although inferential statistics can be applied to them.

Mahalanobis' D^2 is also useful in identifying multivariate outliers, where the distance between each case and the other cases within its group (i.e., the group centroid) can be calculated. This valuable procedure is available through BMD07M and BMDP7M, among other programs (cf. Chapter 4).

9.6.1.2 Stepping Methods Somewhat related to criteria for statistical inference, but not entirely overlapping, is the choice among methods to direct the stepping progression in stepwise DISCRIM. Different methods make the groups maximally different according to different statistical criteria, as indicated in the Purpose column of Table 9.6. A variety of stepping methods, and the computer programs in which they are available, are presented in Table 9.6.

Choice among these methods will depend on the goals of the analysis, available programs, and choice of statistical criterion. For example, if the statistical criterion is to be Wilks' Lambda, it would be beneficial to choose the stepping procedure that minimizes lambda. (In SPSS DISCRIMINANT this is the least expensive method, and is recommended in the absence of contrary reasons.) Or, if the statistical criterion is "change in Rao's V" (cf. Section 8.5.2), the obvious choice would be RAO. With the use of BMDP7M, the stepping criterion is fixed, with the only choice based on whether or not contrasts are specified.

Statistical methods can also be used to modify stepping. For example, the user can specify such values as minimum F for a predictor to enter stepwise selection, minimum F to avoid removal, and so on. Tolerance, or the proportion

TABLE 9.6 METHODS FOR DIRECTING STEPWISE DISCRIMINANT FUNCTION ANALYSIS

Label	Purpose	Program and Option
WILKS	Produces smallest value of Wilks' Lambda (therefore largest multivariate F)	SPSS:WILKS
MAHAL	Produces largest distance (D^2) for two closest groups	SPSS:MAHAL
MAXMINF	Maximizes the smallest F between pairs of groups	SPSS:MAXMINF BMD07M:Col. 44-3
MINRESID	Produces smallest average residual variance ($1 - R^2$) between variables and pairs of groups	SPSS:MINRESID BMD07M:Col. 44-1
WTD MINRESID	Produces smallest average residual variance ($1 - R^2$) between variables and weighted pairs of groups	BMD07M:Col. 44-2
RAO	Produces at each step largest increase in distance between groups as measured by Rao's V	SPSS:RAO
F TO ENTER	At each step picks variable with largest F TO ENTER	BMD07M:Col. 44-0 BMDP7M
CONTRAST	Stepping procedure based on groups as defined by contrasts	BMDP7M:CONTRAST

of variance for a potential predictor that is not already accounted for by other predictors in the equation, can also be modified in all BMD and SPSS stepwise programs. Comparison of programs with respect to these stepwise statistical criteria is available in Table 9.8.

9.6.2 Number of Discriminant Functions

Within a single DISCRIM analysis, a number of orthogonal dimensions are extracted. The maximum number of dimensions extracted is the lesser of either the number of groups minus one or, as in principal components or factor analysis, equal to the number of predictor variables. As in principal components or factor analysis, however, not all of the dimensions may carry worthwhile information. It is frequently the case that the first few discriminant functions account for the lion's share of discriminating power, with no additional information forthcoming from the remaining functions. Only one of the programs, SPSS DISCRIMINANT, offers flexibility with regard to number of discriminant functions. The user can choose the number, the proportion of variance accounted for (with all succeeding discriminant functions dropped once that value is exceeded), or the significance level of additional functions.

SPSS DISCRIMINANT and MANOVA evaluate each of the successive discriminant functions. Referring back to Table 9.2, you will note that eigenvalues, percents of variance, and canonical correlations are given for each discriminant function for the small sample data of Table 9.1. To the right, statistical tests are given for stepwise reduction of the number of functions. With neither of the functions removed, the $\chi^2(8)$ of 20.514 indicates highly significant discriminating power on the basis of both discriminant functions. With the first discriminant function removed, there is still significant discriminating power, indicated by the $\chi^2(3) = 8.484$, $p = .037$. This finding shows that the second discriminant function is also statistically significant and represents an additional dimension in separating groups.

How much of the predictable (between-group) variability is contributed by each discriminant function? The relative sizes of the eigenvalues associated with different discriminant functions indicate the relative proportion of between-group variability contributed by each function. In the small sample example (cf. Table 9.2) 71% of the between-group variability is attributed to the first linear combination of variables, and 29% to the second linear combination.

It should be noted that all dimensions are "used" in the classification functions. Therefore, if classification is the major goal, significance of the individual dimensions is irrelevant, although overall statistical significance (all discriminant functions combined) is still important. It may be worthwhile to delete variables, through hierarchical or stepwise analysis, for more efficient classification, but no useful purpose would be served by using fewer dimensions. Nevertheless, the number of significant discriminant functions may still aid in the decision about how many pairs of discriminant functions are worth plotting to evaluate homogeneity of variance-covariance matrices and to get a notion of the relative separation between groups in the multidimensional space (see the next section).

9.6.3 Interpreting Discriminant Functions

9.6.3.1 Discriminant Function Plots A tool in the interpretation of discriminant functions is to form pairwise plots of group centroids on all significant discriminant functions. It can be recalled from Section 9.4.1 that these centroids are the means of the discriminant scores for each group on each dimension. As these points can only be plotted pairwise, the finding of three significant discriminant functions would require three plots, and so on.

An example of a discriminant function plot is illustrated in Figure 9.1 for the small sample data of Section 9.4. This is simply a plot of the canonical discrimination functions evaluated at group means from Table 9.2. The plot emphasizes the utility of both dimensions in discriminating among the three groups. On the first discriminant function, the MEMORY group is clearly distinguished from the other two groups, but COMMUNICATION and PERCEPTION are not much different from each other. That is, on the abscissa, MEMORY is far from the other two groups, which are closer to each other. It is the second function that highlights differences between COMMUNICATION and the remaining groups. On the basis of both discriminant functions, then, differences among the three groups are clear.

Three programs—BMD07M, BMDP7M, and SPSS DISCRIMINANT— provide plots of discriminant functions (called canonical variates in BMDP7M). In BMD07M and SPSS DISCRIMINANT, cases as well as means are plotted, making interpretation more difficult than with simpler plots, but facilitating evaluation of classification. For BMDP7M, simplified plots of group means as well as plots including cases are available. For all three programs, however,

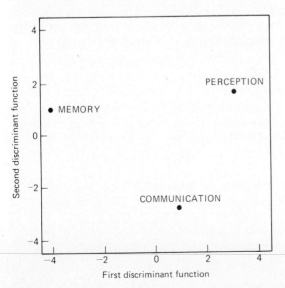

Figure 9.1 Centroids of three learning disability groups on the two discriminant functions derived from sample data of Table 9.1.

only the first pair of discriminant functions (canonical variates) is plotted. Plots including additional significant dimensions would have to be prepared by hand or discriminant scores could be passed to a "plotting" program such as BMDP6D.

With factorial designs (Section 9.6.5) separate sets of plots would be required for each significant main effect and interaction. Main effect plots would have the same format as Figure 9.1, with one data point per marginal mean. Interaction plots would have as many data points as cells in the design.

9.6.3.2 Loading Matrices Another tool in the interpretation of discriminant functions is examination of the loadings of predictor variables on them. Loading matrices are basically factor loading matrices (cf. Chapter 10) that contain correlations between predictor variables and each of the discriminant functions (canonical variates), and can sometimes be useful in naming and interpreting the functions. Mathematically, the loading matrix is the pooled within-group correlation matrix, multiplied by the matrix of standardized discriminant function coefficients.[9]

$$A = R_w D \qquad (9.7)$$

The loading matrix of correlations between predictor variables and discriminant functions, **A,** is found by multiplying the matrix of within-group correlations among predictors, R_w, by a matrix of standardized discriminant function coefficients, **D** (standardized using pooled within-group standard deviations).

For the small sample example of Table 9.1, the loading matrix, as produced by SPSS DISCRIMINANT, appears as the middle matrix in Table 9.2: POOLED WITHIN-GROUPS CORRELATIONS BETWEEN CANONICAL DISCRIMINANT FUNCTIONS AND DISCRIMINATING VARIABLES.

The loading matrix shows that the first discriminant function is correlated most highly with WISC-R Information scores ($r = .23$), while the second is loaded heavily with ITPA Verbal Expression ($r = -.45$). Relating these findings to the plot of Figure 9.1, then, suggests that maximum spread among the three groups is primarily based on Information scores, with the greatest difference between MEMORY and PERCEPTION groups. The COMMUNICATION group is distinguished from the other two primarily on the basis of Verbal Expression.

Caution should be taken, however, in interpreting these loadings. They do not necessarily indicate which variables contribute most heavily to discrimination among groups, after adjustment for remaining variables. If we look at the loadings from the perspective of standard multiple regression (cf. Chapter 5), they are analogous to "raw" correlations rather than semipartial correlations between the DV (a canonical variate in DISCRIM) and the set of predictors.

---

[9] Some texts (e.g., Cooley & Lohnes, 1971) and earlier versions of SPSS use the total correlation matrix to find standardized coefficients rather than the within-group matrix.

For example, Verbal Expression scores might or might not be the heaviest contributor to distinction between the COMMUNICATION group and the other two groups, once the variance due to PERF, INFO, and AGE has been partialled out. Section 9.6.4 deals with methods for interpeting predictors after variance associated with other predictors is removed, if that is desired.

Consensus is lacking regarding how high a correlation in a loading matrix must be to be interpreted. By convention, correlations in excess of .30 (9% of variance) are usually considered eligible and lower ones are not. Guidelines suggested by Comrey (1973) are included in Chapter 10. However, the size of a loading depends both on the "true" value of the correlation in the population *and* on the homogeneity of scores in the sample taken from it. If the sample is unusually homogeneous with respect to some predictor variable, it might be wise to lower the criterion cut (but not below .30) for determining which variables to consider in interpreting a discriminant function.

In some cases, rotation of the loading matrix may facilitate interpretation, as discussed in Chapter 10. But rotation of discriminant function loading matrices is still considered experimental and not recommended for the novice. SPSS DISCRIMINANT and SPSS MANOVA allow rotation of discriminant functions.

9.6.4 Evaluating Predictor Variables

Another tool for evaluating contribution of predictors to discrimination among groups is available through the CONTRAST procedure of BMDP7M. Each group can be contrasted with all other groups combined, to see which predictors are important in isolating that group from the rest. That is, a BMDP7M run is made for each group, contrasting it with all other groups. Each run provides one canonical variable. The F TO ENTER at step 0 is, for each predictor, the univariate F for testing the significance of the difference between the chosen group and the other groups. If a canonical variate simply distinguishes one group from the others, the loadings for the variate correlate perfectly with univariate F's for the predictors in the contrast.

At the last step of each contrast run, with all predictors forced into the analysis, F TO REMOVE indicates the reduction in prediction that would result from removal of each predictor, in turn. Therefore, at step 0, F TO ENTER shows how important a variable is, by itself, in predicting membership in a particular group. At the last step, F TO REMOVE shows the importance of a variable, after adjusting for all other variables.

In order to avoid overinterpretation, it is probably best to consider only variables with F ratios "significant" after adjusting α error for the number of variables in the set. The adjustment can be made on the basis of

$$\alpha = 1 - (1 - \alpha_i)^p \qquad (9.8)$$

The Type I error rate (α) for evaluating contribution of p variables to between-group contrasts is based on the error rate for evaluating each predictor and the number of predictors, $i = 1, 2, \ldots p$.

Each variable, then, is evaluated at α_i. Variables that meet this criterion at the last step can be seen as significantly reducing predictability if they were to be deleted. Even with this adjustment, there is danger of inflation of Type I error rate because of the multiple contrasts (nonorthogonal at that). Further adjustment might be considered (e.g., multiplication of critical F by $k - 1$, where k = number of groups), or the interpretation might simply be made cautiously, deemphasizing statistical justification.

As an example of use of F TO ENTER and F TO REMOVE for interpretation, partial output from the three BMDP7M CONTRAST runs for the small sample data of Table 9.1 is shown in Table 9.7. The contrast between MEMORY and the other two groups appears in Table 9.7(a). PERCEPTION versus the remaining groups appears in Table 9.7(b), and the difference between COMMU-NICATION and the other groups is shown in Table 9.7(c).

With 1 and 3 df at the last step, it is not feasible to evaluate predictors with respect to statistical significance. However, the pattern of F TO ENTER and F TO REMOVE reveals those predictors variables most likely to discriminate each group. Memory deficits are most clearly characterized by low scores on WISC-R Information, as corroborated by the group means shown in Table 9.7(a). For perceptual deficits, no clear pattern is seen at step 0. After adjustment for other variables, however, low WISC-R Performance IQ scores can be seen to distinguish the PERCEP group [Table 7.9(b)]. For the group showing disorders in communication, low ITPA Verbal Expression scores stand out as predominant for prediction [see Table 7.9(c)].

As a measure of strength of association, percent of variance contributed at the last step by significant predictors can be found as a form of a squared semipartial correlation.

$$sr_i{}^2 = \frac{F_i}{df_{res}} (1 - r_c{}^2) \tag{9.9}$$

The squared semipartial correlation ($sr_i{}^2$) between predictor i and the canonical variate representing the difference between two groups (one group vs. all others) is calculated from the F TO REMOVE of the ith predictor at the last step of BMDP7M with all predictors forced, degrees of freedom for error (df_{res}) at that last step, and the $r_c{}^2$ (CANONICAL CORRELA-TIONS,squared).

For example, in evaluating the association between INFO scores and the difference between MEMORY and the other two groups (assuming statistical significance), we obtain

$$sr_{INFO}^2 = \frac{33.373}{3} (1 - .96357^2) = .7958$$

That is, 80% of the variance in INFO scores is shared with the canonical variable separating the MEMORY group from the PERCEP and COMMUN groups.

TABLE 9.7 INPUT AND PARTIAL OUTPUT FOR BMDP7M CONTRAST RUNS ON SMALL SAMPLE EXAMPLE: (A) MEMORY VERSUS OTHER TWO GROUPS; (B) PERCEPTION VERSUS OTHER TWO GROUPS; (C) COMMUNICATION VERSUS OTHER TWO GROUPS

(a)

```
PROGRAM CONTROL INFORMATION

        /PROBLEM      TITLE IS "SMALL SAMPLE DISCRIMINANT EXAMPLE".
        /INPUT        VARIABLES ARE 6.  FORMAT IS "(4X,5F4.0,F4.1)".
        /VARIABLE     NAMES ARE CASENO,GROUP,PERF,INFO,VERBEXP,AGE.
                      LABEL=CASENO.
                      GROUPING IS GROUP.
        /GROUP        CODES(2) ARE 1,2,3.
                      NAMES(2) ARE MEMORY,PERCEP,COMMUN.
        /DISCRIM      LEVEL=0,0,1,1,1,1.  FORCE=1.
                      CONTRAST=2,-1,-1.
        /PLOT         NO CANON.  CONTRAST.
        /PRINT        NO STEP.
        /END
```

```
        MEANS
            GROUP =   MEMORY      PERCEP      COMMUN      ALL GPS.
        VARIABLE
          2 GROUP      1.00000     2.00000     3.00000     2.00000
          3 PERF      98.66667    87.66667   101.33333    95.88889
          4 INFO       7.00000    11.66667     9.66667     9.44444
          5 VERBEXP   36.33333    38.33333    28.33333    34.33333
          6 AGE        7.30000     6.93333     7.63333     7.28889
        COUNTS             3.          3.          3.          9.
```

```
STEP NUMBER   0
    VARIABLE        F TO    FORCE   TOLERANCE   *   VARIABLE        F TO    FORCE   TOLERANCE
                   REMOVE   LEVEL               *                  ENTER   LEVEL
               DF= 1     7                      *               DF= 1     6
                                                *       3 PERF       .162    1       1.000000
                                                *       4 INFO      3.559    1       1.000000
                                                *       5 VERBEXP    .720    1       1.000000
                                                *       6 AGE        .001    1       1.000000
```

```
STEP NUMBER   4
VARIABLE ENTERED   6 AGE
    VARIABLE        F TO    FORCE   TOLERANCE   *   VARIABLE        F TO    FORCE   TOLERANCE
                   REMOVE   LEVEL               *                  ENTER   LEVEL
               DF= 1     3                      *               DF= 1     2
      3 PERF      2.896    1       .106964      *
      4 INFO     33.373    1       .081213      *
      5 VERBEXP   2.722    1       .199297      *
      6 AGE        .345    1       .532781      *
```

```
EIGENVALUES
            12.98000
CUMULATIVE PROPORTION OF TOTAL DISPERSION
            1.00000
CANONICAL CORRELATIONS
            .96357
```

TABLE 9.7 (*Continued*)

(b)

```
PROGRAM CONTROL INFORMATION

          /PROBLEM    TITLE IS "SMALL SAMPLE DISCRIMINANT EXAMPLE".
          /INPUT      VARIABLES ARE 6.  FORMAT IS "(4X,5F4.0,F4.1)".
          /VARIABLE   NAMES ARE CASENO,GROUP,PERF,INFO,VERBEXP,AGE.
                      LABEL=CASENO.
                      GROUPING IS GROUP.
          /GROUP      CODES(2) ARE 1,2,3.
                      NAMES(2) ARE MEMORY,PERCEP,COMMUN.
          /DISCRIM    LEVEL=0,0,1,1,1,1.  FORCE=1.
                      CONTRAST=-1,2,-1.
          /PLOT       NO CANON.   CONTRAST.
          /PRINT      NO STEP.
          /END
```

```
STEP NUMBER   0
   VARIABLE       F TO  FORCE  TOLERANCE  *   VARIABLE       F TO  FORCE  TOLERANCE
                 REMOVE LEVEL             *                 ENTER LEVEL
              DF=  1    7                 *              DF=  1    6
                                          *   3 PERF      1.419   1     1.000000
                                          *   4 INFO      2.941   1     1.000000
                                          *   5 VERBEXP   2.880   1     1.000000
                                          *   6 AGE        .621   1     1.000000

STEP NUMBER   4
VARIABLE ENTERED   5 VERBEXP
   VARIABLE       F TO  FORCE  TOLERANCE  *   VARIABLE       F TO  FORCE  TOLERANCE
                 REMOVE LEVEL             *                 ENTER LEVEL
              DF=  1    3                 *              DF=  1    2
   3 PERF     13.225   1     .106964      *
   4 INFO      3.870   1     .081213      *
   5 VERBEXP    .010   1     .199297      *
   6 AGE        .512   1     .532781      *
```

```
EIGENVALUES
            9.49094
CUMULATIVE PROPORTION OF TOTAL DISPERSION
            1.00000
CANONICAL CORRELATIONS
            .95115
```

TABLE 9.7 (*Continued*)

(c)

```
          PROGRAM CONTROL INFORMATION

              /PROBLEM    TITLE IS "SMALL SAMPLE DISCRIMINANT EXAMPLE".
              /INPUT      VARIABLES ARE 6.  FORMAT IS "(4X,5F4.0,F4.1)".
              /VARIABLE   NAMES ARE CASENO,GROUP,PERF,INFO,VERBEXP,AGE.
                          LABEL=CASENO.
                          GROUPING IS GROUP.
              /GROUP      CODES(2) ARE 1,2,3.
                          NAMES(2) ARE MEMORY,PERCEP,COMMUN.
              /DISCRIM    LEVEL=0,0,1,1,1,1.  FORCE=1.
                          CONTRAST=-1,-1,2.
              /PLOT       NO CANON.   CONTRAST.
              /PRINT      NO STEP.
              /END
```

```
STEP NUMBER    0
   VARIABLE          F TO  FORCE   TOLERANCE  *   VARIABLE         F TO  FORCE    TOLERANCE
                    REMOVE LEVEL              *                   ENTER LEVEL
              DF=   1    7                    *              DF=  1    6
                                             *      3 PERF        .622    1      1.000000
                                             *      4 INFO        .029    1      1.000000
                                             *      5 VERBEXP    6.480    1      1.000000
                                             *      6 AGE         .583    1      1.000000


STEP NUMBER    4
VARIABLE ENTERED   6 AGE
   VARIABLE          F TO  FORCE   TOLERANCE  *   VARIABLE         F TO  FORCE    TOLERANCE
                    REMOVE LEVEL              *                   ENTER LEVEL
              DF=   1    3                    *              DF=  1    2
   3 PERF            .039    1     .106964    *
   4 INFO            .682    1     .081213    *
   5 VERBEXP       12.249    1     .199297    *
   6 AGE            .001    1     .532781     *
U-STATISTIC OR WILKS" LAMBDA     .1400217    DEGREES OF FREEDOM    4    1       6
APPROXIMATE F-STATISTIC            4.606      DEGREES OF FREEDOM    4.00   3.00



EIGENVALUES
              6.14175
CUMULATIVE PROPORTION OF TOTAL DISPERSION
              1.00000
CANONICAL CORRELATIONS
              .92735
```

The procedures detailed in this section are most likely to be useful when the number of groups is fairly small and the separations among them are fairly uniform as displayed on the overall plot of the first two canonical variables. With many groups, some of which may be closely clustered, different sets of contrasts might be suggested by the overall canonical plot. Or with a very large number of groups, the overall loading matrix may have to suffice as a summary device.

Alternatively, contrasts between each group and the others can be attempted from a hierarchical rather than standard approach (cf. Chapter 5).

Instead of each predictor variable being evaluated after adjustment for all other variables, it is adjusted only for higher-priority predictors. This strategy is most easily accomplished through SPSS MANOVA runs, in which the stepdown F's (cf. Chapter 8) are evaluated for each contrast. This might be the procedure of choice if there were some logical basis for assigning priorities to predictors.

All of the procedures suggested for evaluation of DVs in the context of MANOVA apply to evaluation of predictor variables in DISCRIM. Procedures involving stepdown analysis, inspection of univariate F, pooled within-group correlations among predictors, and standardized discriminant function coefficients are equally appropriate, or inappropriate, for DISCRIM as for MANOVA. Therefore the discussion in Section 8.5.3 will not be repeated here.

9.6.5 Design Complexity: Factorial Designs

The notion of placing cases into groups can be easily extended into situations in which groups are formed by differences on more than one dimension. For example, four groups could be formed by cases factorially classified on two dimensions, with two levels of each dimension. An illustration of this would be the large sample example of Section 8.7, where women were classified as high or low femininity, and also as high or low masculinity on the basis of scores on the Bem Sex Role Inventory (BSRI). These two dimensions were factorially combined to form the four groups. On the basis of scores on the six DVs used in that example, women could be assigned, more or less successfully, into the sex role group into which they had been classified by the BSRI.

As long as sample sizes are equal in all cells, factorial problems can easily be resolved through a factorial DISCRIM program, such as BMDP7M. Contrast coding is required to designate a separate run for each main effect and interaction. Discriminant and classification functions are produced for each effect.

With unequal n, classification functions do not vary with adjustment procedure for nonorthogonality, but discriminant functions do vary. Further, DISCRIM programs do not have simple options to deal with unequal n in factorial designs. BMDP7M can be used, but contrast weights have to be chosen so as to reflect decisions about adjustment for unequal n. Therefore a two-stage analysis may be most efficient. First, questions about statistical discrimination among groups can be answered by performing MANOVA. Second, if classification is desired after MANOVA, any of a variety of DISCRIM programs can do the job. Formation of groups for DISCRIM depends on the outcome of MANOVA. If the interaction is statistically significant, groups should be based on the cells of the design. That is, in a two-by-two design, four groups would be formed to be used as the grouping variable in DISCRIM. Note that this grouping will be based on main effects as well as the interaction. The argument can be made, however, that classification of cases into cells would serve most purposes.

If interaction(s) are not statistically significant, classification should be based on significant main effects. For example, for classification into high- and low-femininity groups in the example from Section 8.7, designations of masculinity should be ignored. Classification on the basis of masculinity scores would ignore femininity scores. That is, classification of main effects would be on the basis of marginal groups.

9.6.6 Use of Classification Procedures

The basic technique for classifying cases into groups was outlined in Section 9.4.2. Some canned computer programs, however, offer sophisticated additional features that can be helpful in many classification situations.

9.6.6.1 Cross-Validation and New Cases

Since classification is based on classification coefficients, and these coefficients derive from samples and are only approximations of population parameters, it is often useful to know how well the classification functions perform with a new sample of cases. Cross-validation techniques are especially well developed in the BMDP7M and SPSS DISCRIMINANT programs. If you simply omit information about actual group membership for some cases (hide it from the program), SPSS DISCRIMINANT and BMDP7M will not include those cases in the derivation of classification functions, but will include them in the classification phase. Similarly, with new cases to be classified, omission of group membership will prevent these cases from being used in developing classification functions, but they can be included during the actual classification of cases. One can then examine the accuracy with which the classification functions predict group membership in cases not used to derive them. Cross-validation is demonstrated in Section 9.8.2.

Unfortunately, none of the canned computer programs for DISCRIM allows classification of new cases without repeated entry of the original cases used to derive the classification functions. In practical applications, where classification might be used for diagnostic or personnel purposes for new patients or job applicants, it would be most convenient to input the classification coefficients along with raw data for the new cases to be classified and run the data only through the classification phase. Lacking this, it may be desirable to write your own program based on the classification coefficients to classify cases. Another simple classification technique sometimes feasible is discussed in the next section.

9.6.6.2 Plots

A variety of plots is produced by canned computer programs as part of the classification phase. These plots are typically in the form of scatterplots for the first two discriminant functions (canonical variates) or histograms in the case of a single discriminant function. These plots can be used in interpretation of results as well as for classification purposes (see Section 9.6.3.1). When cases as well as group centroids are shown in the plots so that the spread of cases is displayed as well as the centroids, it is possible to see which groups separate from which other groups. These plots are also useful in identifying outliers. Additional plots may be useful if more than two discriminant functions show statistically significant separation among groups. Group centroids can be easily plotted by hand. When plots of cases are desired, the discriminant scores can be passed to BMDP6D, which can display separate plots for each group and/or all groups on the same plot.

Finally, one plot available only in SPSS DISCRIMINANT shows the placement of all possible cases on the basis of the first two discriminant functions. This territorial map could be used to classify new cases, given their scores on each of the first two discriminant functions (see Section 9.4.1). If only the first two discriminant functions are statistically significant (see Section 9.6.2) and only a few new cases are to be classified, hand calculation of discriminant function scores and comparison with the territorial map would provide a simple, fast classification technique.

9.6.6.3 Jackknifed Classification   A form of bias enters the classification procedure when results of classification are based on the same cases used in developing the classification equations. BMDP7M provides a jackknifed classification scheme that eliminates or reduces this bias. If all variables are forced into the analysis, the bias is eliminated. With a stepwise procedure, the bias is reduced. Each case is classified on the basis of equations developed from all data except the case being classified. The results of jackknifed classification, then, can be viewed as a more realistic estimate of the ability of predictors to discriminate among groups.

Jackknifed classification, however, is very expensive, and with large samples produces little difference in results from the "biased" classification procedures. With more than about 100 cases in the smallest group, it is probably best to forgo jackknifed classification. Applications of jackknifed classification are illustrated in Section 9.8.

9.6.6.4 Evaluating Improvement in Classification   In hierarchical DISCRIM, it can be informative to determine if classification improves as new sets of predictors are added to the analysis. A simple, straightforward test of this improvement is McNemar's repeated-measures chi square test for change. Cases are tabulated as to whether they are correctly or incorrectly classified at an early step and at a later step.

		Early step classification	
		Correct	Incorrect
Later step classification	Correct	(A)	B
	Incorrect	C	(D)

Those cases that have the same result at both steps (cells A and D) are ignored since they show no change. Therefore χ^2 for change is

$$\chi^2 = \frac{(|B - C| - 1)^2}{B + C} \qquad df = 1 \qquad (9.10)$$

Naturally, we would only reject the hypothesis of no improvement if $B >$ C—that is, if more cases became correctly classified after the addition of predictors. So if $B > C$ and χ^2 is greater than the critical value of 3.84, with 1 df at $\alpha = .05$, we can conclude that the addition of predictors at the later step has improved classification. This test is applied to the large sample data of Section 9.8.3.

With very large samples (e.g., thousands of cases) hand tabulation of cases may not be feasible. An alternative, but possibly less desirable, procedure is to test for the significance of the difference between two lambdas, suggested by Frane (personal communication). The Wilks' Lambda from the step with the larger number of variables (Λ_2) is divided by the lambda from the step with the fewer variables (Λ_1), producing a new value of Λ_D.

$$\Lambda_D = \frac{\Lambda_2}{\Lambda_1} \qquad (9.11)$$

Wilks' Lambda for testing the significance of the difference between two lambdas (Λ_D) is calculated by dividing the smaller lambda (Λ_2) by the larger lambda (Λ_1).

The significance of Λ_D is evaluated with the three parameters of degrees of freedom: (1) p, the number of variables in the step with the larger number; (2) $k - 1$, the number of groups, minus 1; and (3) the df_{error} as shown in the step with the larger number of variables. Associated approximate F can be found according to procedures detailed in Section 9.4.1.

For the small sample hierarchical example (Table 9.4), one can test whether INFO and PERF, entered by step 4 ($\Lambda_2 = .0105$), add significantly to discrimination over that achieved by step 2, with AGE and VERBEXP entered in the analysis ($\Lambda_1 = .4436$).

$$\Lambda_D = \frac{.0105}{.4436} = .02367$$

where

$$df_p = 4 = p$$

$$df_{effect} = 2 = k - 1$$

$$df_{error} = 6 = N - k$$

From Section 9.4.1,

$$s = \sqrt{\frac{(4)^2(2)^2 - 4}{(4)^2 + (2)^2 - 5}} = 2$$

$$y = .02367^{1/2} = .15385$$

$$df_1 = (4)(2) = 8$$

$$df_2 = 2\left[6 - \left(\frac{4-2+1}{2}\right)\right] - \left[\frac{4(2)-2}{2}\right] = 6$$

$$\text{approximate } F(8,6) = \left(\frac{1 - .15385}{.15385}\right)\left(\frac{6}{8}\right) = 4.125$$

With critical $F(8, 6) = 4.15$ for $\alpha = .05$, this indicates no significant improvement in discrimination among the three groups with INFO and PERF scores added to AGE and VERBEXP scores.

9.7 COMPARISON OF PROGRAMS

One BMDP program and one SPSS program have been specifically developed for discriminant function analysis, including classification of cases into groups. Also, one of the earlier BMD programs contains options not yet implemented in the BMDP program. In addition, an SPSS program for multivariate analysis of variance can be useful in DISCRIM, although classification of cases is not available. Finally, if the only question is assessment of statistical significance of the discriminating power of a set of predictor variables, any of the MANOVA programs discussed in Chapter 8 is appropriate.

9.7.1 SPSS Programs

The basic program in this package for discriminant function analysis is SPSS DISCRIMINANT, features of which are described in Table 9.8. SPSS DIS-CRIMINANT can be used for direct (standard), hierarchical, or stepwise DIS-CRIM. Strong points include numerous plots, a great deal of classification information, and a great variety of stepwise options. In addition, homogeneity of covariance matrices can be tested, and, should heterogeneity be found, appropriate adjustment can be made in the classification phase. A useful feature is evaluation of successive discriminant functions. The program also directly provides loading matrices (correlations between predictor variables and discriminant functions; cf. Section 9.6.3.2).

A second program is available, SPSS MANOVA, which has some features unobtainable in any of the other DISCRIM programs. This program is partially described in Table 8.9 of Chapter 8, but some aspects especially pertinent to DISCRIM are featured in Table 9.8. There is a variety of statistical criteria available for testing the significance of the discriminating power of the set of predictor variables (cf. Section 9.6.1). Additionally, this program directly provides loading matrices for discriminant functions. There also is a wide variety of matrices that can be printed out and these, along with the determinants

TABLE 9.8 COMPARISON OF PROGRAMS FOR DISCRIMINANT
FUNCTION ANALYSIS

Feature	SPSS DISCRIMINANT	SPSS MANOVA[a]	BMD07M	BMDP7M
Input				
Data must be ordered	No	No	Yes	No
Matrix input optional	Yes	Yes	No	No
Prior probabilities optional	Yes	No	Yes	Yes
Missing data options (cf. Chapter 4)	Yes	No	No	No
More than two groups	Yes	Yes	Yes	Yes
Optional covariance matrix weights	No	No	Yes	No
Factorial arrangement of groups	No	Yes	No	Yes
Suppress intermediate steps	Yes	N.A.	No	Yes
Output				
Mahalanobis' D^2 (between groups)	Yes[b]	No	No	No
χ^2	Yes	No	No	No
Wilks' Lambda	Yes	Yes	Yes	Yes
Approximate F	Yes	Yes	Yes	Yes
Rao's V	Yes[b]	No	No	No
Hotelling's trace criterion	No	Yes	No	No
Roy's gcr	No	Yes	No	No
Pillai's criterion	No	Yes	No	No
Tests of successive discriminant functions (roots)	Yes	Yes	No	No
Univariate F ratios	Yes	Yes	No	Yes[c]
Group means	Yes	Yes	Yes	Yes
Group standard deviations	Yes	Yes	Yes	Yes
Coefficient of variation	No	No	No	Yes
Classification function coefficients	Yes	No	Yes	Yes
Standardized discrimination function—canonical coefficients	Yes	Yes	No	No
Unstandardized discriminant function—canonical coefficients	Yes	Yes	Partial[d]	Yes
Group centroids	Yes	No	Yes	Yes
Within-group covariance matrix	Yes	Yes	Yes	Yes
Within-group covariance matrix inverse	No	No	No	No
Cross-products matrix (within groups)	No	Yes	No	No
Inverse of cross-products matrix (w-g)	No	No	No	No
Cross-product matrices (each group and each hypothesis)	No	Yes	No	No
Within-group correlation matrix	Yes	Yes	Yes	Yes

TABLE 9.8 (*Continued*)

Feature	SPSS DISCRIMINANT	SPSS MANOVA[a]	BMD07M	BMDP7M
Determinant of within-group correlation matrix	No	Yes	No	No
Total covariance matrix	Yes	No	No	No
Group covariance matrices	Yes	Yes	No	No
Determinants of group and within-group covariance matrices	No	Yes	No	No
Equality of group covariance matrices	Yes	Yes	No	No
F matrix, pairwise group comparisons	Yes	No[e]	Yes	Yes
Tests for univariate homogeneity of variance	No	Yes	No	No
Canonical correlations	Yes	Yes	Yes	Yes
Eigenvalues	Yes	Yes	Yes	Yes
Optional number of discriminant functions	Yes	N.A.	No	No
Classification features				
Classification of cases	Yes	No	Yes	Yes
Classification matrix	Yes	No	Yes	Yes
Probability of case classification	Yes	No	No	Yes
Individual discriminant scores	Yes	No	Yes	No
Mahalanobis' D^2 for cases (outliers)	No	No	Yes	Yes
Jackknifed classification matrix	No	No	No	Yes
Classification with a cross-validation sample	Yes	No	No	Yes
Classification information at each step	No	No	Yes	Yes
Classification with separate covariance matrices	Yes	No	No	No
Single-case plot	Yes	No	Yes	Yes
Plot of group centroids alone	No	No	No	Yes
Separate plots by group	Yes	No	No	No
Territorial map	Yes	No	No	No
Rotation of discriminant functions	Yes	Yes	No	No
Correlation of variables with discriminant functions (canonical variates)	Yes	Yes	No	No
Stepwise options				
Optional entry method (cf. Section 9.6.1.2)	Yes	N.A.	Yes	No

TABLE 9.8 *(Continued)*

Feature	SPSS DISCRIMINANT	SPSS MANOVA[a]	BMD07M	BMDP7M
Optional removal method	No	N.A.	No	Yes
Forced entry by level	Yes	N.A.	Yes	Yes
Forced entry overriding criteria	No	N.A.	Yes	Yes
Specify entry by contrasts	No	N.A.	No	Yes
Tolerance	Yes	N.A.	Yes	Yes
Specify maximum number of steps	Yes	N.A.	Yes	Yes
F TO ENTER	Yes	N.A.	Yes	Yes
F TO REMOVE	Yes	N.A.	Yes	Yes
Significance of F TO ENTER[f]	Yes	N.A.	No	No
Significance of F to REMOVE[f]	Yes	N.A.	No	No

[a] Based on Release 9.0 (Hull & Nie, 1981). Additional features reviewed in Section 8.9.

[b] Available as a stepping option.

[c] STEP NUMBER 0, F TO ENTER.

[d] Constant omitted.

[e] Can be obtained through CONTRAST procedure.

[f] Only meaningful at STEP NUMBER 0 (univariate Fs) and at last step if all variables forced.

provided, can be useful for the more sophisticated user. Successive discriminant functions (roots) can be evaluated, as in SPSS DISCRIMINANT.

SPSS MANOVA conveniently provides discriminant functions for the more complex designs, including factorial arrangements with unequal sample sizes. The program is limited, however, in that it includes no classification phase. Further, only standard DISCRIM is available, with no provision for stepwise or hierarchical analysis other than the stepdown analysis described in Chapter 8.

9.7.2 BMD Series

Four BMD series programs for discriminant function analysis are available, with the later P-series program, BMDP7M, incorporating most of the features of the earlier programs. Two of the early programs, BMD04M and BMD05M were specialized programs, but important features are incorporated into BMDP7M. BMD07M, one of the earlier programs, provides stepping options not yet available in BMDP7M. BMD07M was designed specifically for stepwise DISCRIM, but can also be used for direct DISCRIM, by forcing entry of all predictors and ignoring all steps except the final one. With multitudes of predictors, however, output from BMD07M becomes voluminous.

BMDP7M deals well with all varieties of DISCRIM. Although designed as a stepwise program, direct DISCRIM can be efficiently produced by forcing all predictors to enter and suppressing all steps except the last.

BMDP7M and BMD07M have one very substantial advantage over all other DISCRIM programs, namely, identification of outliers through the case classification procedure. Further, an optional JACKKNIFED CLASSIFICATION in BMDP7M identifies cases into groups without using those cases in developing the classification functions. This eliminates or reduces bias in classification, as discussed in Section 9.6.6.3.

The BMDP7M provision of stepping by contrasts allows analysis of factorial designs easily as long as sample sizes are equal among groups. That is, contrasts can be specified to test main effects and interactions, with separate discriminant function analyses run on each effect. This is equivalent to the factorial MANOVA available in SPSS MANOVA. Since sample sizes are equal, and the design orthogonal, there is no need to provide for adjustment of effects. Because this program allows for classification as well as statistical inference, it would frequently be the program of choice for an equal-n factorial design. Unequal-n factorial designs require contrast coefficients that reflect choice of adjustment procedure.

9.8 COMPLETE EXAMPLES OF DISCRIMINANT FUNCTION ANALYSIS

Examples of discriminant function analysis in this section are designed to explore how role-dissatisfied housewives and role-satisfied housewives differ from employed women in attitudes and demography. The sample of 465 women is described in Appendix B. The grouping variable is formed from three groups of women: working women (WORKING), role-satisfied housewives (HAPHOUSE), and role-dissatisfied housewives (UNHOUSE).

For prediction of group membership on the basis of attitudes, predictor variables consist of a measure of internal versus external locus of control (IESCALE), satisfaction with current marital status (ATTSTAT), attitude toward women's role (WOMENROL), and attitude toward homemaking (ATTHOUSE). A fifth attitudinal variable, attitude toward paid work, was dropped from analysis because data were available only for women who had been employed within the past five years. This measure would therefore have involved nonrandom missing values (cf. Chapter 4). The first example of DISCRIM, then, involves a prediction of group membership on the basis of the four attitudinal variables.

The second example involves prediction of group membership on the basis of a hierarchical discriminant function analysis in which demographic variables are entered first, followed by the four attitudinal variables. Demographic variables used as predictors in the second example are marital status (MARITAL), whether or not the women have had children (CHILDREN), religious affiliation (RELIGION), RACE, socioeconomic level (DUN), AGE, and years of education (EDUC).

The two discriminant function analyses allow us to evaluate the distinctions among the three groups on the basis of attitudes alone, demography alone, and demography plus attitudes. We explore the dimensions on which the groups differ, the variables contributing to differences among groups on these dimensions, and the degree to which we can accurately classify members into their own groups. We also evaluate efficiency of classification with a cross-validation sample.

9.8.1 Evaluation of Assumptions

These data will first be evaluated with respect to practical limitations of DISCRIM.

9.8.1.1 Unequal Sample Sizes and Missing Data Twenty-eight of the cases showed data missing on the predictor variables. Missing data were scattered over variables and groups, so that deletion of these cases was deemed appropriate,[10] leaving 437 cases for further evaluation.

For classification, unequal sample sizes were used to modify probabilities with which cases were classified into groups. Assuming a random sample of cases from the population of interest, classification of new cases could be improved by knowledge of the population size for each group of women. For example, knowledge that over half the women work implies that a greater weighting should be given the WORKING group.

9.8.1.2 Linearity An initial inspection of the coding schemes for predictor variables reveals that two of the demographic variables are obviously nonlinear with respect to the remaining variables. Since marital status had originally been coded discretely into three levels—single, married, and broken—there was no reason to expect linear relationships with any of the other predictor variables on the basis of this coding. Therefore the variable was recoded into two dummy variables: single versus other (SINGLE), and married versus other (MARRIED).[11] Similarly, RELIGION was originally coded into four categories of None-or-other, Catholic, Protestant, and Jewish. This variable also would not be expected to be linearly related to any of the other variables. Dummy variable recoding of the RELIGION variable resulted in the creation of three new variables: None-or-other versus all others (NONOTHER), Catholic versus all others (CATHOLIC), and Protestant versus all others (PROTEST). In addition to the 4 attitudinal predictors, then, we have 10 demographic predictors: CHILDREN, RACE, DUN, AGE, EDUC, SINGLE, MARRIED, CATHOLIC, PROTEST, and NONOTHER.

All pairs of 14 predictor variables were evaluated within each group for linearity through SPSS SCATTERGRAM. None of the scatterplots revealed gross deviation from linearity, so no further transformations were felt to be

---

[10] Alternative strategies for dealing with missing data are discussed in Chapter 4.

[11] A variety of sources discuss dummy variable coding (e.g., Cohen & Cohen, 1975).

necessary. (Note that evaluation of scatterplots followed deletion of outliers and cases with missing data, described in the next section.)

9.8.1.3 Outliers To identify univariate outliers, BMDP7D was used to assess the minimum and maximum z score in each of the 14 variables described in Section 9.8.1.2 for each of the three groups separately, after deletion of the 28 cases with missing data. There were some questionable values on EDUC, with one woman reporting 24 years of schooling completed and another reporting completion of 4 years. Since these values were about 4.5 standard deviations from their group means, but still plausible, the decision was made initially to retain those cases and delete them only if they produced multivariate outliers.

In the search for multivariate outliers, the 437 cases were again divided into 3 groups, and outliers for each group were assessed separately through BMDP7M. Two runs were done: In the first pass, all 14 predictors were used; in the second pass, only the 4 attitudinal variables were used. A portion of the BMDP7M output for the WORKING group in the second pass is illustrated in Table 9.9. Outliers are identified as cases with too large a Mahalanobis D^2, evaluated as χ^2 with degrees of freedom equal to the number of predictor variables. In this case, critical χ^2 with 4 df at the 99% level of confidence is 13.28. Therefore, any case with $D^2 >$ than 13.28 is an outlier. In Table 9.9, cases 10, 33, and 36 were identified as outliers for their own group (WORKING). The two checks for outliers identified a total of 13 cases.

For outlying cases, values on all variables were examined. The two cases with extreme scores on EDUC were not among the multivariate outliers, but four cases with extreme values on ATTSTAT were identified. The only other questionable variable noted was RACE, with all three of the outliers on the first multivariate run produced by nonwhite women. Transformation seemed inappropriate for RACE and questionable for ATTSTAT on the basis of only four cases; therefore the decision was made to delete the 13 outliers produced by the two runs. BMDP9R was used to identify the characteristics of all 13 outlying cases, as demonstrated in Section 5.8.1. Together with the deletion of 28 cases with missing data, a final sample of 424 cases was available for analysis.

9.8.1.4 Multivariate Normality Even after identification of missing values and outliers, there were still over 70 cases per group. Further, the BMDP7D run showed no notable skewness for continuous variables, although dichotomous splits were asymmetrical for such variables as RACE. Nevertheless, sample sizes were deemed large enough to suggest normality of sampling distributions of means and there was no reason to doubt multivariate normality.

9.8.1.5 Homogeneity of Variance-Covariance Matrices Examination of sample variances for the 14 predictor variables revealed no gross discrepancies. The ratio of largest to smallest variance for the most discrepant predictor was MARRIED, where the ratio was 6.8:1 for the WORKING versus HAPHOUSE

TABLE 9.9 IDENTIFICATION OF OUTLIERS AMONG SET OF FOUR
ATTITUDINAL VARIABLES (DECK SETUP AND SELECTED
OUTPUT FROM BMDP7M)

```
PROGRAM CONTROL INFORMATION

        /PROBLEM    TITLE IS "LARGE SAMPLE DISCRIM, DIRECT-MULTIVARIATE OUTLIERS".
        /INPUT      VARIABLES ARE 5.  CASE = 465.  UNIT = 99.
                    FORMAT IS "(4X,F4.0,16X,3F4.0,4X,F4.0)".
        /VARIABLES  NAMES ARE WORKSTAT,IESCALE,ATTSTAT,WOMENROL,ATTHOUSE.
                    GROUPING IS WORKSTAT.
        /TRANSFORM  DELETE=37,80,83,95,105,113,118,135,159,196,206,207,208,209,219,
                    265,280,300,303,314,317,341,437,448,457,102,198,253,261,292,299.
        /GROUP      CODES(1) ARE 1 TO 3.
                    NAMES(1) ARE WORKING,HAPHOUSE,UNHOUSE.
                    PRIOR = 0.53, 0.29, 0.18.
        /DISC       LEVEL = 0,1,1,1,1.
                    FORCE = 1.  TOL = 0.001.
        /PLOT       GROUP = 1.  GROUP = 2.  GROUP = 3.
        /PRINT      WITH.  NO STEP.
        /END
```

	INCORRECT CLASSIFICATIONS		MAHALANOBIS D-SQUARE FROM AND POSTERIOR PROBABILITY FOR GROUP –					
GROUP	WORKING		WORKING		HAPHOUSE		UNHOUSE	
CASE								
2			2.4	.516	1.9	.377	3.4	.106
3	HAPHOUSE		2.9	.399	1.5	.433	2.5	.167
6			1.8	.686	3.6	.152	2.5	.162
10			14.6	.632	18.4	.052	13.8	.316
16			2.7	.516	2.1	.390	4.0	.094
18	HAPHOUSE		5.5	.329	3.2	.560	5.5	.111
20			5.6	.459	5.0	.330	5.0	.211
21			5.2	.507	5.1	.298	5.0	.194
22			4.3	.613	6.1	.133	3.9	.254
23			.5	.562	.7	.280	.9	.158
25			3.6	.611	4.6	.203	3.8	.186
26			6.1	.644	8.1	.127	6.0	.229
29			3.9	.443	4.0	.226	2.3	.331
30			6.2	.479	6.6	.221	5.0	.301
31			5.6	.474	5.5	.283	4.8	.243
32			3.3	.701	4.6	.198	5.0	.101
33	HAPHOUSE		13.4	.243	11.1	.403	10.4	.354
36			14.0	.337	12.8	.336	11.9	.327
38			2.9	.633	4.9	.125	2.6	.242
39			4.8	.565	4.4	.372	7.0	.063

groups. That is, there was more variance in marital status for working women
than for role-satisfied housewives. The sample sizes are fairly discrepant, with
almost three and a half times as many working women as role-dissatisfied housewives, but with the use of two-tailed tests and reasonable homogeneity of vari-

ance, DISCRIM procedures could be expected to be robust enough to handle the discrepancies.

Plots of the first two canonical variates for each group were examined for homogeneity of variance-covariance matrices. As can be seen in Figure 9.2 (below and pages 339 and 340), the spread of cases for the three groups is relatively equal. Therefore no direct test of homogeneity of variance-covariance matrices was felt to be necessary.

9.8.1.6 Multicollinearity and Singularity   Since BMDP7M, used for the major analysis, protects against multicollinearity, no formal evaluation was felt to be necessary (cf. Chapter 4).

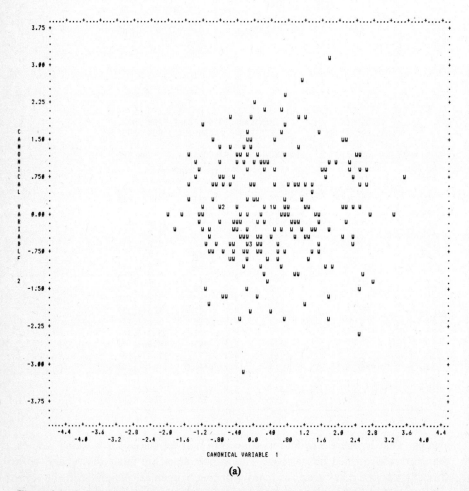

(a)

Figure 9.2   Scatterplots of cases on first two canonical variates for (a) working women, (b) role-satisfied housewives, and (c) role-dissatisfied housewives.

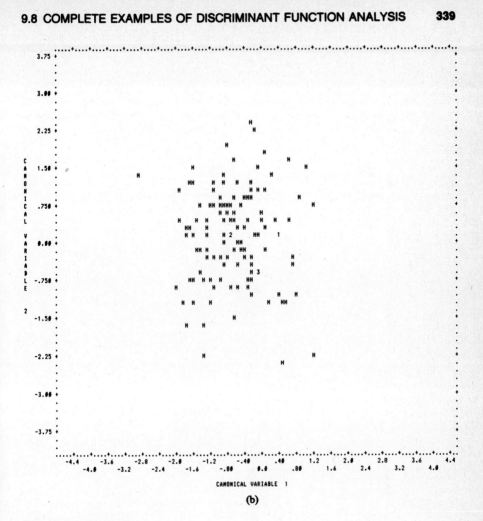

(b)

Figure 9.2 *(Continued)*

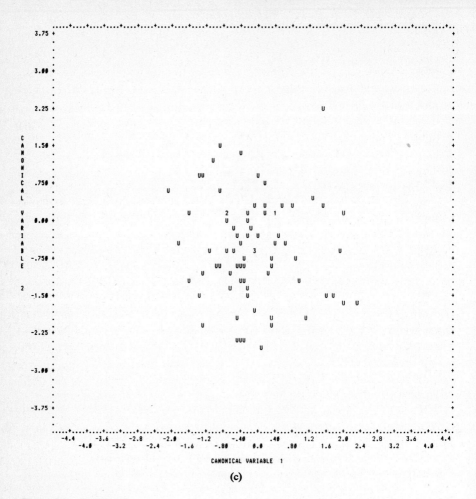

CANONICAL VARIABLE 1

(c)

Figure 9.2   *(Continued)*

9.8.2 Direct Discriminant Function Analysis

Partial output of the direct DISCRIM produced by BMDP7M, with the four attitudinal variables as predictors, appears in Table 9.10. Notice that all attitudinal variables are forced into the analysis. Since this is a direct analysis, stepwise output has been suppressed (NO STEP).

The F TO ENTER values at step 0 are univariate F ratios for the individual predictors. Note that three of the four variables, all except IESCALE, show univariate F ratios that would be significant at the .01 level for discriminating among the three groups.

The approximate F test based on Wilks' Lambda shows highly significant discrimination among the three groups on the basis of all four attitudinal variables combined. Canonical correlations, although small, do not differ dramatically for the two discriminant functions (canonical variables). Note that the COEFFICIENTS FOR CANONICAL VARIABLES are unstandardized discriminant function coefficients.

Table 9.10 shows the classification functions used to classify cases into the three groups (see Eq. 9.3) and the results of that classification, with and without jackknifing (see Section 9.6.6.3). In this case, actual classification was made on the basis of a modified equation in which unequal prior probabilities were set to reflect the unequal group sizes by use of the PRIOR convention, as shown in the deck setup.

An additional run was made for cross-validation, the results of which appear in Table 9.11. The groups for cross-validation were formed by randomly selecting out approximately 25% of the original cases. These were relabeled as to membership into three new groups: NEWWORK (for WORKING), NEWHAP (for HAPHOUSE), and NEWUN (for UNHOUSE). The remaining 323 cases were used to develop the discriminant functions and classification equations. The 101 cases selected out were used only to assess the results of classification. Table 9.11 shows results of classification for both original and cross-validation cases (pages 344 and 345).

Canonical discriminant functions analysis in Table 9.12 (page 345) shows both discriminant functions are statistically significant. The first line (AFTER FUNCTION 0) evaluates both of the discriminant functions combined. The second line (AFTER FUNCTION 1) evaluates the second function alone, with the first function deleted. Since BMDP7M does not provide this analysis, SPSS DISCRIMINANT was used to evaluate the successive discriminant functions.

A plot of placement of the three group centroids (full sample) on the two discriminant functions (canonical variables) appears in Figure 9.3 (page 346). Plotted points are those given in Table 9.10 as CANONICAL VARIABLES EVALUATED AT GROUP MEANS.

Like the dimension reduction analysis, the loading matrix is not given directly by BMDP7M. It could be calculated by hand according to Eq. 9.7 in Section 9.6.3.2. For these data, however, the loading matrix in Table 9.12 was found through SPSS DISCRIMINANT. For interpretation, loadings less than .45 will not be considered.

TABLE 9.10 DECK SETUP AND PARTIAL OUTPUT FROM BMDP7M
DISCRIMINANT FUNCTION ANALYSIS OF FOUR ATTITUDINAL
VARIABLES

```
PROGRAM CONTROL INFORMATION

        /PROBLEM      TITLE IS "LARGE SAMPLE DISCRIM, DIRECT".
        /INPUT        VARIABLES ARE 5.  CASE=424.  UNIT=90.
                      FORMAT IS "(4X,F4.0,16X,3F4.0,4X,F4.0)".
        /VARIABLES    NAMES ARE WORKSTAT,IESCALE,ATTSTAT,WOMENROL,ATTHOUSE.
                      GROUPING IS WORKSTAT.
        /GROUP        CODES(1) ARE 1 TO 3.
                      NAMES(1) ARE WORKING, HAPHOUSE, UNHOUSE.
                      PRIOR = 0.53, 0.29, 0.18.
        /DISC         LEVEL = 0,1,1,1,1.
                      FORCE = 1.  TOL = 0.001.
        /PLOT         GROUP=1. GROUP=2. GROUP=3.
        /PRINT        NO STEP.   WITH.
        /END
  MEANS
        GROUP =    WORKING      HAPHOUSE      UNHOUSE       ALL GPS.
VARIABLE
  1 WORKSTAT     1.00000       2.00000       3.00000       1.64387
  2 IESCALE      6.71300       6.68217       7.01389       6.75472
  3 ATTSTAT     23.03587      20.47287      23.88889      22.40094
  4 WOMENROL    33.78924      37.20155      35.87500      35.18160
  5 ATTHOUSE    23.70852      22.57364      24.75000      23.54009
COUNTS           223.          129.           72.          424.
    STANDARD DEVIATIONS
        GROUP =    WORKING      HAPHOUSE      UNHOUSE       ALL GPS.
VARIABLE
  1 WORKSTAT     0.00000       0.00000       0.00000       0.00000
  2 IESCALE      1.24414       1.31092       1.20437       1.25831
  3 ATTSTAT      8.09452       6.42149       8.44127       7.68794
  4 WOMENROL     6.65328       6.38526       5.62424       6.48082
  5 ATTHOUSE     4.40007       3.87051       3.92446       4.16669
    COEFICIENTS OF VARIATION
        GROUP =    WORKING      HAPHOUSE      UNHOUSE       ALL GPS.
VARIABLE
  1 WORKSTAT     0.00000       0.00000       0.00000       0.00000
  2 IESCALE       .18533        .1961P &*
***********************************************************************************************

STEP NUMBER   0
  VARIABLE        F TO  FORCE  TOLERANCE  *    VARIABLE       F TO  FORCE  TOLERANCE
                 REMOVE LEVEL             *                  ENTER LEVEL
            DF=  2  422                   *              DF= 2  421
                                          *    2 IESCALE     1.864    1   1.000000
                                          *    3 ATTSTAT     6.166    1   1.000000
                                          *    4 WOMENROL   12.092    1   1.000000
                                          *    5 ATTHOUSE    6.688    1   1.000000


STEP NUMBER   4
VARIABLE ENTERED   2 IESCALE
  VARIABLE        F TO  FORCE  TOLERANCE  *    VARIABLE       F TO  FORCE  TOLERANCE
                 REMOVE LEVEL             *                  ENTER LEVEL
            DF=  2  418                   *              DF= 2  417
  2 IESCALE       .953    1   .953621     *
  3 ATTSTAT      3.573    1   .904911     *
  4 WOMENROL    10.972    1   .884189     *
  5 ATTHOUSE     3.596    1   .818146     *
U-STATISTIC OR WILKS" LAMBDA    .9009207      DEGREES OF FREEDOM    4    2     421
APPROXIMATE F-STATISTIC            5.596      DEGREES OF FREEDOM    8.00   836.00
  F - MATRIX        DEGREES OF FREEDOM =   4  418
            WORKING  HAPHOUSE
HAPHOUSE    7.65
UNHOUSE     3.92     4.54
```

TABLE 9.10   (Continued)

```
CLASSIFICATION FUNCTIONS
          GROUP =    WORKING       HAPHOUSE       UNHOUSE
VARIABLE
  2 IESCALE        3.42397        3.46283        3.57931
  3 ATTSTAT         .08297         .04331         .08316
  4 WOMENROL       1.24477        1.32368        1.31609
  5 ATTHOUSE       1.82707        1.82043        1.91746
CONSTANT         -55.77163      -58.41906      -62.59660
```

```
CLASSIFICATION MATRIX
GROUP      PERCENT    NUMBER OF CASES CLASSIFIED INTO GROUP -
           CORRECT
                      WORKING  HAPHOUSE  UNHOUSE
WORKING    87.0        194       28        1
HAPHOUSE   27.1         93       35        1
UNHOUSE     4.2         59       10        3
TOTAL      54.7        346       73        5
JACKKNIFED CLASSIFICATION
GROUP      PERCENT    NUMBER OF CASES CLASSIFIED INTO GROUP -
           CORRECT
                      WORKING  HAPHOUSE  UNHOUSE
WORKING    86.1        192       29        2
HAPHOUSE   23.3         98       30        1
UNHOUSE     1.4         61       10        1
TOTAL      52.6        351       69        4
```

```
EIGENVALUES
                   .07324       .03423
CUMULATIVE PROPORTION OF TOTAL DISPERSION
                   .68147      1.00000
CANONICAL CORRELATIONS
                   .26123       .18193
VARIABLE      COEFFICIENTS FOR CANONICAL VARIABLES

  2 IESCALE       -.06097      -.28840
  3 ATTSTAT        .06481      -.02889
  4 WOMENROL      -.12777      -.08854
  5 ATTHOUSE       .01232      -.18885
CONSTANT          3.16508     10.15588
GROUP       CANONICAL VARIABLES EVALUATED AT GROUP MEANS
  WORKING         .22366       .08517
  HAPHOUSE       -.39052       .08028
  UNHOUSE         .00694      -.40762
```

A summary of interpretative information, as might be included in a table for publication, is provided in Table 9.13 (page 347). Included are the loading matrix, univariate F for each predictor, and pooled within-group correlations among predictor variables, as found through SPSS DISCRIMINANT.

Contrasts were run pitting each group against the other two, using BMDP7M, to determine which variables were most important in distinguishing each group. Tables 9.14, 9.15, and 9.16 (pages 348–350) show the initial F TO ENTER and final F TO REMOVE for each of the three contrast runs. Critical F TO REMOVE for evaluating the four variables at $\alpha_i = .01$ (cf. Section 9.6.4) with 1 and 418 df is approximately 6.8. On the basis of unique variance analyzed by F TO REMOVE, the variable that most clearly distinguishes the WORKING group, then, is WOMENROL. The UNHOUSE group differs most prominently on ATTHOUSE. The HAPHOUSE is not significantly

TABLE 9.11   CROSS-VALIDATION OF CLASSIFICATION OF CASES BY
FOUR ATTITUDINAL VARIABLES (DECK SETUP AND
SELECTED OUTPUT FROM BMDP7M)

PROGRAM CONTROL INFORMATION
```
       /PROBLEM    TITLE IS 'LARGE SAMPLE DISCRIM, DIRECT'.
       /INPUT      VARIABLES ARE 5.
                   FORMAT IS '(4X,F4.0,16X,3F4.0,4X,F4.0)'.
                   CASE = 465.
       /VARIABLES  NAMES ARE WORKSTAT,IESCALE,ATTSTAT,WOMENROL,ATTHOUSE.
                   GROUPING IS WORKSTAT.
       /TRANSFORM
            DELETE=37,80,83,95,105,113,118,135,159,196,206,207,208,209,219,
            265,280,300,303,314,317,341,437,448,457,102,198,253,261,292,299,
            33,166,421,71,279,10,36,85,377,389.
                        PROB = RNDU(872349).
                        NEWGROUP = WORKSTAT + 3.
                        TRUE = PROB LE 0.25.
                        WORKSTAT = NEWGROUP IF TRUE.
       /GROUP      CODES(1) ARE 1 TO 6.
                   NAMES(1) ARE WORKING, HAPHOUSE, UNHOUSE, NEWWORK, NEWHAP,
                   NEWUN.
                   USE = 1 TO 3.
                   PRIOR = 0.53, 0.29, 0.18.
       /DISC       LEVEL = 0,1,1,1,1.
                   FORCE = 1.    TOL = 0.001.
       /PLOT       GROUP = 1.   GROUP = 2.   GROUP = 3.   GROUP = 4.   GROUP = 5.   GROUP = 6.
       /PRINT      WITH.   NO STEP.
       /END
```

CLASSIFICATION FUNCTIONS

GROUP =	WORKING	HAPHOUSE	UNHOUSE
VARIABLE			
2 IESCALE	4.17702	4.16649	4.33770
3 ATTSTAT	0.13447	0.09314	0.13451
4 WOMENROL	1.33560	1.41345	1.40111
5 ATTHOUSE	1.97270	1.97963	2.07843
CONSTANT	−62.24489	−64.85149	−69.27623

CLASSIFICATION MATRIX

GROUP	PERCENT CORRECT	NUMBER OF CASES CLASSIFIED INTO GROUP —		
		WORKING	HAPHOUSE	UNHOUSE
WORKING	88.4	152	20	0
HAPHOUSE	24.5	73	24	1
UNHOUSE	3.8	42	9	2
NEWWORK	0.0	47	3	1
NEWHAP	0.0	26	4	1
NEWUN	0.0	17	1	1
TOTAL	55.1	357	61	6

TABLE 9.11 (*Continued*)

JACKKNIFED CLASSIFICATION

GROUP	PERCENT CORRECT	NUMBER OF CASES CLASSIFIED INTO GROUP —		
		WORKING	HAPHOUSE	UNHOUSE
WORKING	87.2	150	21	1
HAPHOUSE	23.5	73	23	2
UNHOUSE	1.9	43	9	1
NEWWORK	0.0	47	3	1
NEWHAP	0.0	26	4	1
NEWUN	0.0	17	1	1
TOTAL	53.9	356	61	7

TABLE 9.12 PARTIAL DECK SETUP AND SELECTED OUTPUT FROM SPSS DISCRIMINANT FOR PREDICTION OF MEMBERSHIP IN THREE GROUPS ON THE BASIS OF FOUR ATTITUDINAL VARIABLES

```
DISCRIMINANT   GROUPS=WORKSTAT(1,3)/
               VARIABLES=IESCALE,ATTSTAT,WOMENROL,ATTHOUSE/
               ANALYSIS=IESCALE TO ATTHOUSE/
STATISTICS     10
```

CANONICAL DISCRIMINANT FUNCTIONS

FUNCTION	EIGENVALUE	PERCENT OF VARIANCE	CUMULATIVE PERCENT	CANONICAL CORRELATION	AFTER FUNCTION	WILKS LAMBDA	CHI-SQUARED	D.F.	SIGNIFICANCE
					0	.9009207	43.770	8	.0000
1*	.07324	68.15	68.15	.2612255	1	.9669009	14.120	3	.0027
2*	.03423	31.85	100.00	.1819317					

* MARKS THE 2 FUNCTION(S) TO BE USED IN THE REMAINING ANALYSIS.

POOLED WITHIN-GROUPS CORRELATIONS BETWEEN CANONICAL DISCRIMINANT FUNCTIONS AND DISCRIMINATING VARIABLES
VARIABLES ARE ORDERED BY THE FUNCTION WITH LARGEST CORRELATION AND THE MAGNITUDE OF THAT CORRELATION.

	FUNC 1	FUNC 2
WOMENROL	.86467*	.28016
ATTSTAT	-.54651*	.46551
ATTHOUSE	-.44908	.70473*
IESCALE	-.04391	.50455*

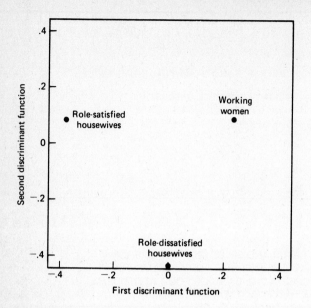

Figure 9.3 Plot of three group centroids on two discriminant functions derived from four attitudinal variables.

TABLE 9.13 RESULTS OF DISCRIMINANT FUNCTION ANALYSIS OF ATTITUDINAL VARIABLES

Predictor variable	Correlations of predictor variables with discriminant functions		Univariate $F(2, 421)$	Pooled within-group correlations among predictors		
	1	2		ATTSTAT	WOMENROL	ATTHOUSE
IESCALE	−.04	.50	1.864	.19	−.02	.14
ATTSTAT	−.55	.47	6.166		−.06	.26
WOMENROL	.86	.28	12.092			−.34
ATTHOUSE	.45	.70	6.688			
Canonical R	.26	.18				
Eigenvalue	.073	.034				

TABLE 9.14 INPUT AND PARTIAL OUTPUT OF BMDP7M CONTRASTING
THE WORKING GROUP WITH THE OTHER TWO GROUPS
FOR ATTITUDINAL VARIABLES

```
PROGRAM CONTROL INFORMATION

        /PROBLEM        TITLE IS "LARGE SAMPLE DISCRIM, DIRECT".
        /INPUT          VARIABLES ARE 5.  CASE=424.  UNIT=90.
                        FORMAT IS "(4X,F4.0,16X,3F4.0,4X,F4.0)".
        /VARIABLES      NAMES ARE WORKSTAT,IESCALE,ATTSTAT,WOMENROL,ATTHOUSE.
                        GROUPING IS WORKSTAT.
        /GROUP          CODES(1) ARE 1 TO 3.
                        NAMES(1) ARE WORKING, HAPHOUSE, UNHOUSE.
                        PRIOR = 0.53, 0.29. 0.18.
        /DISC           LEVEL = 0,1,1,1,1.
                        FORCE = 1.  TOL = 0.001.
                        CONTRAST = -2,1,1.
        /PLOT           NO CANON.  CONTRAST.
        /PRINT          NO STEP.  NO POST.
        /END
```

```
*************************************************************************************************

STEP NUMBER   0
  VARIABLE        F TO  FORCE  TOLERANCE  *   VARIABLE       F TO   FORCE   TOLERANCE
                  REMOVE LEVEL             *                 ENTER  LEVEL
              DF=  1  422                  *             DF=  1   421
                                           *   2 IESCALE    1.164    1     1.000000
                                           *   3 ATTSTAT    1.250    1     1.000000
                                           *   4 WOMENROL  18.596    1     1.000000
                                           *   5 ATTHOUSE    .013    1     1.000000

STEP NUMBER   4
VARIABLE ENTERED    2 IESCALE
  VARIABLE        F TO  FORCE  TOLERANCE  *   VARIABLE       F TO   FORCE   TOLERANCE
                  REMOVE LEVEL             *                 ENTER  LEVEL
              DF=  1  418                  *             DF=  1   417
  2 IESCALE     1.357    1    .953621      *
  3 ATTSTAT     1.989    1    .984911      *
  4 WOMENROL   20.416    1    .884189      *
  5 ATTHOUSE    2.380    1    .818146      *



EIGENVALUES
                .05630
CUMULATIVE PROPORTION OF TOTAL DISPERSION
                1.00000
CANONICAL CORRELATIONS
                .23087
```

TABLE 9.15 INPUT AND PARTIAL OUTPUT OF BMDP7M CONTRASTING
THE HAPHOUSE GROUP WITH THE OTHER TWO GROUPS
FOR ATTITUDINAL VARIABLES

```
PROGRAM CONTROL INFORMATION

        /PROBLEM      TITLE IS "LARGE SAMPLE DISCRIM, DIRECT".
        /INPUT        VARIABLES ARE 5.  CASE=424.  UNIT=90.
                      FORMAT IS "(4X,F4.0,16X,3F4.0,4X,F4.0)".
        /VARIABLES    NAMES ARE WORKSTAT,IESCALE,ATTSTAT,WOMENROL,ATTHOUSE.
                      GROUPING IS WORKSTAT.
        /GROUP        CODES(1) ARE 1 TO 3.
                      NAMES(1) ARE WORKING, HAPHOUSE, UNHOUSE.
                      PRIOR = 0.53, 0.29, 0.18.
        /DISC         LEVEL = 0,1,1,1,1.
                      FORCE = 1.  TOL = 0.001.
                      CONTRAST = 1,-2,1.
        /PLOT         NO CANON.  CONTRAST.
        /PRINT        NO STEP.  NO POST.
        /END
```

```
********************************************************************************

STEP NUMBER   0
  VARIABLE        F TO  FORCE  TOLERANCE  *  VARIABLE        F TO  FORCE  TOLERANCE
                  REMOVE LEVEL             *                  ENTER LEVEL
              DF= 1   422                  *              DF= 1   421
                                           *    2 IESCALE    1.681    1    1.000000
                                           *    3 ATTSTAT   12.248    1    1.000000
                                           *    4 WOMENROL  11.072    1    1.000000
                                           *    5 ATTHOUSE  12.789    1    1.000000

STEP NUMBER   4
VARIABLE ENTERED   2 IESCALE
  VARIABLE        F TO  FORCE  TOLERANCE  *  VARIABLE        F TO  FORCE  TOLERANCE
                  REMOVE LEVEL             *                  ENTER LEVEL
              DF= 1   418                  *              DF= 1   417
    2 IESCALE     .173    1    .953621     *
    3 ATTSTAT    6.512    1    .904911     *
    4 WOMENROL   5.215    1    .884189     *
    5 ATTHOUSE   2.915    1    .818146     *



EIGENVALUES
              .06045
CUMULATIVE PROPORTION OF TOTAL DISPERSION
              1.00000
CANONICAL CORRELATIONS
              .23876
```

TABLE 9.16   INPUT AND PARTIAL OUTPUT OF BMDP7M CONTRASTING
THE UNHOUSE GROUP WITH THE OTHER TWO GROUPS
FOR ATTITUDINAL VARIABLES

```
PROGRAM CONTROL INFORMATION

        /PROBLEM      TITLE IS "LARGE SAMPLE DISCRIM, DIRECT".
        /INPUT        VARIABLES ARE 5.  CASE=424.  UNIT=90.
                      FORMAT IS "(4X,F4.0,16X,3F4.0,4X,F4.0)".
        /VARIABLES    NAMES ARE WORKSTAT,IESCALE,ATTSTAT,WOMENROL,ATTHOUSE.
                      GROUPING IS WORKSTAT.
        /GROUP        CODES(1) ARE 1 TO 3.
                      NAMES(1) ARE WORKING, HAPHOUSE, UNHOUSE.
                      PRIOR = 0.53, 0.29, 0.18.
        /DISC         LEVEL = 0,1,1,1,1.
                      FORCE = 1.  TOL = 0.001.
                      CONTRAST = 1,1,-2.
        /PLOT         NO CANON.  CONTRAST.
        /PRINT        NO STEP.  NO POST.
        /END
```

```
**************************************************************************************

STEP NUMBER   0
   VARIABLE        F TO  FORCE  TOLERANCE  *    VARIABLE        F TO  FORCE  TOLERANCE
                   REMOVE LEVEL            *                    ENTER LEVEL
           DF=  1   422                    *            DF=  1   421
                                           *    2 IESCALE       3.728    1    1.000000
                                           *    3 ATTSTAT       4.548    1    1.000000
                                           *    4 WOMENROL       .207    1    1.000000
                                           *    5 ATTHOUSE      8.798    1    1.000000


STEP NUMBER   4
VARIABLE ENTERED   3 ATTSTAT
   VARIABLE        F TO  FORCE  TOLERANCE  *    VARIABLE        F TO  FORCE  TOLERANCE
                   REMOVE LEVEL            *                    ENTER LEVEL
           DF=  1   418                    *            DF=  1   417
     2 IESCALE     1.585    1    .953621   *
     3 ATTSTAT     1.217    1    .904911   *
     4 WOMENROL    2.099    1    .884189   *
     5 ATTHOUSE    7.183    1    .818146   *



EIGENVALUES
              .03484
CUMULATIVE PROPORTION OF TOTAL DISPERSION
             1.00000
CANONICAL CORRELATIONS
              .18349
```

TABLE 9.17   CHECKLIST FOR DIRECT DISCRIMINANT FUNCTION
ANALYSIS

1. Issues
    a. Unequal sample sizes and missing data
    b. Multivariate normality
    c. Outliers
    d. Linearity
    e. Homogeneity of variance-covariance matrices
    f. Multicollinearity and singularity
2. Major analysis
    a. Significance of discriminant functions. If significant,
        (1) Variance accounted for
        (2) Plot(s) of discriminant functions
        (3) Loading matrix
    b. Variables separating each group
3. Additional analyses
    a. Group means for high-loading variables ($r \geq .45$)
    b. Pooled within-group correlations among predictor variables
    c. Classification results
        (1) Jackknifed classification
        (2) Cross-validation
    d. Change in Rao's V (or stepdown F) plus univariate F for predictors

distinguishable on the basis of any variable after adjustment is made for the remaining variables. Without adjustment (F TO ENTER) all variables except IESCALE would have been significant in discriminating HAPHOUSE from the other two groups.

A checklist for a direct discriminant function analysis appears as Table 9.17. Following is an example of a Results section, in journal format, for the study just described.

RESULTS

A direct discriminant function analysis was performed using the four attitudinal variables as predictors of membership in three groups. Predictor variables were locus of control, attitude toward marital status, attitude toward role of women, and attitude toward homemaking. Groups were working women, role-satisfied housewives, and role-dissatisfied housewives.

Of the original 465 cases, 28 were dropped from analysis owing to missing data. Missing data were scattered over cases and variables, with no evident patterning on the basis of grouping or demographic variables. An additional 13 cases were identified as multivariate outliers with $p < .01$, and were also deleted. Among the outliers, nonwhite women were disproportionately represented ($N = 3$).[12] For the

---

[12] Among the nonwhite women, all worked outside the home and 2 were single, 1 with a child. The third was older and college educated. Among the 10 outliers deleted on the basis of attitudes, 4 were extremely dissatisfied with current marital status. Of these, 2 worked and 2 were role-dissatisfied housewives. Another 4 outliers, all working women, showed unusual combinations of attitudes toward role of women and toward housework. The final 2 outlying cases, both role-dissatisfied housewives, were characterized by near extreme values on 2 variables. Both expressed dissatisfaction with current marital status. In addition, 1 was highly negative toward housework and the other expressed very liberal attitudes toward role of women.

remaining 424 cases, evaluation of assumptions of linearity, normality, multicollinearity or singularity, and homogeneity of variance-covariance matrices revealed no threat to multivariate analysis.

Two discriminant functions were calculated, with a combined $\chi^2(8) = 43.72$, $p < .01$. After removal of the first function, there was still highly significant discriminating power, $\chi^2(3) = 14.12$, $p < .01$. The two discriminant functions accounted for 68% and 32%, respectively, of the between-group variability. [χ^2 **values and percent of variance are from Table 9.12; cf. Section 9.6.2.**] As shown in Figure 9.3, the first discriminant function maximally separates working women from role-satisfied housewives, with role-dissatisfied housewives falling between these two groups. The second discriminant function discriminates role-dissatisfied housewives from the other two groups.

Insert Figure 9.3 about here

A loading matrix of correlations between predictor variables and discriminant functions, as seen in Table 9.13, suggests that the primary variable in distinguishing between working women and role-satisfied housewives (first function) is attitude toward women's role. Role-satisfied housewives have more conservative attitudes toward women's role (mean attitude toward role of women = 37.20) than working women (mean attitude = 33.79). [**Group means are shown in Table 9.10.**]

Insert Table 9.13 about here

Also contributing to discrimination between these two groups of women are attitudes toward current marital status and toward homemaking. Working women are less satisfied with their current marital status (mean attitude toward marital status = 23.04) than role-satisfied housewives (mean = 20.47). Similarly, working women have less positive attitudes toward homemaking (mean attitude toward housework = 23.71) than role-dissatisfied housewives (mean attitude = 22.57). Loadings less than .45 are not interpreted.

After adjustment for all other variables, and keeping overall $\alpha < .05$ for the four variables, only attitude toward women's role significantly separates working women from the other two groups, $F(1, 418) = 20.42$. The squared semipartial correlation between the grouping variable (working women vs. housewives) and attitudes toward women's role is .05. [**Equation 9.9 applied to output in Table 9.14.**] Role-satisfied housewives show no variable that significantly distinguishes them from the remaining groups after adjusting for all other variables, with the strongest variable, attitude toward marital status, producing $F(1, 418) = 6.51$, $.05 > p > .01$.

Three predictors have loadings in excess of .45 in the second discriminant function, which distinguishes between role-dissatisfied housewives and the other two groups. The strongest contribution comes from attitude toward homemaking. Role-dissatisfied housewives are less satisfied with homemaking (mean attitude toward housework = 24.75) than either of the other two groups, means for which have previously been cited. Role-dissatisfied housewives are more likely to attribute control to external sources (mean locus of control = 7.01) than either role-satisfied housewives (mean locus of control = 6.68) or working women (mean = 6.71). They are also less satisfied with current marital status (mean attitude toward marital status = 23.89) than either of the other two groups, means for which have previously been cited.

After adjustment of Type I error rate and adjustment of variables for overlap with each other, only attitude toward housework significantly discriminates role-dissatisfied housewives, $F(1, 418) = 7.18$. The squared semipartial correlation between the grouping of role-dissatisfied housewives versus the other two groups and attitude toward housework is .02. [Equation 9.9 applied to output in Table 9.16.]

Pooled within-group correlations among the four predictors are shown in Table 9.13. Of the six correlations, four would show statistical significance at $\alpha = .01$ if tested individually. There is a small positive relationship between locus of control and attitude toward marital status, with $r(422) = .19$, $p < .01$, indicating that women who are more satisfied with their current marital status are less likely to attribute control of reinforcements to external sources. Attitude toward homemaking is positively correlated with locus of control, $r(422) = .14$, $p < .01$, and attitude toward marital status, $r(422) = .26$, $p < .01$, and negatively correlated with attitude toward women's role, $r(422) = -.34$, $p < .01$. This indicates that women with negative attitudes toward homemaking are likely to attribute control to external sources, to be dissatisfied with their current marital status, and to have more liberal attitudes toward women's role.

With the use of a jackknifed classification procedure for the total usable sample of 424 women, 223 (52.6%) were classified correctly. This classification rate was achieved by classifying a disproportionate number of cases as working women. Although 53% of the women actually were employed, the classification scheme, using sample proportions as prior probabilities, classified 82.8% of the women as employed. This means that the working women were more likely to be correctly classified (86.1% correct classifications) than either the role-satisfied housewives (23.3% correct classifications) or the role-dissatisfied housewives (1.4% correct classifications).

The stability of the classification procedure was checked by a cross-validation run. Approximately 25% of the cases were withheld from calculation of the classification functions in this run. For the 75% of the cases from whom the functions were derived, there was a 53.9% correct classification rate. For the cross-validation cases, 51.5% were correctly classified. This indicates a high degree of consistency in the classification scheme.

9.8.3 Hierarchical Discriminant Function Analysis

Partial output of the hierarchical DISCRIM produced by BMDP7M, using the 10 demographic predictors (after dummy variable coding) followed by the 4 attitudinal predictors, appears in Table 9.18. In DISCRIM, variables are entered individually at each step, however order of entry can be specified. By assigning a lower entry level (1) to demographic variables than to attitudinal variables (2), and then forcing both entry levels, all variables are forced into the analysis, with the demographic variables entering first. (Note that with SPSS DISCRIMINANT, variables with the higher entry level enter first.) The two variables that were dummy variable coded, MARITAL and RELIGION, are ignored by assigning them an entry level of 0.

TABLE 9.18 DECK SETUP AND PARTIAL OUTPUT FROM BMDP7M
HIERARCHICAL DISCRIMINANT FUNCTION ANALYSIS OF
10 DEMOGRAPHIC AND 4 ATTITUDINAL VARIABLES

```
PROGRAM CONTROL INFORMATION

        /PROBLEM      TITLE IS 'LARGE SAMPLE DISCRIM EXAMPLE, HIERARCHICAL'.
        /INPUT        VARIABLES ARE 12.
                      FORMAT IS '(4X,10F4.0,4X,2F4.0)'.
                      CASE = 424.   UNIT=90.
        /VARIABLES    NAMES ARE WORKSTAT,MARITAL,CHILDREN,RELIGION,RACE,IESCALE,ATTSTAT,
                      WOMENROL,DUN,ATTHOUSE,AGE,EDUC,NONOTHER,CATHOLIC,PROTEST,
                      SINGLE,MARRIED.
                      ADD = 5.
                      GROUPING IS WORKSTAT.
        /TRANSFORM    NONOTHER=RELIGION EQ 1.  CATHOLIC=RELIGION EQ 2.
                      PROTEST=RELIGION EQ 3.  SINGLE=MARITAL EQ 1.
                      MARRIED=MARITAL EQ 2.
        /GROUP        CODES(1) ARE 1 TO 3.
                      NAMES(1) ARE WORKING, HAPHOUSE, UNHOUSE.
                      PRIOR = 0.53, 0.29, 0.18.
        /DISC         LEVEL = 0,0,1,0,1,2,2,2,1,2,1,1,1,1,1,1.
                      FORCE = 2.  TOL = 0.001.
        /PRINT        WITHIN.   CLASS = 10,14.
        /PLOT         GROUP = 1.  GROUP = 2.  GROUP = 3.
        /END
```

```
MEANS
         GROUP =   WORKING      HAPHOUSE      UNHOUSE      ALL GPS.
VARIABLE
  1 WORKSTAT     1.00000      2.00000      3.00000      1.64387
  2 MARITAL      2.12556      2.00000      2.09722      2.08255
  3 CHILDREN      .78027       .92248       .88889       .84198
  4 RELIGION     2.56502      2.72093      2.62500      2.62264
  5 RACE         1.10314      1.02326      1.11111      1.08019
  6 IESCALE      6.71300      6.68217      7.01309      6.75472
  7 ATTSTAT     23.03587     20.47287     23.88889     22.40094
  8 WOMENROL    33.78924     37.20155     35.87500     35.18160
  9 DUN         51.73094     56.58140     49.70833     52.86321
 10 ATTHOUSE    23.70852     22.57364     24.75000     23.54009
 11 AGE          4.35426      4.52713      4.11111      4.36557
 12 EDUC        13.60090     13.06977     12.61111     13.27123
 13 NONOTHER      .17937       .16279       .09722       .16038
 14 CATHOLIC      .25561       .20930       .36111       .25943
 15 PROTEST       .38565       .37209       .36111       .37736
 16 SINGLE        .08072       .01550       .02778       .05189
 17 MARRIED       .71300       .96899       .84722       .81368
COUNTS         223.         129.          72.         424.
```

TABLE 9.18 *(Continued)*

```
           STANDARD DEVIATIONS
              GROUP =  WORKING    HAPHOUSE    UNHOUSE   ALL GPS.
        VARIABLE
         1 WORKSTAT   0.00000    0.00000    0.00000    0.00000
         2 MARITAL     .52197     .17678     .38124     .42152
         3 CHILDREN    .41500     .26846     .31648     .36002
         4 RELIGION    .98367    1.02299     .89502     .98158
         5 RACE        .30482     .15130     .31648     .26990
         6 IESCALE    1.24414    1.31092    1.20437    1.25831
         7 ATTSTAT    8.09452    6.42149    8.44127    7.68794
         8 WOMENROL   6.65328    6.38526    5.62424    6.40882
         9 DUN       22.62342   25.01396   23.84187   23.57946
        10 ATTHOUSE   4.40007    3.87051    3.92446    4.16669
        11 AGE        2.14666    2.09940    2.34103    2.16662
        12 EDUC       2.46002    1.95331    1.74061    2.20503
        13 NONOTHER    .38453     .37061     .29834     .36707
        14 CATHOLIC    .43718     .40840     .48369     .43698
        15 PROTEST     .48704     .48525     .48369     .48636
        16 SINGLE      .27301     .12403     .16549     .22045
        17 MARRIED     .45338     .17401     .36230     .37381
```

```
STEP NUMBER   0
   VARIABLE       F TO  FORCE  TOLERANCE  *  VARIABLE      F TO   FORCE  TOLERANCE
               REMOVE  LEVEL              *                ENTER  LEVEL
               DF=  2  422                *              DF=  2   421
                                          *   2 MARITAL    3.678    0    1.000000
                                          *   3 CHILDREN   7.112    1    1.000000
                                          *   4 RELIGION   1.031    0    1.000000
                                          *   5 RACE       4.149    1    1.000000
                                          *   6 IESCALE    1.864    2    1.000000
                                          *   7 ATTSTAT    6.166    2    1.000000
                                          *   8 WOMENROL  12.092    2    1.000000
                                          *   9 DUN        2.505    1    1.000000
                                          *  10 ATTHOUSE   6.688    2    1.000000
                                          *  11 AGE         .858    1    1.000000
                                          *  12 EDUC       6.257    1    1.000000
                                          *  13 NONOTHER   1.367    1    1.000000
                                          *  14 CATHOLIC   2.807    1    1.000000
                                          *  15 PROTEST     .980    1    1.000000
                                          *  16 SINGLE     4.094    1    1.000000
                                          *  17 MARRIED   19.512    1    1.000000
```

TABLE 9.18   (*Continued*)

```
STEP NUMBER 10
VARIABLE ENTERED 13 NONOTHER
    VARIABLE      F TO  FORCE  TOLERANCE  *   VARIABLE     F TO  FORCE  TOLERANCE
                REMOVE LEVEL              *                ENTER LEVEL
            DF= 2  412                    *           DF= 2  411
     3 CHILDREN   3.527   1   .666152     *   2 MARITAL   0.000   0    .000000
     5 RACE       5.008   1   .874110     *   4 RELIGION  0.000   0    .000000
     9 DUN        1.558   1   .816737     *   6 IESCALE   1.243   2    .949470
    11 AGE        1.295   1   .819503     *   7 ATTSTAT   2.424   2    .811931
    12 EDUC       6.966   1   .008877     *   8 WOMENROL 13.492   2    .752779
    13 NONOTHER    .339   1   .644647     *  10 ATTHOUSE  8.581   2    .912750
    14 CATHOLIC   1.288   1   .546897     *
    15 PROTEST     .456   1   .543089     *
    16 SINGLE      .682   1   .612524     *
    17 MARRIED   11.130   1   .731200     *
U-STATISTIC OR WILKS' LAMBDA  .8339961   DEGREES OF FREEDOM  10   2    421
APPROXIMATE F-STATISTIC        3.914     DEGREES OF FREEDOM  20.00  824.00


    F - MATRIX        DEGREES OF FREEDOM =  10  412
                 WORKING  HAPHOUSE
HAPHOUSE    6.54
UNHOUSE     2.41     1.86
CLASSIFICATION FUNCTIONS
        GROUP =  WORKING     HAPHOUSE    UNHOUSE
VARIABLE
     3 CHILDREN    5.09006     6.09374     5.91674
     5 RACE       21.91370    20.42313    21.27392
     9 DUN          .00839      .01833      .01248
    11 AGE          .91628      .87354      .80168
    12 EDUC        3.47295     3.28730     3.26473
    13 NONOTHER    5.90580     5.81808     5.51358
    14 CATHOLIC    5.00773     4.57361     5.29013
    15 PROTEST     7.38006     7.06928     7.29825
    16 SINGLE     11.68057    12.49152    11.85645
    17 MARRIED     7.24586     8.97627     7.89906
CONSTANT         -46.18635   -45.18868   -44.75893
CLASSIFICATION MATRIX
GROUP      PERCENT    NUMBER OF CASES CLASSIFIED INTO GROUP -
           CORRECT
                     WORKING  HAPHOUSE UNHOUSE
WORKING     80.3       179      37        7
HAPHOUSE    38.8        78      50        1
UNHOUSE      0.0        50      22        0
TOTAL       54.0       307     109        8

JACKKNIFED CLASSIFICATION
GROUP      PERCENT    NUMBER OF CASES CLASSIFIED INTO GROUP -
           CORRECT
                     WORKING  HAPHOUSE UNHOUSE
WORKING     77.6       173      43        7
HAPHOUSE    33.3        84      43        2
UNHOUSE      0.0        49      23        0
TOTAL       50.9       306     109        9
```

TABLE 9.18 (*Continued*)

```
STEP NUMBER  14
VARIABLE ENTERED    7 ATTSTAT
   VARIABLE      F TO  FORCE  TOLERANCE  *     VARIABLE      F TO  FORCE  TOLERANCE
                 REMOVE LEVEL            *                  ENTER LEVEL
             DF=  2  408                 *              DF=  2  407
    3 CHILDREN   3.298   1   .653302     *     2 MARITAL  0.000   0    .000000
    5 RACE       7.184   1   .852994     *     4 RELIGION 0.000   0    .000000
    6 IESCALE     .528   2   .898981     *
    7 ATTSTAT     .330   2   .715234     *
    8 WOMENROL  11.233   2   .693285     *
    9 DUN        2.654   1   .772066     *
   10 ATTHOUSE   4.724   2   .766636     *
   11 AGE        2.980   1   .745363     *
   12 EDUC       2.860   1   .726769     *
   13 NONOTHER    .210   1   .629578     *
   14 CATHOLIC   2.914   1   .516727     *
   15 PROTEST    1.686   1   .511764     *
   16 SINGLE      .309   1   .605313     *
   17 MARRIED    7.423   1   .619959     *
U-STATISTIC OR WILKS' LAMBDA    .7548976   DEGREES OF FREEDOM  14   2    421
APPROXIMATE F-STATISTIC         4.399      DEGREES OF FREEDOM  28.00  816.00
F - MATRIX      DEGREES OF FREEDOM =   14  408
              WORKING   HAPHOUSE
HAPHOUSE   6.97
UNHOUSE    2.69     2.80
CLASSIFICATION FUNCTIONS
         GROUP =   WORKING     HAPHOUSE     UNHOUSE
VARIABLE
    3 CHILDREN   1.41949      2.48834      2.08444
    5 RACE      17.06828     15.20182     16.24563
    6 IESCALE    4.15696      4.17283      4.27855
    7 ATTSTAT     .35875       .34950       .36890
    8 WOMENROL   1.60645      1.71149      1.67440
    9 DUN         .09162       .10525       .09929
   10 ATTHOUSE   1.76675      1.73438      1.86201
   11 AGE         .29919       .15763       .15976
   12 EDUC       4.54136      4.44254      4.36990
   13 NONOTHER   9.20652      8.99254      8.93174
   14 CATHOLIC   8.75238      8.06983      9.21798
   15 PROTEST    9.62463      9.00511      9.62845
   16 SINGLE     6.53625      7.07231      6.52021
   17 MARRIED   10.51745     12.12032     11.27240
CONSTANT      -118.48994   -120.63836   -123.20797

CLASSIFICATION MATRIX
GROUP     PERCENT    NUMBER OF CASES CLASSIFIED INTO GROUP -
          CORRECT
                     WORKING  HAPHOUSE  UNHOUSE
WORKING    81.2      181       35        7
HAPHOUSE   52.7       58       68        3
UNHOUSE    12.5       47       16        9
TOTAL      60.8      286      119       19
JACKKNIFED CLASSIFICATION
GROUP     PERCENT    NUMBER OF CASES CLASSIFIED INTO GROUP -
          CORRECT
                     WORKING  HAPHOUSE  UNHOUSE
WORKING    78.5      175       38       10
HAPHOUSE   46.5       65       60        4
UNHOUSE     8.3       50       16        6
TOTAL      56.8      290      114       20
```

TABLE 9.18 *(Continued)*

SUMMARY TABLE

STEP NUMBER	VARIABLE ENTERED	REMOVED	F VALUE TO ENTER OR REMOVE	NUMBER OF VARIABLES INCLUDED	U-STATISTIC	APPROXIMATE F-STATISTIC	DEGREES OF FREEDOM	
1	17 MARRIED		19.5120	1	.9152	19.512	2.00	421.00
2	12 EDUC		5.3006	2	.8926	12.270	4.00	840.00
3	5 RACE		5.6554	3	.8692	10.143	6.00	838.00
4	3 CHILDREN		2.0065	4	.8609	8.126	8.00	836.00
5	14 CATHOLIC		1.9438	5	.8530	6.903	10.00	834.00
6	9 DUN		1.5049	6	.8468	6.010	12.00	832.00
7	11 AGE		1.3939	7	.8412	5.355	14.00	830.00
8	16 SINGLE		.7628	8	.8381	4.778	16.00	828.00
9	15 PROTEST		.6739	9	.8354	4.319	18.00	826.00
10	13 NONOTHER		.3386	10	.8340	3.914	20.00	824.00
11	8 WOMENROL		13.4915	11	.7826	4.872	22.00	822.00
12	10 ATTHOUSE		6.4991	12	.7586	5.862	24.00	820.00
13	6 IESCALE		.6623	13	.7561	4.720	26.00	818.00
14	7 ATTSTAT		.3298	14	.7549	4.399	28.00	816.00

EIGENVALUES
    .24039   .06796
CUMULATIVE PROPORTION OF TOTAL DISPERSION
    .77961   1.00000
CANONICAL CORRELATIONS
    .44023   .25226

VARIABLE	COEFFICIENTS FOR CANONICAL VARIABLES	
3 CHILDREN	-.97437	-.24711
5 RACE	1.68036	-.05618
6 IESCALE	-.02143	-.16462
7 ATTSTAT	.00744	-.02077
8 WOMENROL	-.09591	-.02803
9 DUN	-.01237	-.00198
10 ATTHOUSE	.02231	-.15877
11 AGE	.13225	.10673
12 EDUC	.09694	.18130
13 NONOTHER	.20390	.25355
14 CATHOLIC	.56672	-1.12471
15 PROTEST	.54049	-.41753
16 SINGLE	-.46687	.37963
17 MARRIED	-1.44608	-.02159
CONSTANT	1.43996	4.20870

GROUP	CANONICAL VARIABLES EVALUATED AT GROUP MEANS	
WORKING	.42440	.09950
HAPHOUSE	-.68475	.14750
UNHOUSE	-.08763	-.57246

Although all 14 steps are produced by this analysis, the only ones reproduced in Table 9.18 are those that include entry of all demographic variables (step 10), and entry of all demographic plus attitudinal variables (step 14—last step).

Classification equations and results of classification with and without jackknifing are also shown in Table 9.18. These were requested only for steps 10 and 14. (Jackknifed classification is discussed in Section 9.6.6.3.)

Figure 9.4 shows the centroids for the three groups plotted on the two discriminant functions, based on information from BMDP7M (CANONICAL VARIABLES EVALUATED AT GROUP MEANS).

Information valuable for interpreting discriminant functions and evaluating variables, as produced by SPSS DISCRIMINANT (and discussed in Sections 9.6.3 and 9.6.4) is shown in Table 9.19. Finally, in a form that might be suitable for publication, Table 9.20 (page 360) summarizes information from Tables 9.18 and 9.19; and pooled within-group correlations among predictors (produced by SPSS DISCRIMINANT) are given in Table 9.21 (page 361).

Results of contrasts of each group with the others are shown in Tables 9.22, 9.23, and 9.24 (pages 362–364). Critical F TO REMOVE for evaluating the unique contribution to prediction by variables at $\alpha_i = .003$ (cf. Eq. 9.8) with 1 and 408 df is approximately 9.5.

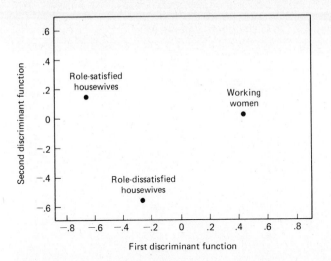

Figure 9.4 Plot of 3 group centroids on 2 discriminant functions derived from 10 demographic and 4 attitudinal variables.

TABLE 9.19   PARTIAL DECK SETUP AND OUTPUT FROM SPSS
            DISCRIMINANT FOR PREDICTION OF MEMBERSHIP IN 3
            GROUPS ON THE BASIS OF 10 DEMOGRAPHIC AND
            4 ATTITUDINAL VARIABLES

```
DISCRIMINANT   GROUPS=WORKSTAT(1,3)/
               VARIABLES=CHILDREN,RACE TO ATTHOUSE,AGE,EDUC,NONOTHER TO MARRIED/
               ANALYSIS=CHILDREN TO MARRIED/
STATISTICS     10
```

CANONICAL DISCRIMINANT FUNCTIONS

FUNCTION	EIGENVALUE	PERCENT OF VARIANCE	CUMULATIVE PERCENT	CANONICAL CORRELATION	-	AFTER FUNCTION	WILKS LAMBDA	CHI-SQUARED	D.F.	SIGNICANCE
					-	0	.7548976	116.55	28	.0000
1*	.24039	77.96	77.96	.4402290	-	1	.9363669	27.252	13	.0115
2*	.06796	22.04	100.00	.2522559	-					

```
* MARKS THE   2 FUNCTION(S) TO BE USED IN THE REMAINING ANALYSIS.
```

POOLED WITHIN-GROUPS CORRELATIONS BETWEEN CANONICAL DISCRIMINANT FUNCTIONS AND DISCRIMINATING VARIABLES
VARIABLES ARE ORDERED BY THE FUNCTION WITH LARGEST CORRELATION AND THE MAGNITUDE OF THAT CORRELATION.

	FUNC 1	FUNC 2
MARRIED	.62009*	-.06183
WOMENROL	.48501*	-.11475
CHILDREN	.36357*	-.17194
SINGLE	-.27316*	.14922
RACE	-.25648*	-.23939
ATTHOUSE	-.22214	-.54124*
EDUC	-.23822	.48648*
CATHOLIC	-.07743	-.41831*
ATTSTAT	-.28406	-.38157*
IESCALE	-.00648	-.36078*
NONOTHER	-.05338	.29237*
DUN	.17420	.26039*
AGE	.06264	.21474*
PROTEST	-.02748	.05415*

TABLE 9.20   RESULTS OF DISCRIMINANT FUNCTION ANALYSIS OF
DEMOGRAPHIC PLUS ATTITUDINAL VARIABLES

Predictor variable	Correlations of predictor variables with discriminant functions		Univariate $F(2, 421)$
	1	2	
Demographic			
CHILDREN	.36	−.17	7.11
RACE	−.26	−.24	4.15
DUN	.17	.26	2.51
AGE	.06	.21	0.86
EDUC	−.24	.49	6.26
NONOTHER	−.05	.29	1.37
CATHOLIC	−.08	−.42	2.81
PROTEST	−.03	.05	0.08
SINGLE	−.27	.15	4.09
MARRIED	.62	−.06	19.51
Attitudinal			
IESCALE	−.01	−.36	1.86
ATTSTAT	−.28	−.38	6.17
WOMENROL	.49	−.11	12.09
ATTHOUSE	−.22	−.54	6.69
Canonical R	.44	.25	
Eigenvalue	.240	.068	

A direct comparison of the two sets of case classifications was possible by comparing output for 10 steps (demography alone) versus output for 14 steps (demography plus attitudes). A portion of the output used for tabulation of the contingency table for McNemar's change test (Eq. 9.10) appears in Table 9.25 (page 365). Note that cases 25 and 40 are among those tabulated in cell B, incorrectly classified at step 10 but correctly classified at step 14. Full comparison of the two runs results in the accompanying contingency table:

		Step 10 classification	
		Correct	Incorrect
Step 14 classification	Correct	(A)	B = 59
	Incorrect	C = 31	(D)

Applying Eq. 9.10, then, we obtain

$$\chi^2 = \frac{(|59 - 31| - 1)^2}{59 + 31} = 8.10$$

TABLE 9.21 POOLED WITHIN-GROUP CORRELATIONS AMONG PREDICTORS FOR DATA OF TABLE 9.20

	Demographic variables									Attitudinal variables			
	RACE	DUN	AGE	EDUC	NON-OTHER	CATH-OLIC	PRO-TEST	SIN-GLE	MAR-RIED	IESCALE	ATTSTAT	WOMENROL	ATTHOUSE
Demographic													
CHILDREN	.13	−.06	.35	−.08	−.05	.11	−.03	−.50	.31	−.02	−.04	.12	.03
RACE		−.13	−.07	−.24	−.03	.25	−.12	−.08	.05	.07	.00	.20	−.08
DUN			.11	.36	.03	−.14	−.02	−.02	.11	−.12	.00	−.23	.01
AGE				.02	−.09	−.02	.10	−.27	.09	−.12	−.05	.25	−.10
EDUC					.09	−.14	−.04	.14	−.09	−.07	−.07	−.35	.15
NONOTHER						−.25	−.34	.14	−.17	−.04	.00	−.09	.08
CATHOLIC							−.46	−.09	.09	.02	.02	.10	−.14
PROTEST								−.03	−.01	−.11	−.04	.13	−.07
SINGLE									−.50	.05	.19	−.06	.02
MARRIED										−.01	−.39	.01	−.02
Attitudinal													
IESCALE											.19	−.02	.14
ATTSTAT												−.06	.26
WOMENROL													−.34

TABLE 9.22 INPUT AND PARTIAL OUTPUT OF BMDP7M CONTRASTING THE WORKING GROUP WITH THE OTHER TWO GROUPS FOR DEMOGRAPHIC AND ATTITUDINAL VARIABLES

```
PROGRAM CONTROL INFORMATION

        /PROBLEM     TITLE IS "LARGE SAMPLE DISCRIM EXAMPLE, HIERARCHICAL".
        /INPUT       VARIABLES ARE 12.
                     FORMAT IS "(4X,10F4.0,4X,2F4.0)".
                     CASE = 424.   UNIT=90.
        /VARIABLES   NAMES ARE WORKSTAT,MARITAL,CHILDREN,RELIGION,RACE,IESCALE,ATTSTAT,
                       WOMENROL,DUN,ATTHOUSE,AGE,EDUC,NONOTHER,CATHOLIC,PROTEST,
                       SINGLE,MARRIED.
                     ADD = 5.
                     GROUPING IS WORKSTAT.
        /TRANSFORM   NONOTHER=RELIGION EQ 1.  CATHOLIC=RELIGION EQ 2.
                     PROTEST=RELIGION EQ 3.  SINGLE=MARITAL EQ 1.
                     MARRIED=MARITAL EQ 2.
        /GROUP       CODES(1) ARE 1 TO 3.
                     NAMES(1) ARE WORKING, HAPHOUSE, UNHOUSE.
                     PRIOR = 0.53, 0.29, 0.18.
        /DISC        LEVEL = 0,0,1,0,1,2,2,2,1,2,1,1,1,1,1,1,1.
                     FORCE = 2.  TOL = 0.001.
                     CONTRAST=-2,1,1.
        /PRINT       NO STEP.    NO POST.
        /PLOT        NO CANON.   CONTRAST.
        /END


*************************************************************************************************
STEP NUMBER   0
   VARIABLE        F TO  FORCE  TOLERANCE  *    VARIABLE        F TO  FORCE   TOLERANCE
                   REMOVE LEVEL             *                   ENTER LEVEL
                 DF=  1  422                *               DF=  1   421
                                            *     2 MARITAL    3.368    0     1.000000
                                            *     3 CHILDREN  12.264    1     1.000000
                                            *     4 RELIGION   1.222    0     1.000000
                                            *     5 RACE       1.794    1     1.000000
                                            *     6 IESCALE    1.164    2     1.000000
                                            *     7 ATTSTAT    1.250    2     1.000000
                                            *     8 WOMENROL  18.596    2     1.000000
                                            *     9 DUN         .363    1     1.000000
                                            *    10 ATTHOUSE    .013    2     1.000000
                                            *    11 AGE         .027    1     1.000000
                                            *    12 EDUC      12.021    1     1.000000
                                            *    13 NONOTHER   1.828    1     1.000000
                                            *    14 CATHOLIC    .464    1     1.000000
                                            *    15 PROTEST     .155    1     1.000000
                                            *    16 SINGLE     7.258    1     1.000000
                                            *    17 MARRIED   27.532    1     1.000000

STEP NUMBER  14
VARIABLE ENTERED    7 ATTSTAT
   VARIABLE        F TO  FORCE  TOLERANCE  *    VARIABLE        F TO  FORCE   TOLERANCE
                   REMOVE LEVEL             *                   ENTER LEVEL
                 DF=  1  408                *               DF=  1   407
    3 CHILDREN    5.347    1    .653302     *     2 MARITAL    0.000    0     .000000
    5 RACE        9.533    1    .852994     *     4 RELIGION   0.000    0     .000000
    6 IESCALE      .558    2    .898981     *
    7 ATTSTAT      .001    2    .715234     *
    8 WOMENROL   18.464    2    .693285     *
    9 DUN         4.077    1    .772066     *
   10 ATTHOUSE    1.094    2    .766636     *
   11 AGE         5.809    1    .745363     *
   12 EDUC        5.423    1    .726769     *
   13 NONOTHER     .421    1    .629578     *
   14 CATHOLIC     .096    1    .516727     *
   15 PROTEST      .954    1    .511764     *
   16 SINGLE       .165    1    .605313     *
   17 MARRIED    10.235    1    .619959     *


EIGENVALUES
              .18110
CUMULATIVE PROPORTION OF TOTAL DISPERSION
              1.00000
CANONICAL CORRELATIONS
              .39158
```

```
PROGRAM CONTROL INFORMATION

        /PROBLEM      TITLE IS "LARGE SAMPLE DISCRIM EXAMPLE, HIERARCHICAL".
        /INPUT        VARIABLES ARE 12.
                      FORMAT IS "(4X,10F4.0,4X,2F4.0)".
                      CASE = 424.   UNIT=90.
        /VARIABLES    NAMES ARE WORKSTAT,MARITAL,CHILDREN,RELIGION,RACE,IESCALE,ATTSTAT,
                      WOMENROL,DUN,ATTHOUSE,AGE,EDUC,NONOTHER,CATHOLIC,PROTEST,
                      SINGLE,MARRIED.
                      ADD = 5.
                      GROUPING IS WORKSTAT.
        /TRANSFORM    NONOTHER=RELIGION EQ 1.  CATHOLIC=RELIGION EQ 2.
                      PROTEST=RELIGION EQ 3.  SINGLE=MARITAL EQ 1.
                      MARRIED=MARITAL EQ 2.
        /GROUP        CODES(1) ARE 1 TO 3.
                      NAMES(1) ARE WORKING, HAPHOUSE, UNHOUSE.
                      PRIOR = 0.53, 0.29, 0.18.
        /DISC         LEVEL = 0,0,1.0,1,2,2,2,1,2,1,1,1,1,1,1,1.
                      FORCE = 2.  TOL = 0.001.
                      CONTRAST=1,-2,1.
        /PRINT        NO STEP.     NO POST.
        /PLOT         NO CANON.    CONTRAST.
        /END
```

`****************************************************************************************************`

```
STEP NUMBER   0
  VARIABLE      F TO  FORCE  TOLERANCE  *   VARIABLE      F TO  FORCE  TOLERANCE
                REMOVE LEVEL            *                 ENTER LEVEL
          DF=  1  422                   *           DF=  1  421
                                        *    2 MARITAL    5.657   0   1.000000
                                        *    3 CHILDREN   4.829   1   1.000000
                                        *    4 RELIGION   1.333   0   1.000000
                                        *    5 RACE       7.821   1   1.000000
                                        *    6 IESCALE    1.681   2   1.000000
                                        *    7 ATTSTAT   12.248   2   1.000000
                                        *    8 WOMENROL  11.072   2   1.000000
                                        *    9 DUN        5.006   1   1.000000
                                        *   10 ATTHOUSE  12.789   2   1.000000
                                        *   11 AGE        1.496   1   1.000000
                                        *   12 EDUC        .022   1   1.000000
                                        *   13 NONOTHER    .361   1   1.000000
                                        *   14 CATHOLIC   4.162   1   1.000000
                                        *   15 PROTEST     .001   1   1.000000
                                        *   16 SINGLE     2.502   1   1.000000
                                        *   17 MARRIED   20.681   1   1.000000

STEP NUMBER  14
VARIABLE ENTERED   6 IESCALE
  VARIABLE      F TO  FORCE  TOLERANCE  *   VARIABLE      F TO  FORCE  TOLERANCE
                REMOVE LEVEL            *                 ENTER LEVEL
          DF=  1  408                   *           DF=  1  407
   3 CHILDREN   3.109   1   .653302     *    2 MARITAL    0.000   0   .000006
   5 RACE       9.036   1   .852994     *    4 RELIGION   0.000   0   .000000
   6 IESCALE     .193   2   .898981     *
   7 ATTSTAT     .584   2   .715234     *
   8 WOMENROL   9.897   2   .693285     *
   9 DUN        2.786   1   .772066     *
  10 ATTHOUSE   5.804   2   .766636     *
  11 AGE        1.217   1   .745363     *
  12 EDUC        .041   1   .726769     *
  13 NONOTHER    .033   1   .629578     *
  14 CATHOLIC   5.631   1   .516727     *
  15 PROTEST    3.165   1   .511764     *
  16 SINGLE      .586   1   .605313     *
  17 MARRIED    8.931   1   .619959     *


EIGENVALUES
                 .16841
CUMULATIVE PROPORTION OF TOTAL DISPERSION
                1.00000
CANONICAL CORRELATIONS
                 .37965
```

TABLE 9.24 INPUT AND PARTIAL OUTPUT OF BMDP7M CONTRASTING
THE UNHOUSE GROUP WITH THE OTHER TWO GROUPS
FOR DEMOGRAPHIC AND ATTITUDINAL VARIABLES

```
PROGRAM CONTROL INFORMATION

        /PROBLEM      TITLE IS "LARGE SAMPLE DISCRIM EXAMPLE, HIERARCHICAL".
        /INPUT        VARIABLES ARE 12.
                      FORMAT IS "(4X,10F4.0,4X,2F4.0)".
                      CASE = 424.   UNIT=90.
        /VARIABLES    NAMES ARE WORKSTAT,MARITAL,CHILDREN,RELIGION,RACE,IESCALE,ATTSTAT,
                      WOMENROL,DUN,ATTHOUSE,AGE,EDUC,NONOTHER,CATHOLIC,PROTEST,
                      SINGLE,MARRIED.
                      ADD = 5.
                      GROUPING IS WORKSTAT.
        /TRANSFORM    NONOTHER=RELIGION EQ 1.  CATHOLIC=RELIGION EQ 2.
                      PROTEST=RELIGION EQ 3.  SINGLE=MARITAL EQ 1.
                      MARRIED=MARITAL EQ 2.
        /GROUP        CODES(1) ARE 1 TO 3.
                      NAMES(1) ARE WORKING, HAPHOUSE, UNHOUSE.
                      PRIOR = 0.53, 0.29, 0.18.
        /DISC         LEVEL = 0,0,1,0,1,2,2,2,1,2,1,1,1,1,1,1,1.
                      FORCE = 2. TOL = 0.001.
                      CONTRAST=1,1,-2.
        /PRINT        NO STEP.    NO POST.
        /PLOT         NO CANON.  CONTRAST.
        /END
```

```
*********************************************************************************

STEP NUMBER   0
  VARIABLE         F TO  FORCE  TOLERANCE  *    VARIABLE         F TO  FORCE  TOLERANCE
                  REMOVE LEVEL             *                    ENTER LEVEL
                  DF=  1  422              *              DF=  1  421
                                           *    2 MARITAL     .394    0   1.000000
                                           *    3 CHILDREN    .641    1   1.000000
                                           *    4 RELIGION    .020    0   1.000000
                                           *    5 RACE       1.859    1   1.000000
                                           *    6 IESCALE    3.728    2   1.000000
                                           *    7 ATTSTAT    4.548    2   1.000000
                                           *    8 WOMENROL    .207    2   1.000000
                                           *    9 DUN        2.099    1   1.000000
                                           *   10 ATTHOUSE   8.798    2   1.000000
                                           *   11 AGE        1.365    1   1.000000
                                           *   12 EDUC       6.365    1   1.000000
                                           *   13 NONOTHER   2.389    1   1.000000
                                           *   14 CATHOLIC   5.115    1   1.000000
                                           *   15 PROTEST     .079    1   1.000000
                                           *   16 SINGLE      .502    1   1.000000
                                           *   17 MARRIED     .016    1   1.000000

STEP NUMBER  14
VARIABLE ENTERED   8 WOMENROL
  VARIABLE         F TO  FORCE  TOLERANCE  *    VARIABLE         F TO  FORCE  TOLERANCE
                  REMOVE LEVEL             *                    ENTER LEVEL
                  DF=  1  408              *              DF=  1  407
   3 CHILDREN     .077    1   .653302      *    2 MARITAL    0.000    0   .000000
   5 RACE         .041    1   .852994      *    4 RELIGION   0.000    0   .000000
   6 IESCALE      .987    2   .898981      *
   7 ATTSTAT      .494    2   .715234      *
   8 WOMENROL     .363    2   .693285      *
   9 DUN          .017    1   .772066      *
  10 ATTHOUSE    9.046    2   .766636      *
  11 AGE          .885    1   .745363      *
  12 EDUC        2.838    1   .726769      *
  13 NONOTHER     .128    1   .629578      *
  14 CATHOLIC    3.468    1   .516727      *
  15 PROTEST      .638    1   .511764      *
  16 SINGLE       .127    1   .605313      *
  17 MARRIED      .010    1   .619959      *


EIGENVALUES
              .06814
CUMULATIVE PROPORTION OF TOTAL DISPERSION
              1.00000
CANONICAL CORRELATIONS
              .25257
```

TABLE 9.25 PARTIAL OUTPUT FROM BMDP7M FOR CLASSIFICATION OF CASES ON THE BASIS OF DEMOGRAPHY ALONE (TOP) AND DEMOGRAPHY PLUS ATTITUDES (BOTTOM)

GROUP	INCORRECT CLASSIFICATIONS WORKING	MAHALANOBIS D-SQUARE FROM AND POSTERIOR PROBABILITY FOR GROUP – WORKING		HAPHOUSE		UNHOUSE	
CASE							
2	HAPHOUSE	3.6	.380	2.2	.419	2.7	.200
3		6.7	.458	6.4	.281	5.6	.262
6		8.2	.477	8.2	.262	7.3	.261
15		13.4	.688	16.2	.093	13.5	.219
17		8.6	.705	11.0	.115	9.2	.180
19	UNHOUSE	8.1	.330	7.0	.306	5.7	.364
20		6.9	.406	6.5	.272	5.2	.322
21		9.7	.463	8.8	.392	9.8	.145
22		3.7	.500	3.1	.360	4.0	.140
24	HAPHOUSE	10.8	.311	8.4	.554	10.3	.134
25	HAPHOUSE	5.6	.345	3.8	.471	4.7	.184
28		12.2	.888	16.8	.050	15.4	.062
29		4.1	.586	4.3	.294	5.1	.120
30		8.9	.583	9.0	.317	10.3	.100
31		3.4	.526	3.2	.314	3.6	.160
34		13.4	.572	15.1	.138	12.6	.290
35		5.6	.427	4.5	.414	5.4	.159
36		6.3	.461	5.3	.413	6.8	.126
37		7.8	.676	8.7	.241	9.8	.084
38		5.4	.424	4.6	.338	4.4	.238
40	HAPHOUSE	8.7	.367	6.9	.484	8.3	.149
41		20.1	.901	24.4	.056	24.0	.044
42		9.2	.369	8.0	.367	7.7	.264
43		5.7	.381	4.6	.359	4.3	.260
44		14.9	.691	15.7	.259	18.0	.050

GROUP	INCORRECT CLASSIFICATIONS WORKING	MAHALANOBIS D-SQUARE FROM AND POSTERIOR PROBABILITY FOR GROUP – WORKING		HAPHOUSE		UNHOUSE	
CASE							
2	HAPHOUSE	5.6	.388	4.0	.482	5.6	.131
3		8.8	.366	7.7	.361	7.3	.274
6		13.5	.708	16.7	.080	13.8	.212
15		15.9	.801	19.3	.079	17.5	.120
17		18.8	.569	19.0	.279	19.3	.152
19	UNHOUSE	14.7	.244	13.0	.320	11.4	.436
20		11.9	.381	11.0	.327	10.3	.292
21		11.4	.483	11.3	.282	10.7	.234
22		4.0	.584	4.4	.266	4.6	.150
24	HAPHOUSE	15.1	.386	13.6	.446	14.6	.168
25		10.5	.507	11.2	.198	9.4	.295
28		16.1	.811	19.8	.068	17.7	.121
29		11.7	.544	12.5	.204	11.1	.253
30		13.9	.489	13.4	.333	13.7	.177
31		6.6	.748	8.7	.145	8.3	.107
34		19.1	.643	23.5	.038	18.3	.319
35		10.4	.574	10.1	.362	12.7	.064
36		8.5	.573	8.4	.323	9.7	.104
37		9.6	.604	9.5	.344	12.4	.052
38		10.0	.372	8.8	.363	8.5	.265
40		14.7	.544	15.3	.212	14.1	.243
41		22.2	.897	26.7	.051	25.7	.052
42		13.1	.465	12.5	.341	12.7	.193
43	HAPHOUSE	9.6	.326	7.9	.413	7.9	.261
44		15.1	.688	15.8	.266	18.4	.046

The obtained value of 8.10 exceeds the critical value of χ^2 with 1 df at the .99 level of confidence. Therefore, we conclude that inclusion of additional variables was appropriate.

Table 9.26 is a checklist for hierarchical discriminant function analysis.

TABLE 9.26 CHECKLIST FOR HIERARCHICAL DISCRIMINANT FUNCTION ANALYSIS

1. Issues
   a. Unequal sample sizes and missing data
   b. Normality
   c. Outliers
   d. Linearity
   e. Homogeneity of variance-covariance matrices
   f. Multicollinearity and singularity
2. Major analysis
   a. Significance of discrimination at each major step
   b. Significance of change at each major step
   c. Number of significant functions at last significant step. If any,
      (1) Variance accounted for
      (2) Plot(s) of discriminant functions
      (3) Loading matrix
   d. Variables separating each group
3. Additional analyses
   a. Group means for high-loading variables ($r \geq .45$)
   b. Pooled within-group correlations among predictor variables
   c. Classification results
      (1) Jackknifed classification
      (2) Cross-validation
   d. Change in Rao's V (or stepdown F) plus univariate F for predictors

An example of a Results section, in journal format, follows.

RESULTS

A hierarchical discriminant function analysis was performed to assess prediction of membership in the 3 groups first from the 10 demographic variables and then from the addition of the 4 attitudinal variables. Groups were working women, role-satisfied housewives, and role-dissatisfied housewives. Demographic variables were children (presence or absence), race, socioeconomic status, age, 3 dummy-coded variables for religion, and 2 dummy-coded variables for marital status. Attitudinal variables were locus of control, attitude toward marital status, attitude toward role of women, and attitude toward homemaking.

Of the original 465 cases, 28 were dropped from analysis owing to missing data. Missing data were randomly scattered over cases and variables, with no patterning on the basis of grouping or demographic variables. An additional 13 cases were identified as multivariate outliers with $p < .01$, and were also deleted. Among the deleted outliers, nonwhite women were disproportionately represented ($N = 4$).[13]

For the remaining 424 cases, evaluation of assumptions of linearity, normality, multicollinearity, and homogeneity of variance-covariance matrices revealed no threat to multivariate analysis.

There was statistically significant discrimination among the 3 groups on the basis of the 10 demographic variables alone, $F(20, 824) = 3.914$, $p < .01$. At this point, based on a jackknife procedure that reduces bias in classification, 50.9% of the women were correctly classified. After adding the 4 attitudinal variables, $F(28, 816) = 4.399$, $p < .01$, indicating again highly significant discriminating power. With all 14 predictor variables, the jackknife procedure correctly classified 56.8% of the women. McNemar's χ^2 test for change in proportion of correct classification with addition of attitudinal variables revealed statistical significance for the gain in classification performance, $\chi^2(1) = 8.10$, $p < .01$.

Sample sizes were used to estimate prior probabilities of group membership. The disproportionate classification of women as employed, however, far exceeded the 53% of the sample who actually were employed. On the basis of demographic variables alone, 72% of all the women were classified as employed. With all 14 variables as predictors, 68% were so classified. With the use of demographic predictors alone, rates of correct classification for the three groups were 77.6%, 33.3%, and 0% for working women, role-satisfied housewives, and role-dissatisfied housewives, respectively. After addition of attitudinal predictors, the correct classification rates increased to 78.5%, 46.5%, and 8.3%, respectively. The gain with attitudinal variables, then, was strongest in properly classifying the two groups of housewives.

On the basis of all 14 predictors, two discriminant functions were calculated, with a combined $\chi^2(28) = 116.55$, $p < .01$. After removal of the first function, statistically significant discriminating power remained, $\chi^2(13) = 27.25$, $p < .05$. The two discriminant functions accounted for 78% and 22%, respectively, of the between-group variability [χ^2 **and percent of variance are from Table 9.19; cf. Section 9.6.2**] in discriminating among groups. As shown in Figure 9.4, the first discriminant function maximally separates working women from role-satisfied housewives, with role-dissatisfied housewives falling between these two groups. The second discriminant function discriminates role-dissatisfied housewives from the other two groups.

Insert Figure 9.4 about here

A loading matrix of correlations between the 14 predictor variables and the two discriminant functions, as seen in Table 9.20, shows that for the first discriminant function, separating working women and role-satisfied housewives, the primary

[13] Among the nonwhite women, all worked outside the home and 2 were single, 1 with a child. The third was older and college educated. Among the 10 outliers deleted on the basis of attitudes, 4 were extremely dissatisfied with current marital status. Of these, 2 worked and 2 were role-dissatisfied housewives. Another 4 outliers, all working women, showed unusual combinations of attitudes toward role of women and toward housework. The final 2 outlying cases, both role-dissatisfied housewives, were characterized by near extreme values on 2 variables. Both expressed dissatisfaction with current marital status. In addition, 1 was highly negative toward housework and the other expressed very liberal attitudes toward the role of women.

predictors (using a loading of .45 or above) are whether or not the woman is married, and attitude toward women's role.

Insert Table 9.20 about here

The findings show that working women are less likely to be currently married (71.3%) than role-satisfied housewives (96.9%). Attitudes toward women's role are more conservative among role-satisfied housewives (mean attitude toward role of women = 37.20) than among working women (mean attitude = 33.79).

The variables separating working women from the other two groups, after all variables are adjusted for each other, and Type I error rate is adjusted for 14 variables, are race, $F(1, 408) = 9.533$, attitude toward women's role, $F(1, 408) = 18.464$, and whether or not the woman is currently married, $F(1, 408) = 10.235$. Squared semipartial correlations between the grouping variable (working women vs. housewives), and race, attitude toward women's role, and whether or not the woman is married are, respectively, 0.02, 0.04, and 0.02. [Equation 9.9 applied to results of Table 9.22.] In regard to race, the only variable not interpreted previously, working women are slightly less likely to be white than housewives, adjusting for the remaining variables.

Role-satisfied housewives are separated from the other two groups on the basis of attitudes toward women's role, $F(1, 408) = 9.897$, with squared semipartial correlation = .02. [Equation 9.9 applied to results of Table 9.23.]

For the second discriminant function, separating role-dissatisfied housewives from the other two groups, the primary predictors are attitude toward homemaking and years of education attained. Role-dissatisfied housewives have less positive attitudes toward homemaking (mean attitude toward housework = 24.75) than the other two groups of women (mean attitude = 23.71 for working women and 22.57 for role-satisfied housewives). Role-dissatisfied women have completed fewer years of education (mean = 12.61 years) than either working women (mean = 13.60 years) or role-satisfied housewives (mean = 13.07 years).

After adjusting all variables for one another and adjusting for inflated Type I error rate with 14 variables, none of the predictors significantly separates role-dissatisfied housewives from the two other groups. The strongest variable is attitude toward homemaking, $F(1, 408) = 9.046$, $.003 < p < .01$.

Many of the pooled within-group correlations, shown in Table 9.21, would reach statistical significance, at the .01 level (all $r > .128$), had such tests been appropriate. Since the emphasis here is not on demography alone, only those correlations involving attitudinal variables or both attitudinal and demographic variables will be discussed. Among attitudinal variables, four of the six within-group correlations show statistical significance, $p < .01$, when tested individually. There is a small positive relationship between locus of control and attitude toward marital status, $r(422) = .19$, indicating that women who are satisfied with their current marital status are less likely to attribute control to external sources. Attitude toward homemaking is positively correlated with locus of control, $r(422) = .14$, and attitude toward current marital status, $r(422) = .26$, and negatively correlated with attitude toward women's role, $r(422) = -.34$. This indicates that women with negative attitudes toward homemaking are likely to attribute control to external sources, to be dissatisfied with their current marital status, and to have more liberal attitudes toward women's role.

Insert Table 9.21 about here

Six of the within-group correlations between attitudinal and demographic variables would reach statistical significance, $p < .01$, tested individually. Attitude toward women's role is positively correlated with race, $r(422) = .20$. This indicates that women with conservative attitudes toward women's role are less likely to be white. Attitude toward women's role is negatively correlated with socioeconomic level, $r(422) = -.23$, and educational level, $r(422) = -.35$, and positively correlated with age, $r(422) = .25$ and being Protestant, $r(422) = .13$. Women with traditional attitudes towards women's role, then, are characterized by lower socioeconomic level, less education, greater age, and Protestant religious affiliation. Additionally, attitude toward homemaking is correlated with whether or not a woman is Catholic, $r(422) = -.14$, with Catholic women showing more positive attitudes toward homemaking than women claiming other religious affiliation. Finally, there is a significant correlation between education and attitude toward homemaking, $r(422) = .15$. The more education a woman has, the less satisfied she is likely to be with housework.

9.9 SOME EXAMPLES FROM THE LITERATURE

Curtis and Simpson (1977) explored predictors for three types of drug users: daily opioid users, less-than-daily opioid users, and nonusers of opioids. A preliminary discriminant function analysis used 31 variables related to demographic characteristics, alcohol use, and drug history to predict differential membership in the three groups. Both discriminant functions were statistically significant and virtually all the predictors showed significant univariate F's for group differences, not surprising with the sample size greater than 23,000. For the second analysis, only the 13 predictor variables with univariate $F > 100$ were selected.

The first discriminant function accounted for 20% of the variance between groups, with the second accounting for 1% of the variance. The major discriminant function distinguished between daily opioid users and the other two groups. (Presumably this interpretation was based on a comparison of group centroids, although they were not reported.) Correlations between predictor variables and discriminant functions (loadings) were used for interpreting the functions. These revealed that daily opioid users were most likely to be black, to be older, to have used illicit drugs for a longer period of time, and to be responsible for slightly more family members, and were more likely to have begun daily use of illicit drugs with a drug other than marijuana.

A series of two-group stepwise discriminant functions was run by Strober and Weinberg (1977) on purchasing versus nonpurchasing families for various types of expenditures. The hypothesis was that variables based on wife's employment status would not predict whether or not the family purchased items considered in the study. Each purchase item (e.g., time-saving durables such as dishwasher, hobby and recreation items, college education) was analyzed in a separate DISCRIM. Predictor variables, in addition to wife's employment status, were family income; net assets; life-cycle stage of the family; whether or not the family recently moved to a different home; and whether or not the family was in the market for the item being analyzed. In none of the analyses did variables related to wife's employment show significant F to enter the stepwise discriminant function analysis for decision to purchase or not purchase the item being analyzed. For the various expenditures, different predictor variables were important. For example, for college education, color TV, and washer, the variable entering the stepwise series first was life-cycle stage of the family. For purchases of furniture, the predictor entering first was net assets.

For those analyses in which sample size was large enough, part of the sample was reserved for cross-validation. Percentage correct prediction was reported for both the analysis and the validity samples, and over the various analyses ranged from 53% to 73%. In all except one analysis in which cross-validation was possible, the difference in correct prediction between analysis and validity samples was less than 2%. For hobby and recreation items, however, the analysis sample provided statistically significant percentage of correct classification, whereas the cross-validation sample did not.

Men involved in midcareer changes were studied by Wiener and Vaitenas (1977). Two personality inventories were used to distinguish midcareer changers from vocationally stable men. Separate discriminant function analyses were run on the two inventories. The first, the Edwards Personal Preference Schedule, consisted of 15 scales. The second analysis was based on the 8 scales of the Gordon Personal Profile and Gordon Personal Inventory. For each of the analyses, Wilks' Lambda showed statistically significant discrimination between groups, with the discriminant functions accounting for 28% and 42% of the variance between groups in the Edwards and Gordon analyses, respectively. Interpretation was based on standardized discriminant function coefficients (rather than the less biased correlations of predictors with discriminant functions). It was concluded that midcareer changers score lower on endurance, dominance, and order scales of the Edwards inventory, and show less responsibility and ascendancy as measured by the Gordon inventory.

Caffrey and Lile (1976) explored differences between psychology and physics students in attitudes toward science. Predictor variables were based on a scientific attitude questionnaire, in which the respondent expressed degree of agreement with 15 quotations from various writers on science. Stepwise discriminant function analysis on upper-division psychology versus physics majors followed preliminary analyses on a random sample of humanities, social science, and nature science majors.

Three items were found to be sufficient to distinguish between psychology and physics majors. Entering the stepwise analysis first was a statement suggesting that the results of scientific knowledge about behavior should be used to change behavior. Psychology majors were more likely to agree with this item than physics majors. In addition, physics majors were more likely to agree with a statement suggesting that humans are free agents, not amenable to scientific prediction (step 2), and with a statement suggesting that Shakespeare conveys more truth about human nature than results of questionnaires (step 3).

Distinction between menopausal and nonmenopausal women in Hawaii was studied by Goodman, Steward, and Gilbert (1977). Predictor variables included medical, gynecological, and obstetrical history; age and age squared; physical measurements; and blood test results. Separate stepwise discriminant function analyses were run on Caucasian and Japanese women, and on a pooled sample. Analyses were run through a stepwise multiple regression program, possible with only two groups to be discriminated. After preliminary runs on the 35 predictors allowed elimination of nonsignficant variables, the stepwise analyses were rerun. In all analyses, age and age squared were forced into the stepwise series first, because of known relationships with the remaining predictors.

Percent of variance accounted for in distinguishing menopausal from nonmenopausal women was 45% and 36% for Caucasian and Japanese women, respectively. After adjustment for effects of age, the only additional predictor variable to be retained in the stepwise analysis for both races was "surgery related to a female disorder." For Caucasian women only, one additional variable, medication, was selected in the stepwise analysis as statistically significant. For age, age squared, surgery, and medication, discriminant function coefficients and their standard errors were given.[14] Finally, variables common to both races were selected and heterogeneity of coefficients between Japanese and Caucasian samples was tested.[15] No significant heterogeneity was found—that is, the two races did not produce significantly different discriminant functions. An alternative strategy for this test would be to do a factorial discriminant function analysis, with race as one of the grouping variables.

---

[14] Since this two-group analysis was run through a multiple regression program, discriminant function coefficients were produced as β weights (cf. Chapter 5) and standard errrors were available.

[15] The test for heterogeneity was reported as that of Rao (1952) as applied by Goodman et al. (1974).

Chapter 10

Principal Components Analysis and Factor Analysis

10.1 GENERAL PURPOSE AND DESCRIPTION

Principal components analysis (PCA) and factor analysis (FA) are statistical techniques that may be applied to a group of variables in which none has been specified as DV or IV. PCA and FA differ from the other multivariate techniques in that intercorrelations among variables in a single set, as opposed to external criteria (e.g., group membership, score on one or more DVs), are analyzed. Further, there are many solutions from which to choose. PCA and FA also differ in that they may reveal relationships between observed and hypothetical variables, while the other techniques are concerned primarily with relationships among observed variables.

In PCA or FA, the researcher is usually interested in discovering which variables in a data set form coherent subgroups that are relatively independent of one another. Inspection of sets of variables that are correlated with one another can reveal a great deal about hypothetical structures or processes that may have generated the combination of outcomes that were measured by the observed variables. The specific goal of analysis may be to summarize patterns of intercorrelations among variables, to reduce a large number of variables to a smaller number of clusters while retaining maximum spread among experimental units, to provide an operational definition (a regression equation) for an unobserved, hypothetical construct by using observed variables, or to test a theory about the nature of underlying variables.

There are two major uses of FA: exploratory and confirmatory. In exploratory FA, one seeks to summarize data by grouping together variables that are intercorrelated. The variables themselves may or may not have been chosen with potential underlying structure in mind. Exploratory FA is usually performed

372

in the early stages of research, when it provides a tool for consolidating variables and for generating hypotheses about relationships in a reduced data set. Confirmatory FA is performed to test hypotheses about the structure of underlying processes. It generally occurs later in the research process, when a theory about structure is to be tested or when hypothesized differences in structure between groups of research units are tested. Variables are specifically chosen to reveal underlying structural processes. Data used in confirmatory FA, then, might be different from those used in exploratory FA.

There are several types of FA, known as R, Q, and P. In the R solution, correlations are computed for pairs of variables across subjects. R solutions reveal intercorrelations among observed variables. In the Q technique, the correlations are computed instead between subjects across variables. The Q solutions reveal intercorrelations among subjects. (An alternative to the Q technique is cluster analysis of cases, frequently thought to produce more valid results.) With the P technique, the unit of analysis is a single person, with correlations computed between variables across occasions (trials). P analysis reveals intercorrelations among variables that tend to change together. By far the most prevalently used technique is the R technique, although Q or P, and others as well (see Comrey, 1973, pp. 212–222), may also be useful, depending on the goals of the researcher.

Steps in a PCA or FA include selecting and measuring a group of variables, preparing the correlation matrix, determining the number of components or factors to be considered, extracting a set of components or factors from the correlation matrix, (probably) rotating the components or factors to increase interpretability, and, finally, interpreting the results. Although there are statistical considerations relevant to most of these steps, the final test of the value of an analysis usually depends on the interpretability of the results: A good PCA or FA "makes sense"; a bad one does not.

One of the problems with FA or PCA is that, after extraction, there are an infinite number of rotations available, all accounting for the same amount of variance in the original data, but each representing factors defined slightly differently. The final choice among the alternatives depends on the researcher's assessment of its scientific utility and interpretability. In the presence of an infinite number of mathematically identical solutions, researchers are bound to differ regarding which is best. Because the differences cannot be resolved by appeal to objective criteria, arguments over the best solution have sometimes become vociferous. However, those who expect a certain amount of ambiguity with respect to the best FA solution will not be surprised when other researchers select a different one. Nor will they be surprised when results are not replicated if different decisions are made at one, or more, of the steps in performing FA.

Another problem with PCA and FA is that they have frequently been used in an attempt to "save" data sets that were ill-conceived originally. The procedures have become associated in the minds of many with the likelihood of sloppy research. In fact, PCA and FA doubtless have provided a mechanism through which results of some poorly designed studies could be interpreted. But the very power of the techniques to create apparent order from real chaos has contributed to their somewhat tarnished reputations as scientific tools. Despite these problems, however, PCA and FA have considerable utility in revealing potential underlying relationships and in reducing a large data set down to a few variables with known properties for, perhaps, input into other analytic procedures.

Mathematically, PCA and FA produce several linear combinations of observed variables called components or factors. When the scores on components or factors for each individual are properly weighted and summed, the individual's scores on observed variables may be closely approximated or reproduced. The number of factors or components required to approximate observed scores, m, is usually fewer than the number of observed variables themselves, p, producing considerable parsimony in summarizing a set of results. Scores on components or factors may, in addition, be more stable than scores on the observed variables from which they were obtained.

But PCA and FA are bootstrap analyses in that the components or factors are, themselves, defined with respect to the observed variables. In particular, the factors are produced by combining scores on observed variables, some of which are correlated, but imperfectly, with each of the factors. A solution is usually more interpretable when a few observed variables are highly correlated with each factor and uncorrelated with all other factors. Interpretation and naming of factors depend on the meaning of the particular combination of observed variables with which that factor is highly correlated.

Before launching a more detailed description of PCA and FA, it will be helpful to define a few terms. The first set of terms involves correlation matrices. The correlation matrix produced by the observed variables as described in Chapter 1 will be called the *observed correlation matrix* **R**. The matrix of correlations among the observed variables produced by the solution will be referred to as the *reproduced correlation matrix*. The difference between the observed and reproduced correlation matrices will be called the *residual correlation matrix*. In a good FA solution, correlations in the residual matrix will be small, indicating a close fit between observed and reproduced matrices.

A second set of terms refers to matrices produced and interpreted as part of the solution. If rotation was *orthogonal* so that all of the factors are uncorrelated with each other, a *loading* matrix is produced. The loading matrix is a matrix of correlations between observed variables and factors. The sizes of the loadings reflect the amount of variance each factor contributes to an observed variable. Most orthogonal FAs are interpreted by looking at the correlations in the loading matrix for each factor.

If rotation was *oblique,* so that the factors themselves could be correlated, several additional matrices are produced. The most straightforward matrix is the *factor correlation* matrix, containing the correlations among the several factors. The loading matrix from orthogonal rotation is split into two matrices: a *structure* matrix of correlations between factors and variables and a *pattern* matrix whose elements reflect the unique variance each factor contributes to the variance of an observed variable. Following oblique rotation, it is common practice to interpret the pattern matrix, although the interpretation is somewhat ambiguous, as would be interpretation of the structure matrix.

Lastly, for both types of rotation there is the *factor-score* coefficients matrix, which is a matrix of coefficients used to generate scores on factors from scores on observed variables for each individual.

FA produces *factors,* while PCA produces *components.* However, the processes are similar except in preparation of the correlation matrix of observed variables for extraction. The primary difference between the two major procedures, PCA and FA, is the variance that is analyzed. In PCA, all the variance in the observed variables contributes to the solution. In FA, only the variance a variable shares with other variables is considered relevant to the solution, and attempts are made to disregard variance due to error and variance that is unique to each variable. The term *factor* will be used henceforth to refer to both components and factors unless the distinction is critical, in which case the appropriate term will be used.

10.2 KINDS OF RESEARCH QUESTIONS

The goal of research using PCA or FA is to reduce a large number of variables down into a smaller set, or to concisely describe (and perhaps understand) the pattern of interrelationships among observed variables (exploratory FA), or to test theory about underlying structure (confirmatory FA). Sometimes the results of exploratory FA may seem to reflect underlying structure as well, even though discovery of structure was not the primary goal of analysis.

Often FA is used in conjunction with test construction when the researcher aspires to an inventory to assess some aspect of personality or performance. Preliminary lists of items are factor analyzed and refined until a reliable and sensitive instrument measuring several factors is constructed. In many research programs, scores on factors (see Section 10.6.6) are tested against treatment differences in experimental settings to establish validity for the factor structure.

Although PCA and FA have many uses, some of the questions that are frequently asked are presented in Sections 10.2.1 through 10.2.7.

10.2.1 Nature of Latent Structure

Which set of latent variables most adequately summarizes the pattern of correlations found among observed variables? That is, which set of factors or components has the greatest scientific utility? These questions, and others like them, refer to the ability of PCA and FA to uncover patterns in relationships among variables. Using PCA and FA to describe patterns involves rotating the solution for maximum interpretability (see Section 10.6.3) and making sense of the resulting solution (see Section 10.6.5). For instance, a researcher might be interested in generating an instrument to assess self-esteem. Possible items might cover a wide range of issues: self-esteem in a social setting, in a work setting, in the family, with regard to intellectual prowess, and so forth. FA of the items follow-ing their administration to a large group of persons might reveal that self-esteem is composed of several latent variables, identified by the pattern of correlations between the factors and the test items.

10.2.2 Number of Latent Structures

What is the minimum number of latent variables needed to approximate the pattern of correlations found in **R,** the observed correlation matrix? This question involves reducing the set of observed variables to the smallest set of latent variables that still preserves the information in **R.** Strategies for choosing an appropriate number of factors and for assessing the correspondence between observed and reproduced correlation matrices are discussed in Section 10.6.2.

For instance, the researcher in self-esteem might find that the pattern of correlations in **R** can be reproduced to some extent by three factors, but to successively greater extents by four and five factors. However, the fifth factor, although contributing variance to the solution, may not be interpretable. The researcher may conclude that the four-factor solution is the one to be reported.

10.2.3 Importance of Factors and Solutions

Which factors account for the most variance in the solution? How much variabil-ity in the data set is accounted for by the chosen factor solution? Methods for addressing these questions are found in Section 10.6.4. The researcher may discover that the first three factors in the solution account for large amounts of spread in self-esteem while the fourth factor (and the possible fifth) accounts for less. The first three are, in some sense, more important than the fourth. The factor solution as a whole may account for, say, 60% of the total variabil-ity or spread in scores, leaving 40% unaccounted for by this particular solu-tion.

10.2.4 Testing Theory in FA

How well does the obtained factor solution fit an expected factor solution? Tests of theory in FA are addressed, in a preliminary form, in both Sections 10.6.2 and 10.6.7. The researcher may, at the outset, have generated hypotheses regarding both the number and the nature of the factors thought to underlie self-esteem. Comparisons between the proposed structure and the four-factor solution provide a test of hypotheses.

10.2.5 Comparing Structure in Different Groups

How similar is the underlying structure for persons with different characteristics or different experiences? Consistency in structure between two groups may be assessed, in preliminary fashion, using techniques described in Section 10.6.7. It seems likely that the structure of self-esteem differs between, say, women and men or between members of different ethnic groups. Comparisons among structures for different groups are possible and frequently enlightening. Alternatively, the researcher may seek to change the structure of self-esteem by application of a treatment to an experimental group. Comparisons between control and experimental groups are also possible provided sufficiently large numbers of persons are studied.

10.2.6 Distinguishing Among Subjects

How can individual differences among research units be maximized? The goal here is to find a dimension that spreads subjects apart as much as possible. The first principal component provides a single dimension along which subjects are maximally distinguishable. The first set of m principal components provides m dimensions along which subjects are maximally different. These advantages, which may be useful in some research settings, are described further in Section 10.5.1.2. For instance, the researcher in self-esteem may have been interested not so much in the structure underlying that concept as in discovering the linear combination of all self-esteem items that distinguishes maximally among respondents. The first principal component provides such a combination, which can then be used as, say, a DV in ANOVA or MANOVA.

10.2.7 Estimating Scores on Latent Variables

Had factors been measured directly, what scores would each of my subjects have received on each of them? Estimation of factor scores is the topic of Section 10.6.6. Consolidation of many scores on observed variables to a few scores on factors may vastly simplify interpretation of results of other analyses in which factor scores are used as DVs or IVs. For instance, the researcher interested in self-esteem, using the equations from the four-factor solution, could generate four scores on factors from scores on observed variables for each respondent. Respondents could in turn have been subjected to experimental manipulation designed to alter one or more factors of self-esteem used as DVs in MANOVA.

10.3 LIMITATIONS

10.3.1 Theoretical Issues

Most applications of PCA or FA are exploratory in nature when FA is used as a tool for reducing the number of variables or examining patterns of correlations among variables without a serious intent to test theory. Under these circumstances, both the theoretical and the practical limitations to FA may be relaxed in favor of a frank exploration of the data. Decisions about number of factors and rotational scheme are based on pragmatic rather than theoretical criteria.

The research project that is designed specifically to be factor analyzed, however, differs from other projects in several important respects. Among the best detailed discussions of the differences is the one found in Comrey (1973, pp. 189–211), from which some of the following discussion was taken.

The first task of the researcher is to generate hypotheses about factors believed to underlie the domain of interest. Statistically, it is important to make the research inquiry broad enough to include five or six hypothesized factors so that the solution is stable. Logically, in order to reveal the structure underlying a research area, all relevant factors have to be included. Failure to measure some important factor may distort the apparent relationships among measured factors. Inclusion of all relevant factors poses a logical, but not statistical, problem to the researcher.

Next, one selects variables to observe. For each hypothesized factor, five or six variables, thought to be relatively pure measures of the factor, should be included. Pure measures are called marker variables. Marker variables are useful because they define clearly the nature of a factor; marker variables are highly correlated with one and only one factor, and load on it regardless of extractional or rotational technique. Adding potential variables to a factor to round it out is much more meaningful if the factor is unambiguously defined by marker variables to begin with.

The complexity of the variables is also relevant to the factor solution. Complexity is related to the number of factors with which a variable is associated. A pure variable is associated with only one factor, whereas a complex variable is associated with several. If variables differing in complexity are all included in an analysis, those with similar complexity levels may "catch" each other in factors that have little to do with underlying structures. Variables with similar complexity may be correlated with each other *because of their complexity* and not because they are related to the same factor. Estimating the complexity of variables is part of generating hypotheses about factors and selecting variables to measure them.

Several other considerations are required of the researcher planning a factor analytic study. It is important, for instance, that the sample chosen exhibit spread in scores with respect to the variables and the factors they measure. If all subjects achieve about the same score on some factor, correlations among the observed variables may be low and the factor may not emerge in analysis. Selection of research units that are thought to differ on the underlying factors is important in designing the study.

One should also be leery about pooling the results of several samples, or the same sample with measures repeated in time, for factor analytic purposes. First, it may be that samples that are known to be different with respect to some criteria (e.g., socioeconomic status) also differ in factor structure. Analyses of just such differences may enrich a research area. Second, underlying factor structure may shift in time for the same subjects with learning or experience in an experimental setting and, again, the differences may be quite revealing. Pooling results from diverse groups in FA may obscure differences rather than illuminate them.

On the other hand, if different samples do produce the same structure, pooling them is desirable because of increase in sample size. For example, if men and women produce the same factor structure of strategies for coping with stress, the samples should be combined and the results of the single FA reported. Strategies for evaluating differences in factor structure among groups are discussed in Section 10.6.7.

Because FA and PCA are exquisitely sensitive to the sizes of correlations in **R,** it is critical that honest, reliable correlations be employed. Review Chapter 4 before conducting research using FA.

10.3.2 Practical Issues

Sensitivity to outlying cases, problems created by missing data, and degradation of correlations between poorly distributed variables all plague FA and PCA. A review of the issues in Chapter 4 (Sections 4.2, 4.3, and 4.4) is especially important to FA or PCA. Thoughtful solutions to some of the problems, including data transformations, may markedly enhance structural analyses, whether performed for exploratory or confirmatory purposes. However, the limitations apply with greater force to confirmatory FA.

10.3.2.1 Outliers Among Cases As in all multivariate techniques, cases may be outliers either on individual original variables (univariate) or on combinations of original variables (multivariate). Consult Chapter 4 for methods of detecting and reducing the effects of both univariate and multivariate outliers.

10.3.2.2 Sample Size Correlation coefficients tend to be less reliable when estimated from small samples. Therefore, it is important that sample size be large enough that correlations are reliably estimated. Comrey (1973) gives as a guide sample sizes of 50 as very poor, 100 as poor, 200 as fair, 300 as good, 500 as very good, and 1000 as excellent. Others suggest that *a sample size of 100–200 is good enough for most purposes, particularly when subjects are homogeneous and number of variables is not too large.* The required sample size depends also on magnitude of population correlation and number of factors. If there are strong, reliable correlations and a few, distinct factors, a sample size of 50 may even be adequate, as long as there are notably more cases than factors.

10.3.2.3 Factorability of R A matrix that is factorable should include several sizable correlations. The expected size depends, to some extent, on N (larger sample sizes tend to produce smaller correlations), but if no correlation exceeds .30, use of FA might be questionable. If the observed variables are uncorrelated with one another, there is probably nothing to factor analyze. *Inspect* **R** *for correlations in excess of* .30 and, if none are found, reconsider use of FA, except in its most exploratory and pragmatic sense.

A notoriously sensitive test of the hypothesis that the correlations in a matrix represent random fluctuations around true values of zero was offered by Bartlett (1954). Because of its sensitivity and its dependence on N, Bartlett's test will be significant with samples of substantial size even if linkages among variables are slight, and is therefore not recommended.

10.3.2.4 Multicollinearity and Singularity In PCA, multicollinearity is not a problem because there is no need to invert a matrix. For most forms of FA (i.e., those that require matrix inversion) and for estimation of factor scores in any form of FA, extreme multicollinearity and/or singularity must be avoided. *For FA, if the determinant of* **R** *and eigenvalues associated with some factors approach 0, multicollinearity is present. It may be necessary to eliminate some variables,* as recommended in Chapter 4.

10.3.2.5 Outliers Among Variables In both exploratory and confirmatory FA, *variables that are unrelated to others in the set should be identified,* but what one does with them depends on the goal of analysis. *A variable with a low squared multiple correlation with all other variables or low correlations with all factors does not share variance with either variables or factors and has not participated in the analysis.* In exploratory FA, this outlier among variables might be retained as sole representative of an important dimension in other analysis or deleted, as pragmatic considerations dictate. In confirmatory FA, the dimension represented by the outlying variable may represent either a promising lead for future work or (probably) error variance, but its interpretation awaits clarification by more research.

Similarly, factors that are defined by only one or two variables are potentially unreliable and should be interpreted with great caution or not at all (cf. Mulaik, 1972, p. 195; Harman, 1967, p. 134). The problem with two-variable factors is complicated, but it boils down to the fact that there are more potential correlations between pairs of variables than there are degrees of freedom to define them adequately. Therefore one never knows which two-variable factor is "real." Suggestions for determining reliability of factors defined by one or two variables are discussed in Section 10.6.2.

10.3.2.6 Outliers with Respect to the Solution In FA and PCA, cases may be unusual with respect to their scores on the factors. These are cases that have unusually large or small scores on the factors as estimated from the regression equation applied to scores on observed variables. Probably, the deviant scores are from cases for which the factor solution is inadequate. If BMDP4M is used, *outliers with respect to the solution will have large Mahalanobis distances, estimated as chi square values, from the location of the case in the space defined by the factors to the centroid of all cases in the same space.* If scatterplots between pairs of factors are requested, these cases will appear along the borders. Although these cases could be dropped from analysis, it might also be informative to examine them for consistency along some dimension as a guide to the kinds of cases for which the FA is *not* appropriate.

10.3.2.7 Normality and Linearity As long as PCA and FA are used to describe a sample, or as convenient ways to summarize the relationships in a large set of observed variables, assumptions regarding the distributions of variables are not required. If variables are normally distributed with linear relationships between them, the solution is enhanced. To the extent that normality and linearity fail, the solution is degraded, but may still be worthwhile.

However, when statistical inference to a population is to be made about the number of factors, it is assumed that the distribution of variables is multivariate normal. Morrison (1967, pp. 80–94) provides a detailed discussion of multivariate normality. Essentially it is the assumption that all variables, and all linear combinations of variables, are normally distributed. *Normality among single variables may be assessed by skewness* (see Chapter 4). *Normality, and implied linearity, among pairs of variables may be assessed through inspection of scatterplots.*

10.4 FUNDAMENTAL EQUATIONS FOR ANALYSES OF STRUCTURE

Because of the variety and complexity of the calculations involved in preparing the correlation matrix, extracting factors, and rotating them, and because, in our judgment, little insight is produced by demonstrations of some of these procedures, this section will not show the basic calculations. Instead, the relationships between some of the more important matrices will be shown, with an assist from SPSS FACTOR for underlying calculations.

Table 10.1 lists many of the important matrices that the user of analyses of structure will encounter. Although the list is lengthy, it is composed mostly of *matrices of correlations* (between variables, between factors, and between variables and factors), *matrices of standard scores* (on variables and on factors), and *matrices of regression weights* (for producing scores on factors from scores on variables and for estimating the unique contribution of each factor to the variance in a variable).

Two additional matrices included are the matrix of eigenvalues and the matrix of their corresponding eigenvectors. Eigenvalues and eigenvectors themselves are also covered, albeit scantily, because of their importance in factor extraction, the frequency with which one encounters the terminology, and the close association between eigenvalues and variance.

A data set that is appropriate for analysis of structure consists of numerous cases or subjects, each of whom has been measured on several variables. One such grossly inadequate data set is shown in Table 10.2 (page 384), which illustrates some of the relationships in structural analyses. Five respondents who were trying on ski boots late one Friday night in January were asked about the importance of each of four variables to their selection of a ski resort. The variables were cost of ski ticket (COST), kind of ski lift (LIFT), depth of snow (DEPTH), and kind of snow (POWDER). Larger numbers indicate greater importance. The researcher desired to investigate the pattern of relationships among the variables (see **R**) in an effort to understand better the dimensions underlying choice of ski area.

TABLE 10.1 COMMONLY ENCOUNTERED MATRICES IN STRUCTURAL ANALYSES

Label	Name	Rotation	Size[a]	Description
R	Correlation matrix	Both orthogonal and oblique	$p \times p$	Matrix of correlations between variables.
Z	Variable matrix	Both orthogonal and oblique	$N \times p$	Matrix of standardized observed variable scores.
F	Factor-score matrix	Both orthogonal and oblique	$N \times m$	Matrix of standard scores on factors or components.
A	Factor loading matrix Pattern matrix	Orthogonal Oblique	$p \times m$	Matrix of regressionlike weights used to estimate the unique contribution of each factor to the variance in a variable. If orthogonal, also correlations between variables and factors.
B	Factor-score coefficients matrix	Both orthogonal and oblique	$p \times m$	Matrix of regression weights used to generate factor scores from variables.

C	Structure matrix[b]	Oblique	$p \times m$	Matrix of correlations between variables and (correlated) factors
Φ	Factor correlation matrix	Oblique	$m \times m$	Matrix of correlations among factors
L	Eigenvalue matrix[c]	Both orthogonal and oblique	$m \times m$	Diagonal matrix of eigenvalues, one per factor[e]
V	Eigenvector matrix[d]	Both orthogonal and oblique	$p \times m$	Matrix of eigenvectors, one vector per eigenvalue

[a] Row by column dimensions where
p = number of variables
N = number of subjects
m = number of factors or components

[b] In most textbooks, the structure matrix is labeled S. However, we have used S to represent the sum-of-squares and cross-products matrix elsewhere and will use C for the structure matrix here.

[c] Also called characteristic roots or latent roots.

[d] Also called characteristic vectors or latent vectors.

[e] If the matrix is of full rank, there are actually p rather than m eigenvalues and eigenvectors. Only m are of interest, however, so the remaining $p - m$ are not displayed.

TABLE 10.2 SMALL SAMPLE OF HYPOTHETICAL DATA
FOR ILLUSTRATION OF FACTOR ANALYSIS

Skiers	Variables			
	COST	LIFT	DEPTH	POWDER
S_1	32	64	65	67
S_2	61	37	62	65
S_3	59	40	45	43
S_4	36	62	34	35
S_5	62	46	43	40

Correlation matrix

	COST	LIFT	DEPTH	POWDER
COST	1.000	−.953	−.055	−.130
LIFT	−.953	1.000	−.091	−.036
DEPTH	−.055	−.091	1.000	.990
POWDER	−.130	−.036	.990	1.000

Notice the pattern of correlations in the matrix as set off by the vertical and horizontal lines. The strong correlations in the upper left and lower right quadrants show that scores on COST and LIFT are related, as are scores on DEPTH and POWDER. The upper right and lower left quadrants show small cross-correlations among the variables. That is, scores on DEPTH and LIFT are unrelated, as are scores on POWDER and LIFT, and so on.

An important theorem from matrix algebra indicates that, under certain conditions, matrices can be diagonalized. Correlation and covariance matrices are among those that can be diagonalized. When a matrix is diagonalized, it is transformed into a matrix with numbers in the positive diagonal and zeros everywhere else. The numbers in the positive diagonal represent variance from the original matrix that has been repackaged by manipulating it as follows:

$$\mathbf{L} = \mathbf{V'RV} \tag{10.1}$$

Diagonalization of $\mathbf{R}$ is accomplished by post- and premultiplying it by the matrix $\mathbf{V}$ and its transpose.

The columns in $\mathbf{V}$ are called eigenvectors, and the values in the main diagonal of $\mathbf{L}$ are called eigenvalues. The first eigenvector corresponds to the first eigenvalue, and so forth.

Eigenvalues and eigenvectors were calculated for the example in Table 10.2 with an assist from SPSS FACTOR. They are shown in Table 10.3.

Because there were four variables, four eigenvalues and their corresponding eigenvectors were available for analysis. However, because the goal of FA is

TABLE 10.3 EIGENVECTORS AND CORRE-SPONDING EIGENVALUES FOR THE EXAMPLE	Eigenvector 1	Eigenvector 2
	−.283	.651
	.177	−.685
	.658	.252
	.675	.207
	Eigenvalue 1	Eigenvalue 2
	2.00	1.91

to summarize a pattern of correlations with as few factors as possible, and because each eigenvalue corresponds to a different potential factor, usually only large eigenvalues and their corresponding eigenvectors are retained. These duplicate the correlation matrix as faithfully as possible with the fewest factors. In this case, eigenvalues of 2.02, 1.94, .04, and .00 were initially identified corresponding to each of the four possible factors. Only the first two factors, with values over 1.00, were retained in subsequent analyses. These two alone had eigenvalues of 2.00 and 1.91, respectively, as indicated in the table.

Using Eq. 10.1 and inserting the values from the example, we obtain

$$\mathbf{L} = \begin{bmatrix} -.283 & .177 & .658 & .675 \\ .651 & -.685 & .252 & .207 \end{bmatrix} \begin{bmatrix} 1.000 & -.953 & -.055 & -.130 \\ -.953 & 1.000 & -.091 & -.036 \\ -.055 & -.091 & 1.000 & .990 \\ -.130 & -.036 & .990 & 1.000 \end{bmatrix} \begin{bmatrix} -.283 & .651 \\ .177 & -.685 \\ .658 & .252 \\ .675 & .207 \end{bmatrix}$$

$$\mathbf{L} = \begin{bmatrix} 2.00 & .00 \\ .00 & 1.91 \end{bmatrix}$$

(All values agree with computer output. Hand calculation may produce discrepancies due to rounding error.)

The matrix of eigenvectors premultiplied by its transpose produces the identity matrix with ones in the positive diagonal and zeros elsewhere:

$$\mathbf{V'V = I} \tag{10.2}$$

From the example:

$$\begin{bmatrix} -.283 & .177 & .658 & .675 \\ .651 & -.685 & .252 & .207 \end{bmatrix} \begin{bmatrix} -.283 & .651 \\ .177 & -.685 \\ .658 & .252 \\ .675 & .207 \end{bmatrix} = \begin{bmatrix} 1.000 & .000 \\ .000 & 1.000 \end{bmatrix}$$

The important point is that because correlation matrices meet requirements for diagonalizability, it is possible to bring to bear on them the insights of the matrix algebra of eigenvectors and eigenvalues, with FA as one result. When a matrix is diagonalized, the information contained in it is changed from one form to another. In FA, the variance in the correlation matrix is condensed into the eigenvalues. The factor with the largest eigenvalue contains the most variance and so on down to factors with small or negative eigenvalues, which are usually omitted from solutions.

The calculations involved in finding eigenvectors and eigenvalues are extremely laborious and not particularly enlightening (although they are illustrated in Appendix A on a small matrix). They require solving p equations in p unknowns with additional side constraints. The calculations are rarely performed by hand. Once the eigenvalues and eigenvectors are known, however, the rest of FA (or PCA) more or less "falls out," as will become clear from Eqs. 10.3 to 10.6.

Equation 10.1 can be reorganized as follows:

$$\mathbf{R} = \mathbf{VLV'} \qquad (10.3)$$

The correlation matrix can be decomposed into a product of three matrices—the matrices of eigenvalues and corresponding eigenvectors.

$$\mathbf{R} = \mathbf{V}\sqrt{\mathbf{L}}\,\sqrt{\mathbf{L}}\,\mathbf{V'} \qquad (10.4)$$

or

$$\mathbf{R} = (\mathbf{V}\sqrt{\mathbf{L}})(\sqrt{\mathbf{L}}\,\mathbf{V'})$$

If $\mathbf{V}\sqrt{\mathbf{L}}$ is called $\mathbf{A}$, and $\sqrt{\mathbf{L}}\,\mathbf{V'}$ is $\mathbf{A'}$, then

$$\mathbf{R} = \mathbf{AA'} \qquad (10.5)$$

The correlation matrix can also be decomposed into a product of two matrices, each a combination of eigenvectors and the square root of eigenvalues.

This equation is frequently called the fundamental equation for FA.[1] It represents the assertion that the correlation matrix can be decomposed into the product of the factor loading matrix, $\mathbf{A}$, and its transpose.

Equations 10.4 and 10.5 also reveal that the major work of FA (and PCA) is calculation of the eigenvalues and eigenvectors. Once they are known, the unrotated factor loading matrix can be found by straightforward matrix multiplication, as follows.

$$\mathbf{A} = \mathbf{V}\sqrt{\mathbf{L}} \qquad (10.6)$$

For the example:

$$\mathbf{A} = \begin{bmatrix} -.283 & .651 \\ .177 & -.685 \\ .658 & .252 \\ .675 & .207 \end{bmatrix} \begin{bmatrix} \sqrt{2.00} & 0 \\ 0 & \sqrt{1.91} \end{bmatrix} = \begin{bmatrix} -.400 & .900 \\ .251 & -.947 \\ .932 & .348 \\ .956 & .286 \end{bmatrix}$$

---

[1] Note that Eqs. 10.4 and 10.5 presume that all p eigenvalues and eigenvectors are used, not just m of them.

The factor loading matrix contains correlations between factors and variables. The first column contains correlations between the first factor and each variable in turn, COST (−.400), LIFT (.251), DEPTH (.932), and POWDER (.956). The second contains correlations between the second factor and each variable in turn, COST (.900), LIFT (−.947), DEPTH (.348), and POWDER (.286). Interpretation of a factor is interpretation of the set of variables that are highly correlated with it, that is, that have high loadings on it. Thus the first factor is primarily a snow conditions factor (DEPTH and POWDER), while the second reflects resort conditions (COST and LIFT). (The negative correlation indicates that more attractive lifts are also more costly.)

Notice, however, that all the variables are correlated with both factors to a considerable extent. Interpretation is still fairly clear for this hypothetical example, but most likely would not be so for real data. Usually a factor is most interpretable when a few variables load highly on it and the rest do not.

Rotation is ordinarily used after extraction to maximize high correlations and minimize low ones. Several criteria for rotation are available (see Section 10.5.2) but the most commonly used technique, and the one illustrated here, is called varimax. The goal of varimax rotation is to maximize the variance of the factor loadings for each factor by making high loadings higher and low ones lower.

This goal is accomplished by means of a transformation matrix Λ (as defined in Eq. 10.8), where

$$\mathbf{A}_{\text{unrotated}} \, \Lambda = \mathbf{A}_{\text{rotated}} \qquad (10.7)$$

The unrotated factor loading matrix is multiplied by the transformation matrix to produce the rotated loading matrix.

For the example:

$$\mathbf{A} = \begin{bmatrix} -.400 & .900 \\ .251 & -.947 \\ .932 & .348 \\ .956 & .286 \end{bmatrix} \quad \begin{bmatrix} .946 & -.325 \\ .325 & .946 \end{bmatrix} = \begin{bmatrix} -.086 & .981 \\ -.071 & -.977 \\ .994 & .026 \\ .997 & -.040 \end{bmatrix}$$

Compare the rotated and unrotated loading matrices. Notice that in the rotated matrix the low correlations are around zero and the high ones are higher still. Emphasizing differences in correlations usually facilitates interpretation of factors by making unambiguous the pattern of variables that correlates with a factor.

The numbers in the transformation matrix have a spatial interpretation.

$$\Lambda = \begin{bmatrix} \cos \Psi & -\sin \Psi \\ \sin \Psi & \cos \Psi \end{bmatrix} \qquad (10.8)$$

The transformation matrix is a matrix of sines and cosines of an angle Ψ.

TABLE 10.4 RELATIONSHIPS AMONG LOADINGS, COMMUNALITIES,
SSLs, VARIANCE, AND COVARIANCE OF
ORTHOGONALLY ROTATED FACTORS

	Factor 1	Factor 2	Communalities (h^2)
COST	−.086	.981	$\Sigma a^2 =$ .970
LIFT	−.071	−.977	$\Sigma a^2 =$ .960
DEPTH	.994	.026	$\Sigma a^2 =$ .989
POWDER	.997	−.040	$\Sigma a^2 =$ .996
SSLs	$\Sigma a^2 = 1.994$	$\Sigma a^2 = 1.919$	3.915
Proportion of variance	.50	.48	.98
Proportion of covariance	.51	.49	

For the example, the angle is approximately 19°. That is, cos 19 ≅ .946 and
sin 19 ≅ .325. Geometrically, this corresponds to a 19° swivel of the factor
axes about the origin. Section 10.5.2.3 gives greater detail regarding the geometric
meaning of rotation.

Once the rotated loading matrix is available, other relationships can be
demonstrated, as shown in Table 10.4. The communality for a variable is the
sum of squared loadings (SSL) within a variable across factors. The communality
is also the squared multiple correlation of the variable as predicted from the
factors. The proportion of variance *in the original variables* accounted for by
a factor is the SSL for the factor divided by the number of variables (if rotation
is orthogonal).[2] The proportion of covariance accounted for by a factor, the
proportion of variance *in the solution* that the factor accounts for, is the SSL
for the factor divided by the sum of the communalities (or, equivalently, the
sum of the SSLs).

The reproduced correlation matrix can be generated using Eq. 10.5. For
the example:

$$\overline{\mathbf{R}} = \begin{bmatrix} -.086 & .981 \\ -.071 & -.977 \\ .994 & .026 \\ .997 & -.040 \end{bmatrix} \begin{bmatrix} -.086 & -.071 & .994 & .997 \\ .981 & -.977 & .026 & -.040 \end{bmatrix}$$

$$= \begin{bmatrix} .970 & -.953 & -.059 & -.125 \\ -.953 & .962 & -.098 & -.033 \\ -.059 & -.098 & .989 & .990 \\ -.125 & -.033 & .990 & .996 \end{bmatrix}$$

[2] Note that the SSL is equal to the eigenvalue for unrotated factors only. That is, in an
unrotated factor mattrix, Σa^2 (summed over the variables within a factor) is the eigenvalue. In a
rotated matrix, the term SSL is used in place of eigenvalue.

Notice that the reproduced correlation matrix differs slightly from the original correlation matrix. In fact,

$$\mathbf{R}_{res} = \mathbf{R} - \overline{\mathbf{R}} \tag{10.9}$$

The residual correlation matrix is the difference between the observed correlation matrix and the reproduced correlation matrix.

For the example, with communalities inserted in the positive diagonal of $\mathbf{R}$:

$$
\mathbf{R}_{res} =
\begin{bmatrix}
.970 & -.953 & -.055 & -.130 \\
-.953 & .960 & -.091 & -.036 \\
-.055 & -.091 & .989 & .990 \\
-.130 & -.036 & .990 & .996
\end{bmatrix}
-
\begin{bmatrix}
.970 & -.953 & -.059 & -.125 \\
-.953 & .962 & -.098 & -.033 \\
-.059 & -.098 & .989 & .990 \\
-.125 & -.033 & .990 & .996
\end{bmatrix}
$$

$$
=
\begin{bmatrix}
.000 & .000 & .004 & -.005 \\
.000 & -.002 & .007 & -.003 \\
.004 & .007 & .000 & .000 \\
-.005 & -.003 & .000 & .000
\end{bmatrix}
$$

In a "good" FA, the numbers in the residual correlation matrix will be small, reflecting few differences between the original correlation matrix and the correlation matrix generated from factor loadings.

Because FA is essentially circular, scores on factors can be predicted for each case once the loading matrix is available. The first step involves calculation of regression coefficients for weighting variable scores to produce factor scores. Since $\mathbf{R}^{-1}$ is the inverse of the matrix of correlations among variables and $\mathbf{A}$ is the matrix of correlations between factors and variables, the next equation can be recognized as similar to Eq. 5.6, the equation for regression coefficients in multiple regression.

$$\mathbf{B} = \mathbf{R}^{-1}\mathbf{A} \tag{10.10}$$

Regression coefficients for producing factor scores from variable scores are a product of the inverse of the correlation matrix and the factor loading matrix.

For the example:[3]

---

[3] The numbers in $\mathbf{B}$ are different from factor-score coefficients generated by computer for the small data set. The difference is due to rounding error following inversion of a multicollinear correlation matrix.

$$\mathbf{B} = \begin{bmatrix} 25.485 & 22.689 & -31.655 & 35.479 \\ 22.689 & 21.386 & -24.831 & 28.312 \\ -31.655 & -24.831 & 99.917 & -103.950 \\ 35.479 & 28.312 & -103.950 & 109.567 \end{bmatrix} \begin{bmatrix} -.086 & .981 \\ -.072 & -.978 \\ .994 & .027 \\ .997 & -.040 \end{bmatrix}$$

$$= \begin{bmatrix} 0.082 & 0.537 \\ 0.054 & -0.461 \\ 0.190 & 0.087 \\ 0.822 & -0.074 \end{bmatrix}$$

To generate a score on the first factor, then, one weights the standardized score on the first variable, COST by 0.082, LIFT by 0.054, DEPTH by 0.190, and POWDER by 0.822. In matrix form,

$$\mathbf{F} = \mathbf{ZB} \tag{10.11}$$

Factor scores are a product of standardized scores on variables and regression coefficients.

For the example:

$$\mathbf{F} = \begin{bmatrix} -1.22 & 1.14 & 1.15 & 1.14 \\ 0.75 & -1.02 & 0.92 & 1.01 \\ 0.61 & -0.78 & -0.36 & -0.47 \\ -0.95 & 0.98 & -1.20 & -1.01 \\ 0.82 & -0.30 & -0.51 & -0.67 \end{bmatrix} \begin{bmatrix} 0.082 & 0.537 \\ 0.054 & -0.461 \\ 0.190 & 0.087 \\ 0.822 & -0.074 \end{bmatrix}$$

$$= \begin{bmatrix} 1.12 & -1.16 \\ 1.01 & 0.88 \\ -0.45 & 0.69 \\ -1.08 & -0.99 \\ -0.60 & 0.58 \end{bmatrix}$$

That is, the first subject is predicted to have a standardized score of 1.12 on the first factor and −1.16 on the second factor, and so on for all five subjects. Note that the sum of standardized scores across subjects for a single factor is zero.

Predicting scores on variables from scores on factors is also possible. The equation for doing so is

$$\mathbf{Z} = \mathbf{FA'} \tag{10.12}$$

Standardized scores on variables may be predicted as a product of scores on factors weighted by factor loadings.

For the example:

$$\mathbf{Z} = \begin{bmatrix} 1.12 & -1.16 \\ 1.01 & 0.88 \\ -0.45 & 0.69 \\ -1.08 & -0.99 \\ -0.60 & 0.58 \end{bmatrix} \begin{bmatrix} -.086 & -.072 & .994 & .997 \\ .981 & -.978 & .027 & -.040 \end{bmatrix}$$

$$= \begin{bmatrix} -1.23 & 1.05 & 1.08 & 1.16 \\ 0.78 & -0.93 & 1.03 & 0.97 \\ 0.72 & -0.64 & -0.43 & -0.48 \\ -0.88 & 1.05 & -1.10 & -1.04 \\ 0.62 & -0.52 & -0.58 & -0.62 \end{bmatrix}$$

That is, the first subject (row one of $\mathbf{Z}$) is predicted to have a standardized score of -1.23 on COST, 1.05 on LIFT, 1.08 on DEPTH, and 1.16 on POWDER. Like the reproduced correlation matrix, these values will be similar to the observed values if the FA captures the relationship among the variables.

Expansion of the products may prove enlightening. For example, for the first subject,

$$-1.23 = -.086(1.12) + .981(-1.16)$$
$$1.05 = -.072(1.12) - .978(-1.16)$$
$$1.08 = .994(1.12) + .027(-1.16)$$
$$1.16 = .997(1.12) - .040(-1.16)$$

Or, in algebraic form,

$$z_{\text{COST}} = a_{11}F_1 + a_{12}F_2$$

$$z_{\text{LIFT}} = a_{21}F_1 + a_{22}F_2$$

$$z_{\text{DEPTH}} = a_{31}F_1 + a_{32}F_3$$

$$z_{\text{POWDER}} = a_{41}F_1 + a_{42}F_3$$

That is, a score on an observed a variable may be conceptualized as a properly weighted and summed combination of the scores on factors that underlie it.

All the relationships mentioned thus far are based on a rotational scheme in which factors remain orthogonal (uncorrelated). Most of the complexities of orthogonal rotation remain and several others are added when oblique (correlated) rotation is used. Consult Table 10.1 for a listing of additional matrices and a hint of the discussion to follow.

SPSS FACTOR was run on the data from Table 10.2 using the default option for oblique rotation (cf. Section 10.5.2.2). Values for the pattern matrix, $\mathbf{A}$, were calculated along with factor-score coefficients, $\mathbf{B}$.

In oblique rotation, the loading matrix becomes the pattern matrix. Values in the pattern matrix when squared represent the unique contribution of each factor to the variance of each variable but do not include segments of variance that come from overlapping variance between correlated factors. For the example the pattern matrix following oblique rotation was

$$
A = \begin{bmatrix} -.079 & .981 \\ -.078 & -.978 \\ .994 & .033 \\ .997 & -.033 \end{bmatrix}
$$

Factor-score coefficients following oblique rotation were also identified, as follows:

$$
B = \begin{bmatrix} 0.104 & 0.584 \\ 0.081 & -0.421 \\ 0.159 & -0.020 \\ 0.856 & 0.034 \end{bmatrix}
$$

Applying Eq. 10.11 to produce factor scores results in the following values:

$$
F = \begin{bmatrix} -1.22 & 1.14 & 1.15 & 1.14 \\ 0.75 & -1.02 & 0.92 & 1.01 \\ 0.61 & -0.78 & -0.36 & -0.47 \\ -0.95 & 0.98 & -1.20 & -1.01 \\ 0.82 & -0.30 & -0.51 & -0.67 \end{bmatrix} \begin{bmatrix} 0.104 & 0.584 \\ 0.081 & -0.421 \\ 0.159 & -0.020 \\ 0.856 & 0.034 \end{bmatrix}
$$

$$
= \begin{bmatrix} 1.12 & -1.18 \\ 1.01 & 0.88 \\ -0.46 & 0.68 \\ -1.07 & -0.98 \\ -0.59 & 0.59 \end{bmatrix}
$$

Once the factor scores are determined, correlations among factors can be obtained. Among the equations used for that purpose is

$$
\Phi = \left(\frac{1}{N-1} \right) F'F \tag{10.13}
$$

Correlations among factors may sometimes be obtained by producing a matrix of cross products of standardized factor scores and dividing the results by the number of cases minus one.

The matrix of correlations between factors is a standard part of computer output following oblique rotation. For the example:

$$\Phi = \frac{1}{4}\begin{bmatrix} 1.12 & 1.01 & -0.46 & -1.07 & -0.59 \\ -1.18 & 0.88 & 0.68 & -0.98 & 0.59 \end{bmatrix} \begin{bmatrix} 1.12 & -1.18 \\ 1.01 & 0.88 \\ -0.46 & 0.68 \\ -1.07 & -0.98 \\ -0.59 & 0.59 \end{bmatrix}$$

$$= \begin{bmatrix} 1.00 & -0.01 \\ -0.01 & 1.00 \end{bmatrix}$$

Notice that the correlation between the first and second factor is quite low, −.01. For the example, there is almost no relationship between the two factors, although considerable correlation could have been produced had it been warranted. Ordinarily one would use orthogonal rotation in a case like this because complexities introduced by oblique rotation would not be warranted by such low correlations among factors.

However, if oblique rotation is used, then the structure matrix, **C**, is composed of correlations between variables and factors. These correlations represent both the unique relationship between the variable and the factor (shown in the pattern matrix) and the relationship between the variable and parts of the variance of the factor that it shares with other factors. The equation for the structure matrix is

$$\mathbf{C} = \mathbf{A}\Phi \tag{10.14}$$

The structure matrix is a product of the pattern matrix and the matrix of correlations among factors.

For example:

$$\mathbf{C} = \begin{bmatrix} -.079 & .981 \\ -.078 & -.978 \\ .994 & .033 \\ .997 & -.033 \end{bmatrix} \begin{bmatrix} 1.00 & -.01 \\ .01 & 1.00 \end{bmatrix} = \begin{bmatrix} -.069 & .982 \\ -.088 & -.977 \\ .994 & .023 \\ .997 & -.043 \end{bmatrix}$$

Thus variables COST, LIFT, DEPTH, and POWDER correlate −.069, −.088, .994, and .997 with the first factor and .982, −.977, .023, and −.043 with the second factor, respectively (without Kaiser's normalization).

There is some debate as to whether one should interpret and report the pattern matrix or the structure matrix following oblique rotation. The structure matrix of correlations between factors and variables is appealing because it is readily understood. However, the sizes of the correlations between variables and any given factor are likely to be inflated by the overlap between that factor and other factors (with which the variables also correlate). The problem becomes more severe as the correlations among factors increase. For this reason, the set of variables that composes the factor may be hard to see. On the other hand, the pattern matrix contains values representing the unique contributions of each factor to the variance in the variables. Shared variance is omitted (as it is with standard multiple regression), but the set of variables that composes a factor is usually easier to see. For this reason, most researchers interpret and report the pattern matrix rather than the structure matrix. However, if the researcher reports either the structure or pattern matrix and also Φ, then the interested reader can generate the other using Eq. 10.14 as desired.

In oblique rotation, $\overline{R}$ is produced as follows:

$$\overline{R} = CA' \qquad (10.15)$$

The reproduced correlation matrix is generated by multiplying the structure matrix by the transpose of the pattern matrix.

Once the reproduced correlation matrix is available, Eq. 10.9 can be used to generate the residual correlation matrix, a matrix that is frequently useful for diagnosing adequacy of fit in FA.

10.5 MAJOR TYPES OF ANALYSIS OF STRUCTURE

Numerous procedures for factor extraction and rotation are available. However, only those procedures available in the BMDP and SPSS packages will be summarized here. Other extraction and rotational techniques are described in Mulaik (1972) and Harman (1967).

10.5.1 Factor Extraction Techniques

Among the extraction techniques available in the BMDP and SPSS packages are principal components (PCA), principal factors, maximum likelihood factoring (including Rao's canonical factoring), image factoring, and alpha factoring (see Table 10.5). Of these, PCA and principal factors are the most commonly used.

All of the extraction techniques calculate a set of orthogonal components or factors that, in combination, reproduce **R**. Criteria that are used to generate the factors (e.g., maximize variance, minimize residual correlations) differ from technique to technique. But the differences between the solutions they produce may be small with a good data set and similar communality estimates. In fact, one test of the stability of a FA solution is its similarity following different extraction procedures. A stable solution usually tends to appear regardless of which extraction technique was employed. Table 10.6 (page 396) shows results from the same data set after extraction with the use of several techniques followed by varimax rotation. Similarities among the solutions are obvious.

None of the extraction techniques routinely provides an interpretable solution without rotation. All results of extraction may be rotated by any of the procedures described in Section 10.5.2, except Kaiser's Second Little Jiffy Extraction, which has its own rotational procedure.

Lastly, when using FA the researcher should hold in abeyance well-learned

TABLE 10.5 SUMMARY OF EXTRACTION PROCEDURES

Extraction technique	Program	Goal of analysis	Special features
Principal components	SPSS BMDP4M	Maximize variance extracted by orthogonal components.	Mathematically determined, empirical solution with common, unique, and error variance mixed into components.
Principal factors	SPSS BMDP4M	Maximize variance extracted by orthogonal factors.	Estimates communalities to attempt to eliminate unique and error variance from factors.
Image factoring	SPSS BMDP4M (Second Little Jiffy)		Uses SMCs between each variable and all others as communalities to generate a mathematically determined solution with error variance and unique variance eliminated.
Maximum likelihood factoring	BMDP4M SPSS (Rao's canonical factoring)	Estimate factor loadings for population that maximize the likelihood of sampling the observed correlation matrix.	BMDP4M uses newer, more efficient computing algorithm.
Alpha factoring	SPSS	Maximize the generalizability of orthogonal factors.	

proscriptions regarding data snooping. It is quite common to use PCA as a preliminary extraction technique, followed by one or more of the other procedures, perhaps varying number of factors, communality estimates, and rotational methods at each step. Analysis terminates when the researcher decides on the preferred solution.

10.5.1.1 PCA Versus FA One of the most important decisions is the choice between PCA and FA. Mathematically, the difference involves the contents of the positive diagonal in the correlation matrix (the diagonal that contains the correlation between a variable and itself). In either PCA or FA, the variance that is available to be analyzed is the sum of the values in the positive diagonal. If 1s are present in the diagonal (as in PCA), there is as much variance to be redistributed among components as there are observed variables: Each variable contributes a unit of variance by contributing a 1 to the positive diagonal of the correlation matrix. All the variance is redistributed, including error variance and variance unique to each observed variable. So if all components are retained, PCA duplicates exactly the standard scores of the observed variables by a linear combination of components.

In FA, on the other hand, only the variance that each observed variable

TABLE 10.6   RESULTS OF DIFFERENT EXTRACTION METHODS
ON SAME DATA SET

Unrotated Factor Loadings

	Factor 1				Factor 2			
Variables	PCA	PFa	Rao	Alpha	PCA	PFa	Rao	Alpha
1	.58	.63	.70	.54	.68	.68	−.54	.76
2	.51	.48	.56	.42	.66	.53	−.47	.60
3	.40	.38	.48	.29	.71	.55	−.50	.59
4	.69	.63	.55	.69	−.44	−.43	.54	−.33
5	.64	.54	.48	.59	−.37	−.31	.40	−.24
6	.72	.71	.63	.74	−.47	−.49	.59	−.40
7	.63	.51	.50	.53	−.14	−.12	.17	−.07
8	.61	.49	.47	.50	−.09	−.09	.15	−.03

Rotated Factor Loadings (Varimax)

	Factor 1				Factor 2			
Variables	PCA	PFa	Rao	Alpha	PCA	PFa	Rao	Alpha
1	.15	.15	.15	.16	.89	.91	.87	.92
2	.11	.11	.10	.12	.83	.71	.72	.73
3	−.02	.01	.02	.00	.81	.67	.69	.66
4	.82	.76	.78	.76	−.02	−.01	−.03	.01
5	.74	.62	.62	.63	.01	.04	.03	.04
6	.86	.86	.87	.84	.04	−.02	−.01	−.03
7	.61	.49	.48	.50	.20	.18	.21	.17
8	.57	.46	.45	.46	.23	.20	.20	.19

Note: The largest difference in communality estimates for a single variable between extraction
techniques was 0.08.

a Principal factors.

shares with other observed variables is made available for analysis. Exclusion
of error and unique sources of variability from FA is based on the belief that
such variability only confuses the structural picture that emerges from an analysis
of underlying processes. The common variance is estimated by *communalities,*
values between 0 and 1 that are inserted in the positive diagonal of the correlation
matrix.[4] The solution in FA concentrates on those variables with high communality values. The sum of the communalities (sum of the SSLs) is the variance
that is distributed among factors and is less than the total variance in the set
of observed variables. Because unique and error variances are omitted, a linear
combination of factors will approximate, but not duplicate, scores on observed
variables.

   The goal of PCA is to extract the maximum variance from the data set
by identification of a few orthogonal components. The goal of FA is to reproduce

---

   [4] Maximum likelihood extraction manipulates off-diagonal elements rather than values in
the diagonal.

the correlation matrix as closely as possible with the fewest number of factors. PCA offers a unique mathematical solution, whereas most forms of FA do not. PCA may be thought of as analyzing variance, FA as analyzing covariance (communality).

The choice between PCA and FA depends on your assessment of the fit between the common factor model, the data set, and the goals of the research. If you are interested in an inferred, hypothetical solution uncontaminated by unique and error variability, FA should be the choice. If, on the other hand, you desire an empirical summary of the data set, PCA is the better choice.

10.5.1.2 Principal Components Extraction of principal components is possible through BMDP4M and SPSS FACTOR. The goal of PCA is to extract maximum variance from the data set with each component. The first principal component is the linear combination of observed variables that maximally separates subjects by maximizing the variance of their component scores. The second component is formed from variability remaining in the data set after the variance associated with the first component is removed; it is the linear combination of observed variables that extracts maximum variability uncorrelated with the first component. Subsequent components also extract maximum variability from residual correlations and are orthogonal to all previously extracted components.

A set of principal components will be ordered, with the first component extracting the most variance and the last component the least variance. The solution is mathematically unique and, if all components are retained, will reproduce exactly the observed correlation matrix. Further, since the components are orthogonal, their use in other analyses may greatly facilitate interpretation of results.

PCA is the solution of choice for the researcher who is primarily interested in reducing a large number of variables down to a smaller number. PCA is also recommended as the first step toward a more detailed FA. PCA can reveal a great deal about probable number and nature of common factors even though it may not provide the solution that is finally interpreted.

10.5.1.3 Principal Factors Principal factors extraction differs from PCA in that estimates of communality, instead of 1s, are inserted in the positive diagonal of the observed correlation matrix. These estimates are derived through an iterative procedure, with SMCs (squared multiple correlations of each variable with all other variables) used as the starting values in the iteration. The goal of analysis, like that for PCA, is to extract maximum orthogonal variance from the data set with each succeeding factor. Advantages to principal factors extraction are that it is widely used (and commonly understood) and that it conforms to the factor analytic model in which common variance is analyzed independent of unique and error variance.

Because the goal is to maximize variance extraction, however, principal factors may not be as good as other extraction procedures in reproducing the correlation matrix. Also, communalities must be estimated and the solution will, to some extent, be determined by the values of those estimates.

10.5.1.4 Image Factor Extraction   Image factoring is available through both SPSS FACTOR and BMDP4M (as Kaiser's Second Little Jiffy). The technique is called image factoring because the analysis partitions among factors the variance of an observed variable that is *reflected* by the other variables, the SMC. Image factor extraction provides an interesting compromise between the characteristics of PCA and those of principal factors. Like PCA, image extraction provides a mathematically unique solution by the expedient of using fixed values in the positive diagonal of **R**. Like principal factors, the values in the diagonal are communalities with unique and error variability excluded. The compromise is struck by using the squared multiple correlation (SMC or R^2) of each variable with all others as the communality for that variable. With a hefty sample size and in excess of 10 to 15 observed variables, SMCs become stable and, depending on the adequacy of sampling among *variables,* provide a decent estimate of communality.

10.5.1.5 Maximum Likelihood Factor Extraction   The maximum likelihood method of factor extraction was developed originally by Lawley in the 1940s (see Lawley & Maxwell, 1963). Although the method was intuitively appealing, the number and difficulty of the required calculations made use of the method impractical. Now, however, an apparently efficient algorithm for maximum likelihood extraction is available through BMDP4M. Frane and Hill (1974) assert that the algorithm now used in BMDP4M usually converges quickly and accurately on population values. Maximum likelihood extraction is also available through Rao's canonical factoring in SPSS but with an inefficient and potentially quite expensive algorithm.

Maximum likelihood extraction estimates population values for factor loadings by calculating the set of factor loadings that maximizes the probability of sampling the observed correlation matrix from a population. Or, within constraints imposed by the correlations among variables, population estimates for factor loadings are calculated that have the greatest probability of yielding a sample with the observed correlation matrix.

10.5.1.6 Alpha Factoring   Alpha factor extraction, available through SPSS FACTOR, grew out of psychometric considerations in which researchers were interested in discovering which common factors would be found consistently when repeated samples of *items* were taken from a population of items. The problem is logically similar to identifying differences that would be consistently found among samples of subjects taken from a population of subjects—a question at the heart of most univariate and multivariate statistical procedures. In alpha factoring, however, the concern is with the reliability of the common factors rather than with the reliability of group differences. Coefficient alpha is a measure derived in psychometrics for the reliability (also called generalizability) of a score taken in a variety of situations. In alpha factoring, communalities are calculated, using iterative procedures, to maximize coefficient alpha for the factors.

Probably the greatest advantage to the procedure is that it focuses the researcher's attention squarely on the problem of sampling items from the domain of items of interest. Disadvantages stem from the relative unfamiliarity of most researchers with the procedure and the reasons for it.

10.5.2 Rotation

The results of factor extraction, unaccompanied by rotation, are likely to be uninterpretable regardless of which extraction technique is used. After extraction, rotation is used to improve the interpretability and scientific utility of the solution. It is *not* used to improve the quality of the mathematical fit between the observed and reproduced correlation matrices because, in fact, all (orthogonally) rotated solutions are mathematically equivalent to one another and to the initial solution.

A fundamental decision is required between an orthogonal and an oblique rotational scheme. In orthogonal rotation, the factors are uncorrelated with one another. Orthogonal solutions offer ease of description and interpretation of results; yet they may strain "reality" unless the researcher is convinced that the underlying structures producing the results—the factors—actually are operating almost independently of one another. The researcher unprepared to believe that common factors come as uncorrelated packages may wish to employ an oblique rotation. In oblique rotations the common factors may be correlated with one another, with conceptual advantages and practical disadvantages in interpreting and describing the results.

Among the tens of rotational techniques that have been proposed, only those that are available in the BMDP and SPSS packages are included in this discussion (see Table 10.7). The reader who wishes to know more about these or other techniques could consult Harman (1967) and Mulaik (1972). A helpful presentation of rotation-by-hand is available in Comrey (1973, pp. 109–145).

10.5.2.1 Orthogonal Rotation
Three orthogonal rotational techniques are available in both SPSS FACTOR and BMDP4M: varimax, quartimax, and equimax. Varimax rotation is easily the most commonly used of all the rotations available.

Just as the extraction procedures have slightly different goals, both logically and mathematically, so also the rotational procedures were derived to perform different tasks. The goal of varimax rotation is to make the factors as simple as possible by maximizing the variance of the loadings across variables within factors. In this way, loadings tend to become higher for those variables with high correlations with a factor and smaller for the other variables. Interpreting a factor is facilitated because the composite of variables with which a factor is correlated is obvious. Varimax also tends to reapportion the variance among factors so that they become relatively equal in importance, accomplishing this

TABLE 10.7 SUMMARY OF ROTATIONAL TECHNIQUES IN SPSS AND BMDP

Rotational technique	Program	Type	Goal of analysis	Comments
Varimax	BMDP4M SPSS	Orthogonal	Minimize complexity of factors (simplify columns of loading matrix) by maximizing variance of loadings on each factor.	Most commonly used rotation; recommended as default option (in BMDP, Γ [gamma] = 1)
Quartimax	BMDP4M SPSS	Orthogonal	Minimize complexity of variables (simplify rows of loading matrix) by maximizing variance of loadings on each variable.	First factor tends to be general with others subclusters of variables (in BMDP, $\Gamma = 0$)
Equimax	BMDP4M SPSS	Orthogonal	Simplify both variables and factors (rows and columns); compromise between quartimax and varimax.	May behave erratically ($\Gamma = \frac{1}{2}$)
Orthogonal with gamma	BMDP4M	Orthogonal	Simplify either factors or variables, depending on value of gamma (Γ).	Gamma (Γ) continuously variable
Direct oblimin	BMDP4M SPSS	Oblique	Simplify factors by minimizing cross products of loadings.	Continuous values of gamma, Γ (BMDP) or delta, δ (SPSS), available; allows wide range of factor intercorrelations
(Direct) quartimin	BMDP4M SPSS	Oblique	Simplify factors by minimizing sum of cross products of squared loadings in pattern matrix.	Permits fairly high correlations among factors. Recommended oblique rotation by BMDP series; in BMDP, $\Gamma = 0$. Achieved in SPSS by setting $\delta = 0$
Orthoblique	BMDP4M	Both orthogonal and oblique	Rescale factor loadings to yield orthogonal solution; nonrescaled loadings may be correlated.	Accompanies Kaiser's Second Little Jiffy (image) extraction only.

by taking variance from factors first extracted and distributing it among the later ones.

Quartimax does for the variables what varimax does for the factors: It simplifies them by increasing the dispersion of the loadings within variables across factors. Unlike varimax, which operates on the columns of the loading matrix, quartimax operates on the rows. Quartimax is not nearly as popular as varimax because one is usually more interested in simple factors than in simple variables.

Equimax is a hybrid between varimax and quartimax that tries simultaneously to simplify the factors and the variables. Mulaik (1972) reports that equimax tends to behave erratically unless the researcher can specify the number of factors with confidence.

Although varimax rotation simplifies the factors, quartimax the variables, and equimax both, they do so in BMDP4M by setting levels on a simplicity criterion of 1, 0, and ½, respectively. The criterion, called Γ (gamma), can also be continuously varied between 0 (variables simplified) and 1 (factors simplified) by using orthogonal with Γ rotation, which allows the user to specify any desired Γ level.

For most applications, the varimax rotation is probably the rotation of choice; it is recommended as the default option by both SPSS and BMDP.

10.5.2.2 Oblique Rotation

An *embarasse de richesse* awaits the researcher who decides to employ oblique rotation (see Table 10.7). Oblique rotations offer a continuous range of correlations between factors. The amount of correlation that is permitted between factors is determined by a variable called delta, δ, by SPSS FACTOR and gamma, Γ, by BMDP4M.[5] The values of delta and gamma determine the maximum amount of correlation permitted among factors. When gamma or delta is less than zero, solutions are increasingly orthogonal. When either gamma or delta is zero, solutions are allowed to be fairly highly correlated. Gamma or delta values near 1 can produce factors that are very highly correlated. Although there is a positive relationship between size of delta or gamma and size of correlation, the exact amount of maximum correlation for various sizes of gamma and delta cannot be specified independent of a data set.

The family of procedures used for oblique rotation with varying degrees of correlation in SPSS and BMDP is direct oblimin. In the special case where Γ or δ = 0 (the default option for both SPSS and BMDP), the procedure is called direct quartimin. Values of gamma or delta greater than zero permit high correlations among factors, and the researcher should take care that the correct number of factors is chosen. Otherwise highly correlated factors may not remain distinguishable one from the other. Some trial and error, coupled with inspection of the scatterplots of relationships between pairs of factors may be required to determine the most useful size of gamma or delta. Or, one might

---

[5] Note that Γ is used by BMDP to indicate nature of simplicity in orthogonal rotation and amount of obliquity in oblique rotation.

simply trust to the default values for maximum flexibility with minimum ambiguity.

Orthoblique rotation is designed to accompany Kaiser's Second Little Jiffy (Image) Factor extraction, and does so automatically through BMDP4M. Orthoblique rotation uses the quartimax algorithm to produce an orthogonal solution *on rescaled factor loadings;* therefore the solution may be oblique with respect to the original factor loadings.

10.5.2.3 Geometric Interpretation   Recall from algebra or geometry that a set of points can be represented in two-dimensional space by listing their coordinates with respect to X and Y axes, as illustrated in Figure 10.1 for the four variables of Section 10.4. Points representing COST (−.400, .900), LIFT (.251, −.947), DEPTH (.932, .348), and POWDER (.956, .286) in Figure 10.1(a) can also be accurately located with respect to different axes, as shown in Figure 10.1(b). Note that the position of points has not changed, although their coordinates have changed in the new axis system where COST has become (−.086, .981), LIFT (−.071, −.977), DEPTH (.994, .026), and POWDER (.997, −.040), as shown in Figure 10.1(b). The effect of rotation has been to amplify high loadings while minimizing low ones. Spatially, the effect has been to rotate the axes so that they "shoot through" the variable clusters more closely.

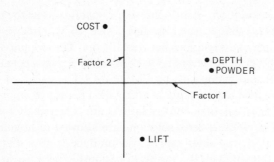

(a) Location of COST, LIFT, DEPTH, and POWDER after extraction (before rotation)

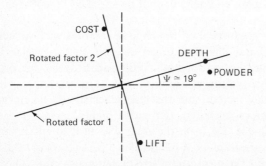

(b) Location of COST, LIFT, DEPTH, and POWDER vis-à-vis rotated axes

Figure 10.1 Illustration of rotation of axes to provide a better definition of factors vis-à-vis the variables with which they correlate.

Factor extraction yields a solution in which observed variables are represented as vectors (whose lengths are the same as the communalities of the variables) that terminate at the points indicated by the coordinate system. The coordinate system is formed from the common factors that are axes for the system. The coordinates of each point are the weights from the loading matrix. If there are three common factors, then the space has three dimensions and the point for each observed variable will be positioned in the space by three coordinates.

If the factors are orthogonal—that is, the common factor axes are all at right angles to one another—then the coordinates of the variable points also represent the correlations between the common factors and the observed variables. Correlations (factor loadings) can be read directly from the graphs by projecting perpendicular lines from each point to each of the common factor axes.

One of the primary goals of PCA or FA, and the motivation behind extraction, is to discover the minimum number of common factor axes that is needed to reproduce adequately the scores on the observed variables. This goal is akin to that of finding the minimum number of dimensions by which the points can be reliably positioned.

A second major goal, and the motivation behind rotation, is the discovery of the meaning attached to the common factors that underlie responses to observed variables. This goal is akin to that of interpreting the common factor axes that are used to define the space. Factor rotation is conducted to reposition common factor axes so as to make them maximally interpretable. Repositioning the axes will change the coordinates of the variable points but not the positions of the points with respect to each other.

Common factors are usually considered interpretable when some observed variables load highly on them and the rest do not. This allows interpretation of the factor in terms of the high-loading variables. Ideally, each variable would be highly correlated with one, and only one, common factor. In graphic terms this means that the points representing each variable in the common factor space would lie far out along one common factor axis but near the origin with respect to the other axes, or that the coordinates of the point would be large for one axis and near zero for the other axes.

If you had only one observed variable, of course, it would be trivial to position the common factor axis to meet these criteria (variable point and axis would overlap in a space of one dimension). However, with many variables and several common factor axes, compromises are required in positioning the common factor axes. The variables become a "swarm" in which variables that are correlated with one another form a cluster of points that are near one another. With luck, the clusters of points will be about 90° away from one another so that an orthogonal solution is indicated. And with lots of luck, the variables will also cluster in groups with empty spaces between them so that the common factor axes are nicely defined.

In oblique rotation the situation is slightly more complicated. Since common factors may be correlated with one another, common factor axes will not necessarily be at right angles. And, though it is easier to position axes so that loadings are either high or low, it may be harder to visualize and interpret the solution. For practical suggestions of ways to use graphic techniques to judge the adequacy of rotation, see Section 10.6.3.

10.5.3 Some Practical Recommendations

Although an almost overwhelming combination of extraction and rotation techniques is available, in practice differences among them may be slight. With a large number of variables, several high-loading variables per factor, with the same, well-chosen number of factors, and with similar values for communality, the results of extraction will be similar regardless of which extraction method is used. Further, differences that are apparent after extraction tend to disappear after rotation.

Most researchers begin their FA by using principal components extraction and varimax rotation.[6] From the results, one can estimate the rank of the observed correlation matrix (Section 10.3.2.4), the number of factors (Section 10.6.2), and variables that might be excluded from subsequent analyses (Section 10.3.2.5).

During the next few runs, researchers experiment with different numbers of factors, different extraction techniques, and both orthogonal and oblique rotations. Some number of factors with some combination of extraction and rotation will produce the result with the greatest scientific utility, consistency, and meaning; presumably this will be the solution that is interpreted.

10.6 SOME IMPORTANT ISSUES

Resolution of several of the issues raised in this section may be accomplished through a variety of methods. Usually the results of application of different methods will be consistent; occasionally they will not. When they are not (for instance, when one method tells you that enough factors have been extracted, whereas another method indicates the presence of one or more additional factors), you must rely on judgment regarding which answer to attend, remembering all the while that interpretability and replicability are the final tests of any FA.

10.6.1 Estimates of Communalities

FA differs from PCA in that communality values (numbers between 0 and 1) replace 1s in the positive diagonal of **R** in preparation for factor extraction. Communality values are used in an effort to produce a solution in which factor structure is uncontaminated by the unique and error variabilities associated with each observed variable. But communality values have to be estimated, and there is considerable dispute regarding how that should be done.

---

[6] They might also profitably begin with a cluster analysis, as described in Section 10.6.2.

The SMC of each variable with the others in the sample is usually used as the starting estimate of communality. As the solution develops, iterative procedures (which can be directed by the researcher) are used to adjust communality values to provide the best fit between observed and reproduced correlation matrices with the smallest number of factors. Iteration continues until successive communality values hardly differ from one another. Final estimates of communality are also SMCs, but now between each variable and the factors. That is, communality represents the proportion of variance in a variable that is predictable from the factors underlying it. Communality estimates do not change with orthogonal rotation.

Image extraction and maximum likelihood extraction are slightly different. In image extraction, estimates of communality, which are the SMCs of each variable with all others, are used as the communality values throughout. In maximum likelihood extraction, number of factors instead of communality values are estimated and off-diagonal correlations are "rigged" so as to produce the best fit between observed and reproduced matrices, given the estimated number of factors.

BMDP4M offers SMCs, user-specified values, or maximum row values (which are lower bounds for SMCs) as initial communality estimates. SPSS FACTOR permits any of these when using principal factor extraction without iteration but otherwise uses SMCs. Fewer iterations are usually required when starting from SMCs.

The seriousness with which estimates of communality should be regarded depends on the number of observed variables. If the number of variables exceeds, say, 20, sample SMCs probably provide reasonable estimates of communality. Furthermore, with 20 or more variables, the elements in the positive diagonal are few compared to the total number of elements in **R,** and their sizes will not influence the solution very much. Actually, if the communality values for all variables are of approximately the same magnitude, results of PCA and FA are also likely to be very similar.

If communality values exceed 1, however, problems with the solution are indicated. Probably the number of factors extracted is wrong, and addition or deletion of one or two factors will reduce the communality below 1. Very low communality values, on the other hand, indicate that the variables with them are outliers that might be deleted (Section 10.3.2.5).

10.6.2 Adequacy of Extraction and Number of Factors

Because inclusion of more factors in a solution improves the fit between observed and reproduced correlation matrices, these two issues are inseparable. The more factors one permits, the better the fit and the greater the percent of variance in the data "explained" by the factor solution. However, the greater the number of factors included, the less parsimonious the solution. To account for all the variance (PCA) or covariance (FA) in a data matrix, one would normally have to include as many factors as observed variables. It is clear, then, that a trade-off is required: One wants to include enough factors for an adequate fit, but not so many that parsimony is lost. The question then becomes, What is adequate?

Selection of the number of factors is probably more critical than selection of extraction and rotational techniques or communality values. In confirmatory FA, selection of the number of factors is really selection of the theoretical framework underlying a research area, although interpretation of the factors is also important. You can partially check the consistency between a theoretical factor framework and an obtained solution by asking if the theoretical number of factors is adequate to fit the data. Hypothesized underlying factors may have to be added or dropped for adequate fit.

You can assess adequacy of extraction for a given number of factors in several ways, some of which are presented next. Because determination of adequacy involves judgment, and because the BMDP and SPSS systems differ in the kinds of information available, you may want to perform only some of the tests that will be described here.

Using SPSS, you can get a first quick estimate of the number of factors from the sizes of the eigenvalues reported as part of an initial run with PCA extraction. Eigenvalues represent variance. Because the variance that each standardized observed variable contributes is 1 (or less), any factor with an eigenvalue less than 1 is not as important, from a variance perspective, as an observed variable. The number of factors with eigenvalues greater than 1 is an estimate of the maximum number of factors.

As a second estimate, you can perform the scree test (Cattell, 1966) on the percent of variance accounted for by each of the factors in the solution. The percent of variance is given directly, both for each factor separately and for the factors cumulatively. To perform the test, all the potential factors, in descending order, are arranged along an abscissa with percent of variance as the ordinate. Usually the resulting curve is negatively decreasing—percent of variance is highest for the first factor and moderate but decreasing for the next few factors before reaching an asymptotically small level for the last several factors. You look for a point of diminishing returns, where inclusion of more factors adds very little to the variance accounted for by the solution, as illustrated in Figure 10.2. Probably you will want to experiment in future analyses by adding or subtracting one or two factors around the point of diminishing returns. Just where the point is, for you, is a matter of judgment.

At a practical level, then, still within the SPSS framework, you perform an initial run with PCA extraction and varimax rotation to determine both the number of factors with eigenvalues greater than 1, and, within that group, the number of factors accounting for meaningful percents of variance. In the default option only factors with eigenvalues greater than 1 are rotated. A look at the SSLs and percents (or proportions) of variance obtained *after* rotation may provide more information regarding the number of reliable factors.[7] Usually

---

[7] In SPSS, SSLs need to be calculated by squaring and summing the loadings for each rotated factor, as illustrated in Table 10.4. The table and its accompanying narrative also show how proportion of variance is found.

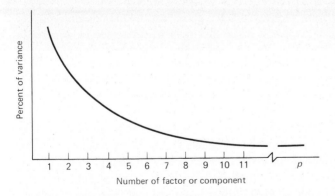

Number of factor or component

Figure 10.2　Cattell's scree test for estimating the number of meaningful factors.

a few of the factors with eigenvalues greater than 1 before rotation have SSLs less than 1 after it. The number of factors remaining with SSLs greater than 1 after rotation is probably a good estimate of the number of reliable factors.

At this point it becomes important to look at the rotated loading matrix to determine the number of variables used to define each factor (see Section 10.6.5). If only one variable loads highly on a factor, the factor is insufficiently defined. If two variables define a factor, then whether or not it is reliable depends on the pattern of correlations of those two variables with each other and with other variables in **R**. If the two variables are highly correlated with each other (say, $r > .70$) and relatively uncorrelated with others, the factor may be reliable. If this pattern in correlations is observed, you should then inspect the two items themselves, to make sure they are distinct and not really just one variable.

If you have access to BMDP, you can perform all the tests just described for number of factors and adequacy of extraction and, like many researchers, you may wish to use them. However, a couple of less complicated procedures are available, one through cluster analysis using BMDP1M. As illustrated in Section 10.8, number of reliable factors is estimated by subjecting the group of variables to cluster analysis and then inspecting the sorted and shaded correlations (as optionally available through the SHADE command in the PRINT paragraph). The number of densely shaded triangles is a good estimate of the number of reliable factors.

Once the FA has been performed using BMDP4M, adequacy of extraction may be assessed through inspection of the residual correlation matrix, as indicated in Section 10.4. If all the residual correlations are less than, say, .10, then the reproduced correlation matrix is very similar to the original correlation matrix and extraction is adequate. If not, more factors may be required.

There is some debate concerning whether it is better to retain too many or too few factors if the number is ambiguous. Sometimes a researcher may want to rotate, but not interpret, marginal factors for statistical purposes (e.g., to keep all communality values ≤ 1). Other times the last few factors, even though of marginal stability, may represent the most interesting and unexpected findings in a research area. These, then, are good reasons for retaining factors of marginal reliability. However, a researcher may be interested in using only demonstrably reliable factors; in this case, retention of the fewest possible factors is recommended.

10.6.3 Adequacy of Rotation and Simple Structure

The decision between orthogonal and oblique rotation is an important one that should be made as soon as the number of reliable factors is reasonably apparent. In most factor analytic situations, it seems more likely that factors are correlated than that they are not, since clusters of variables in the "real world" are more likely than not related to one another. However, reporting the results of oblique rotation requires reporting the elements of the pattern matrix (**A**) as well as the factor correlation matrix (**Φ**), whereas reporting orthogonal rotation requires only the loading matrix (**A**). Thus simplicity of reporting results would favor orthogonal rotation.

Further, if factor scores or factorlike scores (Section 10.6.6) are to be used as IVs or DVs in other analyses, or if a goal of analysis is comparison of factor structure in groups (Section 10.6.7), then orthogonal rotation has distinct advantages. However, for all its conceptual simplicity, orthogonal rotation is *not* an advantage if it distorts the true relationships among factors by imposing independence on correlated dimensions.

Perhaps the best way to decide between orthogonal and oblique rotations is to request oblique rotation with the desired number of factors specified and inspect the size of the correlations among factors. The default options for both SPSS FACTOR and BMDP4M will calculate factors that are fairly highly correlated if the data so indicate. However, if the data do not indicate the presence of correlated factors, those factors will not be correlated. If some correlations in **Φ** exceed .30, they indicate a 10% (or more) overlap in variance among factors, enough variance to be taken seriously unless there are compelling reasons to do otherwise. In other words, we recommend interpreting and reporting oblique rotation if correlations among factors exceed .30 (or thereabouts) unless the desire to compare structure in groups, a need for orthogonal factors in other analyses, or theory dictate otherwise.

The *adequacy* of rotation may be assessed through several considerations. Perhaps the simplest test is inspection of the observed correlation matrix for patterns of correlations among variables. Are the patterns represented in the rotated solution? Do variables that are highly correlated tend to load on the same factor? If you included marker variables, do they load on factors as predicted?

Another criterion is simple structure, an idea developed by Thurstone (1947). If simple structure is present (and factors are reasonably uncorrelated), only a few variables correlate highly with each factor and only one factor correlates highly with each variable. In other words, the columns of **A**, which define factors, should have several high and many low values while the rows of **A**, which define variables vis-à-vis factors, should have only one high value.

Rows with several high correlations correspond to variables that are said to be complex, that is, to simultaneously reflect two or more dimensions or ideas defined by factors. At least in the initial stages of a research program, complex variables should be avoided (Section 10.3.1). Whereas rows reflect the complexity of the variables, columns reflect the simplicity of the factors. Various orthogonal rotational methods (see Section 10.5.2) have been designed to simplify either the variables or the factors (or both). The most popular orthogonal rotational method, varimax, tends to produce simple factors.

Adequacy of rotation may also be ascertained through inspection of the Plot of Rotated Factors in SPSS (for orthogonal rotations *only*) or the Rotated Factor Loadings in BMDP (available for both kinds of rotation). In the figures, factors are considered two at a time. One should consider the *distance, clustering,* and *direction* of the points representing variables relative to the factors as axes in the figures. The distance of a variable point from the origin reflects the size of factor loadings; variables that correlate highly with a factor are far out on the factor axis. Ideally, each variable point would be far out on one axis and near the origin on all others. Clusters of variable points reveal the simplicity of factors. One likes to see a cluster of several points near the ends of each axis and all other points near the origin. A smattering of points at various distances from the origin indicates a factor that is not simple, while a cluster of points midway between two axes may reflect the presence of another factor. Finally, the direction of the clusters (after orthogonal rotation) may indicate the need for oblique rotation. If data points do not distribute themselves close to factor axes after orthogonal rotation, the angle between clusters with respect to the origin will not be 90°. That is, the optimal fit of data points is around axes that are not orthogonal. Another analysis with oblique rotation may reveal substantial correlations among factors. Several of these relationships are depicted in Figure 10.3.

10.6.4 Importance and Internal Consistency of Factors

The importance of a factor (or a set of factors) is probably best evaluated by the proportion (or percent) of variance or covariance explained by the factor after rotation. The proportion of variance attributable to individual components or factors may differ before and after rotation because rotation tends to redistribute variance among factors somewhat. Ease of ascertaining proportions of variance depends on whether rotation was orthogonal or oblique.

After orthogonal rotation, the importance of individual factors is related to the sizes of their SSLs. Factors with SSLs greater than 1 can be converted to proportion of variance (by dividing its SSL by p, the number of variables) or proportion of covariance (by dividing the SSL by the sum of SSLs or, equivalently, sum of communalities). Table 10.4 in Section 10.4 illustrates the calculations for the example. The proportion of variance accounted for by a factor indicates the amount of variance in the original variables (where each has contributed a unit of variance) that has been condensed into the factor. The proportion of covariance accounted for by a factor indicates the relative importance

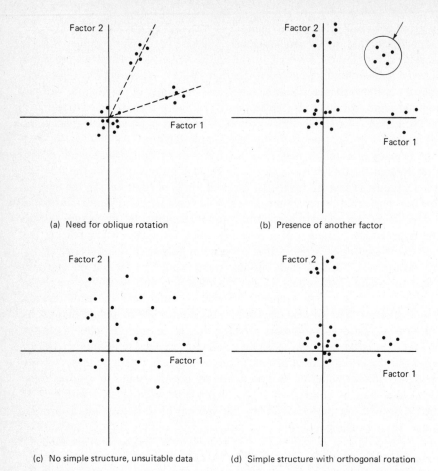

Figure 10.3 Pairwise plots of factor loadings following orthogonal rotation and indicating: (a) need for oblique rotation; (b) presence of another factor; (c) unsuitable data; and (d) simple structure.

of the factor to the total variance accounted for by all factors combined. All factors combined are, of course, likely to account for only a fraction of the total variance in the original solution.

Although the proportions of variance for an obliquely rotated solution can be obtained *before* rotation by the methods just described, they will be only rough indicators of the proportions of variance and covariance of factors in the rotated solution. Because factors are correlated, they share overlapping variability so that assignment of variance to individual factors or components is ambiguous. After oblique rotation the size of the SSLs associated with individual factors may be taken as a rough approximation to their importance—bigger ones are more important—but proportions of variance cannot be specified directly.

An estimate of the internal consistency of the solution—the certainty with which common factor axes are fixed in the variable space—is given by the squared multiple correlations of the factor scores predicted from scores on the observed variables. A high SMC (say, .70 or better) means that the observed variables account for a substantial variance in the factor scores. A low SMC means the factors are poorly defined by the observed variables included in the research. The larger the SMCs, the more stable the factors and the greater the confidence with which interpretations may be assigned to them.

BMDP4M prints these SMCs as the positive diagonal of the factor-score covariance matrix. The values could be generated in SPSS by calculating factor scores with FACTOR and then using REGRESSION to get SMCs between the observed variables and the factor scores.

10.6.5 Interpretations of Factors

To interpret a factor, one tries to understand the underlying dimension that unifies the group of variables defining it. In both orthogonal and oblique rotations, the information regarding which variables define a factor is obtained from the loading matrix, **A**, but the meaning of the loadings is different for the two rotations.

After orthogonal rotations, the values in the loading matrix are correlations between the variables and the factors. Thus one decides which size of correlation is meaningful (usually a correlation at least larger than .30), collects together the variables with loadings in excess of the criterion, and searches for a concept that unifies them, with greater attention to variables having higher loadings.

After oblique rotation, the process is the same, but the interpretation of the values in **A**, the pattern matrix, is no longer straightforward. In this case the loading may be taken as a measure of the unique contribution of each factor to the variance in a variable, but it is not a correlation. Because the factors are correlated, the correlations between the variables and factors (available in the structure matrix, **C**) are inflated by the amount of overlap between factors. Thus a variable may correlate with one factor through its correlation with another factor rather than directly. The elements in the pattern matrix have had variance overlap among factors "partialled out," but at the expense of conceptual simplicity. Actually, the reason for interpretation of the pattern matrix rather than the structure matrix is probably pragmatic—it's easier. The difference between high values and low values is more apparent in the pattern matrix than in the structure matrix.

As a rule of thumb, loadings in excess of .30 are eligible for interpretation, whereas lower ones are not, because a factor loading of .30 indicates at least a 9% overlap in variance between the variable and the factor. The greater the overlap between a variable and a factor, the more that variable is a pure measure of the factor. Comrey (1973) suggests that loadings in excess of .71 (50% variance) are considered excellent, .63 (40%) very good, .55 (30%) good, .45 (20%) fair, and .32 (10% of variance) poor. Choice of the cutoff of size of loading to be interpreted is a matter of researcher preference.

Because the size of loadings reflects, to some extent, the homogeneity of scores in the sample, if homogeneity is suspected, interpretation of lower loadings may be warranted. That is, if the sample does not produce a wide spread of scores on observed variables, factor interpretation may require inspection of lower loadings.

At some point, a researcher generally tries to characterize a factor by assigning it a name or a label, a process that involves art as well as science. Rummel (1970) provides numerous helpful hints on interpreting and naming factors. If BMDP4M is available, interpretation of factors is facilitated by output of the matrix of sorted and shaded correlations that groups variables by their correlations with factors.

The replicability, utility, and complexity of factors are also considerations in interpretation. Is the factor structure replicable in time and/or with different groups? Is it a useful addition to scientific thinking in a research area? Where does the factor fit in the hierarchy of "explanations" about a phenomenon? Is it sufficiently complex so as to be intriguing without being so complex that it is uninterpretable?

10.6.6 Factor Scores

Among the more potentially useful outcomes of PCA or FA are factor scores. These scores represent estimates of the scores subjects would have received on each of the latent variables had they been measured directly.

Because there are normally fewer factors than observed variables, and because factor scores are nearly independent if factors are orthogonal (cf. Figure 10.3), input of factor scores to subsequent analyses may facilitate both calculation and interpretation. Multicollinear matrices can be reduced to orthogonal components using PCA, for instance. Or one could use FA to reduce a large number of DVs to a small number of factors for use as DVs in MANOVA. Alternatively, one could reduce a large number of IVs to a small number of factors for purposes of predicting a DV in multiple regression or group membership in discriminant analysis. If factors are fewer in number, more stable, and more meaningful than observed variables, their use enhances subsequent analyses.

Procedures for estimating factor scores range between simple-minded (but frequently adequate) and reasonably sophisticated. The method described in Section 10.4 (especially Eqs. 10.10 and 10.11) is the one used by BMDP and SPSS to generate factor scores by applying sophisticated regression techniques. This regression method, like all others (see Chapter 5), may capitalize on chance relationships among variables so that factor-score estimates are overdetermined.

Several rather simple-minded techniques exist for estimating factor scores (Comrey, 1973). Perhaps the simplest is to sum scores on variables that load highly on the same factor. This procedure weights the importance of variables in producing factor scores by their variability, a problem that is alleviated if variable scores are standardized or if the variables have roughly equal dispersion to begin with. For many research purposes, a "quick and dirty" estimate like this one may be entirely adequate.

Several limitations to use of factor scores are worthy of note. First, they are *estimates* only. Second, changes in extraction or rotation may change the estimate considerably. Third, factor-score estimates (but not component-score estimates) may be somewhat correlated even following orthogonal rotation. Lastly, the internal consistency with which factors have been derived is a major consideration to use of factor scores. Consult Section 10.6.4 for a discussion of internal consistency and its assessment.

10.6.7 Comparisons Among Solutions and Groups

Frequently a researcher is interested in deciding whether or not two groups that differ in experience or characteristics share the same latent structure. These comparisons may involve the *pattern* of the correlations between variables and factors, or both the *pattern and magnitude* of the correlations between them. Rummel (1970), Levine (1977), and Pinneau and Newhouse (1964) give excellent summaries of several comparisons that might be of interest. Only the easier comparison techniques will be mentioned here.

It is important to note that theory can be tested in FA using these procedures. Theory regarding underlying structure is used to generate one set of loadings to be compared with a set derived from a sample. Estimation of the magnitude of factor loadings for variables from theory does not have to be very precise: 1s can be used as loadings for variables that are expected to load on a factor, while zeros are used as loadings for the other variables. Comparisons between loadings from theory and loadings from sample data are then conducted in confirmatory FA.

The first step in comparing factors from two different samples is to generate them. For purposes of comparison, it is critical that similar procedures be employed at the various stages of analysis with the data sets to be compared. Similar variables and, if possible, similar marker variables should be included during data collection. Similar procedures for handling missing data and outliers should be employed. Variable transformations, if used, should be applied toward the same goals with the same variables in both data sets. Extractional and rotational techniques should be the same, as should the criterion for determining number of factors. If factor scores are to be compared, they should be generated by the same procedures.

Once data sets have been factor analyzed, the next decision involves which pairs of factors to compare. Comparing all possible pairs of factors may result in spuriously significant results by capitalizing on chance relationships. Presence of marker variables will vastly simplify this stage of comparison.

At this point careful inspection of the loading matrices for both groups may reveal similarities or differences in factor structure sufficiently obvious as to dispel the need for more formal procedures. When the same criteria were used, did both groups generate the same number of factors? If not, there is an obvious difference in overall structure. Do almost the same variables load highly on the different factors for the two groups? Could you reasonably use the same labels to name factors for both groups? If all three questions are answered in the affirmative, it may be unnecessary to proceed to statistical comparisons.

Most of the more formal, numerical comparisons are performed on either the loading matrix or the pattern matrix. An important decision is whether to compare just the pattern of correlations or both the pattern and magnitude of the correlations among variables and factors in the data sets. Comparisons involving magnitude are somewhat more stringent than those involving just pattern. However, the magnitude of loadings may be influenced by extraneous features of data collection (such as homogeneity of a sample for factors being compared). Cattell's salient similarity index, s (Cattell & Baggaley, 1960; Cattell, 1957), is sensitive to pattern of loadings, while the Pearson product-moment correlation coefficient, r, is sensitive to both pattern and magnitude of loadings.

To illustrate methods of comparison, the two loading matrices in Table 10.8 will be used. Both are products of overactive imagination but do illustrate a typical problem in factor comparison, namely, that factor 1 in Set 1 is similar to factor 2 (rather than to factor 1) in Set 2. Sometimes it is hard to decide which factors to compare. Marker variables can be of help.

In calculating s, the first step is to construct a two-way frequency table with pairs of loadings for each variable on each factor contributing a single tally to the table according to whether the loadings are positively salient (*PS*), negatively salient (*NS*), or neither (hyperplane or *HP*) on each of the factors being compared. Cattell used a cut of .10 for determining salience; loadings at or above .10 were salient while lower ones were not. But as a cut in loadings of .32 or better seems more appropriate, it will be employed here.

Once the frequency table is constructed, s is calculated as follows:

$$s = \frac{c_{11} + c_{33} - c_{13} - c_{31}}{c_{11} + c_{33} + c_{13} + c_{31} + .5(c_{12} + c_{21} + c_{23} + c_{32})} \tag{10.16}$$

The c values in the equation represent frequency counts in cells in the frequency table. Application of the equation to comparison of factor 1 in Set 1 and factor 2 in Set 2 is illustrated in Table 10.9.

Although exact probability statements about the value of s are not currently available, estimates of them are provided by Cattell and colleagues (1969) and reproduced in Appendix C, Table C.7. Probabilities must be assessed with respect to both the number of variables, p, and the percentage of cases that fall into the hyperplane for the pair of factors being compared: 60%, 70%, 80%, or

TABLE 10.8 LOADING MATRICES FROM TWO HYPOTHETICAL DATA SETS

	Set 1		Set 2	
	Factor 1	Factor 2	Factor 1	Factor 2
COST	−.086	.981	.732	.265
LIFT	−.072	−.978	.649	.537
DEPTH	.994	.027	.211	.874
POWDER	.997	−.040	.189	.796

TABLE 10.9 CALCULATION OF CATTELL'S SALIENT SIMILARITY INDEX s

		Set 1		
		PS	HP	NS
	PS	c_{11}	c_{12}	c_{13}
Set 2	HP	c_{21}	c_{22}	c_{23}
	NS	c_{31}	c_{32}	c_{33}

For the example:

		Set 1		
		PS	HP	NS
	PS	2	1	
Set 2	HP		1	
	NS			

$$s = \frac{2+0-0-0}{2+0+0+0+.5(1+0+0+0)}$$

$$= \frac{2}{2.5}$$

$$= .80$$

90%. If a value of s exceeds that of v_s for some hyperplane percentage and number of variables, then the factors are similar at the corresponding level of probability. For instance, if the hyperplane count is 60% and 40 variables are being compared, an s value in excess of .26 indicates a relationship between the factors at the .041 significance level. For the example, as only 25% of the loadings were in the hyperplane and only four variables were included, the significance of s cannot be determined.

A second method of comparing factors involves calculation of Pearson r (see Chapter 2, Eq. 2.29) using as data the loadings on the factors to be compared for two groups. For the same comparison of factors as shown in Table 10.9, $r = .91$. That is, the loadings for factor 1 in Set 1 and factor 2 in Set 2 from Table 10.8 correlate .91. Although calculating r for loadings is a straightforward procedure and the meaning of r is widely understood, this method of comparing factors has drawbacks. If a large number of variables is in the set used to define the factors, it is possible for r to be large even though none of the variables with large loadings correspond between the two factors. The correlation will be sizable because of the numerous variables with small loadings that are not used to define either factor. Thus caution is urged in interpretation of r used to compare factors.

Another method, somewhat experimental at this stage, involves generating pairs of factor scores for a group by using both the factor-score coefficients for that group and those for the other group and then correlating the pairs of factor scores. If correlation is high, it implies that there is good correspondence between factor scores generated by the two different groups and that factor structure is therefore similar.

10.7 COMPARISON OF PROGRAMS

There is one BMDP program and one SPSS program for PCA and FA. In both programs, options are available for methods of extraction as well as methods of rotation. The programs are different, however, both in flexibility and in the amount of supplemental information provided.

10.7.1 SPSS Package

The program in the SPSS package developed for PCA and FA is FACTOR. Features of the program are described in Table 10.10.

TABLE 10.10   COMPARISON OF FACTOR ANALYSIS PROGRAMS

Feature	SPSS Factor	BMDP4M
Input		
Correlation matrix	Yes	Yes
About origin	No	Yes
Covariance matrix	No	Yes
About origin	No	Yes
Factor loadings	Yes	Yes
Factor-score coefficients	No	Yes
Specify maximum number of factors	Yes	Yes
Extraction method (see Table 10.5)		
PCA	Yes	Yes
PFA	Yes	Yes
Image	Yes	Yes
Maximum likelihood (Rao's canonical)	Yes	Yes
Alpha	Yes	No
Specify communalities	Yes	Yes
Specify minimum eigenvalue	Yes	Yes
Specify maximum number of iterations	Yes	Yes
Specify tolerance	Yes	Yes
Rotation method (see Table 10.7)		
Varimax	Yes	Yes
Quartimax	Yes	Yes
Equimax	Yes	Yes
Direct oblimin	Yes	Yes
Direct quartimin	Yes[a]	Yes
Indirect oblimin	No	Yes
Orthoblique	No	Yes
Optional Kaiser's normalization	Normalized only	Yes
Maximum number of factors	No limit	10[b]
Differential case weighting	No	Yes
Output		
Means and standard deviations	Yes	Yes

TABLE 10.10 (*Continued*)

Feature	SPSS Factor	BMDP4M
Number of cases per variable (missing data)	Yes	No
Coefficients of variation	No	Yes
Minimums and maximums	No	Yes
z for minimums and maximums	No	Yes
First case for minimums and maximums	No	Yes
Correlation matrix	Yes	Yes
Covariance matrix	No	Yes
Initial communalities	Yes	Yes
Iteration for communalities	No	No
Final communalities	Yes	Yes
Eigenvalues	Yes	Yes
Percent of covariance	Yes	No
Cumulative percent of covariance	Yes	No
Cumulative proportion of total variance	No	Yes
Unrotated factor loadings	Yes	Yes
Variance explained by factors	No	Yes[c]
Simplicity criterion, each iteration	Oblique only	Yes
Rotated factor loadings (pattern)	Yes	Yes
Rotated factor loadings (structure)	Yes	Yes
Transformation matrix	Yes	No
Factor-score coefficients	Yes	Yes
Factor scores	Data file only	Yes
Plots of unrotated factor loadings	No	Yes
Plots of rotated factor loadings	Orthogonal rotation only	Yes
Plots of factor scores	No	Yes
Sorted rotated factor loadings	No	Yes
Inverse of correlation matrix	Yes	Yes
Determinant of correlation matrix	Yes	No
Factor-score covariance matrices	No	Yes
Mahalanobis distance of cases	No	Yes
Standard scores—each variable for each case	No	Yes
Partial correlations (anti-image matrix)	No	Yes
Residual correlation matrix	No	Yes
Shaded correlations	No	Yes

[a] $\delta = 0$.

[b] Can be increased with reduced number of variables.

[c] For all loading matrices.

SPSS FACTOR does a PCA or FA on a correlation matrix or a factor loading matrix. Several extraction methods and a variety of orthogonal rotation methods are available. Oblique rotation is done using direct oblimin, one of the best methods currently available (see Section 10.5.2.2).

Univariate output is limited to means, standard deviations, and number of cases per variable, so that any search for univariate outliers must be done through other programs in the SPSS package. Similarly, information about multivariate outliers among cases is not readily available. Output of extraction and rotation information is extensive. However, diagnostic output for judging adequacy of extraction and rotation is quite limited. FACTOR is the only program reviewed that, under conditions requiring matrix inversion, prints out the determinant of the correlation matrix, which may be helpful in assessing multicollinearity and singularity (Sections 10.3.2.4 and 4.4.4).

Factor scores (Section 10.6.6) are available, but only as external output. They are not given as part of regular printed output. In addition, information about outliers with respect to the solution is not available.

10.7.2 BMD Series

There are four programs in the BMD series: BMD01M, BMD03M, BMD08M, and BMDP4M. The original BMD programs for FA were BMD01M and BMD03M. These, however, were replaced by BMD08M, which was then replaced by BMDP4M. Features of BMDP4M are described in Table 10.10.

BMDP4M is flexible in its input, performing PCA and FA on either correlation or covariance matrices, with an option for analysis to proceed around the origin (setting all variable means to zero).

Several varieties of factor extraction and both orthogonal and oblique rotation methods are available. Information is given for all varieties of outlier detection—variables, univariate and multivariate outliers among cases, and outliers in the solution. About the only feature missing is the transformation matrix, but it is generally of limited interest.

A large variety of plots can be obtained. Both rotated and unrotated factor loadings can be plotted, as well as factor scores. Particularly helpful is a table of sorted rotated factor loadings, in which variables that load highly on each factor can be easily identified visually. (Loadings below a particular cutoff, specified by the user, are printed out as 0.0.) As an aid to evaluating adequacy of extraction, the residual correlation matrix may be printed (see Section 10.6.2). The factor-score covariance matrix is useful for assessing internal consistency and interpretability of factors. With these and other features listed in Table 10.10, BMDP4M is clearly the FA program of choice.

10.8 COMPLETE EXAMPLE OF FA

During the second year of the panel study described in Appendix B, male and female respondents completed the Bem Sex Role Inventory (BSRI; Bem, 1974). Female respondents were 369 middle-class, English-speaking women between the ages of 21 and 60 who were interviewed in person. One hundred sixty-two men who were "close to" the women also filled out and returned a written version of the BSRI when requested to do so by letter several months after the women were interviewed. Most of the men were husbands; a few were fathers, sons, boyfriends, and so forth.

The BSRI is a 45-item inventory, consisting of 20 items judged as indicative of femininity, 20 items indicative of masculinity,[8] and 5 items measuring social desirability. Respondents attribute traits (e.g., "gentle," "shy," "dominant") to themselves by assigning numbers between 1 ("never or almost never true of me") and 7 ("always or almost always true of me") to each of the items. Responses are summed to get separate masculine and feminine scores. Masculinity and femininity are conceived to be orthogonal dimensions of personality with both, one, or neither descriptive of any given individual.

Previous factor analytic work had indicated the presence of between three and five factors underlying the items of the BSRI, with potentially different factors for men and women. Comparison of factor structures between the sexes is a goal of this analysis.

10.8.1 Evaluation of Limitations

Because the BSRI was neither developed through nor designed for factor analytic work, it meets only marginally the requirements listed in Section 10.3.1. For instance, marker variables are not included and variables from the feminine scale differ in social desirability as well as in meaning (e.g., "tender" and "gullible"), so some of these variables are likely to be complex.

Nonetheless, the widespread use of the instrument and its convergent validity with some behavioral measures make factor structure, and the comparison of structure for men and women, important research questions. Because the BSRI is already published and in use, no deletion of variables or transformation of them was performed. Missing and out-of-range values were replaced with the mean response for the variable in question for the appropriate sex.

10.8.1.1 Outliers Among Cases Multivariate outliers among men and women were identified separately and deleted from subsequent analysis.

Selected portions of BMDP4M output relevant for identifying outliers are shown in Table 10.11. The first column, labeled CHISQ/DF, contains Mahalanobis distances, evaluated as χ^2s, of each case from the centroid of the cases for the original data. Because χ^2 has been divided by degrees of freedom in the column, one must look up critical χ^2 at the desired α level and divide it by degrees of freedom to get the value against which the numbers in the column are compared. With 44 df, the critical χ^2 value at $\alpha = .01$ is 68.71. Therefore, numbers in excess of 1.56 (or 68.71/44) in the column indicate outliers. Using this criterion, 41 women and 18 men were deleted from subsequent analyses as outliers.

Because of the large number of outliers and variables, a case-by-case analysis (cf. Chapter 4) was not deemed feasible. Instead, for each gender group, a discriminant function analysis (BMDP7M, cf. Chapter 9) was used to identify those variables that significantly discriminated between two subgroups: outliers and nonoutliers. On the last step of the discriminant analysis for women, only

---

[8] Owing to clerical error, one of the items, "aggression," was omitted from our questionnaires.

TABLE 10.11 SELECTED PORTIONS OF BMDP4M OUTPUT USED TO IDENTIFY OUTLYING CASES

ESTIMATED FACTOR SCORES AND MAHALANOBIS DISTANCES (CHI-SQUARES) FROM EACH CASE TO THE CENTROID OF ALL CASES
FOR ORIGINAL DATA (44 D.F.) FACTOR SCORES (11 D.F.) AND THEIR DIFFERENCE (33 D.F.)
EACH CHI-SQUARE HAS BEEN DIVIDED BY ITS DEGREE OF FREEDOM.
CASE NUMBERS BELOW REFER TO DATA MATRIX BEFORE DELETION OF MISSING DATA.

CASE LABEL NO.	CHISQ/DF FACTOR 44	11	CHISQ/DF 11	CHISQ/DF 33	FACTOR 1	FACTOR 2	FACTOR 3	FACTOR 4	FACTOR 5	FACTOR 6	FACTOR 7	FACTOR 8
1	1.695	0.154	0.933	1.949	-0.966	0.934	0.703	1.021	1.131	1.631	-1.164	-1.063
2	1.164	0.379	0.916	1.246	-1.221	-0.937	-1.009	-0.703	0.628	-2.195	0.786	0.142
3	0.656	-1.008	0.585	0.679	-0.677	-0.509	0.211	0.045	-1.155	0.272	-0.159	-1.403
4	0.429	1.065	0.259	0.485	-0.537	-0.066	-0.159	0.491	-0.265	0.207	-0.363	0.142
5	0.562	0.438	1.097	0.384	1.316	1.122	1.100	-0.493	1.689	0.523	-2.034	0.112
6	0.951	1.346	1.159	0.882	-1.215	-1.040	0.444	1.828	-1.659	-0.640	0.753	-0.684
7	1.033	0.695	1.783	0.783	-1.044	0.583	-0.230	-0.693	-2.441	-0.200	-0.282	0.090
8	0.402	-0.768	0.542	0.355	0.207	-0.260	0.869	1.549	-0.379	-0.261	0.324	0.156
9	0.848	0.654	0.617	0.926	0.980	-0.737	-0.296	0.383	-0.504	-1.909	0.413	0.221
10	0.993	-0.584	1.183	0.929	0.271	-0.814	2.293	-0.772	-0.844	1.009	-0.734	1.216
11	1.060	-0.692	1.064	1.058	0.341	1.175	0.913	0.870	-2.329	-0.084	-0.860	1.394
12	1.587	-1.120	1.995	1.450	-0.842	1.931	0.888	-1.737	-1.144	0.169	-0.532	2.426
13	0.568	0.874	0.839	0.478	1.151	1.011	0.529	0.811	1.522	1.207	1.003	-0.543
14	1.020	-1.188	1.223	0.952	1.355	-0.754	0.705	-0.030	-2.316	-0.387	0.347	0.512
15	1.932	-1.630	1.171	2.186	0.734	-0.185	-0.452	1.226	0.623	-1.075	-0.303	0.335
16	0.729	0.153	0.280	0.879	0.090	0.058	-0.286	-0.187	-0.937	0.970	-0.562	-0.858
17	0.650	1.018	0.529	0.690	0.511	-0.086	0.613	1.102	0.567	-0.135	0.553	-1.327
18	1.136	-0.675	0.996	1.183	-1.638	-1.590	-0.750	0.901	0.259	-1.418	0.195	-0.537
19	0.879	0.703	0.917	0.866	-0.356	-1.963	0.992	0.842	-0.576	-1.179	-0.302	0.771
20	0.766	-1.503	1.148	0.639	1.289	0.212	-0.740	0.275	-0.132	0.258	0.711	0.231
21	1.169	1.177	0.888	1.263	-0.489	0.958	-2.167	0.712	-0.857	-0.251	0.664	0.835
22	1.461	1.931	1.166	1.559	-0.251	0.913	-1.800	0.632	-1.257	-0.623	-1.350	-0.326
23	0.764	-1.369	0.714	0.780	-0.541	-0.880	-0.756	0.797	-1.294	-0.279	0.746	-0.493
24	0.593	0.090	0.636	0.579	1.451	0.107	0.585	-0.229	1.417	-0.951	-0.081	0.453
25	1.294	1.056	1.501	1.226	0.225	-0.199	1.286	0.617	1.296	0.397	-2.904	-1.482
26	1.604	1.626	0.977	1.813	-1.456	-0.227	0.347	0.947	-0.561	0.309	-1.337	0.359

one variable was found to discriminate outliers as a group. For the men, outliers differed on five variables.

10.8.1.2 Sample Size   With outlying cases deleted, FAs were conducted on responses of 328 women and 144 men. Using the guidelines of Section 10.3.2.2, and anticipating not more than five factors, we retained a good sample size for women and an adequate one for men.

10.8.1.3 Factorability of R   Inspection of correlation matrices among the 44 items for both women and men produced by BMDP4M revealed numerous correlations in excess of .30 and some considerably higher. Patterns in responses to variables were therefore anticipated.

10.8.1.4 Multicollinearity and Singularity   Inspection of the original non-rotated PCA runs for women and men revealed that the smallest eigenvalues (the ones associated with the factor 44) were 0.146 for women and 0.101 for men, neither dangerously close to zero. Simultaneously it was observed that SMCs between variables where each in turn serves as DV for the others, did not approach 1 (see Table 10.12). The largest SMC among the variables for women was .73 and for men was .79. So multicollinearity and singularity were no problem in these data sets.

10.8.1.5 Outliers Among Variables   SMCs among variables as shown in Table 10.12 may also be used to evaluate outliers among the variables, as discussed in Section 10.3.2.5. The lowest SMC among variables was .11 for women and .10 for men. However, these SMCs were obtained for different variables. It was therefore decided to retain all 44 variables in both FAs, although several were known to be largely unrelated to others in the set. (In fact, 43% of the 44 variables in the analysis for women and 34% of the 44 variables in the analysis for men had loadings too low on all the factors to aid in interpreting the factors in the final solution.)

10.8.1.6 Outliers with Respect to the Solution   The table used to uncover outlying cases in the initial run is the same one used in the final run to find cases that do not fit the solution very well. The second column of CHISQ/ DF (see Table 10.13, page 423) shows Mahalanobis distance of each case from the centroid of the factor scores evaluated as χ^2/df. Cases with large values are deviant cases in the space of the factor solution. How large must the value be? Again, critical χ^2 divided by the chosen number of factors as degrees of freedom provides the cutoff value. At $\alpha = .01$, the cutoff for four factors is 3.32 (or 13.28/4). Cases with values larger than this in the column are outliers. With the use of this criterion, three cases among the men and nine among the women were found to be outliers in the solution space. No attempt was made to discover unifying characteristics, if any, between cases that fit and those that do not, but in some research such an attempt might prove fruitful.

TABLE 10.12   SELECTED PORTIONS OF BMDP4M OUTPUT USED TO
ASSESS MULTICOLLINEARITY AMONG RESPONSES OF
WOMEN

SQUARED MULTIPLE CORRELATIONS (SMC) OF EACH VARIABLE WITH ALL OTHER VARIABLES

		SMC
1	HELPFUL	0.34886
2	RELIANT	0.44520
3	DEFBEL	0.39426
4	YIELDING	0.21225
5	CHEERFUL	0.47777
6	INDPT	0.48678
7	ATHLET	0.27458
8	SHY	0.31712
9	ASSERT	0.51961
10	STRPERS	0.60038
11	FORCEFUL	0.56104
12	AFFECT	0.57104
13	FLATTER	0.28074
14	LOYAL	0.37668
15	ANALYT	0.23412
16	FEMININE	0.38845
17	SYMPATHY	0.43333
18	MOODY	0.36025
19	SENSITIV	0.43589
20	UNDSTAND	0.53911
21	COMPASS	0.58764
22	LEADERAB	0.72746
23	SOOTHE	0.45266
24	RISK	0.37705
25	DECIDE	0.48222
26	SELFSUFF	0.59919
27	CONSCIEN	0.38084
28	DOMINANT	0.55596
29	MASCULIN	0.36035
30	STAND	0.54778
31	HAPPY	0.51838
32	SOFTSPOK	0.37306
33	WARM	0.56035
34	TRUTHFUL	0.31162
35	TENDER	0.56922
36	GULLIBLE	0.25057
37	LEADACT	0.72462
38	CHILDLIK	0.29580
39	INDIVID	0.36492
40	FOULLANG	0.10507
41	LOVECHIL	0.22962
42	COMPETE	0.45442
43	AMBITIOU	0.42281
44	GENTLE	0.56367

TABLE 10.13   SELECTED BMDP4M OUTPUT USED TO EVALUATE
OUTLYING CASES IN THE SPACE OF THE SOLUTION
FOR WOMEN (TOP) AND MEN (BOTTOM) (SETUP SHOWN
IN TABLE 10.17)

ESTIMATED FACTOR SCORES AND MAHALANOBIS DISTANCES (CHI-SQUARES) FROM EACH CASE TO THE CENTROID OF ALL CASES
FOR ORIGINAL DATA (44 D.F.) FACTOR SCORES (4 D.F.) AND THEIR DIFFERENCE (40 D.F.).
EACH CHI-SQUARE HAS BEEN DIVIDED BY ITS DEGREES OF FREEDOM.
CASE NUMBERS BELOW REFER TO DATA MATRIX BEFORE DELETION OF MISSING DATA.

CASE LABEL	NO.	CHISQ/DF 44	CHISQ/DF 4	CHISQ/DF 40	FACTOR 1	FACTOR 2	FACTOR 3	FACTOR 4
	116	1.035	1.068	1.032	1.356	1.592	.658	.167
	117	.843	1.235	.803	1.698	-.099	-.979	.658
	118	.568	.334	.591	.632	-.157	-.331	-.613
	119	.717	.916	.697	-.095	1.391	1.422	.282
	120	.856	1.907	.751	.287	.414	.530	-2.266
	121	1.842	3.490	1.678	-.533	1.321	-2.081	.943
	122	.785	.923	.771	.503	1.617	1.259	.194
	123	.565	.097	.611	.335	-.124	.336	.164
	124	1.223	1.611	1.184	-.373	1.083	1.778	1.140
	125	.792	1.909	.681	.994	-2.185	-.786	.169
	127	1.478	2.379	1.388	.266	.701	-2.170	.355
	128	.667	1.343	.600	-.550	1.060	-.115	-1.625
	129	1.087	.476	1.148	.804	.468	.984	-.175
	130	.807	1.778	.709	-.141	-2.186	-1.296	.959
	132	1.049	.666	1.087	-.710	.633	-.742	.062
	133	.832	3.517	.563	-1.584	-3.069	-.463	.700

ESTIMATED FACTOR SCORES AND MAHALANOBIS DISTANCES (CHI-SQUARES) FROM EACH CASE TO THE CENTROID OF ALL CASES
FOR ORIGINAL DATA (44 D.F.) FACTOR SCORES (3 D.F.) AND THEIR DIFFERENCE (41 D.F.).
EACH CHI-SQUARE HAS BEEN DIVIDED BY ITS DEGREES OF FREEDOM.
CASE NUMBERS BELOW REFER TO DATA MATRIX BEFORE DELETION OF MISSING DATA.

CASE LABEL	NO.	CHISQ/DF 44	CHISQ/DF 3	CHISQ/DF 41	FACTOR 1	FACTOR 2	FACTOR 3
	112	1.391	1.347	1.394	-1.294	-1.219	.616
	113	1.458	.899	1.498	-.201	.747	1.301
	114	1.178	.379	1.236	-.510	-.859	.142
	115	.712	1.342	.666	-1.030	1.422	-.692
	116	1.172	.585	1.215	-.515	-.031	1.071
	117	1.099	1.715	1.054	1.003	-.197	-1.778
	118	1.183	.066	1.264	.026	-.419	.028
	119	1.031	.404	1.076	.977	-.110	-.350
	120	.642	.613	.644	-.389	-1.059	-.642
	121	.697	.576	.705	-.432	-1.176	-.029
	122	1.085	2.272	.999	-2.160	1.066	.647
	123	.638	.807	.626	.845	.163	-1.115
	124	.815	.052	.870	-.173	.075	-.306
	125	1.000	.839	1.012	.252	-1.020	.988
	126	.461	.051	.491	.065	.037	.346
	127	1.697	.160	1.809	.423	.297	-.368
	128	.936	.868	.941	-1.419	.110	.552
	129	1.273	1.813	1.233	-1.287	1.646	-.700

10.8.1.7 Normality and Linearity   Distributions of the 44 variables were
examined for skewness and, for a few pairs, linearity (through bivariate scatterplots). Most of the variables were well enough behaved that distortions of the
solution were not anticipated.

10.8.2 Principal Factors Extraction with Varimax Rotation: Comparison of Two Groups

Although PCA with varimax rotation was used in initial runs with women
and men separately to evaluate assumptions and limitations, PCA was not se-

TABLE 10.14 COMMUNALITY VALUES FOR WOMEN (FOUR FACTORS) AND MEN (THREE FACTORS), PRINCIPAL FACTORS EXTRACTION WITH VARIMAX ROTATION: SELECTED PORTIONS OF BMDP4M OUTPUT

(a) Women			(b) Men		
COMMUNALITIES OBTAINED FROM 4 FACTORS AFTER 5 ITERATIONS. THE COMMUNALITY OF A VARIABLE IS ITS SQUARED MULTIPLE CORRELATION (CO-VARIANCE) WITH THE FACTORS.			COMMUNALITIES OBTAINED FROM 3 FACTORS AFTER 4 ITERATIONS. THE COMMUNALITY OF A VARIABLE IS ITS SQUARED MULTIPLE CORRELATION (CO-VARIANCE) WITH THE FACTORS.		
1	HELPFUL	0.2965	1	HELPFUL	.3870
2	RELIANT	0.4095	2	RELIANT	.3352
3	DEFBEL	0.2718	3	DEFBEL	.2505
4	YIELDING	0.1867	4	YIELDING	.3856
5	CHEERFUL	0.3893	5	CHEERFUL	.4059
6	INDPT	0.5134	6	INDPT	.3830
7	ATHLET	0.1667	7	ATHLET	.0371
8	SHY	0.1399	8	SHY	.1254
9	ASSERT	0.4780	9	ASSERT	.4281
10	STRPERS	0.5257	10	STRPERS	.5374
11	FORCEFUL	0.4670	11	FORCEFUL	.5857
12	AFFECT	0.5501	12	AFFECT	.4935
13	FLATTER	0.1924	13	FLATTER	.0620
14	LOYAL	0.2535	14	LOYAL	.2498
15	ANALYT	0.1630	15	ANALYT	.0773
16	FEMININE	0.2213	16	FEMININE	.0918
17	SYMPATHY	0.4940	17	SYMPATHY	.3524
18	MOODY	0.2945	18	MOODY	.3347
19	SENSITIV	0.4744	19	SENSITIV	.4559
20	UNDSTAND	0.5749	20	UNDSTAND	.5398
21	COMPASS	0.6497	21	COMPASS	.5035
22	LEADERAB	0.5715	22	LEADERAB	.4621
23	SOOTHE	0.4005	23	SOOTHE	.5664
24	RISK	0.2586	24	RISK	.3595
25	DECIDE	0.3658	25	DECIDE	.3769
26	SELFSUFF	0.8188	26	SELFSUFF	.3587
27	CONSCIEN	0.3482	27	CONSCIEN	.2322
28	DOMINANT	0.5659	28	DOMINANT	.5319
29	MASCULIN	0.2092	29	MASCULIN	.2620
30	STAND	0.4533	30	STAND	.5195
31	HAPPY	0.4902	31	HAPPY	.3938
32	SOFTSPOK	0.2342	32	SOFTSPOK	.1769
33	WARM	0.5850	33	WARM	.6151
34	TRUTHFUL	0.1260	34	TRUTHFUL	.2675
35	TENDER	0.5360	35	TENDER	.6268
36	GULLIBLE	0.1859	36	GULLIBLE	.2293
37	LEADACT	0.5462	37	LEADACT	.4506
38	CHILDLIK	0.1684	38	CHILDLIK	.2904
39	INDIVID	0.2557	39	INDIVID	.2033
40	FOULLANG	0.0307	40	FOULLANG	.1275
41	LOVECHIL	0.1389	41	LOVECHIL	.1995
42	COMPETE	0.2959	42	COMPETE	.1969
43	AMBITOU	0.2475	43	AMBITIOU	.2557
44	GENTLE	0.4613	44	GENTLE	.5450

lected to be the reported solution. Instead, a common factor solution was chosen to eliminate the effects of unique and error variability. Because principal factors extraction is most widely reported, its use is illustrated here. A check on the stability of the solutions using maximum likelihood extraction revealed that all four factors for women and at least the first two for men were stable. Principal factors extraction with oblique rotation was also employed, as discussed below, as a check on adequacy of rotation and simple structure. Altogether, 10 analyses were performed, 5 for women and 5 for men, before final solutions were selected.

Because 44 variables were involved, sizes of estimated communalities were not critically important to the solution (Section 10.6.1). The size of communality does indicate, however, the percent of variance in a variable that overlaps variance in the factors. As seen in Table 10.14, communality values for a number of variables for both women and men were quite low (e.g., ATHLET, FLATTER). In other words, one would be unable to predict scores on many of the variables from scores on factors by using the solutions chosen here. Nine of the variables for women and nine of those for men have communality values lower than .2. Heterogeneity among the variables is indicated by this outcome. It should be recalled, however, that factorial purity was not a criterion used to devise the BSRI.

As a preliminary estimate of the number of factors, data for men and women separately were subjected to cluster analysis through BMDP1M. The sorted and shaded correlation matrices (see Figure 10.4) were inspected for number of triangular patterns along the diagonal that might each represent a factor. For men, three such triangles were seen, one associated with the variables HELPFUL through SYMPATHY, a second with DEFBEL through DECIDE, and a third with RELIANT through SELFSUFF. For women, four such triangles were visible. The first corresponded with the variables RELIANT through DECIDE, the second with ASSERT through LEADACT, the third with AFFECT through GENTLE, and the fourth with SYMPATHY through SENSITIV. Because of the shaded area that connects clusters 3 and 4 for women, it remains to be determined whether or not factors corresponding to these clusters are really distinct.

Adequacy of extraction and numbers of factors were evaluated in several additional analyses for both women and men. Maximum numbers of factors were determined in initial unrotated PCA runs by the number of components with eigenvalues larger than 1. Eleven of the components for women and 12 for men had large enough eigenvalues for consideration. After the first run with principal factors extraction, inspection of variance extracted by the first several components (see Table 10.15, page 428) revealed expected declines. Components 1–6 represented, in order, 17.2, 10.8, 4.4, 3.5, 2.3, and 1.6 percents of variance extracted for men. The same six factors for women extracted, respectively, 16.4, 13.3, 5.5, 3.3, 2.4, and 1.2 percents of variance. As seen in Table 10.15, not only percents of variance but also eigenvalues drop off precipitously after about the first four factors. Principal factors extraction with varimax rotation of four factors for women and three for men were therefore requested.

(a) Women

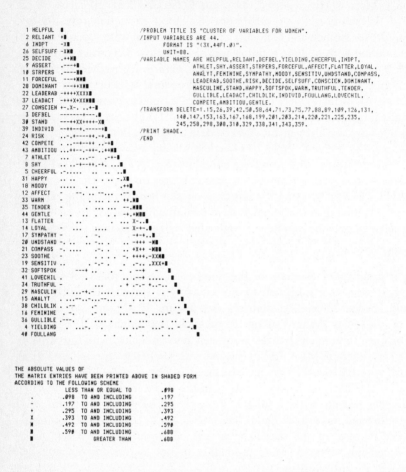

THE ABSOLUTE VALUES OF
THE MATRIX ENTRIES HAVE BEEN PRINTED ABOVE IN SHADED FORM
ACCORDING TO THE FOLLOWING SCHEME

	LESS THAN OR EQUAL TO	.098
.	.098 TO AND INCLUDING	.197
-	.197 TO AND INCLUDING	.295
+	.295 TO AND INCLUDING	.393
X	.393 TO AND INCLUDING	.492
M	.492 TO AND INCLUDING	.590
■	.590 TO AND INCLUDING	.688
■	GREATER THAN	.688

Figure 10.4 Deck setup and selected output of partial and shaded correlation matrix from BMDP1M: estimate of number of factors for women and men.

Scatterplots for this solution with pairs of rotated factors as axes and variables as points are shown in Figure 10.5 (pages 429–430). Ideally, variable points would appear only at the origin (unmarked middle of figures) or in clusters at the ends of factor axes. Scatterplots between factor 1 and factor 2 for both sexes, and of factor 2 with factor 4 for women, seem reasonably clear. The scatterplots between several other pairs of factors for women show evidence of correlation among factors requiring further investigation. (Note that an axis rotated 45° from vertical between factors 1 and 4, for instance, would provide

(b) Men

ABSOLUTE VALUES OF CORRELATIONS IN SORTED AND SHADED FORM

THE ABSOLUTE VALUES OF
THE MATRIX ENTRIES HAVE BEEN PRINTED ABOVE IN SHADED FORM
ACCORDING TO THE FOLLOWING SCHEME

```
               LESS THAN OR EQUAL TO          .099
   .           .099  TO AND INCLUDING         .198
   -           .198  TO AND INCLUDING         .297
   +           .297  TO AND INCLUDING         .396
   X           .396  TO AND INCLUDING         .495
   M           .495  TO AND INCLUDING         .593
   ▣           .593  TO AND INCLUDING         .692
   ■                 GREATER THAN             .692
```

Figure 10.4 (*Continued*)

a far better fit.) Otherwise, the scatterplots are disappointing but consistent with other evidence of heterogeneity among the variables in the BSRI.

As a final test of adequacy of extraction, it was noted that most values in the residual correlation matrices for the four-factor orthogonal solution for women and the three-factor solution for men are near zero. Therefore it was decided that the most reasonable number of factors for women was four and for men was three.

TABLE 10.15   EIGENVALUES AND PROPORTIONS OF VARIANCE FOR
FIRST EIGHT COMPONENTS, MEN AND WOMEN:
SELECTED BMDP4M OUTPUT

(a) Men

FACTOR	VARIANCE EXPLAINED	CUMULATIVE PROPORTION OF TOTAL VARIANCE
1	7.561870	0.171861
2	4.738637	0.279557
3	1.967550	0.324274
4	1.538152	0.359232
5	1.003113	0.382030
6	0.727920	0.398574
7	0.607152	0.412373
8	0.484983	0.423395

(b) Women

FACTOR	VARIANCE EXPLAINED	CUMULATIVE PROPORTION OF TOTAL VARIANCE
1	7.194941	0.163521
2	5.866262	0.296846
3	2.411947	0.351662
4	1.463652	0.384927
5	1.043182	0.408636
6	0.888051	0.420833
7	0.838563	0.447891
8	0.764920	0.465275

Adequacy of rotation and simple structure were assessed using a variety of techniques mentioned in Section 10.6.3. Oblique rotation after principal factor extraction of four factors for women and three for men was requested. Following the recommendations in Dixon (1981) for BMDP4M, direct quartimin (DQUART) was the oblique method employed. The highest correlation found was between factors 2 and 3 for women, $r = .30$ (see Table 10.16, page 430). The generally oblong shape of the scatterplot of factor scores between these two factors (see Figure 10.6, page 431) confirms the correlation. However, all other correlations between factors were low for both sexes. Interpretation of factors did not change with oblique rotation for either women or men. Because interpretation was substantially the same, because of complexities added in reporting results with oblique rotation, and because of the desire to compare factors for the two sexes, orthogonal rotation was considered adequate.

Simplicity of structure (Section 10.6.3) in factor loadings following orthogonal rotation was assessed from the table of ROTATED FACTOR LOADINGS (see Table 10.17, pages 432–433). In each column are a few high and many low correlations between variables and factors. Although not as clear as one might hope, the factors appear to be distinguishable (correlated with different variables) if messy, a condition that is also obvious from the scatterplots in Figure 10.5.

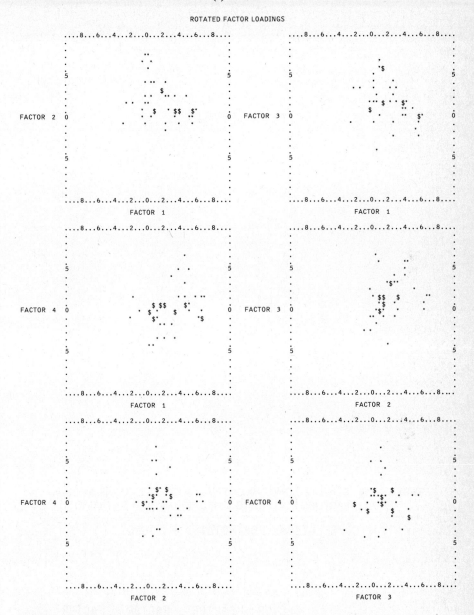

Figure 10.5 Selected BMDP4M output showing scatterplots of variable loadings with factors as axes for (a) women and (b) men.

(b) Men

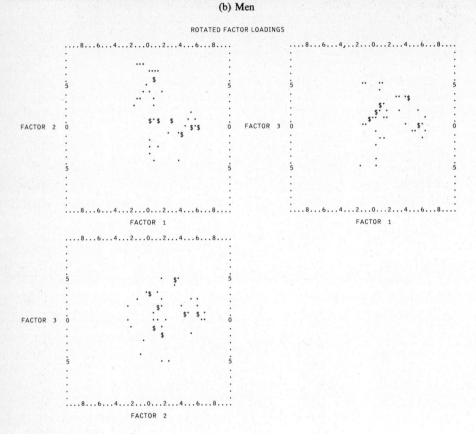

Figure 10.5   (*Continued*)

TABLE 10.16   SELECTED BMDP4M OUTPUT OF CORRELATIONS AMONG
FACTORS FOLLOWING DIRECT QUARTIMIN ROTATION

FACTOR CORRELATIONS FOR ROTATED FACTORS

		FACTOR 1	FACTOR 2	FACTOR 3	FACTOR 4
FACTOR	1	1.000			
FACTOR	2	.141	1.000		
FACTOR	3	.073	.304	1.000	
FACTOR	4	.121	.019	.018	1.000

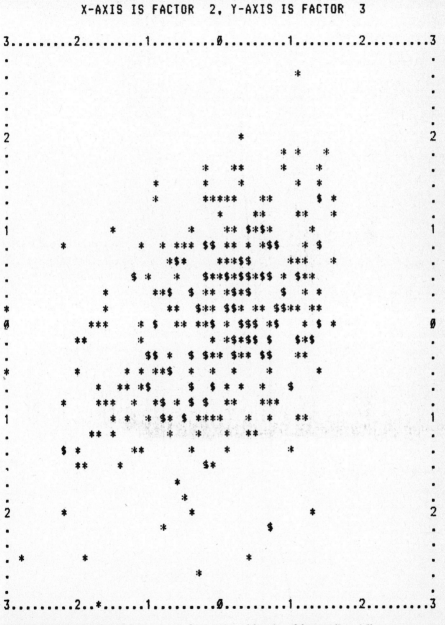

Figure 10.6 Scatterplot of factor scores for women with pairs of factors (2 and 3) as axes following oblique rotation.

TABLE 10.17   FACTOR LOADINGS FOR PRINCIPAL FACTORS EXTRACTION
AND VARIMAX ROTATION OF THREE FACTORS FOR MEN
AND FOUR FOR WOMEN: SELECTED BMDP4M OUTPUT

(a) Men

```
PROGRAM CONTROL INFORMATION

      /PROBLEM TITLE IS "FACTOR ANALYSIS FOR MEN".
      /INPUT VARIABLES ARE 44.
            FORMAT IS "(3X,44F1.0)".
            UNIT=87.
      /VARIABLE NAMES ARE HELPFUL,RELIANT,DEFBEL,YIELDING,CHEERFUL,INDPT,
                  ATHLET,SHY,ASSERT,STRPERS,FORCEFUL,AFFECT,FLATTER,LOYAL,
                  ANALYT,FEMININE,SYMPATHY,MOODY,SENSITIV,UNDSTAND,COMPASS,
                  LEADERAB,SOOTHE,RISK,DECIDE,SELFSUFF,CONSCIEN,DOMINANT,
                  MASCULIN,STAND,HAPPY,SOFTSPOK,WARM,TRUTHFUL,TENDER,
                  GULLIBLE,LEADACT,CHILDLIK,INDIVID,FOULLANG,LOVECHIL,
                  COMPETE,AMBITIOU,GENTLE.
      /TRANSFORM DELETE=4,22,24,25,39,45,57,70,78,109,110,140,142,151,153,
            157,158,160.
      /FACTOR  METHOD=PFA. NUMBER=3.
      /PRINT SHADE. RESIDUAL. FSTR. NO CORR.
      /END
```

ROTATED FACTOR LOADINGS (PATTERN)

		FACTOR 1	FACTOR 2	FACTOR 3
HELPFUL	1	.499	-.046	.368
RELIANT	2	-.079	.325	.473
DEFBEL	3	-.002	.487	.117
YIELDING	4	.483	-.390	-.021
CHEERFUL	5	.539	-.119	.317
INDPT	6	-.131	.329	.508
ATHLET	7	.045	.102	.157
SHY	8	.041	-.344	-.072
ASSERT	9	.038	.635	.152
STRPERS	10	-.034	.730	.053
FORCEFUL	11	-.154	.749	.013
AFFFCT	12	.701	.040	-.021
FLATTER	13	.157	.096	-.168
LOYAL	14	.398	.042	.299
ANALYT	15	.090	.117	.236
FEMININE	16	.027	-.141	-.267
SYMPATHY	17	.591	.022	-.053
MOODY	18	-.211	.217	-.493
SENSITIV	19	.657	.105	-.114
UNDSTAND	20	.704	.038	.208
COMPASS	21	.695	.137	.039
LEADERAB	22	.175	.649	.102
SOOTHE	23	.726	.077	-.182
RISK	24	.107	.573	-.142
DECIDE	25	.089	.547	.264
SELFSUFF	26	.087	.287	.518
CONSCIEN	27	.372	-.059	.301
DOMINANT	28	-.118	.720	-.002
MASCULIN	29	.118	.315	.386
STAND	30	.123	.656	.272
HAPPY	31	.530	-.044	.333
SOFTSPOK	32	.116	-.380	.137
WARM	33	.783	-.011	.050
TRUTHFUL	34	.167	.083	.482
TENDER	35	.787	.003	-.082
GULLIBLE	36	.050	-.221	-.422
LEADACT	37	.204	.631	.102
CHILDLIK	38	.038	.099	-.528
INDIVID	39	.067	.429	.123
FOULLANG	40	.149	-.237	.222
LOVECHIL	41	.409	.045	.174
COMPETE	42	-.030	.429	.108
AMBITIOU	43	.199	.424	.191
GENTLE	44	.737	-.006	.038
	VP	6.458	5.776	3.035

THE VP FOR EACH FACTOR IS THE SUM OF THE SQUARES OF THE ELEMENTS OF THE COLUMN OF THE FACTOR PATTERN MATRIX
CORRESPONDING TO THAT FACTOR. WHEN THE ROTATION IS ORTHOGONAL, THE VP IS THE VARIANCE EXPLAINED BY THE FACTOR.

TABLE 10.17 (*Continued*)

(b) Women

PROGRAM CONTROL INFORMATION

```
/PROBLEM TITLE IS 'PFA ANALYSIS FOR WOMEN—VARIMAX ROTATION'.
/INPUT VARIABLES ARE 44.
        FORMAT IS '(3X, 44F1.0)'.
/VARIABLES NAMES ARE HELPFUL, RELIANT, DEFBEL, YIELDING, CHEERFUL, INDPT, ATHLET,
        SHY, ASSERT, STRPERS, FORCEFUL, AFFECT, FLATTER, LOYAL, ANALYT,
        FEMININE, SYMPATHY, MOODY, SENSITIV, UNDSTAND, COMPASS, LEADERAB,
        SOOTHE, RISK, DECIDE, SELFSUFF, CONSCIEN, DOMINANT, MASCULIN,
        STAND, HAPPY, SOFTSPOK, WARM, TRUTHFUL, TENDER, GULLIBLE, LEADACT,
        CHILDLIK, INDIVID, FOULLANG, LOVECHIL, COMPETE, AMBITIOU, GENTLE.
/TRANSFORM DELETE=1,15,26,39,42,50,58,64,71,73,75,77,88,89,109,126,131,
        140,1147,153,163,167,168,199,201,203,214,220,221,225,235,
        245,258,298,308,310,329,338,341,343,359.
/FACTOR METHOD IS PFA.
 NUMBER IS 4.
/PRINT PART.
        RESI.
        SHADE.
/PLOT FINAL=4. SCORE=4.
/END
```

ROTATED FACTOR LOADINGS (PATTERN)

		FACTOR 1	FACTOR 2	FACTOR 3	FACTOR 4
HELPFUL	1	0.328	0.229	0.300	0.162
RELIANT	2	0.364	0.104	0.129	0.499
DEFBEL	3	0.427	0.290	−0.022	0.071
YIELDING	4	−0.205	0.188	0.330	0.025
CHEERFUL	5	0.165	0.061	0.586	0.120
INDPT	6	0.498	0.051	0.001	0.512
ATHLET	7	0.324	−0.109	0.221	−0.024
SHY	8	−0.362	−0.064	−0.055	−0.046
ASSERT	9	0.680	0.098	−0.058	−0.057
STRPERS	10	0.714	0.064	−0.036	−0.104
FORCEFUL	11	0.659	0.012	−0.172	−0.055
AFFECT	12	0.285	0.376	0.437	−0.370
FLATTER	13	0.141	0.119	0.203	−0.342
LOYAL	14	0.227	0.310	0.300	−0.127
ANALYT	15	0.316	0.237	−0.074	0.035
FEMININE	16	0.020	0.254	0.384	0.095
SYMPATHY	17	−0.035	0.692	0.118	0.006
MOODY	18	0.038	0.127	−0.413	−0.326
SENSITIV	19	0.095	0.676	0.034	0.086
UNDSTAND	20	0.016	0.731	0.167	0.112
COMPASS	21	0.059	0.785	0.174	0.019
LEADERAB	22	0.726	0.115	0.024	0.174
SOOTHE	23	0.067	0.556	0.286	−0.074
RISK	24	0.479	0.119	0.114	0.046
DECIDE	25	0.466	0.076	0.133	0.354
SELFSUFF	26	0.431	0.105	0.127	0.637
CONSCIEN	27	0.232	0.309	0.205	0.396
DOMINANT	28	0.695	−0.083	−0.272	0.039
MASCULIN	29	0.323	−0.199	−0.256	−0.018
STAND	30	0.602	0.268	0.058	0.126
HAPPY	31	0.108	0.013	0.686	0.090
SOFTSPOK	32	−0.281	0.170	0.313	0.169
WARM	33	0.147	0.455	0.569	−0.182
TRUTHFUL	34	0.145	0.290	0.132	0.058
TENDER	35	0.107	0.425	0.556	−0.186
GULLIBLE	36	−0.056	0.058	0.098	−0.412
LEADACT	37	0.724	−0.010	0.028	0.146
CHILDLIK	38	0.005	−0.083	−0.123	−0.382
INDIVID	39	0.446	0.108	0.051	0.206
FOULLANG	40	−0.037	0.034	0.166	0.025
LOVECHIL	41	−0.013	0.195	0.293	−0.121
COMPETE	42	0.520	−0.068	0.140	−0.030
AMBITIOU	43	0.450	0.029	0.176	0.116
GENTLE	44	0.023	0.411	0.529	−0.106
VP		6.164	4.016	3.369	2.258

THE VP FOR EACH FACTOR IS THE SUM OF THE SQUARES OF THE ELEMENTS OF THE COLUMN OF
THE FACTOR PATTERN MATRIX CORRESPONDING TO THAT FACTOR. WHEN THE ROTATION IS
ORTHOGONAL, THE VP IS THE VARIANCE EXPLAINED BY THE FACTOR.

TABLE 10.18 PERCENTS OF VARIANCE AND COVARIANCE
EXPLAINED BY EACH OF THE ROTATED
ORTHOGONAL FACTORS FOR MEN AND
WOMEN

	Factors			
	1	2	3	4
Men				
SSL	6.46	5.78	3.04	
Percent of variance	14.68	13.14	6.91	
Percent of covariance	42.28	37.83	19.90	
Women				
SSL	6.16	4.02	3.37	2.26
Percent of variance	14.00	9.14	7.66	5.14
Percent of covariance	38.96	25.43	21.32	14.29

The complexity of variables is assessed by examination of loadings across factors for a variable. For a loading cut of .45 (see Section 10.6.5 and later in the present section), only two variables for women, WARM and INDPT, and none for men load on more than one factor.

The importance of each of the factors to the solutions for both sexes was calculated as percents of variance and covariance they represent, following the procedures demonstrated in Table 10.4. SSLs, called VP at the bottom of each column of loadings shown in Table 10.17, were used in the calculations. For orthogonal rotation, it is important to use SSLs from rotated, not unrotated, factors, because the distribution of variance among factors changes with rotation.[9] Proportion of variance for a factor is simply SSL for the factor divided by number of variables. Proportion of covariance is SSL divided by sum of SSLs. Results, converted to percent, are shown in Table 10.18. Each of the factors for both sexes accounts for between 6 and 15% of the variance in the variables, not an outstanding performance. The first two factors represent most of the covariance (communality) for men, while covariance is more evenly distributed among the first three factors for women.

Internal consistency of the factors is shown as SMCs in the diagonal of the FACTOR SCORE COVARIANCE matrix. Factors serve as DVs with variables as IVs. Factors that are well defined by the variables have high SMCs, whereas poorly defined factors have low SMCs. As can be seen in Table 10.19, all factors are internally consistent for both women and men. The off-diagonal elements in these matrices are correlations between factors. Although uniformly low, the values are not zero. That is, low correlations among scores on factors may be obtained even with orthogonal rotation.

Factors are interpreted through their factor loadings. The process is facilitated with BMDP4M by requesting output of SORTED ROTATED FACTOR LOADINGS and SORTED AND SHADED LOADINGS. In both these ma-

---

[9] Variance is redistributed after oblique rotation, too, but the amount accounted for by each of several correlated factors is ambiguous, as mentioned in Section 10.5.2.2.

TABLE 10.19 SMCs FOR FACTORS WITH VARIABLES AS IVs AND
CORRELATIONS AMONG FACTORS: SELECTED BMDP4M
OUTPUT

(a) Men

FACTOR SCORE COVARIANCE (COMPUTED FROM FACTOR STRUCTURE AND FACTOR SCORE
COEFFICIENTS)

	FACTOR 1	FACTOR 2	FACTOR 3
FACTOR 1	.927		
FACTOR 2	−.001	.907	
FACTOR 3	.021	.034	.802

THE DIAGONAL OF THE ABOVE MATRIX CONTAINS THE SQUARED MULTIPLE CORRELA-
TIONS OF EACH FACTOR WITH THE VARIABLES.

(b) Women

FACTOR SCORE COVARIANCE (COMPUTED FROM FACTOR STRUCTURE AND FACTOR SCORE
COEFFICIENTS)

	FACTOR 1	FACTOR 2	FACTOR 3	FACTOR 4
FACTOR 1	.907			
FACTOR 2	.021	.850		
FACTOR 3	.001	.081	.812	
FACTOR 4	.041	−.005	.000	.787

THE DIAGONAL OF THE ABOVE MATRIX CONTAINS THE SQUARED MULTIPLE CORRELA-
TIONS OF EACH FACTOR WITH THE VARIABLES.

trices, variables are grouped by factors and reordered by size of loading. Examples
from the data for women and men are shown in Table 10.20. Interpretation
of factors comes from meaning associated with variables that are clustered together for each factor.

Although a default cut of .25 is used for inclusion of variables in factors
by BMDP4M, from a variance perspective the cutoff values proposed by Comrey
(1973) and reported in Section 10.6.5 make somewhat better sense. It was decided
for these examples that a loading of .45, which represents a 20% variance
overlap, would be required for inclusion of a variable in definition of a factor.
With the use of that criterion, Table 10.21 (pages 438–439) was generated as
a further assist to interpretation. Variables are clustered by factors and put in
ascending order by factor loading with the largest loadings on top. In interpreting

TABLE 10.20   SORTED AND SORTED AND SHADED FACTOR LOADINGS FROM BMDP4M

(a) Women

ABSOLUTE VALUES OF CORRELATIONS IN SORTED AND SHADED FORM

```
22  LEADERAB  ▓
37  LEADACT   ▓▓
10  STRPERS   XX▓
28  DOMINANT  X▓X▓
 9  ASSERT    XXX▓
11  FORCEFUL  XX▓▓N▓
30  STAND     XXX+X+▓
42  COMPETE   X+-+--▓
21  COMPASS   . . . . ▓
20  UNDSTAND  . .   ▓▓
17  SYMPATHY  . .   ▓▓▓
19  SENSITIV  -  .  ▓▓▓▓
23  SOOTHE    .   XXX+▓
31  HAPPY     . . -. ▓
 5  CHEERFUL  . . -. . X▓
33  WARM      . . . X++++++▓
35  TENDER    . . X+--++-▓▓
44  GENTLE    . X++-++-▓▓▓
26  SELFSUFF  ++----+-. . . . ▓
 6  INDPT     X++++-+. .  X▓
15  ANALYT    -. .--+.-. . . . ▓
 1  HELPFUL   ++-. . +.---------. ▓
13  FLATTER   .  . -. . ▓
24  RISK      ++----X+. . . . . . . ▓
25  DECIDE    +X.--++-. . . . . N+.- X▓
12  AFFECT    .. - .+-.+-.+--NN+  .++. ▓
27  CONSCIEN  +-. . +.-----.-.--X-.+ .. ▓
 4  YIELDING  . . .-. . . . -. . . .  . ▓
29  MASCULIN  --.+.-. . .-. . . -. . . ▓
 7  ATHLET    --. . . .+  .--.-. . ▓
14  LOYAL     .. -  .---.---++-. .-- .X. ▓
32  SOFTSPOK  .+--  -. . . .- . . . .+ ▓
16  FEMININE  .  -. .----+. . . . .-.+ -▓
34  TRUTHFUL  .  -. . . .- --- .-.+  + ▓
 3  DEFBEL    ----+-X.-. . . .---- ---- . .▓
36  GULLIBLE  .  .  .-. . . . -. . ▓
 2  RELIANT   +----. .    NN + .+ -  ▓
38  CHILDLIK  . .  . .   . . .-. . .▓
39  INDIVID   +-+--.X-. .    ++-- -. . -.+ ▓
40  FOULLANG  .    . ▓
41  LOVECHIL  .   . . . . .---+ .  ▓
 8  SHY       ++-+--. .    .  ▓
43  AMBITIOU  +----.-▓  .. .----- ++.--  - +. ▓
18  MOODY     .       X+. . . .  . .  ..· ▓
```

THE ABSOLUTE VALUES OF
THE MATRIX ENTRIES HAVE BEEN PRINTED ABOVE IN SHADED FORM
ACCORDING TO THE FOLLOWING SCHEME

	LESS THAN OR EQUAL TO	0.103
.	0.103 TO AND INCLUDING	0.205
-	0.205 TO AND INCLUDING	0.308
+	0.308 TO AND INCLUDING	0.411
X	0.411 TO AND INCLUDING	0.514
N	0.514 TO AND INCLUDING	0.616
▓	0.616 TO AND INCLUDING	0.719
▓	GREATER THAN	0.719

SORTED ROTATED FACTOR LOADINGS (PATTERN)

		FACTOR 1	FACTOR 2	FACTOR 3	FACTOR 4
LEADERAB	22	0.726	0.0	0.0	0.0
LEADACT	37	0.724	0.0	0.0	0.0
STRPERS	10	0.714	0.0	0.0	0.0
DOMINANT	28	0.695	0.0	−0.272	0.0
ASSERT	9	0.680	0.0	0.0	0.0
FORCEFUL	11	0.659	0.0	0.0	0.0
STAND	30	0.602	0.268	0.0	0.0
COMPETE	42	0.520	0.0	0.0	0.0
COMPASS	21	0.0	0.785	0.0	0.0
UNDSTAND	20	0.0	0.731	0.0	0.0
SYMPATHY	17	0.0	0.692	0.0	0.0
SENSITIV	19	0.0	0.676	0.0	0.0
SOOTHE	23	0.0	0.556	0.286	0.0
HAPPY	31	0.0	0.0	0.686	0.0
CHEERFUL	5	0.0	0.0	0.586	0.0
WARM	33	0.0	0.455	0.569	0.0
TENDER	35	0.0	0.425	0.556	0.0
GENTLE	44	0.0	0.411	0.529	0.0
SELFSUFF	26	0.431	0.0	0.0	0.637
INDPT	6	0.498	0.0	0.0	0.512
ANALYT	15	0.316	0.0	0.0	0.0
HELPFUL	1	0.328	0.269	0.300	0.0
FLATTER	13	0.0	0.0	0.0	−0.342
RISK	24	0.479	0.0	0.0	0.0
DECIDE	25	0.466	0.0	0.0	0.354
AFFECT	12	0.285	0.376	0.437	−0.370
CONSCIEN	27	0.0	0.309	0.0	0.396
YIELDING	4	0.0	0.0	0.330	0.0
MASCULIN	29	0.323	0.0	−0.256	0.0
ATHLET	7	0.324	0.0	0.0	0.0
LOYAL	14	0.0	0.310	0.300	0.0
SOFTSPOK	32	−0.281	0.0	0.313	0.0
FEMININE	16	0.0	0.254	0.384	0.0
TRUTHFUL	34	0.0	0.290	0.0	0.0
DEFBEL	3	0.427	0.290	0.0	0.0
GULLIBLE	36	0.0	0.0	0.0	−0.412
RELIANT	2	0.364	0.0	0.0	0.499
CHILDLIK	38	0.0	0.0	0.0	−0.382
INDIVID	39	0.446	0.0	0.0	0.0
FOULLANG	40	0.0	0.0	0.0	0.0
LOVECHIL	41	0.0	0.0	0.293	0.0
SHY	8	−0.362	0.0	0.0	0.0
AMBITIOU	43	0.450	0.0	0.0	0.0
MOODY	18	0.0	0.0	−0.413	−0.326
	VP	6.164	4.016	3.369	2.258

TABLE 10.20 (*Continued*)

(b) Men

ABSOLUTE VALUES OF CORRELATIONS IN SORTED AND SHADED FORM

SORTED ROTATED FACTOR LOADINGS (PATTERN)

```
35 TENDER   ▮
33 WARM     ▮▮
44 GENTLE   ▮▮▮
23 SOOTHE   ▮XX▮
20 UNDSTAND ▮XXX▮
12 AFFECT   ▮▮▮▮X▮
21 COMPASS  ▮▮▮▮▮X▮
19 SENSITIV XXX▮▮X▮▮
17 SYMPATHY X+XXX+▮▮▮
 5 CHEERFUL +X+-XX+.-▮
31 HAPPY    +XX+X+---▮▮
11 FORCEFUL ...  .  ..▮
10 STRPERS         ▮▮
28 DOMINANT .       . ▮X▮
30 STAND     . - .. .X▮+▮
22 LEADERAB ...-.... .XXX▮▮
 9 ASSERT        .   ▮▮XXX▮
37 LEADACT  ...-.... ..+XXX▮X▮
24 RISK      . .-. .. .+X+++++▮
25 DECIDE     ..    -X+X+XXMX▮
38 CHILDLIK   ..    ..    ..▮
26 SELFSUFF  ..   .-.----.-.+-▮
 6 INDPT      . .-  ------..-.▮▮
 8 SHY        ..   .+------..-▮
13 FLATTER  ... . .         -  ▮
16 FEMININE  . -....-+- . ▮
27 CONSCIEN .---+.--+--     .  ▮
14 LOYAL    ----X--++--  . .. -. .-▮
29 MASCULIN ..  ... ...+--- ..-- ▮.-▮
15 ANALYT   . ...  .  . .   . .--.▮
 4 YIELDING +++X++--.+-X+X- -  .. .-. ▮
32 SOFTSPOK ..   ..+X-..+..  + . -▮
 2 RELIANT   . . -------.+.▮▮. . - ▮
34 TRUTHFUL  -.. .. -    --- ..+-- -▮
 1 HELPFUL  +X-+X+X++++. ..   .-. X+... .+▮
36 GULLIBLE   . ..-...+.-.-. ...-▮
18 MOODY    .. - . X+.. ..+.  -. . .---▮
 7 ATHLET   ..   . .. ...  . . .▮
39 INDIVID  . . .. ++++.---. ++-  .. .  . ▮
40 FOULLANG .. . +-..  ...  . .-. ▮
41 LOVECHIL -X+--X--.-+  ... .--. - -..▮
42 COMPETE      +------++ .. -  -.... +. ▮
43 AMBITIOU .-..... .-.---X+X-- --  . .+  -- . .-..X▮
 3 DEFBEL      . +X+▮-+--. .-.  + +.-. ..-  . ▮
```

		FACTOR 1	FACTOR 2	FACTOR 3
TENDER	35	.787	0.000	0.000
WARM	33	.783	0.000	0.000
GENTLE	44	.737	0.000	0.000
SOOTHE	23	.726	0.000	0.000
UNDSTAND	20	.704	0.000	0.000
AFFECT	12	.701	0.000	0.000
COMPASS	21	.695	0.000	0.000
SENSITIV	19	.657	0.000	0.000
SYMPATHY	17	.591	0.000	0.000
CHEERFUL	5	.539	0.000	.317
HAPPY	31	.530	0.000	.333
FORCEFUL	11	0.000	.749	0.000
STRPERS	10	0.000	.730	0.000
DOMINANT	28	0.000	.720	0.000
STAND	30	0.000	.656	.272
LEADERAB	22	0.000	.649	0.000
ASSERT	9	0.000	.635	0.000
LEADACT	37	0.000	.631	0.000
RISK	24	0.000	.573	0.000
DECIDE	25	0.000	.547	.264
CHILDLIK	38	0.000	0.000	-.528
SELFSUFF	26	0.000	.287	.518
INDPT	6	0.000	.329	.508
SHY	8	0.000	-.344	0.000
FLATTER	13	0.000	0.000	0.000
FEMININE	16	0.000	0.000	-.267
CONSCIEN	27	.372	0.000	.301
LOYAL	14	.398	0.000	.299
MASCULIN	29	0.000	.315	.386
ANALYT	15	0.000	0.000	0.000
YIELDING	4	.483	-.390	0.000
SOFTSPOK	32	0.000	-.380	0.000
RELIANT	2	0.000	.325	.473
TRUTHFUL	34	0.000	0.000	.482
HELPFUL	1	.499	0.000	.368
GULLIBLE	36	0.000	0.000	-.422
MOODY	18	0.000	0.000	-.493
ATHLET	7	0.000	0.000	0.000
INDIVID	39	0.000	.429	0.000
FOULLANG	40	0.000	0.000	0.000
LOVECHIL	41	.489	0.000	0.000
COMPETE	42	0.000	.429	0.000
AMBITIOU	43	0.000	.424	0.000
DEFBEL	3	0.000	.487	0.000
VP		6.458	5.776	3.035

THE ABSOLUTE VALUES OF
THE MATRIX ENTRIES HAVE BEEN PRINTED ABOVE IN SHADED FORM
ACCORDING TO THE FOLLOWING SCHEME

```
            LESS THAN OR EQUAL TO        .100
   .     .100  TO AND INCLUDING          .199
   -     .199  TO AND INCLUDING          .299
   +     .299  TO AND INCLUDING          .399
   X     .399  TO AND INCLUDING          .498
   ▮     .498  TO AND INCLUDING          .598
   ▮     .598  TO AND INCLUDING          .698
   ▮                  GREATER THAN       .698
```

TABLE 10.21 ORDER (BY SIZE OF LOADINGS) IN WHICH VARIABLES CONTRIBUTE TO FACTORS FOR WOMEN AND MEN

(a) Women

Factor 1: *Dominance*	Factor 2: *Empathy*	Factor 3: *Positive Affect*	Factor 4: *Independence*
Has leadership abilities	Compassionate	Happy	Self-sufficient
Acts as a leader	Understanding	Cheerful	Independent
Strong personality	Sympathetic	Warm	Self-reliant
Dominant	Sensitive to needs of others	Tender	
Assertive	Eager to soothe hurt feelings	Gentle	
Forceful	Warm		
Willing to take a stand			
Competitive			
Independent			
Willing to take risks			
Makes decisions easily			
Ambitious			
Individualistic			

Note: Most important variables are near top of columns. Proposed labels are in italics.

(b) Men

Factor 1: *Positive Affect and Empathy*	Factor 2: *Dominance*	Factor 3: *Mature Independence*
Tender	Forceful	Not childlike
Warm	Strong personality	Self-sufficient
Gentle	Dominant	Independent
Eager to soothe hurt feelings	Willing to take a stand	Not moody
Understanding	Has leadership abilities	Truthful
Affectionate	Assertive	Self-reliant
Compassionate	Acts as a leader	
Sensitive to needs of others	Willing to take risks	
Sympathetic	Decisive	
Cheerful	Defends own beliefs	
Happy		
Helpful		
Yielding		

a factor, items near the top of the columns are given somewhat greater weight. Labels for the factors are suggested in the table and variable names are written out in greater detail.

Comparisons between factor structures for women and men were made using visual inspection, the salient similarity index s, and r. It was first noted that the optimal number of factors was different for the two sexes, precluding an exact correspondence in structure. Visual inspection revealed the near-identity between factor 1 for women and factor 2 for men, both labeled Dominant. Of the 10 variables defining the factor for men, and the 13 for women, 8 were the same variables. It was further noted that the variables that were different tended to be those with lower loadings.

Visual inspection also revealed the similarity between factor 1 for men and factors 2 and 3 for women, interpreted as Empathy and Positive Affect. Ten of 13 variables for men (in factor 1) and 10 of 10 variables for women (in factors 2 and 3 combined) were the same. Again, the variables that were different tended to have lower loadings. Taken at face value, this result would indicate that women differentiate between Empathy and Positive Affect while men do not. However, it should be recalled that the cluster analysis (Figure 10.4) revealed some correlation between the variables in factors 2 and 3 for women, and that when oblique rotation was performed, factors 2 and 3 for women correlated .30. Thus, although this sample of women differentiated between Empathy and Positive Affect, they did not do so completely.

The relationship between factor 4 for women and factor 3 for men is not as obvious. Although the same three variables that define the factor for women also load on the factor for men, the factor for men contains three additional variables. Because visual inspection was insufficient to remove doubts about the similarity of the factors, s was calculated to compare their loadings, as shown in Table 10.22. Loadings for the comparisons were taken from Table 10.17. As evaluated against Table C.7 in Appendix C, with a percentage in the hyperplane of .86, and 44 variables, the probability of getting an s value of .67 by chance alone is less than .001.

A correlation coefficient was also calculated between the third factor loadings for men and the fourth factor loadings for women in Table 10.17. The r value was .72.

TABLE 10.22 COMPARISON OF FACTOR 3 FOR MEN WITH FACTOR 4 FOR WOMEN THROUGH THE SALIENT SIMILARITY INDEX, s

		Factor 4, Women		
		PS	*HP*	*NS*
	PS	3	1	
Factor 3, Men	*HP*		38	
	NS		2	

$$s = \frac{3+0-0-0}{3+0+0+0+.5(1+0+0+2)}$$

$$= .67$$

TABLE 10.23 CHECKLIST FOR FACTOR
ANALYSIS

1. Issues
   a. Outliers among cases
   b. Sample size
   c. Factorability of **R**
   d. Multicollinearity and singularity
   e. Outliers among variables
   f. Outliers in the solution
   g. Normality and linearity of variables
2. Major analyses
   a. Number of factors
   b. Nature of factors
   c. Adequacy of rotation
   d. Importance of factors
3. Additional analyses
   a. Factor scores
   b. Distinguishability and simplicity of factors
   c. Complexity of variables
   d. Internal consistency of factors
   e. Comparison of factors
      (1) Between groups
      (2) With theory

Table 10.23 provides a checklist for FA. A Results section in standard journal format follows for the data analyzed in this section.

RESULTS

Principal factors extraction with varimax rotation was performed through BMDP4M on 44 items from the BSRI separately for women and for men. Principal factors extraction followed principal components extraction that was used to estimate number of factors, presence of outliers, absence of multicollinearity, and factorability of the correlation matrices. With an $\alpha = .01$ cutoff level, 41 of 369 women and 18 of 162 men were identified as outliers and deleted from principal factors extraction.[10]

Number of factors extracted was four for women and three for men. All factors were distinguishable and well defined for both groups. As indicated by SMCs, all factors for both groups are internally consistent and well defined by the variables; the lowest of the SMCs for factors from variables was .787. [Information on SMCs is from Table 10.19.] The reverse is not true, however. Variables are by and large not well defined by this factor solution. Communality values, as seen in Table 10.24, tend to be low. With a cut of .45 for inclusion of a variable in interpretation of a factor, 19 and 15 of the 44 variables for the solutions of women and men, respectively,

---

[10] Outliers for each gender were compared as a group to nonoutlying cases through discriminant function analysis. As a group, the 41 women outliers were more ambitious ($p < .05$) than the remaining female sample. The group of 18 outliers among men differed on several variables from the nonoutlying sample. They were moodier and had less leadership ability ($p < .001$), were more feminine and less childlike ($p < .01$), and were less individualistic ($p < .05$).

were not included in the solutions. However, only two of the variables in the solution for women, "warm" and "independent," were complex. Failure of numerous variables to load on a factor reflects heterogeneity of items on the BSRI.

Insert Table 10.24 about here

Orthogonal rotation was retained because of conceptual simplicity and ease of description. There was evidence, however, that factors 2 and 3 for women correlated .30, with 9% overlap in variance. Correlation was revealed when oblique rotation was requested. However, because interpretation of the factors was near-identical for both oblique and orthogonal rotation and because the highest interfactor correlation was .30, with all other correlations much lower, orthogonal rotation was deemed adequate.

Loadings of variables on factors, communalities, and percents of variance and covariance are shown in Table 10.24. Variables have been ordered and grouped by size of loading to facilitate interpretation. Loadings under .45 (20% of variance) have been replaced by zeros. Interpretive labels are suggested for each factor at the bottom of the columns.

Inspection of the pattern of loadings in Table 10.24 reveals the strong similarity of the Dominance factors (factor 1 for women and factor 2 for men) for both sexes. Factor 1 for men appears to be a composite of factors 2 and 3 for women, which were labeled Empathy and Positive Affect. Although these results might suggest that women distinguish between Empathy and Positive Affect whereas men do not, it should be pointed out that the two factors for women correlate .30 with each other in oblique rotation. Empathy and Positive Affect are correlated concepts, it would appear, for both sexes but more so for men.

Factor 3 for men (Mature Independence) and 4 for women (Independence) were defined by the same three variables, but, for men, an additional three variables loaded on the factor. Because the correspondence between these two factors was less obvious, Cattell's Salient Similarity Index, s (Cattell & Baggaley, 1960; Cattell, 1957), and r were calculated from the full set of factor loadings. The s value was .67, which exceeded the value expected by chance at $p < .001$. The correlation among the loadings was .72. Both results indicate similarity between factor 4 for women and factor 3 for men.

TABLE 10.24 FACTOR LOADINGS, COMMUNALITIES (h^2), PERCENTS OF VARIANCE AND COVARIANCE FOR FOUR-FACTOR PRINCIPAL FACTORS EXTRACTION, AND VARIMAX ROTATION FOR WOMEN AND MEN ON BSRI ITEMS

Item	Women					Men			
	F_1	F_2	F_3	F_4	h^2	F_1	F_2	F_3	h^2
Leadership ability	.73				.57	.00	.65	.00	.46
Acts as leader	.72				.55	.00	.63	.00	.45
Strong personality	.71				.53	.00	.73	.00	.54
Dominant	.70				.57	.00	.72	.00	.53
Assertive	.68				.48	.00	.64	.00	.43
Forceful	.66				.47	.00	.75	.00	.59
Takes stand	.60				.45	.00	.66	.00	.52
Competitive	.52				.30	.00	.00	.51	.20
Independent	.50			.51	.51	.00	.00	.00	.38
Takes risks	.48				.26	.00	.57	.00	.36
Makes decisions	.47				.37	.00	.55	.00	.38
Ambitious	.45				.25	.00	.00	.00	.26
Individualistic	.45				.26	.70	.00	.00	.20
Compassionate		.78			.65	.70	.00	.00	.50
Understanding		.73			.57	.59	.00	.00	.54
Sympathetic		.69			.49	.66	.00	.00	.35
Sensitive		.68			.47	.73	.00	.00	.46
Eager to soothe		.56	.57		.40	.78	.00	.00	.57
Warm		.46	.69		.58	.53	.00	.00	.62
Happy			.69		.49	.54	.00	.00	.39
Cheerful			.59		.39	.79	.00	.00	.41
Tender			.56		.54	.74	.00	.00	.63
Gentle			.53		.46	.70	.00	.00	.55
Affectionate					.55	.50	.00	.00	.49
Helpful					.30	.48	.00	.00	.39
Yielding					.19				.39
Self-sufficient				.64	.62	.00	.00	.52	.36
Self-reliant				.50	.41	.00	.00	.47	.34
Truthful					.13	.00	.00	.48	.27
Not childlike					.17	.00	.00	.53	.29
Not moody					.29	.00	.00	.49	.33
Percent of variance	14.00	9.14	7.66	5.14	35.94	14.68	13.14	6.91	34.73
Percent of covariance	38.96	25.43	21.32	14.29		42.27	37.83	19.90	
Label	Dominance	Empathy	Positive affect	Independence		Empathy and positive affect	Dominance	Mature independence	

Note: The following variables had loadings under .45 on all factors in both solutions: defends beliefs, athletic, shy, flatterable, loyal, analytical, feminine, conscientious, masculine, softspoken, gullible, does not use harsh language, and loves children.

10.9 SOME EXAMPLES FROM THE LITERATURE

McLeod, Brown, and Becker (1977) used principal components extraction with varimax rotation to collapse responses to 17 items into 4 factors for use in a study of the relationship between perceived locus of blame for the Watergate scandal and several different indicators of political orientation and behaviors. Perceived locus of blame for Watergate among the 617 respondents was composed of 4 factors accounting for a total of 48.7% of the variance and labeled the Media, the Regime (Republican and Democratic Parties, Congress), Nixon and Entourage, and the System (party politics, the political system, the economic system). Factor scores were generated for 181 respondents who were part of a panel study; scores on each factor were correlated with such indicators of political behavior as political trust, party affiliation, campaign participation, and voting direction. Results were somewhat different for those who had or had not supported Nixon in 1972 and indicated that deleterious effects of Watergate were short-term and should be considered in the context of a more general and longer-term decline in political trust and participation.

Menozzi, Piazza, and Cavalli-Sforza (1978) studied the relationship between distributions of genetic characteristics and the expansion of early farming in Europe. The overall goal of analysis was to determine, if possible, if early farmers themselves had migrated or if farming techniques had been adopted by widening circles of local hunters and gatherers. PCA was used to collapse data from 10 genetic loci and 38 alleles (for blood group substances and the like) for 400 map locations into three components. Gradients (or clines) were then formed of the densities of each component over the geographical area of Europe and the Mid-East. The first component showed greatest density in the southeast (the Mid-East) and least in the northwest (Northern Europe and Scandinavia). This map was in remarkable agreement with a map showing archaeological evidence of spread of early farming; and both argue, but not exclusively, for a spread of farmers rather than a spread of only their technology. The second and third components show changes in genetic densities corresponding to east-west gradients and northeast to southwest gradients, respectively.

In a study of willingness to donate human body parts (Pessemier, Bemmoar & Hanssens, 1977), FA was used to collapse 10 items measuring willingness to donate into 3 factors: donation of blood, skin, marrow; donations upon death; and donation of kidney. FA was performed after a Guttman scale analysis revealed that willingness to donate was not unidimensional. Responses to items composing factors were summed to yield three sum-score variables for each respondent that served as DVs in separate analyses of variance.[11] Middle-aged female respondents with above-average incomes were most likely to donate regenerative tissue or parts after death.

To investigate the relationship between certain attitudinal constructs and willingness to donate, an attempt was made to reduce 36 attitudinal questions to a smaller set of underlying variables using FA. When the determinant of

---

[11] The authors might well have used MANOVA instead of ANOVA.

the covariance matrix was near-zero (indicating extreme multicollinearity), FA was abandoned in favor of inspection of partial correlations between the 36 variables, guided by face validity of the items. Of the 36 items, 25 were retained, composing 9 attitudinal constructs for which responses were summed. A high-medium-low split was used on each attitudinal construct to generate groups for IVs in ANOVA, with the willingness-to-donate variables used as DVs in three separate analyses of each IV. Significant differences were found between most attitudinal variables (e.g., liberalism, interest in physical attractiveness) and one or more of the DVs. However, because of the number of analyses performed among both IVs and DVs that are likely to be correlated, it is difficult to have much faith in any one of the significant differences.

A study by Wong and Allen (1976) combined categories of drugs with three major cognitive dimensions in FAs of semantic differentials to study the relationships between 18 drugs and various attitudes. The three major dimensions were "pleasantness," "strength," and "dangerousness." Four hundred and fifty respondents evaluated each of 18 drugs on each dimension using a Likert-type scale with 10 gradations (from "very safe" to "very dangerous," for instance). For the set of 18 drugs, principal factors extraction of 4 factors followed by rotation were performed separately for each attitudinal dimension plus usage. Only drugs loading .60 or higher with the factors were retained. Usage and "pleasantness" tended to have similar factor structures, as did "dangerousness" and "strength." Other analyses revealed that usage is inversely related to both perceived dangerousness and strength, but directly related to pleasantness. Alcoholic beverages and marijuana tended to show deviant patterns from the rest.

The meaning of money was investigated in a study by Wernimont and Fitzpatrick (1972). Principal components extraction and varimax rotation were used to identify 7 components underlying patterns of responses of 533 subjects to 40 adjective pairs, each rated in a 7-point scale. The 7 components were labeled "shameful failure," "social acceptability," "pooh-pooh attitude," "moral evil," "comfortable security," "social unacceptability," and "conservative business attitudes." Following PCA for the group as a whole, subjects were subdivided into 11 groups on the basis of occupation, which served as IVs in one-way ANOVAs, with component scores as DVs. Separate ANOVAs were performed for each component. The groups differed significantly in their component scores, with a break usually found between the meaning of money to the unemployed (hard-core trainees, college persons) versus the employed (scientists, managers, salespersons, and the like). In general, it was concluded that money means very different things to different segments of the population.

Chapter 11

An Overview of the General Linear Model

11.1 LINEARITY AND THE GENERAL LINEAR MODEL

The emphasis in this book, up to this point, has been on distinctions among statistical methods in order to facilitate choice of the most useful technique to answer any given research question. As has been hinted throughout the book, however, all of these techniques are special applications of the same statistical model—the general linear model (GLM). The goal of this chapter is to introduce the GLM and to fit the various techniques into the model. In addition to the aesthetic pleasure provided by insight into the GLM, an understanding of the model can provide a great deal of flexibility in data analysis as well as allowing use of more sophisticated statistical techniques and computer programs. Most data sets can be fruitfully analyzed by several alternative techniques, and frequently more than one technique will be useful in making sense of the data. Section 11.3 represents an example of the use of alternative research strategies.

The GLM deals with relationships among variables. Linearity is relevant to the relationship because the shape of the relationship between each pair of variables is assumed by the model to be linear; that is, any relation that exists between two variables may be represented by a straight line. For example, body weight, at least in the mid ranges, typically is linearly related to height.

Additivity is also relevant to the GLM. If one variable is to be predicted by a set of other variables, the effects of the variables within the set are additive in the prediction equation. The second variable in the set adds predictability to the first one, the third adds to the first two, and so on. That is, the equation relating sets of variables is composed of a series of weighted terms, each added to the others.

These assumptions, however, do *not* preclude the inclusion of variables with curvilinear or multiplicative relationships. As was discussed throughout this book, variables can be dichotomized, transformed, or recoded so that even complex relationships can be evaluated within the GLM.

11.2 BIVARIATE TO MULTIVARIATE STATISTICS AND OVERVIEW OF TECHNIQUES

11.2.1 Bivariate Form

The GLM is based on prediction or, in jargon, regression. A regression equation represents the value of a DV, Y, associated with one or a combination of IVs, X. The simplest case of the GLM, then, is the familiar bivariate regression:

$$A + BX + e = Y \qquad (11.1)$$

where B is the change in Y associated with a one-unit change in X; A is a constant representing the value of Y when X is zero; and e is a random variable representing error of prediction.

If X and Y are converted to standard z scores, the constant A automatically becomes zero. That is, the standard normal deviate transformation of X and Y produces two variables, z_X and z_Y, which are measured on the same scale and cross at the point where both z scores equal zero. Because, after standardization of variances, slope is measured in equal units (rather than the possibly unequal units of X and Y raw scores), it now also represents strength of the relationship between X and Y. In bivariate regression with standardized variables, β is equal to the Pearson product-moment correlation coefficient. The closer β is to 1.00 or -1.00, the better the prediction of Y from X (or X from Y). Equation 11.1 is then simplified to

$$\beta z_X + e = z_Y \qquad (11.2)$$

As discussed in Chapters 1 and 2, one distinction that is sometimes important in statistics is whether data are continuous or discrete.[1] We can therefore distinguish three forms of bivariate regression (actually correlation) for those situations in which X and Y are (1) both continuous, in which case we deal with Pearson product-moment correlation, (2) mixed, with X dichotomous and Y continuous, dealt with as point biserial correlation, and (3) both dichotomous, dealt with as phi coefficient. In fact, there is no true distinction among these three forms of correlation. Given zero or one coding of the dichotomous variables,

---

[1] A further distinction between discrete and dichotomous variables is useful for our purposes. Discrete variables with more than two levels can be recoded into a series of dichotomous variables. Whereas discrete variables may produce complex, nonlinear relationships with other variables, dichotomous variables cannot. In this section, when we speak of statistical techniques using discrete variables (rather than specifying dichotomous variables), we imply that recoding is unnecessary or is handled internally in computer programs designed for the particular analysis.

TABLE 11.1   OVERVIEW OF TECHNIQUES IN THE GENERAL
LINEAR MODEL

A. Bivariate form (Eq. 11.2)
    1. Pearson product-moment correlation: X continuous, Y continuous
    2. Point biserial correlation: X dichotomous, Y continuous
    3. Phi coefficient: X dichotomous, Y dichotomous
B. Simple multivariate form (Eq. 11.3)
    1. Multiple regression: all X's continuous, Y continuous
    2. ANOVA: all X's discrete, Y continuous
    3. ANCOVA: some X's continuous and some discrete, Y continuous
    4. Two-group discriminant function analysis: all X's continuous, Y dichotomous
C. Full multivariate form (Eq. 11.4)
    1. Canonical correlation: all X's continuous, all Y's continuous
    2. MANOVA: all X's discrete, all Y's continuous
    3. MANCOVA: some X's continuous and some discrete, all Y's continuous
    4. Discriminant function analysis: all X's continuous, all Y's discrete
    5. Factor analysis (FA)/Principal components analysis (PCA): all Y's continuous, all X's latent

---

any of the correlations can be calculated using the equation for Pearson product-moment correlation. For example, coding one category of X as 1 and the other category as 0, and using the same coding for the two categories of Y, application of the Pearson product-moment correlation will give the same answer as application of the equation for phi coefficient. Table 11.1 compares the three bivariate forms of the GLM.

11.2.2 Simple Multivariate Form

The first generalization of the simple bivariate form of the GLM is to increase the number of variables, X, used to predict Y. It is here that the additivity of the model first becomes evident. In standardized form;

$$\sum_{i=1}^{k} \beta_i z_{X_i} + e = z_Y \tag{11.3}$$

That is, Y is predicted by a weighted *sum* of X variables. The weights, β, no longer reflect the simple strength of relation between Y and any given X, since they may be affected by relationships among X variables as well. Here, again, there are special statistical techniques associated with whether all X's are continuous, or not, as seen in Table 11.1. Here also, with appropriate coding, the most general form of the equation can be used to solve all the special cases.

    If Y and all X's are continuous, the associated statistical technique is multiple regression. Indeed, as was seen in Chapter 5, Eq. 11.3 is used to describe the multiple regression problem. If Y is continuous but all X's are discrete, we have the special case of regression known as analysis of variance. The values of X represent "groups," and the emphasis is on finding how groups differ in their average scores on Y, rather than on predicting Y, but the basic equation is the same. A significant difference between groups suggests that knowledge of X can be used to predict differential performance on Y.

As implied before, analysis of variance problems can be solved through application of multiple regression computer programs. If there are more than two groups, they can be coded to yield as many X's as there are degrees of freedom (cf. Chapter 2). For example, three groups can be coded into two dichotomous X's, one representing the first group versus the other two, and the second representing the second group versus the other two. The third group is represented by those who are not in either of the other two groups. Inclusion of a third component would be completely redundant since that component would be perfectly predictable from the combination of the other two.

If IVs (groups) are factorially combined, coding can still be used to form a series of dichotomous X variables. For example, one IV might be anxiety level divided into three groups, while a second IV might be task difficulty, divided into two groups. In addition to the two X components representing anxiety level, there could be an X component representing whether or not the task was difficult. An additional two X components could be developed to represent the interaction between anxiety level and task difficulty in their effect on performance. The five X components representing the 5 df could then be recombined and used to test the comparisons typically of interest in ANOVA: each of the two main effects and the interaction. Or the five components could be tested individually, if those comparisons were of interest. Since there are a number of excellent books (e.g., Cohen & Cohen, 1975) describing analysis of variance using multiple regression techniques, that fascinating topic has not been pursued in this book.

If some X's are continuous and others are discrete, with Y continuous, we have what is typically labeled analysis of covariance (Chapter 7). The continuous X's are labeled covariates, and the discrete ones are labeled IVs. The effects of IVs on Y are then assessed after adjustments are made for the effects of the covariates on Y. Actually, the GLM can deal with combinations of continuous and discrete X in much more general ways than traditional analysis of covariance, as was discussed in Chapters 5 and 7.

Finally, if Y is dichotomous, with X's continuous, we have two-group discriminant function analysis[2] (Chapter 9). Here the aim is to predict membership in one of the two groups (represented by the Y dichotomy) on the basis of the set of X. Notice the reversal in terminology here between ANOVA and discriminant function analysis. In ANOVA the groups are represented by X, while in discriminant function analysis the groups are represented by Y. The distinction, although confusing, is trivial within the GLM. As will be seen in forthcoming sections, *all* the special techniques are simply special cases of the full form of the GLM.

---

[2] In discriminant function analysis, dichotomous X's may be used if they meet the requirement of a normal sampling distribution of means. As shown by the central limit theorem, with large sample sizes this condition is met with dichotomous variables. In such cases, the dichotomous variables may be used as if they were "continuous and normally distributed." Similarly, dichotomous Y's that meet this requirement may be used in all forms of ANOVA, including the multivariate form to be discussed in the next section.

11.2.3 Full Multivariate Form

The GLM takes a major leap when the Y side of the equation is expanded as well as the X side. It now may take more than one equation to relate the two sets of variables:

Root

$$1: \quad \sum_{i=1}^{k} \beta_{i1} z_{X_{i1}} = \sum_{j=1}^{p} \gamma_{j1} z_{Y_{j1}}$$

$$2: \quad \sum_{i=1}^{k} \beta_{i2} z_{X_{i2}} = \sum_{j=1}^{p} \gamma_{j2} z_{Y_{j2}}$$

$$\vdots$$

$$(11.4)$$

$$m: \quad \sum_{k=1}^{k} \beta_{im} z_{X_{im}} + e = \sum_{j=1}^{p} \gamma_{jm} z_{Y_{jm}}$$

where m equals k or p, whichever is smaller, and γ equals regression weights for the standardized Y variables.

As a matter of fact, relating the two sets of variables can take as many equations as the number of X or Y variables, whichever is smaller. As long as there is only one Y, all X's can be combined in such a way that they produce a single straight-line relationship with Y. That is, X's can be combined into a single value, and that value forms a linear relationship with Y. Once there is another Y, however, it may not combine with the first in such a way that the relationship between combined Y's and combined X's is linear but rather may require another dimension, or linear combination of X's, to form a linear relationship with combined Y's.

Let's take the simplest example of Y—a dichotomous situation where Y represents two groups. The X's have been combined so as to form a single combined variable that best separates the two group means, $\overline{Y}_1$ and $\overline{Y}_2$. This is illustrated in Figure 11.1(a). We can simply draw a line parallel to an imaginary line connecting the two means, and let the parallel line represent the linear combination of X's, or first root of X. Once a third group is added, however, it may not fall along that same line. To maximally separate the three groups (representing 2 df, as you recall from Chapter 2), it may be necessary to add a second linear combination of X's, or second root. In the example in Figure 11.1(b), the first root separates the means of the first group, $\overline{Y}_1$, from those of the other two groups, $\overline{Y}_2$ and $\overline{Y}_3$, but does not distinguish between groups 2 and 3. The second root separates out the mean of the third group, $\overline{Y}_3$, but does not distinguish between the first two groups.

Root can be called by other names, depending on the special statistical technique in which they are developed. For example, roots are called discriminant functions, principal components, canonical variates, and so forth. The full multi-

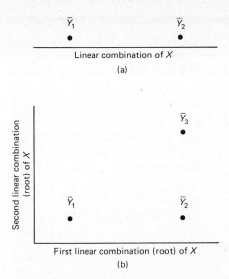

Figure 11.1 (a) Plot of two group means, $\overline{Y}_1$ and $\overline{Y}_2$, on a scale representing the linear combination of X. (b) Plot of two linear combinations of X required to distinguish among three group means: $\overline{Y}_1$, $\overline{Y}_2$, and $\overline{Y}_3$.

variate techniques refer to the need for multidimensional space to describe relationships among variables. In the example given, two dimensions are needed. With four groups, up to three dimensions might be needed.

With three groups, however, it might be the case that all groups could be separated by a single straight line [i.e., $\overline{Y}_3$ of Figure 11.1(b) would also fall along the line representing the first root of X]. Then only the first root would be needed to describe the relationship. That is, the number of roots *necessary* to describe the relationship between two sets of variables may not be the number of roots maximally available. For this reason, the error term in Eq. 11.4 might not necessarily be associated with the mth root. It would be associated with the last necessary root, with "necessary" statistically or psychometrically defined.

As with the bivariate and simple multivariate forms of the GLM, specialized statistical techniques are associated with whether or not variables are continuous (cf. Table 11.1). The most general form, the noble ancestor of the GLM, is canonical correlation, in which all X's and Y's are continuous (Chapter 6). With appropriate recoding, all problems in bivariate and multivariate statistics (with the exceptions of PCA and FA) could be solved through application of canonical correlation. Practically, however, computer programs labeled "canonical correlation" tend not to be rich in the kind of information frequently desired when one or more of the X or Y variables is discrete. Programs labeled "multivariate general linear model" tend to be rich, but extremely difficult to use.

With all X's discrete and all Y's continuous, we have a situation usefully thought of in terms of multivariate analysis of variance (MANOVA, Chapter 8). The discrete X variables represent groups, and the combination of Y variables is examined to see how they differ as a function of group membership. If the question of interest is how groups differ on the DVs, the answer is best found through MANOVA. If some of the X's are continuous, they can be analyzed as covariates, just as in ANCOVA. MANCOVA is used to discover how groups differ on Y's after adjustment is made for the effects of covariates.

If questions about prediction for groups and variables are turned around, so that one investigates how well group membership is predicted from a set of predictor variables, the answer is best sought through discriminant function analysis. Here the discrete Y variables represent groups (as illustrated in Figure 11.1) and the continuous X variables represent the measurements taken on units within the groups.

Finally, there is a family of procedures—FA and PCA—in which only the Y variables are measured. A set of Y's is measured empirically, but the set of X variables is latent. It is assumed that a set of roots underlie the Y's; the purpose of the analysis is to uncover this set of roots or factors, or X's.

11.3 ALTERNATIVE RESEARCH STRATEGIES

For most data sets, there will be more than one appropriate research strategy, depending on such considerations as how the data are interrelated, your preference for interpreting statistics associated with certain techniques, and the audience you intend to address.

A data set for which alternative strategies are appropriate might consist of groups of people distinguished on the basis of type of treatment received—behavior modification, short-term psychotherapy, and a waiting-list control group. Suppose there are a great many variables: self-reports of symptoms and moods, reports of family members, therapist reports, and a host of tests measuring personality, pathology, and attitudes. The major goal of analysis would probably be to find out if, and on which variables, the groups differ after treatment.

The obvious strategy seems to be MANOVA, but a likely problem is that number of variables may exceed clients in each group, leading to singularity (cf. Chapters 4 and 8). Further, with so many variables, some probably will be highly related to combinations of others. You might decide to choose among them or combine them on some rational basis, or you might want to first look at empirical relationships among them.

A first step in dealing with the data, then, might be a look at the squared multiple correlations of each variable with all the others. These could be obtained by a series of runs through a regression program (cf. Chapter 5) or, more simply, through a FA program (Chapter 10). The SMCs might or might not provide sufficient information for a judicious decision about which variables to delete and/or combine. If not, the next step might be a PCA on the pooled within-cells correlation matrix (cf. Chapters 8 and 10).

The usual procedures for deciding number of components and type of rotation can be followed. Out of this analysis would come scores for each client on each component, and some idea of what each component represents in terms of the variables that load highly on it. Depending on the outcome, subsequent strategies might differ. If the principal components solution is orthogonal, the component scores can each serve as the DV in turn, in a series of univariate ANOVAs, with adjustment for experiment-wise Type I error (cf. Section 8.5.3.2).

If the PCA provides an oblique solution, then MANOVA can proceed, using component scores as DVs. The stepdown hierarchy might well correspond to that of the components; that is, the scores on the first component enter first, and so on.

In addition, if you are inclined to think about the dimensionality underlying group differences, you might want to view the component scores through a discriminant function analysis (cf. Chapter 9). That is, it might be of interest to learn that differences between behavior modification and short-term psychotherapy are most notable on components loaded heavily with attitudes and self-reports, while differences between treated groups and the control group are associated with components loaded with therapist reports and personality measures.

You could, in fact, structure the entire problem as a discriminant function analysis. Protection against multicollinearity and singularity is built into discriminant function programs by setting a tolerance level so that variables that are highly predicted by the other variables are not allowed to participate in the solution.

These strategies are all "legitimate" and simply represent different ways of getting to the same goal. In the immortal words spoken one Tuesday night in the Jacuzzi by Sanford A. Fidell, "You mean you only know one thing, but you have a dozen different names for it?"

A Skimpy Introduction to Matrix Algebra

The purpose of this appendix is to provide readers with sufficient background to follow, and duplicate as desired, calculations illustrated in the fourth sections of Chapters 5 through 10. The purpose is not to provide a thorough review of matrix algebra or even to facilitate an in-depth understanding of it. The reader who is interested in more than calculational rules has several excellent discussions available, particularly those in Tatsuoka (1971) and Rummel (1970).

Most of the algebraic manipulations with which the reader is familiar—addition, subtraction, multiplication, and division—have counterparts in matrix algebra. In fact, the algebra that most of us learned is a special case of matrix algebra involving only a single number, a scalar, instead of an ordered array of numbers, a matrix. Some generalizations from scalar algebra to matrix algebra seem "natural" (i.e., matrix addition and subtraction) while others (multiplication and division) are convoluted. Nonetheless, matrix algebra provides an extremely powerful and compact method for manipulating sets of numbers to arrive at desirable statistical products.

The matrix calculations illustrated here are calculations performed on square matrices. Square matrices have the same number of rows as columns. Sums-of-squares and cross-products matrices, variance-covariance matrices, and correlation matrices are all square. In addition, these three very commonly encountered matrices are symmetrical, having the same value in row 1, column 2, as in column 1, row 2, and so forth. Symmetrical matrices are mirror images of themselves about the main diagonal (the diagonal going from top left to bottom right in the matrix).

There is a more complete matrix algebra that includes nonsquare matrices as well. However, once one proceeds from the data matrix, which has as many rows as research units (subjects) and as many columns as variables, to the sum-of-squares and cross-products matrix, as illustrated in Section 1.6, most calculations illustrated in this book involve square, symmetrical matrices. A further restriction on this appendix was to limit the discussion to only those manipulations used in the fourth sections of Chapters 5 through 10. For purposes of numerical illustration, two very simple matrices, square, but not symmetrical (to eliminate any uncertainty regarding which elements are involved in calculations), will be defined as follows:

$$\mathbf{A} = \begin{bmatrix} a & b & c \\ d & e & f \\ g & h & i \end{bmatrix} = \begin{bmatrix} 3 & 2 & 4 \\ 7 & 5 & 0 \\ 1 & 0 & 8 \end{bmatrix}$$

$$\mathbf{B} = \begin{bmatrix} r & s & t \\ u & v & w \\ x & y & z \end{bmatrix} = \begin{bmatrix} 6 & 1 & 0 \\ 2 & 8 & 7 \\ 3 & 4 & 5 \end{bmatrix}$$

A.1 ADDITION OR SUBTRACTION OF A CONSTANT TO A MATRIX

If one has a matrix, **A**, and wants to add or subtract a constant, k, to the elements of the matrix, one simply adds (or subtracts) the constant to every element in the matrix.

$$\mathbf{A} + k = \begin{bmatrix} a & b & c \\ d & e & f \\ g & h & i \end{bmatrix} + k$$

$$= \begin{bmatrix} a+k & b+k & c+k \\ d+k & e+k & f+k \\ g+k & h+k & i+k \end{bmatrix} \tag{A.1}$$

If $k = -3$, then

$$\mathbf{A} + k = \begin{bmatrix} 0 & -1 & 1 \\ 4 & 2 & -3 \\ -2 & -3 & 5 \end{bmatrix}$$

A.2 MULTIPLICATION OR DIVISION OF A MATRIX BY A CONSTANT

Multiplication or division of a matrix by a constant is a straightforward process.

$$k\mathbf{A} = k \begin{bmatrix} a & b & c \\ d & e & f \\ g & h & i \end{bmatrix}$$

$$k\mathbf{A} = \begin{bmatrix} ka & kb & kc \\ kd & ke & kf \\ kg & kh & ki \end{bmatrix} \tag{A.2}$$

and

$$\frac{1}{k}\mathbf{A} = \begin{bmatrix} \dfrac{a}{k} & \dfrac{b}{k} & \dfrac{c}{k} \\[2ex] \dfrac{d}{k} & \dfrac{e}{k} & \dfrac{f}{k} \\[2ex] \dfrac{g}{k} & \dfrac{h}{k} & \dfrac{i}{k} \end{bmatrix} \tag{A.3}$$

Numerically, if $k = 2$, then

$$k\mathbf{A} = \begin{bmatrix} 6 & 4 & 8 \\ 14 & 10 & 0 \\ 2 & 0 & 16 \end{bmatrix}$$

A.3 ADDITION AND SUBTRACTION OF TWO MATRICES

These procedures are straightforward, as well as useful. If matrices **A** and **B** are as defined at the beginning of this appendix, one simply performs the addition or subtraction of corresponding elements:

$$\mathbf{A} + \mathbf{B} = \begin{bmatrix} a & b & c \\ d & e & f \\ g & h & i \end{bmatrix} + \begin{bmatrix} r & s & t \\ u & v & w \\ x & y & z \end{bmatrix}$$

$$= \begin{bmatrix} a+r & b+s & c+t \\ d+u & e+v & f+w \\ g+x & h+y & i+z \end{bmatrix} \tag{A.4}$$

and

$$\mathbf{A} - \mathbf{B} = \begin{bmatrix} a-r & b-s & c-t \\ d-u & e-v & f-w \\ g-x & h-y & i-z \end{bmatrix} \tag{A.5}$$

For the numerical example:

$$\mathbf{A} + \mathbf{B} = \begin{bmatrix} 3 & 2 & 4 \\ 7 & 5 & 0 \\ 1 & 0 & 8 \end{bmatrix} + \begin{bmatrix} 6 & 1 & 0 \\ 2 & 8 & 7 \\ 3 & 4 & 5 \end{bmatrix}$$

$$\mathbf{A} + \mathbf{B} = \begin{bmatrix} 9 & 3 & 4 \\ 9 & 13 & 7 \\ 4 & 4 & 13 \end{bmatrix}$$

Calculation of a difference between two matrices is required when, for instance, one desires a residuals matrix, the matrix obtained by subtracting a reproduced matrix from an obtained matrix (as in factor analysis, Chapter 10). Or, if the matrix that is subtracted happens to consist of columns with appropriate means of variables inserted in every slot, then the difference between it and a matrix of raw scores produces a deviation matrix.

A.4 MULTIPLICATION, TRANSPOSES, AND SQUARE ROOTS OF MATRICES

Matrix multiplication is both unreasonably complicated and undeniably useful. Note that the ijth element of the resulting matrix is a function of row i of the first matrix and column j of the second.

$$\mathbf{AB} = \begin{bmatrix} a & b & c \\ d & e & f \\ g & h & i \end{bmatrix} \begin{bmatrix} r & s & t \\ u & v & w \\ x & y & z \end{bmatrix}$$

$$= \begin{bmatrix} ar + bu + cx & as + bv + cy & at + bw + cz \\ dr + eu + fx & ds + ev + fy & dt + ew + fz \\ gr + hu + ix & gs + hv + iy & gt + hw + iz \end{bmatrix} \tag{A.6}$$

Numerically,

$$\mathbf{AB} = \begin{bmatrix} 3 & 2 & 4 \\ 7 & 5 & 0 \\ 1 & 0 & 8 \end{bmatrix} \begin{bmatrix} 6 & 1 & 0 \\ 2 & 8 & 7 \\ 3 & 4 & 5 \end{bmatrix}$$

$$= \begin{bmatrix} 3 \cdot 6 + 2 \cdot 2 + 4 \cdot 3 & 3 \cdot 1 + 2 \cdot 8 + 4 \cdot 4 & 3 \cdot 0 + 2 \cdot 7 + 4 \cdot 5 \\ 7 \cdot 6 + 5 \cdot 2 + 0 \cdot 3 & 7 \cdot 1 + 5 \cdot 8 + 0 \cdot 4 & 7 \cdot 0 + 5 \cdot 7 + 0 \cdot 5 \\ 1 \cdot 6 + 0 \cdot 2 + 8 \cdot 3 & 1 \cdot 1 + 0 \cdot 8 + 8 \cdot 4 & 1 \cdot 0 + 0 \cdot 7 + 8 \cdot 5 \end{bmatrix}$$

$$= \begin{bmatrix} 34 & 35 & 34 \\ 52 & 47 & 35 \\ 30 & 33 & 40 \end{bmatrix}$$

Regrettably, $\mathbf{AB} \neq \mathbf{BA}$ in matrix algebra. Thus

$$\mathbf{BA} = \begin{bmatrix} 6 & 1 & 0 \\ 2 & 8 & 7 \\ 3 & 4 & 5 \end{bmatrix} \begin{bmatrix} 3 & 2 & 4 \\ 7 & 5 & 0 \\ 1 & 0 & 8 \end{bmatrix}$$

$$\mathbf{BA} = \begin{bmatrix} 25 & 17 & 24 \\ 69 & 44 & 64 \\ 42 & 26 & 52 \end{bmatrix}$$

If another concept of matrix albegra is introduced, some useful statistical properties of matrix algebra can be shown. The transpose of a matrix is indicated by a prime (′) and stands for a rearrangement of the elements of the matrix such that the first row becomes the first column, the second row the second column, and so forth. Thus

$$\mathbf{A}' = \begin{bmatrix} a & d & g \\ b & e & h \\ c & f & i \end{bmatrix}$$

$$= \begin{bmatrix} 3 & 7 & 1 \\ 2 & 5 & 0 \\ 4 & 0 & 8 \end{bmatrix} \tag{A.7}$$

When transposition is used in conjunction with multiplication, then some advantages of matrix multiplication become clear, namely,

$$\mathbf{AA}' = \begin{bmatrix} a & b & c \\ d & e & f \\ g & h & i \end{bmatrix} \begin{bmatrix} a & d & g \\ b & e & h \\ c & f & i \end{bmatrix}$$

$$= \begin{bmatrix} a^2+b^2+c^2 & ad+be+cf & ag+bh+ci \\ ad+be+cf & d^2+e^2+f^2 & dg+eh+fi \\ ag+bh+ci & dg+eh+fi & g^2+h^2+i^2 \end{bmatrix} \tag{A.8}$$

The elements in the main diagonal are the sums of squares while those off the diagonal are cross products.

Had **A** been multiplied by itself, rather than by a transpose of itself, a different result would have been achieved.

$$\mathbf{AA} = \begin{bmatrix} a^2+bd+cg & ab+be+ch & ac+bf+ci \\ da+ed+fg & db+e^2+fh & dc+ef+fi \\ ga+hd+ig & gb+he+ih & gc+hf+i^2 \end{bmatrix}$$

If $\mathbf{AA} = \mathbf{C}$, then $\mathbf{C}^{1/2} = \mathbf{A}$. That is, there is a parallel in matrix algebra to squaring and taking the square root of a scalar, but it is a complicated business because of the complexity of matrix multiplication. If, however, one has a matrix **C** from which a square root is desired (as in canonical correlation, Chapter 6), one searches for a matrix, **A**, which, when multiplied by itself, produces **C**. If, for example,

$$C = \begin{bmatrix} 27 & 16 & 44 \\ 56 & 39 & 28 \\ 11 & 2 & 68 \end{bmatrix}$$

then

$$C^{1/2} = \begin{bmatrix} 3 & 2 & 4 \\ 7 & 5 & 0 \\ 1 & 0 & 8 \end{bmatrix}$$

A.5 MATRIX "DIVISION" (INVERSES AND DETERMINANTS)

If you liked matrix multiplication, you'll love matrix inversion. Logically, the process is analogous to performing division for single numbers by finding the reciprocal of the number and multiplying by the reciprocal: if $a^{-1} = 1/a$, then $(a)(a^{-1}) = a/a = 1$. That is, the reciprocal of a scalar is a number that, when multiplied by the number itself, equals 1. Both the concepts and the notation are similar in matrix algebra, but they are complicated by the fact that a matrix is an array of numbers.

To determine if the reciprocal of a matrix has been found, one needs the matrix equivalent of the 1 as employed in the preceding paragraph. The identity matrix, I, a matrix with 1s in the main diagonal and zeros elsewhere, is such a matrix. Thus

$$I = \begin{bmatrix} 1 & 0 & 0 \\ 0 & 1 & 0 \\ 0 & 0 & 1 \end{bmatrix} \tag{A.9}$$

Matrix division, then, becomes a process of finding A^{-1} such that

$$A^{-1}A = AA^{-1} = I \tag{A.10}$$

One way of finding A^{-1} requires a two-stage process, the first of which consists of finding the determinant of A, noted $|A|$. The determinant of a matrix is sometimes said to represent the generalized variance of the matrix, as most readily seen in a 2×2 matrix. Thus we define a new matrix as follows:

$$D = \begin{bmatrix} a & b \\ c & d \end{bmatrix}$$

where

$$|D| = ad - bc \tag{A.11}$$

If **D** is a variance-covariance matrix where a and d are variances while b and c are covariances, then $ad - bc$ represents variance minus covariance. It is this property of determinants that makes them useful for hypothesis testing (see, for example, Chapter 8, Section 8.4, where Wilks' Lambda is used in MANOVA).

Calculation of determinants becomes rapidly more complicated as the matrix gets larger. For example, in our 3 by 3 matrix,

$$|\mathbf{A}| = a(ei - fh) + b(fg - di) + c(dh - eg) \qquad (A.12)$$

Should the determinant of **A** equal zero, then the matrix cannot be inverted because the next operation in inversion would involve division by zero. Multicollinear or singular matrices (those with variables that are linear combinations of one another, as discussed in Chapter 4) have zero determinants that prohibit inversion.

A full inversion of **A** is

$$\mathbf{A}^{-1} = \begin{bmatrix} a & b & c \\ d & e & f \\ g & h & i \end{bmatrix}^{-1}$$

$$= \frac{1}{|\mathbf{A}|} \begin{bmatrix} ei - fh & ch - bi & bf - ce \\ fg - di & ai - cg & cd - af \\ dh - eg & bg - ah & ae - bd \end{bmatrix} \qquad (A.13)$$

Please recall that because **A** is not a variance-covariance matrix, a negative determinant is possible, even somewhat likely. Thus, in the numerical example,

$$|\mathbf{A}| = 3(5 \cdot 8 - 0 \cdot 0) + 2(0 \cdot 1 - 7 \cdot 8) + 4(7 \cdot 0 - 5 \cdot 1)$$

$$= -12$$

and

$$\mathbf{A}^{-1} = \frac{1}{-12} \begin{bmatrix} 5 \cdot 8 - 0 \cdot 0 & 4 \cdot 0 - 2 \cdot 8 & 2 \cdot 0 - 4 \cdot 5 \\ 0 \cdot 1 - 7 \cdot 8 & 3 \cdot 8 - 4 \cdot 1 & 4 \cdot 7 - 3 \cdot 0 \\ 7 \cdot 0 - 5 \cdot 1 & 2 \cdot 1 - 3 \cdot 0 & 3 \cdot 5 - 2 \cdot 7 \end{bmatrix}$$

$$= \begin{bmatrix} 40/{-12} & -16/{-12} & -20/{-12} \\ -56/{-12} & 20/{-12} & 28/{-12} \\ -5/{-12} & 2/{-12} & 1/{-12} \end{bmatrix}$$

$$= \begin{bmatrix} -3.33 & 1.33 & 1.67 \\ 4.67 & -1.67 & -2.33 \\ .42 & -.17 & -.08 \end{bmatrix}$$

Confirm that, within rounding error, Eq. A.10 is true. Once the inverse of **A** is found, "division" by it is accomplished whenever required by using the inverse and performing matrix multiplication.

A.6 EIGENVALUES AND EIGENVECTORS: PROCEDURES FOR CONSOLIDATING VARIANCE FROM A MATRIX

We promised you a demonstration of computation of eigenvalues and eigenvectors for a matrix, so here it is. Like a Chinese dinner, however, you may well find that this discussion satisfies your appetite for only a couple of hours. During that time, round up Tatsuoka (1971), get the cat off your favorite chair, and prepare for an intelligible, if somewhat lengthy, description of the same subject.

Most of the multivariate procedures rely on eigenvalues and their corresponding eigenvectors (also called characteristic roots and vectors) in one way or another because they consolidate the variance in a matrix (the eigenvalue) while providing the linear combination of variables (the eigenvector) to do it. The coefficients applied to variables to form linear combinations of variables in all the multivariate procedures are rescaled elements from eigenvectors. The variance that the solution "accounts for" is associated with the eigenvalue, and is sometimes called so directly.

Calculation of eigenvalues and eigenvectors is best left up to a computer with any realistically sized matrix. For illustrative purposes, a 2×2 matrix will be used here. The logic of the process is also somewhat difficult, involving several of the more abstract notions and relationships in matrix algebra, including the equivalence between matrices, systems of linear equations with several unknowns, and roots of polynomial equations.

Solution of an eigenproblem involves solution of the following equation:

$$(\mathbf{D} - \lambda \mathbf{I})V = 0 \qquad (A.14)$$

where λ is the eigenvalue and V the eigenvector to be sought. Expanded, this equation becomes

$$\left[\begin{bmatrix} a & b \\ c & d \end{bmatrix} - \lambda \begin{bmatrix} 1 & 0 \\ 0 & 1 \end{bmatrix} \right] \begin{bmatrix} v_1 \\ v_2 \end{bmatrix} = 0$$

or

$$\left[\begin{bmatrix} a & b \\ c & d \end{bmatrix} - \begin{bmatrix} \lambda & 0 \\ 0 & \lambda \end{bmatrix} \right] \begin{bmatrix} v_1 \\ v_2 \end{bmatrix} = 0$$

or, by applying Eq. A.5,

$$\begin{bmatrix} a - \lambda & b \\ c & d - \lambda \end{bmatrix} \begin{bmatrix} v_1 \\ v_2 \end{bmatrix} = 0 \tag{A.15}$$

If one considers the matrix **D**, whose eigenvalues are sought, a variance-covariance matrix, one can see that a solution is desired to "capture" the variance in **D** while rescaling the elements in **D** by v_1 and v_2 to do so.

It is obvious from Eq. A.15 that a solution is always available when v_1 and v_2 are zero. A nontrivial solution may also be available when the determinant of the leftmost matrix in Eq. A.15 is zero.[1] That is, if (following Eq. A.11)

$$(a - \lambda)(d - \lambda) - bc = 0 \tag{A.16}$$

then there may exist values of λ and values of v_1 and v_2 that satisfy the equation and are not zero. However, expansion of Eq. A.16 gives a polynomial equation, in λ, of degree 2:

$$\lambda^2 - (a + d)\lambda + ad - bc = 0 \tag{A.17}$$

Solving for the eigenvalues, λ, requires solving for the roots of this polynomial. If the matrix has certain properties (see footnote 1), there will be as many positive roots to the equation as there are rows (or columns) in the matrix.

If Eq. A.17 is rewritten as $x\lambda^2 + y\lambda + z = 0$, the roots may be found by applying the following equation:

$$\lambda = \frac{-y \pm \sqrt{y^2 - 4xz}}{2x} \tag{A.18}$$

For a numerical example, consider the following matrix.

$$\mathbf{D} = \begin{bmatrix} 5 & 1 \\ 4 & 2 \end{bmatrix}$$

Applying Eq. A.17, we obtain

$$\lambda^2 - (5 + 2)\lambda + 5 \cdot 2 - 1 \cdot 4 = 0$$

or

$$\lambda^2 - 7\lambda + 6 = 0$$

The roots to this polynomial may be found by Eq. A.18 as follows:

[1] Read Tatsuoka (1971); a matrix is said to be positive definite when all $\lambda_i > 0$, positive semidefinite when all $\lambda_i \geq 0$, and ill-conditioned when some $\lambda_i < 0$.

$$\lambda_1 = \frac{-(-7) + \sqrt{(-7)^2 - 4 \cdot 1 \cdot 6}}{2 \cdot 1}$$

$$= 6$$

and

$$\lambda_2 = \frac{-(-7) - \sqrt{(-7)^2 - 4 \cdot 1 \cdot 6}}{2 \cdot 1}$$

$$= 1$$

(The roots could also be found by factoring to get $[\lambda - 6][\lambda - 1]$.)

Once the roots are found, they may be used in Eq. A.15 to find v_1 and v_2, the eigenvector. There will be one set of eigenvectors for the first root and a second set for the second root. Both solutions require solving sets of two simultaneous equations in two unknowns, to wit, for the first root, 6, and applying Eq. A.15:

$$\begin{bmatrix} 5 - 6 & 1 \\ 4 & 2 - 6 \end{bmatrix} \begin{bmatrix} v_1 \\ v_2 \end{bmatrix} = 0$$

or

$$\begin{bmatrix} -1 & 1 \\ 4 & -4 \end{bmatrix} \begin{bmatrix} v_1 \\ v_2 \end{bmatrix} = 0$$

so that

$$-1v_1 + 1v_2 = 0$$

and

$$4v_1 - 4v_2 = 0$$

When $v_1 = 1$ and $v_2 = 1$, a solution is found.

For the second root, 1, the equations become

$$\begin{bmatrix} 5 - 1 & 1 \\ 4 & 2 - 1 \end{bmatrix} \begin{bmatrix} v_1 \\ v_2 \end{bmatrix} = 0$$

or

$$\begin{bmatrix} 4 & 1 \\ 4 & 1 \end{bmatrix} \begin{bmatrix} v_1 \\ v_2 \end{bmatrix} = 0$$

so that

$$4v_1 + 1v_2 = 0$$

and

$$4v_1 + 1v_2 = 0$$

When $v_1 = -1$ and $v_2 = 4$, a solution is found. Thus the first eigenvalue is 6, with [1, 1] as a corresponding eigenvector, while the second eigenvector is 1, with [−1, 4] as a corresponding eigenvector.

Because the matrix was 2×2, the polynomial for eigenvalues was quadratic and there were two equations in two unknowns to solve for eigenvectors. Imagine the joys of a matrix 15×15, a polynomial with terms to the 15th power for the first half of the solution and 15 equations in 15 unknowns for the second half. A little more appreciation for the computer center, please, next time you visit!

Research Design for Complete Examples

Data used in all the large sample examples were collected with the aid of a grant from the National Institute on Drug Abuse (#DA 00847) to L. S. Fidell and J. E. Prather in 1974–1976. Methods of collecting the data and references to the measures included in the study are described here approximately as they have been previously reported (Hoffman & Fidell, 1979).

METHOD

A structured interview, containing a variety of health, demographic, and attitudinal measures was given to a randomly selected group of 465 female, 20- to 59-year-old, English-speaking residents of the San Fernando Valley, a suburb of Los Angeles, in February 1975. A second interview, focusing primarily on health variables but also containing the Bem Sex Role Inventory (BSRI, 1974) and the Eysenck Personality Inventory (EPI, 1963), was conducted with 369 (79.4%) of the original respondents in February 1976.

The 1975 target sample of 703 names was approximately a .003 probability sample of appropriately aged female residents of the San Fernando Valley, and was randomly drawn from lists prepared by listers during the weeks immediately preceding the sample selection. Lists were prepared for census blocks that had been randomly drawn (proportional to population) from 217 census tracts, which were themselves randomly drawn after they were stratified by income and assigned probabilities proportional to their populations. Respondents were contacted after first receiving a letter soliciting their cooperation. Substitutions were not allowed. A minimum of four callbacks was required before the attempt to obtain an interview was terminated. The completion rate for the target sample was 66.1%, with a 26% refusal rate and a 7.9% "unobtainable" rate.

The demographic characteristics of the 465 respondents who cooperated in 1975 confirmed the essentially white, middle- and working-class composition of the San Fernando Valley, and agreed, for the most part, with the profile of characteristics of women in the valley that was calculated from 1970 Census Bureau data. The final sample was 91.2% white, with a median family income (before taxes) of $17,000 per year and an average Duncan scale (Featherman, 1973) socioeconomic level (DUN) rating of 51. Respondents were also well educated (13.2 years of school completed, on average), and predominantly Protestant (38%), with 26% Catholic, 20% Jewish, and the remainder "None" or "Other." A total of 52.9% worked (either full-time—33.5%—or part-time—19.4%). Seventy-eight percent were living with husbands at the time of the first interview, with 9% divorced, 6% single, 3% separated, 3% widowed, and fewer than 1% "living together." Altogether, 82.4% of the women had had children; the average number of children was 2.7, with 2.1 children, on the average, still living in the same house as the respondent.

Of the original 465 respondents, 369 (79.4%) were reinterviewed a year later. Of the 96 respondents who were not reinterviewed, 51 refused, 36 had moved and could not be relocated, 8 were known to be in the Los Angeles area but were not contacted after a minimum of 5 attempts, and 1 was deceased. Those who were and were not reinterviewed were similar (by analyses of variance) on health and attitudinal variables. They differed, however, on some demographic measures. Those who were reinterviewed tended to be higher-DUN, higher-income white women who were better-educated, were older, and had experienced significantly fewer life change units (Rahe, 1974) in 1975.

The 1975 interview schedule was composed of items assessing a number of demographic, health, and attitudinal characteristics (see Table B.1). Insofar as possible, previously tested and validated items and measures were used, although time constraints prohibited including all items from some measures. Coding on most items was prearranged so that responses given large numbers reflected increasingly unfavorable attitudes, dissatisfaction, poorer health, lower income, increasing stress, increasing use of drugs, and so forth.

The 1976 interview schedule repeated many of the health items, with a shorter set of items assessing changes in marital status and satisfaction, changes in work status and satisfaction, and so forth. The BSRI and EPI were also included, as previously mentioned. The interview schedules for both 1975 and 1976 took 75 minutes on average to administer and were conducted in respondents' homes by experienced and trained interviewers.

To obtain median values for the masculine and feminine scores of the BSRI for a comparable sample of men, the BSRI was mailed to the 369 respondents who cooperated in 1976, with instructions to ask a man near to them (husband, friend, brother, etc.) to fill out and return it. The completed BSRI was received from 162 (46%) men, of whom 82% were husbands, 8.6% friends, 3.7% fiancés, 1.9% brothers, 1.2% sons, 1.2% ex-husbands, 0.6% brothers-in-law, and 0.6% fathers. Analyses of variance were used to compare the demo-

TABLE B.1 DESCRIPTION OF SOME OF THE VARIABLES AVAILABLE FROM 1975–1976 INTERVIEWS

Variable	Abbreviation	Brief description	Source
Demographic variables			
Socioeconomic level	DUN	Measure of deference accorded employment categories	Featherman (1973), update of Duncan scale
Education	EDUC	Number of years completed	
Income	INCOME	Total family income before taxes	
Age	AGE	Chronological age in 5-year categories	
Marital status	MARITAL	A categorical variable assessing current marital status	
Parenthood	CHILDREN	A categorical variable assessing whether or not one has had children	
Ethnic group membership	RACE	A categorical variable assessing ethnic affiliation	
Employment status	WORKING or CUEMPST	A categorical variable assessing whether or not one is currently employed	
Happy housewives	HAPHOUSE	WORKING (unemployed) plus a negative response to the question of whether or not employment is desired	
Unhappy housewives	UNHOUSE	WORKING (unemployed) plus a positive response to the question of whether or not employment is desired	
Religious affiliation	RELIGION	A categorical variable assessing category of religious affiliation	
Time utilization variables			
Time spent on housework	AMHOUSE	Hours spent per week doing housework	
Time spent on self	SELFIND	Hours spent doing leisure activities, napping; willingness to carry on when ill	

Variable	Code	Description	Source
Attitudinal variables			
Attitudes toward housework	ATTHOUSE	Frequency of experiencing various favorable and unfavorable attitudes toward home-making	Derived from Johnson (1955)
Attitudes toward paid work	PAIDWORK	Frequency of experiencing various favorable and unfavorable attitudes toward paid work	Johnson (1955)
Attitudes toward role of women	WOMENROL	Measure of conservative or liberal attitudes toward role of women	Spence and Helmreich (1972)
Locus of control	IESCALE	Measures of control ideology; internal or external	Rotter (1966)
Attitudes toward marital status	HAPSTAT or ATTSTAT	Satisfaction with current marital status	From Burgess and Locke (1960); Locke and Wallace (1959); and Rollins and Feldman (1970)
Religiosity	REL	Measure of involvement with religion	Glock, Ringer, and Babbie (1967)
Personality variables			
Self-esteem	SELFESTE	Measures of self-esteem and confidence in various situations	Rosenberg (1965)
Neuroticism-stability index	NEUROTIC	A scale derived from factor analysis to measure neuroticism versus stability	Eysenck and Eysenck, 1963
Introversion-extraversion index	INTEXT	A scale derived from factor analysis to measure introversion versus extraversion	Eysenck and Eysenck, 1963
Health variables			
Mental health	MENHEALT	Frequency count of mental health problems (feeling somewhat apart, can't get along, etc.)	Langner (1962)
Physical health	PHYHEALT	Frequency count of problems with various body systems (circulation, digestion, etc.); general description of health	

TABLE B.1 (*Continued*)

Variable	Abbreviation	Brief description	Source
Number of visits to health professionals	TIMEDHS	Frequency count of visits to physical and mental health professionals	Balter and Levine (1971)
Use of psychotropic drugs	DRUGUSE	A frequency, recency measure of involvement with prescription and nonprescription major and minor tranquilizers, sedatives-hypnotics, antidepressants, and stimulants	
Use of psychotropic and over-the-counter drugs	ALLSUM	DRUGUSE plus a frequency, recency measure of over-the-counter mood-modifying drug intake	
Attitudes toward medication	ATTMED	Items concerning attitudes toward use of medication	
Use of substances	SUBSUSE	Involvement with caffeine, vitamins, nicotine, alcohol, and nonprescription, nonpsychotropic remedies	
Life change units	RAHE	Weighted items reflecting number and importance of changes in life situation	Rahe (1974)

graphic characteristics of the men who returned the BSRI with those who did not (insofar as such characteristics could be determined by responses of the women to questions in the 1975 interview). The two groups differed in that, as with the reinterviewed women, the men who responded presented an advantaged socioeconomic picture relative to those who did not. Respondents had higher DUN ratings, were better educated, and enjoyed higher income. The unweighted averages of the men's and women's median masculine scores and median feminine scores were used to split the sample of women into those who were feminine, masculine, androgynous, and undifferentiated.

Statistical Tables

TABLE C.1 NORMAL CURVE AREAS **473**

TABLE C.1 NORMAL CURVE AREAS

z	.00	.01	.02	.03	.04	.05	.06	.07	.08	.09
0.0	.0000	.0040	.0080	.0120	.0160	.0199	.0239	.0279	.0319	.0359
0.1	.0398	.0438	.0478	.0517	.0557	.0596	.0636	.0675	.0714	.0753
0.2	.0793	.0832	.0871	.0910	.0948	.0987	.1026	.1064	.1103	.1141
0.3	.1179	.1217	.1255	.1293	.1331	.1368	.1406	.1443	.1480	.1517
0.4	.1554	.1591	.1628	.1664	.1700	.1736	.1772	.1808	.1844	.1879
0.5	.1915	.1950	.1985	.2019	.2054	.2088	.2123	.2157	.2190	.2224
0.6	.2257	.2291	.2324	.2357	.2389	.2422	.2454	.2486	.2517	.2549
0.7	.2580	.2611	.2642	.2673	.2704	.2734	.2764	.2794	.2823	.2852
0.8	.2881	.2910	.2939	.2967	.2995	.3023	.3051	.3078	.3106	.3133
0.9	.3159	.3186	.3212	.3238	.3264	.3289	.3315	.3340	.3365	.3389
1.0	.3413	.3438	.3461	.3485	.3508	.3531	.3554	.3577	.3599	.3621
1.1	.3643	.3665	.3686	.3708	.3729	.3749	.3770	.3790	.3810	.3830
1.2	.3849	.3869	.3888	.3907	.3925	.3944	.3962	.3980	.3997	.4015
1.3	.4032	.4049	.4066	.4082	.4099	.4115	.4131	.4147	.4162	.4177
1.4	.4192	.4207	.4222	.4236	.4251	.4265	.4279	.4292	.4306	.4319
1.5	.4332	.4345	.4357	.4370	.4382	.4394	.4406	.4418	.4429	.4441
1.6	.4452	.4463	.4474	.4484	.4495	.4505	.4515	.4525	.4535	.4545
1.7	.4554	.4564	.4573	.4582	.4591	.4599	.4608	.4616	.4625	.4633
1.8	.4641	.4649	.4656	.4664	.4671	.4678	.4686	.4693	.4699	.4706
1.9	.4713	.4719	.4726	.4732	.4738	.4744	.4750	.4756	.4761	.4767
2.0	.4772	.4778	.4783	.4788	.4793	.4798	.4803	.4808	.4812	.4817
2.1	.4821	.4826	.4830	.4834	.4838	.4842	.4846	.4850	.4854	.4857
2.2	.4861	.4864	.4868	.4871	.4875	.4878	.4881	.4884	.4887	.4890
2.3	.4893	.4896	.4898	.4901	.4904	.4906	.4909	.4911	.4913	.4916
2.4	.4918	.4920	.4922	.4925	.4927	.4929	.4931	.4932	.4934	.4936
2.5	.4938	.4940	.4941	.4943	.4945	.4946	.4948	.4949	.4951	.4952
2.6	.4953	.4955	.4956	.4957	.4959	.4960	.4961	.4962	.4963	.4964
2.7	.4965	.4966	.4967	.4968	.4969	.4970	.4971	.4972	.4973	.4974
2.8	.4974	.4975	.4976	.4977	.4977	.4978	.4979	.4979	.4980	.4981
2.9	.4981	.4982	.4982	.4983	.4984	.4984	.4985	.4985	.4986	.4986
3.0	.4987	.4987	.4987	.4988	.4988	.4989	.4989	.4989	.4990	.4990

Source: Abridged from Table 1 of *Statistical Tables and Formulas,* by A. Hald. Copyright ©
1952, John Wiley & Sons, Inc. Reprinted by permission of John Wiley & Sons, Inc.

TABLE C.2   CRITICAL VALUES OF THE
              t DISTRIBUTION FOR
              $\alpha = .05$ AND .01, TWO-
              TAILED TEST

Degrees of freedom	.05	.01
1	12.706	63.657
2	4.303	9.925
3	3.182	5.841
4	2.776	4.604
5	2.571	4.032
6	2.447	3.707
7	2.365	3.499
8	2.306	3.355
9	2.262	3.250
10	2.228	3.169
11	2.201	3.106
12	2.179	3.055
13	2.160	3.012
14	2.145	2.977
15	2.131	2.947
16	2.120	2.921
17	2.110	2.898
18	2.101	2.878
19	2.093	2.861
20	2.086	2.845
21	2.080	2.831
22	2.074	2.819
23	2.069	2.807
24	2.064	2.797
25	2.060	2.787
26	2.056	2.779
27	2.052	2.771
28	2.048	2.763
29	2.045	2.756
30	2.042	2.750
40	2.021	2.704
60	2.000	2.660
120	1.980	2.617
∞	1.960	2.576

Source: Abridged from Table 9 in *Biometrika Tables for Statisticians,* vol. 1, 2d ed., edited by E. S. Pearson and H. O. Hartley (New York: Cambridge University Press, 1958). Reproduced with the permission of the trustees of *Biometrika.*

TABLE C.3 CRITICAL VALUES OF THE *F* DISTRIBUTION　　　　475

TABLE C.3 CRITICAL VALUES OF THE *F* DISTRIBUTION

df_2		df_1 1	2	3	4	5	6	8	12	24	∞
1	0.1%	405284	500000	540379	562500	576405	585937	598144	610667	623497	636619
	0.5%	16211	20000	21615	22500	23056	23437	23925	24426	24940	25465
	1 %	4052	4999	5403	5625	5764	5859	5981	6106	6234	6366
	2.5%	647.79	799.50	864.16	899.58	921.85	937.11	956.66	976.71	997.25	1018.30
	5 %	161.45	199.50	215.71	224.58	230.16	233.99	238.88	243.91	249.05	254.32
	10 %	39.86	49.50	53.59	55.83	57.24	58.20	59.44	60.70	62.00	63.33
2	0.1	998.5	999.0	999.2	999.2	999.3	999.3	999.4	999.4	999.5	999.5
	0.5	198.50	199.00	199.17	199.25	199.30	199.33	199.37	199.42	199.46	199.51
	1	98.49	99.00	99.17	99.25	99.30	99.33	99.36	99.42	99.46	99.50
	2.5	38.51	39.00	39.17	39.25	39.30	39.33	39.37	39.42	39.46	39.50
	5	18.51	19.00	19.16	19.25	19.30	19.33	19.37	19.41	19.45	19.50
	10	8.53	9.00	9.16	9.24	9.29	9.33	9.37	9.41	9.45	9.49
3	0.1	167.5	148.5	141.1	137.1	134.6	132.8	130.6	128.3	125.9	123.5
	0.5	55.55	49.80	47.47	46.20	45.39	44.84	44.13	43.39	42.62	41.83
	1	34.12	30.81	29.46	28.71	28.24	27.91	27.49	27.05	26.60	26.12
	2.5	17.44	16.04	15.44	15.10	14.89	14.74	14.54	14.34	14.12	13.90
	5	10.13	9.55	9.28	9.12	9.01	8.94	8.84	8.74	8.64	8.53
	10	5.54	5.46	5.39	5.34	5.31	5.28	5.25	5.22	5.18	5.13
4	0.1	74.14	61.25	56.18	53.44	51.71	50.53	49.00	47.41	45.77	44.05
	0.5	31.33	26.28	24.26	23.16	22.46	21.98	21.35	20.71	20.03	19.33
	1	21.20	18.00	16.69	15.98	15.52	15.21	14.80	14.37	13.93	13.46
	2.5	12.22	10.65	9.98	9.60	9.36	9.20	8.98	8.75	8.51	8.26
	5	7.71	6.94	6.59	6.39	6.26	6.16	6.04	5.91	5.77	5.63
	10	4.54	4.32	4.19	4.11	4.05	4.01	3.95	3.90	3.83	3.76
5	0.1	47.04	36.61	33.20	31.09	29.75	28.84	27.64	26.42	25.14	23.78
	0.5	22.79	18.31	16.53	15.56	14.94	14.51	13.96	13.38	12.78	12.14
	1	16.26	13.27	12.06	11.39	10.97	10.67	10.29	9.89	9.47	9.02
	2.5	10.01	8.43	7.76	7.39	7.15	6.98	6.76	6.52	6.28	6.02
	5	6.61	5.79	5.41	5.19	5.05	4.95	4.82	4.68	4.53	4.36
	10	4.06	3.78	3.62	3.52	3.45	3.40	3.34	3.27	3.19	3.10
6	0.1	35.51	27.00	23.70	21.90	20.81	20.03	19.03	17.99	16.89	15.75
	0.5	18.64	14.54	12.92	12.03	11.46	11.07	10.57	10.03	9.47	8.88
	1	13.74	10.92	9.78	9.15	8.75	8.47	8.10	7.72	7.31	6.88
	2.5	8.81	7.26	6.60	6.23	5.99	5.82	5.60	5.37	5.12	4.85
	5	5.99	5.14	4.76	4.53	4.39	4.28	4.15	4.00	3.84	3.67
	10	3.78	3.46	3.29	3.18	3.11	3.05	2.98	2.90	2.82	2.72
7	0.1%	29.22	21.69	18.77	17.19	16.21	15.52	14.63	13.71	12.73	11.69
	0.5%	16.24	12.40	10.88	10.05	9.52	9.16	8.68	8.18	7.65	7.08
	1 %	12.25	9.55	8.45	7.85	7.46	7.19	6.84	6.47	6.07	5.65
	2.5%	8.07	6.54	5.89	5.52	5.29	5.12	4.90	4.67	4.42	4.14
	5 %	5.59	4.74	4.35	4.12	3.97	3.87	3.73	3.57	3.41	3.23
	10 %	3.59	3.26	3.07	2.96	2.88	2.83	2.75	2.67	2.58	2.47
8	0.1	25.42	18.49	15.83	14.39	13.49	12.86	12.04	11.19	10.30	9.34
	0.5	14.69	11.04	9.60	8.81	8.30	7.95	7.50	7.01	6.50	5.95
	1	11.26	8.65	7.59	7.01	6.63	6.37	6.03	5.67	5.28	4.86
	2.5	7.57	6.06	5.42	5.05	4.82	4.65	4.43	4.20	3.95	3.67
	5	5.32	4.46	4.07	3.84	3.69	3.58	3.44	3.28	3.12	2.93
	10	3.46	3.11	2.92	2.81	2.73	2.67	2.59	2.50	2.40	2.29
9	0.1	22.86	16.39	13.90	12.56	11.71	11.13	10.37	9.57	8.72	7.81
	0.5	13.61	10.11	8.72	7.96	7.47	7.13	6.69	6.23	5.73	5.19
	1	10.56	8.02	6.99	6.42	6.06	5.80	5.47	5.11	4.73	4.31
	2.5	7.21	5.71	5.08	4.72	4.48	4.32	4.10	3.87	3.61	3.33
	5	5.12	4.26	3.86	3.63	3.48	3.37	3.23	3.07	2.90	2.71
	10	3.36	3.01	2.81	2.69	2.61	2.55	2.47	2.38	2.28	2.16
10	0.1	21.04	14.91	12.55	11.28	10.48	9.92	9.20	8.45	7.64	6.76
	0.5	12.83	9.43	8.08	7.34	6.87	6.54	6.12	5.66	5.17	4.64
	1	10.04	7.56	6.55	5.99	5.64	5.39	5.06	4.71	4.33	3.91
	2.5	6.94	5.46	4.83	4.47	4.24	4.07	3.85	3.62	3.37	3.08
	5	4.96	4.10	3.71	3.48	3.33	3.22	3.07	2.91	2.74	2.54
	10	3.28	2.92	2.73	2.61	2.52	2.46	2.38	2.28	2.18	2.06

TABLE C.3   (*Continued*)

df₂ \ df₁		1	2	3	4	5	6	8	12	24	∞
11	0.1	19.69	13.81	11.56	10.35	9.58	9.05	8.35	7.63	6.85	6.00
	0.5	12.23	8.91	7.60	6.88	6.42	6.10	5.68	5.24	4.76	4.23
	1	9.65	7.20	6.22	5.67	5.32	5.07	4.74	4.40	4.02	3.60
	2.5	6.72	5.26	4.63	4.28	4.04	3.88	3.66	3.43	3.17	2.88
	5	4.84	3.98	3.59	3.36	3.20	3.09	2.95	2.79	2.61	2.40
	10	3.23	2.86	2.66	2.54	2.45	2.39	2.30	2.21	2.10	1.97
12	0.1	18.64	12.97	10.80	9.63	8.89	8.38	7.71	7.00	6.25	5.42
	0.5	11.75	8.51	7.23	6.52	6.07	5.76	5.35	4.91	4.43	3.90
	1	9.33	6.93	5.95	5.41	5.06	4.82	4.50	4.16	3.78	3.36
	2.5	6.55	5.10	4.47	4.12	3.89	3.73	3.51	3.28	3.02	2.72
	5	4.75	3.88	3.49	3.26	3.11	3.00	2.85	2.69	2.50	2.30
	10	3.18	2.81	2.61	2.48	2.39	2.33	2.24	2.15	2.04	1.90
13	0.1%	17.81	12.31	10.21	9.07	8.35	7.86	7.21	6.52	5.78	4.97
	0.5%	11.37	8.19	6.93	6.23	5.79	5.48	5.08	4.64	4.17	3.65
	1 %	9.07	6.70	5.74	5.20	4.86	4.62	4.30	3.96	3.59	3.16
	2.5%	6.41	4.97	4.35	4.00	3.77	3.60	3.39	3.15	2.89	2.60
	5 %	4.67	3.80	3.41	3.18	3.02	2.92	2.77	2.60	2.42	2.21
	10 %	3.14	2.76	2.56	2.43	2.35	2.28	2.20	2.10	1.98	1.85
14	0.1	17.14	11.78	9.73	8.62	7.92	7.43	6.80	6.13	5.41	4.60
	0.5	11.06	7.92	6.68	6.00	5.56	5.26	4.86	4.43	3.96	3.44
	1	8.86	6.51	5.56	5.03	4.69	4.46	4.14	3.80	3.43	3.00
	2.5	6.30	4.86	4.24	3.89	3.66	3.50	3.29	3.05	2.79	2.49
	5	4.60	3.74	3.34	3.11	2.96	2.85	2.70	2.53	2.35	2.13
	10	3.10	2.73	2.52	2.39	2.31	2.24	2.15	2.05	1.94	1.80
15	0.1	16.59	11.34	9.34	8.25	7.57	7.09	6.47	5.81	5.10	4.31
	0.5	10.80	7.70	6.48	5.80	5.37	5.07	4.67	4.25	3.79	3.26
	1	8.68	6.36	5.42	4.89	4.56	4.32	4.00	3.67	3.29	2.87
	2.5	6.20	4.77	4.15	3.80	3.58	3.41	3.20	2.96	2.70	2.40
	5	4.54	3.68	3.29	3.06	2.90	2.79	2.64	2.48	2.29	2.07
	10	3.07	2.70	2.49	2.36	2.27	2.21	2.12	2.02	1.90	1.76
16	0.1	16.12	10.97	9.00	7.94	7.27	6.81	6.19	5.55	4.85	4.06
	0.5	10.58	7.51	6.30	5.64	5.21	4.91	4.52	4.10	3.64	3.11
	1	8.53	6.23	5.29	4.77	4.44	4.20	3.89	3.55	3.18	2.75
	2.5	6.12	4.69	4.08	3.73	3.50	3.34	3.12	2.89	2.63	2.32
	5	4.49	3.63	3.24	3.01	2.85	2.74	2.59	2.42	2.24	2.01
	10	3.05	2.67	2.46	2.33	2.24	2.18	2.09	1.99	1.87	1.72
17	0.1	15.72	10.66	8.73	7.68	7.02	6.56	5.96	5.32	4.63	3.85
	0.5	10.38	7.35	6.16	5.50	5.07	4.78	4.39	3.97	3.51	2.98
	1	8.40	6.11	5.18	4.67	4.34	4.10	3.79	3.45	3.08	2.65
	2.5	6.04	4.62	4.01	3.66	3.44	3.28	3.06	2.82	2.56	2.25
	5	4.45	3.59	3.20	2.96	2.81	2.70	2.55	2.38	2.19	1.96
	10	3.03	2.64	2.44	2.31	2.22	2.15	2.06	1.96	1.84	1.69
18	0.1	15.38	10.39	8.49	7.46	6.81	6.35	5.76	5.13	4.45	3.67
	0.5	10.22	7.21	6.03	5.37	4.96	4.66	4.28	3.86	3.40	2.87
	1	8.28	6.01	5.09	4.58	4.25	4.01	3.71	3.37	3.00	2.57
	2.5	5.98	4.56	3.95	3.61	3.38	3.22	3.01	2.77	2.50	2.19
	5	4.41	3.55	3.16	·2.93	2.77	2.66	2.51	2.34	2.15	1.92
	10	3.01	2.62	2.42	2.29	2.20	2.13	2.04	1.93	1.81	1.66
19	0.1%	.15.08	10.16	8.28	7.26	6.61	6.18	5.59	4.97	4.29	3.52
	0.5%	10.07	7.09	5.92	5.27	4.85	4.56	4.18	3.76	3.31	2.78
	1 %	8.18	5.93	5.01	4.50	4.17	3.94	3.63	3.30	2.92	2.49
	2.5%	5.92	4.51	3.90	3.56	3.33	3.17	2.96	2.72	2.45	2.13
	5 %	4.38	3.52	3.13	2.90	2.74	2.63	2.48	2.31	2.11	1.88
	10 %	2.99	2.61	2.40	2.27	2.18	2.11	2.02	1.91	1.79	1.63
20	0.1	14.82	9.95	8.10	7.10	6.46	6.02	5.44	4.82	4.15	3.38
	0.5	9.94	6.99	5.82	5.17	4.76	4.47	4.09	3.68	3.22	2.69
	1	8.10	5.85	4.94	4.43	4.10	3.87	3.56	3.23	2.86	2.42
	2.5	5.87	4.46	3.86	3.51	3.29	3.13	2.91	2.68	2.41	2.09
	5	4.35	3.49	3.10	2.87	2.71	2.60	2.45	2.28	2.08	1.84
	10	2.97	2.59	2.38	2.25	2.16	2.09	2.00	1.89	1.77	1.61

TABLE C.3 CRITICAL VALUES OF THE F DISTRIBUTION 477

TABLE C.3 (*Continued*)

df_2 \ df_1		1	2	3	4	5	6	8	12	24	∞
21	0.1	14.59	9.77	7.94	6.95	6.32	5.88	5.31	4.70	4.03	3.26
	0.5	9.83	6.89	5.73	5.09	4.68	4.39	4.01	3.60	3.15	2.61
	1	8.02	5.78	4.87	4.37	4.04	3.81	3.51	3.17	2.80	2.36
	2.5	5.83	4.42	3.82	3.48	3.25	3.09	2.87	2.64	2.37	2.04
	5	4.32	3.47	3.07	2.84	2.68	2.57	2.42	2.25	2.05	1.81
	10	2.96	2.57	2.36	2.23	2.14	2.08	1.98	1.88	1.75	1.59
22	0.1	14.38	9.61	7.80	6.81	6.19	5.76	5.19	4.58	3.92	3.15
	0.5	9.73	6.81	5.65	5.02	4.61	4.32	3.94	3.54	3.08	2.55
	1	7.94	5.72	4.82	4.31	3.99	3.76	3.45	3.12	2.75	2.31
	2.5	5.79	4.38	3.78	3.44	3.22	3.05	2.84	2.60	2.33	2.00
	5	4.30	3.44	3.05	2.82	2.66	2.55	2.40	2.23	2.03	1.78
	10	2.95	2.56	2.35	2.22	2.13	2.06	1.97	1.86	1.73	1.57
23	0.1	14.19	9.47	7.67	6.69	6.08	5.65	5.09	4.48	3.82	3.05
	0.5	9.63	6.73	5.58	4.95	4.54	4.26	3.88	3.47	3.02	2.48
	1	7.88	5.66	4.76	4.26	3.94	3.71	3.41	3.07	2.70	2.26
	2.5	5.75	4.35	3.75	3.41	3.18	3.02	2.81	2.57	2.30	1.97
	5	4.28	3.42	3.03	2.80	2.64	2.53	2.38	2.20	2.00	1.76
	10	2.94	2.55	2.34	2.21	2.11	2.05	1.95	1.84	1.72	1.55
24	0.1	14.03	9.34	7.55	6.59	5.98	5.55	4.99	4.39	3.74	2.97
	0.5	9.55	6.66	5.52	4.89	4.49	4.20	3.83	3.42	2.97	2.43
	1	7.82	5.61	4.72	4.22	3.90	3.67	3.36	3.03	2.66	2.21
	2.5	5.72	4.32	3.72	3.38	3.15	2.99	2.78	2.54	2.27	1.94
	5	4.26	3.40	3.01	2.78	2.62	2.51	2.36	2.18	1.98	1.73
	10	2.93	2.54	2.33	2.19	2.10	2.04	1.94	1.83	1.70	1.53
25	0.1%	13.88	9.22	7.45	6.49	5.88	5.46	4.91	4.31	3.66	2.89
	0.5%	9.48	6.60	5.46	4.84	4.43	4.15	3.78	3.37	2.92	2.38
	1 %	7.77	5.57	4.68	4.18	3.86	3.63	3.32	2.99	2.62	2.17
	2.5%	5.69	4.29	3.69	3.35	3.13	2.97	2.75	2.51	2.24	1.91
	5 %	4.24	3.38	2.99	2.76	2.60	2.49	2.34	2.16	1.96	1.71
	10 %	2.92	2.53	2.32	2.18	2.09	2.02	1.93	1.82	1.69	1.52
26	0.1	13.74	9.12	7.36	6.41	5.80	5.38	4.83	4.24	3.59	2.82
	0.5	9.41	6.54	5.41	4.79	4.38	4.10	3.73	3.33	2.87	2.33
	1	7.72	5.53	4.64	4.14	3.82	3.59	3.29	2.96	2.58	2.13
	2.5	5.66	4.27	3.67	3.33	3.10	2.94	2.73	2.49	2.22	1.88
	5	4.22	3.37	2.98	2.74	2.59	2.47	2.32	2.15	1.95	1.69
	10	2.91	2.52	2.31	2.17	2.08	2.01	1.92	1.81	1.68	1.50
27	0.1	13.61	9.02	7.27	6.33	5.73	5.31	4.76	4.17	3.52	2.75
	0.5	9.34	6.49	5.36	4.74	4.34	4.06	3.69	3.28	2.83	2.29
	1	7.68	5.49	4.60	4.11	3.78	3.56	3.26	2.93	2.55	2.10
	2.5	5.63	4.24	3.65	3.31	3.08	2.92	2.71	2.47	2.19	1.85
	5	4.21	3.35	2.96	2.73	2.57	2.46	2.30	2.13	1.93	1.67
	10	2.90	2.51	2.30	2.17	2.07	2.00	1.91	1.80	1.67	1.49
28	0.1	13.50	8.93	7.19	6.25	5.66	5.24	4.69	4.11	3.46	2.70
	0.5	9.28	6.44	5.32	4.70	4.30	4.02	3.65	3.25	2.79	2.25
	1	7.64	5.45	4.57	4.07	3.75	3.53	3.23	2.90	2.52	2.06
	2.5	5.61	4.22	3.63	3.29	2.06	2.90	2.69	2.45	2.17	1.83
	5	4.20	3.34	2.95	2.71	2.56	2.44	2.29	2.12	1.91	1.65
	10	2.89	2.50	2.29	2.16	2.06	2.00	1.90	1.79	1.66	1.48
29	0.1	13.39	8.85	7.12	6.19	5.59	5.18	4.64	4.05	3.41	2.64
	0.5	9.23	6.40	5.28	4.66	4.26	3.98	3.61	3.21	2.76	2.21
	1	7.60	5.42	4.54	4.04	3.73	3.50	3.20	2.87	2.49	2.03
	2.5	5.59	4.20	3.61	3.27	3.04	2.88	2.67	2.43	2.15	1.81
	5	4.18	3.33	2.93	2.70	2.54	2.43	2.28	2.10	1.90	1.64
	10	2.89	2.50	2.28	2.15	2.06	1.99	1.89	1.78	1.65	1.47
30	0.1	13.29	8.77	7.05	6.12	5.53	5.12	4.58	4.00	3.36	2.59
	0.5	9.18	6.35	5.24	4.62	4.23	3.95	3.58	3.18	2.73	2.18
	1	7.56	5.39	4.51	4.02	3.70	3.47	3.17	2.84	2.47	2.01
	2.5	5.57	4.18	3.59	3.25	3.03	2.87	2.65	2.41	2.14	1.79
	5	4.17	3.32	2.92	2.69	2.53	2.42	2.27	2.09	1.89	1.62
	10	2.88	2.49	2.28	2.14	2.05	1.98	1.88	1.77	1.64	1.46

TABLE C.3 *(Continued)*

df₂		1	2	3	4	5	6	8	12	24	∞
40	0.1%	12.61	8.25	6.60	5.70	5.13	4.73	4.21	3.64	3.01	2.23
	0.5%	8.83	6.07	4.98	4.37	3.99	3.71	3.35	2.95	2.50	1.93
	1 %	7.31	5.18	4.31	3.83	3.51	3.29	2.99	2.66	2.29	1.80
	2.5%	5.42	4.05	3.46	3.13	2.90	2.74	2.53	2.29	2.01	1.64
	5 %	4.08	3.23	2.84	2.61	2.45	2.34	2.18	2.00	1.79	1.51
	10 %	2.84	2.44	2.23	2.09	2.00	1.93	1.83	1.71	1.57	1.38
60	0.1	11.97	7.76	6.17	5.31	4.76	4.37	3.87	3.31	2.69	1.90
	0.5	8.49	5.80	4.73	4.14	3.76	3.49	3.13	2.74	2.29	1.69
	1	7.08	4.98	4.13	3.65	3.34	3.12	2.82	2.50	2.12	1.60
	2.5	5.29	3.93	3.34	3.01	2.79	2.63	2.41	2.17	1.88	1.48
	5	4.00	3.15	2.76	2.52	2.37	2.25	2.10	1.92	1.70	1.39
	10	2.79	2.39	2.18	2.04	1.95	1.87	1.77	1.66	1.51	1.29
120	0.1	11.38	7.31	5.79	4.95	4.42	4.04	3.55	3.02	2.40	1.56
	0.5	8.18	5.54	4.50	3.92	3.55	3.28	2.93	2.54	2.09	1.43
	1	6.85	4.79	3.95	3.48	3.17	2.96	2.66	2.34	1.95	1.38
	2.5	5.15	3.80	3.23	2.89	2.67	2.52	2.30	2.05	1.76	1.31
	5	3.92	3.07	2.68	2.45	2.29	2.17	2.02	1.83	1.61	1.25
	10	2.75	2.35	2.13	1.99	1.90	1.82	1.72	1.60	1.45	1.19
∞	0.1	10.83	6.91	5.42	4.62	4.10	3.74	3.27	2.74	2.13	1.00
	0.5	7.88	5.30	4.28	3.72	3.35	3.09	2.74	2.36	1.90	1.00
	1	6.64	4.60	3.78	3.32	3.02	2.80	2.51	2.18	1.79	1.00
	2.5	5.02	3.69	3.12	2.79	2.57	2.41	2.19	1.94	1.64	1.00
	5	3.84	2.99	2.60	2.37	2.21	2.09	1.94	1.75	1.52	1.00
	10	2.71	2.30	2.08	1.94	1.85	1.77	1.67	1.55	1.38	1.00

Source: Abridged from Table 18 in *Biometrika Tables for Statisticians,* vol. 1, 2d ed., edited by E. S. Pearson and H. O. Hartley (New York: Cambridge University Press, 1958). Reproduced with the permission of the trustees of *Biometrika.*

df	0.250	0.100	0.050	0.025	0.010	0.005	0.001
1	1.32330	2.70554	3.84146	5.02389	6.63490	7.87944	10.828
2	2.77259	4.60517	5.99147	7.37776	9.21034	10.5966	13.816
3	4.10835	6.25139	7.81473	9.34840	11.3449	12.8381	16.266
4	5.38527	7.77944	9.48773	11.1433	13.2767	14.8602	18.467
5	6.62568	9.23635	11.0705	12.8325	15.0863	16.7496	20.515
6	7.84080	10.6446	12.5916	14.4494	16.8119	18.5476	22.458
7	9.03715	12.0170	14.0671	16.0128	18.4753	20.2777	24.322
8	10.2188	13.3616	15.5073	17.5346	20.0902	21.9550	26.125
9	11.3887	14.6837	16.9190	19.0228	21.6660	23.5893	27.877
10	12.5489	15.9871	18.3070	20.4831	23.2093	25.1882	29.588
11	13.7007	17.2750	19.6751	21.9200	24.7250	26.7569	31.264
12	14.8454	18.5494	21.0261	23.3367	26.2170	28.2995	32.909
13	15.9839	19.8119	22.3621	24.7356	27.6883	29.8194	34.528
14	17.1170	21.0642	23.6848	26.1190	29.1413	31.3193	36.123
15	18.2451	22.3072	24.9958	27.4884	30.5779	32.8013	37.697
16	19.3688	23.5418	26.2962	28.8454	31.9999	34.2672	39.252
17	20.4887	24.7690	27.5871	30.1910	33.4087	35.7185	40.790
18	21.6049	25.9894	28.8693	31.5264	34.8053	37.1564	42.312
19	22.7178	27.2036	30.1435	32.8523	36.1908	38.5822	43.820
20	23.8277	28.4120	31.4104	34.1696	37.5662	39.9968	45.315
21	24.9348	29.6151	32.6705	35.4789	38.9321	41.4010	46.797
22	26.0393	30.8133	33.9244	36.7807	40.2894	42.7956	48.268
23	27.1413	32.0069	35.1725	38.0757	41.6384	44.1813	49.728
24	28.2412	33.1963	36.4151	39.3641	42.9798	45.5585	51.179
25	29.3389	34.3816	37.6525	40.6465	44.3141	46.9278	52.620
26	30.4345	35.5631	38.8852	41.9232	45.6417	48.2899	54.052
27	31.5284	36.7412	40.1133	43.1944	46.9630	49.6449	55.476
28	32.6205	37.9159	41.3372	44.4607	48.2782	50.9933	56.892
29	33.7109	39.0875	42.5569	45.7222	49.5879	52.3356	58.302
30	34.7998	40.2560	43.7729	46.9792	50.8922	53.6720	59.703
40	45.6160	51.8050	55.7585	59.3417	63.6907	66.7659	73.402
50	56.3336	63.1671	67.5048	71.4202	76.1539	79.4900	86.661
60	66.9814	74.3970	79.0819	83.2976	88.3794	91.9517	99.607
70	77.5766	85.5271	90.5312	95.0231	100.425	104.215	112.317
80	88.1303	96.5782	101.879	106.629	112.329	116.321	124.839
90	98.6499	107.565	113.145	118.136	124.116	128.299	137.208
100	109.141	118.498	124.342	129.561	135.807	140.169	149.449

ource: Abridged from Table 8 in *Biometrika Tables for Statisticians,* vol. 1, 2d ed., edited by E. S. Pearson and . O. Hartley (New York: Cambridge University Press, 1958). Reproduced with the permission of the trustees f *Biometrika.*

TABLE C.5 CRITICAL VALUES FOR SQUARED MULTIPLE CORRELATION (R^2) IN FORWARD STEPWISE SELECTION

$\alpha = .05$

k	M	$N - k - 1$															
		10	12	14	16	18	20	25	30	35	40	50	60	80	100	150	200
2	1	38	33	29	26	24	22	18	15	13	12	9	8	6	5	3	2
2	2	45	39	35	31	28	26	21	18	16	14	11	10	7	6	4	3
4	1	39	35	31	28	26	24	20	17	15	14	11	9	7	6	4	3
4	2	50	44	40	37	34	31	26	23	20	18	14	12	9	7	5	4
4	3	55	49	45	41	37	34	29	25	22	19	16	13	10	8	5	4
4	4	58	52	47	43	39	36	31	26	23	21	17	14	11	9	6	5
6	1	38	34	31	29	26	25	21	18	16	14	12	10	8	6	4	3
6	2	50	45	41	38	35	33	28	25	22	20	16	14	11	9	6	5
6	3	57	51	47	43	40	38	32	28	25	22	19	16	12	10	7	6
6	4	61	55	51	47	43	40	34	30	27	24	20	17	13	11	7	6
6	6	66	60	55	51	47	44	37	33	29	26	22	18	14	12	8	6
8	1	36	33	30	28	26	24	21	18	16	15	12	11	8	7	5	4
8	2	49	45	41	38	36	34	29	26	23	21	18	15	12	10	7	6
8	3	57	52	48	45	42	39	34	30	27	24	20	18	14	11	8	7
8	4	62	57	52	49	45	43	37	33	29	26	22	19	15	12	8	7
8	6	68	62	57	53	50	47	40	36	32	29	24	21	16	13	9	7
8	8	71	66	61	56	53	49	43	38	34	30	25	22	17	14	10	7
10	1	35	32	29	27	26	24	21	18	16	15	13	11	8	7	5	4
10	2	48	44	41	38	36	34	30	26	24	21	18	16	12	10	7	6
10	3	56	52	48	45	42	40	35	31	28	25	22	19	15	12	9	7

N	M																
10	4	61	57	53	50	47	44	38	34	31	28	24	21	16	14	10	8
10	6	69	64	59	55	52	49	43	38	34	31	26	23	18	15	11	9
10	8	72	67	63	59	55	52	45	40	36	33	28	24	19	16	11	9
10	10	75	70	65	61	57	54	47	42	38	34	29	25	20	16	11	9
15	1	31	29	27	25	24	23	20	18	16	15	13	11	9	7	5	4
15	2	44	41	39	36	35	33	29	26	24	22	19	16	13	11	8	7
15	3	53	50	47	44	42	40	35	32	29	27	23	20	16	14	10	8
15	4	59	56	53	50	47	45	40	36	33	30	26	23	18	16	11	9
15	6	68	64	60	57	54	52	46	42	38	35	30	26	21	18	13	11
15	8	74	70	66	62	59	56	50	45	41	38	33	29	23	20	14	11
15	10	78	74	69	66	62	59	53	48	43	40	34	30	24	21	15	11
15	15	81	77	73	69	65	62	56	50	46	42	36	31	25	21	15	11
20	1	27	26	24	23	22	21	19	17	16	14	12	11	9	7	5	4
20	2	40	38	36	34	33	31	28	25	23	21	18	16	13	11	8	6
20	3	50	47	45	42	40	38	35	32	29	27	23	21	17	14	10	8
20	4	57	54	51	48	46	44	40	36	33	31	27	24	19	16	12	10
20	6	66	63	60	57	54	52	47	43	39	37	32	28	23	20	14	12
20	8	73	69	66	63	60	57	52	47	44	40	35	31	26	22	16	13
20	10	78	74	70	67	64	61	56	51	47	43	38	33	27	23	17	14
20	15	85	81	77	74	70	67	61	55	51	47	41	37	30	25	18	14
20	20	85	81	77	74	71	68	62	56	52	48	42	37	30	25	18	14

Note: Decimals are omitted; k = number of predictors; M = number of predictors selected; N = sample size.

Source: Adapted from Tables 1 and 2 in "Tests of significance in stepwise regression," by L. Wilkinson, Psychological Bulletin, 1979, **86**(1), 168–174. Copyright 1979 by the American Psychological Association. Reprinted by permission.

TABLE C.5 (*Continued*)

$\alpha = .01$

								$N - k - 1$									
k	M	10	12	14	16	18	20	25	30	35	40	50	60	80	100	150	200
2	1	53	47	42	38	35	32	27	23	20	18	14	12	9	8	5	4
2	2	60	54	48	44	40	37	31	26	23	21	17	14	11	9	6	5
4	1	52	47	42	39	36	33	28	24	22	19	16	14	10	9	6	4
4	2	63	57	51	47	44	41	35	30	27	24	20	17	13	11	7	5
4	3	69	62	56	51	48	44	38	33	29	26	22	19	15	12	8	6
4	4	71	64	59	54	50	47	40	35	31	28	23	20	15	12	8	6
6	1	49	45	41	38	36	33	28	25	22	20	17	14	11	9	6	5
6	2	61	56	52	48	45	42	36	32	29	26	22	19	15	12	8	7
6	3	68	62	57	54	50	47	40	35	32	29	24	21	16	14	9	8
6	4	72	66	61	57	53	49	43	38	34	30	26	22	17	14	10	8
6	6	76	71	66	61	57	54	47	41	37	33	28	24	19	15	10	8
8	1	47	43	40	37	35	32	28	25	22	20	17	14	11	9	6	5
8	2	60	55	51	48	45	42	37	33	29	27	23	20	15	13	9	7
8	3	67	62	58	54	51	48	42	37	33	30	26	22	18	15	10	8
8	4	72	66	62	58	54	51	45	40	36	32	27	24	19	16	11	9
8	6	78	72	67	63	59	55	48	43	39	35	30	26	21	17	12	9
8	8	80	75	70	66	62	59	52	46	41	37	32	27	22	18	12	9
10	1	44	41	38	36	34	32	28	24	22	20	17	15	12	9	7	5
10	2	58	54	50	47	44	42	37	33	30	27	23	20	16	13	9	7
10	3	66	61	57	54	51	48	42	38	34	31	27	23	18	15	11	9

10	12	17	20	25	29	34	37	41	46	52	55	58	62	66	71	4	10
11	14	19	22	28	32	37	40	45	50	57	60	63	67	72	77	6	10
11	14	20	23	29	33	39	43	47	53	60	63	67	71	76	81	8	10
11	14	20	24	30	35	41	45	50	56	63	66	70	74	78	83	10	10
5	7	10	12	15	17	19	21	23	26	29	31	32	34	36	39	1	15
7	9	13	16	20	23	27	29	32	36	40	42	45	47	50	53	2	15
9	11	16	19	24	27	32	35	38	42	47	50	52	55	58	62	3	15
11	13	18	22	27	30	35	38	42	47	52	55	57	60	64	68	4	15
13	15	21	25	30	34	40	43	47	52	58	61	64	67	71	75	6	15
14	17	23	27	33	37	43	46	51	56	62	65	68	72	76	80	8	15
14	18	24	28	35	39	45	49	53	59	66	69	72	76	80	84	10	15
14	18	25	30	37	42	49	53	57	63	70	73	76	80	83	87	15	15
5	7	10	12	14	16	19	20	22	24	27	28	30	31	33	35	1	20
7	9	13	16	20	22	26	28	31	34	38	40	42	44	47	49	2	20
9	11	16	19	24	27	32	34	37	41	46	48	50	52	55	58	3	20
11	13	19	22	27	31	36	38	42	46	51	53	55	58	61	64	4	20
14	16	22	26	32	36	41	44	48	52	58	60	63	66	69	72	6	20
16	18	24	29	35	39	45	48	52	57	63	65	68	71	74	78	8	20
16	20	26	31	37	42	48	51	55	60	66	69	72	75	79	82	10	20
16	21	29	34	41	46	53	57	61	67	73	76	79	82	86	89	15	20
16	21	29	35	42	48	54	58	63	68	75	77	80	83	87	90	20	20

TABLE C.6 CRITICAL VALUES FOR FMAX (S^2_{max}/S^2_{min}) DISTRIBUTION FOR $\alpha = .05$ AND .01

$\alpha = .05$

df \ k	2	3	4	5	6	7	8	9	10	11	12
4	9.60	15.5	20.6	25.2	29.5	33.6	37.5	41.1	44.6	48.0	51.4
5	7.15	10.8	13.7	16.3	18.7	20.8	22.9	24.7	26.5	28.2	29.9
6	5.82	8.38	10.4	12.1	13.7	15.0	16.3	17.5	18.6	19.7	20.7
7	4.99	6.94	8.44	9.70	10.8	11.8	12.7	13.5	14.3	15.1	15.8
8	4.43	6.00	7.18	8.12	9.03	9.78	10.5	11.1	11.7	12.2	12.7
9	4.03	5.34	6.31	7.11	7.80	8.41	8.95	9.45	9.91	10.3	10.7
10	3.72	4.85	5.67	6.34	6.92	7.42	7.87	8.28	8.66	9.01	9.34
12	3.28	4.16	4.79	5.30	5.72	6.09	6.42	6.72	7.00	7.25	7.48
15	2.86	3.54	4.01	4.37	4.68	4.95	5.19	5.40	5.59	5.77	5.93
20	2.46	2.95	3.29	3.54	3.76	3.94	4.10	4.24	4.37	4.49	4.59
30	2.07	2.40	2.61	2.78	2.91	3.02	3.12	3.21	3.29	3.36	3.39
60	1.67	1.85	1.96	2.04	2.11	2.17	2.22	2.26	2.30	2.33	2.36
∞	1.00	1.00	1.00	1.00	1.00	1.00	1.00	1.00	1.00	1.00	1.00

α = .01

df \ k	2	3	4	5	6	7	8	9	10	11	12
4	23.2	37	49	59	69	79	89	97	106	113	120
5	14.9	22	28	33	38	42	46	50	54	57	60
6	11.1	15.5	19.1	22	25	27	30	32	34	36	37
7	8.89	12.1	14.5	16.5	18.4	20	22	23	24	26	27
8	7.50	9.9	11.7	13.2	14.5	15.8	16.9	17.9	18.9	19.8	21
9	6.54	8.5	9.9	11.1	12.1	13.1	13.9	14.7	15.3	16.0	16.6
10	5.85	7.4	8.6	9.6	10.4	11.1	11.8	12.4	12.9	13.4	13.9
12	4.91	6.1	6.9	7.6	8.2	8.7	9.1	9.5	9.9	10.2	10.6
15	4.07	4.9	5.5	6.0	6.4	6.7	7.1	7.3	7.5	7.8	8.0
20	3.32	3.8	4.3	4.6	4.9	5.1	5.3	5.5	5.6	5.8	5.9
30	2.63	3.0	3.3	3.4	3.6	3.7	3.8	3.9	4.0	4.1	4.2
60	1.96	2.2	2.3	2.4	2.4	2.5	2.5	2.6	2.6	2.7	2.7
∞	1.00	1.0	1.0	1.0	1.0	1.0	1.0	1.0	1.0	1.0	1.0

Note: S_{max}^2 is the largest and S_{min}^2 the smallest in a set of k independent mean squares, each based on degrees of freedom (df).

Source: Abridged from Table 31 in *Biometrika Tables for Statisticians*, vol. 1, 2d ed., edited by E. S. Pearson and H. O. Hartley (New York: Cambridge University Press, 1958). Reproduced with the permission of the trustees for *Biometrika*.

TABLE C.7 PROBABILITIES FOR THE SALIENT VARIABLE SIMILARITY INDEX, s
Probabilities $s \geq v_s$ for Hyperplane Count 60%

Number of variables, p

10	v_s:	.76	.51	.26	.01	.00						
	p:	.001	.020	.138	.364	.500						
20	v_s:	.63	.51	.26	.13	.01	.00					
	p:	.000	.004	.086	.207	.393	.500					
30	v_s:	.59	.51	.42	.34	.26	.17	.09	.01	.00		
	p:	.000	.001	.005	.019	.054	.131	.256	.411	.500		
40	v_s:	.51	.44	.38	.32	.26	.19	.13	.07	.01	.00	
	p:	.000	.001	.005	.016	.041	.090	.169	.282	.426	.500	
50	v_s:	.46	.41	.36	.31	.26	.21	.16	.11	.06	.01	.00
	p:	.000	.001	.004	.011	.026	.061	.115	.200	.301	.428	.500
60	v_s:	.42	.38	.34	.30	.26	.21	.17	.13	.09	.05	.01
	p:	.000	.001	.003	.008	.018	.039	.078	.132	.215	.318	.440
80	v_s:	.35	.32	.29	.26	.22	.19	.16	.13	.10	.07	.04
	p:	.000	1.001	.003	.008	.018	.034	.063	.107	.170	.243	.338
	v_s:	.00										
	p:	.500										
100	v_s:	.36	.31	.28	.26	.23	.21	.18	.16	.13	.11	.08
	p:	.000	.001	.003	.006	.011	.019	.034	.058	.091	.136	.196
	v_s:	.03	.01	.00								
	p:	.353	.449	.500								

Probabilities $s \geq v_s$ for Hyperplane Count 70%

Number of variables, p													
10	v_s:	.67	.34	.01	.00								
	p:	.002	.052	.316	.500								
20	v_s:	.67	.51	.34	.17	.01	.00						
	p:	.000	.002	.027	.135	.357	.500						
30	v_s:	.56	.45	.34	.23	.12	.01	.00					
	p:	.000	.002	.016	.064	.190	.383	.500					
40	v_s:	.51	.42	.34	.26	.17	.09	.00					
	p:	.000	.002	.007	.034	.098	.222	.500					
50	v_s:	.47	.41	.34	.27	.21	.14	.07	.01	.00			
	p:	.000	.001	.004	.018	.052	.123	.247	.407	.500			
60	v_s:	.39	.34	.28	.23	.17	.12	.06	.01	.00			
	p:	.000	.002	.007	.025	.066	.142	.262	.415	.500			
80	v_s:	.38	.34	.30	.26	.21	.17	.13	.09	.05	.01	.00	
	p:	.000	.001	.002	.007	.019	.046	.092	.174	.283	.425	.500	
100	v_s:	.34	.31	.27	.24	.21	.17	.14	.11	.07	.04	.01	.00
	p:	.000	.001	.002	.006	.016	.037	.069	.125	.207	.318	.438	.500

Source: Adapted from Tables 1–4 in "Factor matching procedures: An improvement of the s index; with tables," by R. B. Cattell, K. R. Balcar, J. L. Horn, and J. R. Nesselroade, Educational and Psychological Measurement, 1969, 29, 781–792. Reproduced with permission of the publisher and authors.

TABLE C.7 (*Continued*)
Probabilities $s \geq v_s$ for Hyperplane Count 80%

Number of variables, p

10	v_s:	.51	.01	.00					
	p:	.012	.187	.500					

20	v_s:	.76	.51	.26	.01	.00			
	p:	.000	.003	.041	.279	.500			

30	v_s:	.67	.51	.34	.17	.01	.00		
	p:	.000	.001	.011	.083	.316	.500		

40	v_s:	.51	.38	.26	.13	.01	.00		
	p:	.000	.003	.024	.115	.347	.500		

50	v_s:	.51	.41	.31	.21	.11	.01	.00	
	p:	.000	.001	.006	.036	.142	.361	.500	

60	v_s:	.42	.34	.26	.17	.09	.01	.00	
	p:	.000	.002	.012	.052	.164	.369	.500	

80	v_s:	.38	.32	.26	.19	.13	.07	.01	.00
	p:	.000	.001	.004	.022	.076	.199	.301	.500

100	v_s:	.31	.26	.21	.16	.11	.06	.01	.00
	p:	.000	.002	.010	.036	.105	.220	.402	.500

Probabilities $s \geq v_s$ for Hyperplane Count 90%

Number of variables, p

10	v_s:	.01	.00			
	p:	.052	.500			

20	v_s:	.51	.01	.00		
	p:	.003	.099	.500		

30	v_s:	.67	.34	.01	.00	
	p:	.000	.007	.133	.500	

40	v_s:	.51	.26	.01	.00	
	p:	.000	.012	.167	.500	

50	v_s:	.41	.21	.01	.00	
	p:	.000	.018	.198	.500	

60	v_s:	.51	.34	.17	.01	.00
	p:	.000	.002	.029	.217	.500

80	v_s:	.38	.26	.13	.01	.00
	p:	.000	.004	.045	.251	.500

100	v_s:	.31	.21	.11	.01	.00
	p:	.000	.007	.061	.286	.500

References

Asher, H. B. *Causal Modeling*. Sage University Paper Series on Quantitative Applications in the Social Sciences, Series No. 07-003. Beverly Hills and London: Sage Publications, 1976.

Balter, M. D., and Levine, J. Character and extent of psychotropic drug usage in the United States. Paper presented at the Fifth World Congress on Psychiatry, Mexico City, 1971.

Bartlett, M. S. The statistical significance of canonical correlations. *Biometrika*, 1941, **32**, 29–38.

Bartlett, M. S. A note on the multiplying factors for various chi-square approximations. *Journal of the Royal Statistical Society*, 1954, **16** (Series B), 296–298.

Beatty, J. R. Identifying decision-making policies in the diagnosis of learning disabilities. *Journal of Learning Disabilities*, 1977, **10**(4), 201–209.

Bem, S. L. The measurement of psychological androgyny. *Journal of Consulting and Clinical Psychology*, 1974, **42**, 155–162.

Bentler, P. M. Multivariate analysis with latent variables: Causal modeling. *Annual Review of Psychology*, 1980, **31**, 419–456.

Bock, R. D. Contributions of multivariate experimental designs to educational research. In Cattell, R. B., ed., *Handbook of Multivariate Experimental Psychology*. Chicago: Rand McNally, 1966.

Bock, R. D. *Multivariate Statistical Methods in Behavioral Research*. New York: McGraw-Hill, 1975.

Bock, R. D., and Haggard, E. A. The use of multivariate analysis of variance in behavioral research. In Whitla, D. K., ed., *Handbook of Measurement and Assessment in Behavioral Sciences*. Reading, Mass.: Addison-Wesley, 1968.

Box, G. E. P., and Cox, D. R. An analysis of transformations. *Journal of the Royal Statistical Society*, 1964, **26** (Series B), 211–243.

Bradley, R. H., and Gaa, J. P. Domain specific aspects of locus of control: Implications for modifying locus of control orientation. *Journal of School Psychology,* 1977, **15**(1), 18–24.

Brown, R. D., Braskamp, L. A., and Newman, D. L. Evaluator credibility as a function of report style. *Evaluation Quarterly,* 1978, **2**(2), 331–334.

Browne, M. W. Predictive validity of a linear regression equation. *British Journal of Mathematical and Statistical Psychology,* 1975, **28**, 79–87.

Burgess, E., and Locke, H. *The Family.* 2nd ed. New York: American Book, 1960.

Caffrey, B., and Lile, S. Similarity of attitudes toward science on the part of psychology and physics students. *Teaching of Psychology,* 1976, **3**(1), 24–26.

Campbell, D. R., and Stanley, J. C. *Experimental and Quasi-experimental Designs for Research.* New York: Rand McNally, 1966.

Cattell, R. B. *Personality and Motivation Structures and Measurement.* Yonkers-on-Hudson, N.Y.: World Book, 1957.

Cattell, R. B. The scree test for the number of factors. *Multivariate Behavioral Research,* 1966, **1**, 245–276.

Cattell, R. B., and Baggaley, A. R. The salient variable similarity index for factor matching. *British Journal of Statistical Psychology,* 1960, **13**, 33–46.

Cattell, R. B., Balcar, K. R., Horn, J. L., and Nesselroade, J. R. Factor matching procedures: An improvement of the *s* index; with tables. *Educational and Psychological Measurement,* 1969, **29**, 781–792.

Cattin, P. Note on the estimation of the squared cross-validated multiple correlation of a regression model. *Psychological Bulletin,* 1980, **87**(1), 63–65.

Cohen, E., and Burns, P. SPSS-MANOVA. Document No. 413 (Rev. A). Evanston: Northwestern University, Vogelback Computing Center, 1977.

Cohen, J., and Cohen, P. *Applied Multiple Regression/Correlation Analysis for the Behavioral Sciences.* New York: Lawrence Erlbaum Associates, 1975.

Cohen P., Gaughran, E., and Cohen, J. Age patterns of childbearing: A canonical analysis. *Multivariate Behavioral Research,* 1979, **14**, 75–89.

Comrey, A. L. *A First Course in Factor Analysis.* New York: Academic Press, 1973.

Cooke, T. D., and Campbell, D. T. *Quasi-experimentation: Design and Analysis Issues for Field Settings.* Chicago: Rand McNally College Publishing Co.; Boston: Houghton Mifflin, 1979.

Cooley, W. W., and Lohnes, P. R. *Multivariate Data Analysis.* New York: Wiley, 1971.

Cornbleth, T. Effects of a protected hospital ward area on wandering and nonwandering geriatric patients. *Journal of Gerontology,* 1977, **32**(5), 573–577.

Curtis, B., and Simpson, D. D. Differences in background and drug use history among three types of drug users entering drug therapy programs. *Journal of Drug Education,* 1977, **7**(4), 369–379.

Dixon, W. J., ed. *BMD Biomedical Computer Programs.* Los Angeles: University of California Press, 1970.

Dixon, W. J., ed. *BMD X-Series Supplement.* Los Angeles: University of California Press, 1972.

Dixon, W. J., ed. *BMD Biomedical Computer Programs.* Los Angeles: University of California Press, 1974.

Dixon, W. J., ed. *BMDP Statistical Software*. Berkeley: University of California Press, 1981.

Edwards, A. L. *An Introduction to Linear Regression and Correlation*. San Francisco: Freeman, 1976.

Eysenck, H. J., and Eysenck, S. B. G. *The Eysenck Personality Inventory*. San Diego, Calif.: Educational and Industrial Testing Service; London: University of London Press, 1963.

Featherman, D. Metrics of occupational status reconciled to the 1970 Bureau of Census Classification of Detailed Occupational Titles (based on Census Technical Paper No. 26, "1970 Occupation and Industry Classification Systems in Terms of Their 1960 Occupation and Industry Elements"). Washington, D.C.: Government Printing Office, 1973. (Update of Duncan's socioeconomic status metric described in Reiss, J., et al., *Occupational and Social Status*. New York: Free Press of Glencoe, 1961.)

Fidell, S. Nationwide urban noise survey. *Journal of the Acoustical Society of America*, 1978, **16**(1) 198–206.

Fleming, J. S., and Pinneau, S. R., Sr. Measuring the stability of linear weighting systems via correlated scoring functions. In preparation, 1980.

Fornell, C. External single-set components analysis of multiple criterion/multiple predictor variables. *Multivariate Behavioral Research*, 1979, **14**, 323–338.

Frane, J. W. Personal communication. Health Sciences Computing Facility, University of California, Los Angeles, August 3, 1977.

Frane, J. W. The univariate approach to repeated measures—foundation, advantages, and caveats. BMD Technical Report No. 69. Health Sciences Computing Facility, University of California, Los Angeles, 1980.

Frane, J. W., and Hill, M. Annotated computer output for factor analysis using Professor Jarvik's smoking questionnaire. Technical Report No. 8. Health Sciences Computing Facility, University of California, Los Angeles, 1974.

Gaito, J. Measurement scales and statistics: Resurgence of an old misconception. *Psychological Bulletin*, 1980, **87**(3), 564–567.

Games, P. A. Type IV errors revisited. *Psychological Bulletin*, 1973, **80**(4), 304–307.

Giarrusso, R. The effects of attitude similarity and physical attractiveness on romantic attraction and time perception. Master's thesis, California State University, Northridge, 1977.

Glock, C., Ringer, B., and Babbie, E. *To Comfort and to Challenge*. Berkeley: University of California Press, 1967.

Goodman, M. J., Chung, C. S., and Gilbert, F. Racial variation in diabetes mellitus in Japanese and Caucasians living in Hawaii. *Journal of Medical Genetics*, 1974, **11**, 328–338.

Goodman, M. J., Steward, C. J., and Gilbert, F., Jr. Patterns of menopause: A study of certain medical and physiological variables among Caucasian and Japanese women living in Hawaii. *Journal of Gerontology*, 1977, **32**(3), 291–298.

Hakstian, A. R., Roed, J. C., and Lind, J. C. Two-sample T^2 procedure and the assumption of homogeneous covariance matrices. *Psychological Bulletin*, 1979, **86**, 1255–1263.

Hall, S. M., Hall, R. G., DeBoer, G., and O'Kulitch, P. Self and external management compared with psychotherapy in the control of obesity. *Behavior Research and Therapy*, 1977, **15**, 89–95.

Harman, H. H. *Modern Factor Analysis.* University of Chicago Press, 1967.

Harris, R. J. *A Primer of Multivariate Statistics.* New York: Academic Press, 1975.

Heise, D. R. *Causal Analysis.* New York: Wiley, 1975.

Hoffman, D., and Fidell, L. S. Characteristics of androgynous, undifferentiated, masculine and feminine middle class women. *Sex Roles,* 1979, **5**(6), 765–781.

Hull, C. H., and Nie, N. H. *SPSS Update 7–9.* New York: McGraw-Hill, 1981.

Jakubczak, L. F. Age differences in the effects of palatability of diet on regulation of calorie intake and body weight of rats. *Journal of Gerontology,* 1977, **32**(1), 49–57.

Johnson, G. An instrument for the assessment of job satisfaction. *Personnel Psychology,* 1955, **8**, 27–37.

Jöreskog, K. G., and Sörbom, D. *LISREL IV Users Guide.* Chicago: National Educational Research, 1978.

Keppel, G. *Design and Analysis: A Researcher's Handbook.* Englewood Cliffs, N.J.: Prentice-Hall, 1973.

Kim, J., and Kohout, F. J. Multiple regression analysis: Subprogram regression. In Nie, N. H., Hull, C. H., Jenkins, J. G., Steinbrenner, K., and Bent, D. H., eds., *SPSS: Statistical Package for the Social Sciences.* New York: McGraw-Hill, 1970.

Kirk, R. E. *Experimental Design: Procedures for the Behavioral Sciences.* Monterey, Calif.: Brooks/Cole, 1968.

Kish, L. *Survey Sampling.* New York: Wiley, 1965.

Langer, T. A 22-item screening score of psychiatric symptoms indicating impairment. *Journal of Health and Human Behavior,* 1962, **3**, 269–276.

Larzelere, R. E., and Mulaik, S. A. Single-sample tests for many correlations. *Psychological Bulletin,* 1977, **84**(3), 557–569.

Lawley, D. N., and Maxwell, A. E. *Factor Analysis as a Statistical Method.* London: Butterworths, 1963.

Lee, W. *Experimental Design and Analysis.* San Francisco: Freeman, 1975.

Levin, J. R., and Marascuilo, L. A. Type IV errors and interactions. *Psychological Bulletin,* 1972, **78**, 368–374.

Levin, J. R., and Marascuilo, L. A. Type IV errors and games. *Psychological Bulletin,* 1973, **80**(4), 308–309.

Levine, M. S. *Canonical Analysis and Factor Comparison.* Sage University Paper Series in Quantitative Applications in the Social Sciences, Series No. 07–006. Beverly Hills and London: Sage Publications, 1977.

Linton, M., and Gallo, P. *The Practical Statistician: Simplified Handbook of Statistics.* Monterey, Calif.: Brooks/Cole, 1975.

Locke, H., and Wallace, K. Short marital-adjustment and prediction tests: Their reliability and validity. *Marriage and Family Living,* 1959, **21**, 251–255.

McLeod, J. M., Brown, J. D., and Becker, L. B. Watergate and the 1974 congressional elections. *Public Opinion Quarterly,* 1977, **41**, 181–195.

McNeil, K. S., Kelly, F. J., and McNeil, J. T. *Testing Research Hypotheses Using Multiple Linear Regression.* Carbondale: Southern Illinois University Press, 1975.

Maki, J. E., Hoffman, D. M., and Berk, R. A. A time series analysis of the impact of a water conservation campaign. *Evaluation Quarterly,* 1978, **2**(1), 107–118.

Mardia, K. V. The effect of nonnormality on some multivariate tests and robustness to nonnormality in the linear model. *Biometrika,* 1971, **58**(1), 105–121.

Menozzi, P., Piazza, A., and Cavalli-Sforza, L. Synthetic maps of human gene frequencies in Europeans. *Science,* 1978, **201,** 786–792.

Merrill, P. F., and Towle, J. J. The availability of objectives and performance in a computer-managed graduate course. *Journal of Experimental Education,* 1976, **45**(1), 12–29.

Miller, J. K., and Farr, S. D. Bimultivariate redundancy: A comprehensive measure of interbattery relationship. *Multivariate Behavioral Research,* 1971, **6,** 313–324.

Morrison, D. F. *Multivariate Statistical Methods.* New York: McGraw-Hill, 1967.

Moser, C. A., and Kalton, G. *Survey Methods in Social Investigation.* New York: Basic Books, 1972.

Mosteller, F., and Tukey, J. W. *Data Analysis and Regression.* Reading, Mass.: Addison-Wesley, 1977.

Mulaik, S. A. *The Foundation of Factor Analysis.* New York: McGraw-Hill, 1972.

Myers, J. L. *Fundamentals of Experimental Design.* 3rd ed. Boston: Allyn and Bacon, 1979.

Nicholson, W., and Wright, S. R. Participants' understanding of the treatment in policy experimentation. *Evaluation Quarterly,* 1977, **1**(2), 245–268.

Nie, N. H., Hull, C. H., Jenkins, J. G., Steinbrenner, K., and Bent, D. H. *Statistical Package for the Social Sciences.* 2nd ed. New York: McGraw-Hill, 1975.

Norušis, M. J. *SPSS Statistical Algorithms.* Chicago: SPSS, 1978.

O'Kane, J. M., Barenblatt, L., Jensen, P. K., and Cochran, L. T. Anticipatory socialization and male Catholic adolescent sociopolitical attitudes. *Sociometry,* 1977, **40**(1), 67–77.

Olson, C. L. On choosing a test statistic in multivariate analysis of variance. *Psychological Bulletin,* 1976, **83**(4), 579–586.

Olson, C. L. Practical considerations in choosing a MANOVA test statistic: A rejoinder to Stevens. *Psychological Bulletin,* 1979, **86,** 1350–1352.

Overall, J. E., and Spiegel, D. K. Concerning least squares analysis of experimental data. *Psychological Bulletin,* 1969, **72**(5), 311–322.

Overall, J. E., and Woodward, J. A. Nonrandom assignment and the analysis of covariance. *Psychological Bulletin,* 1977, **84**(3), 588–594.

Pessemier, E. A., Bemmoar, A. C., and Hanssens, D. M. Willingness to supply human body parts: Some empirical results. *Journal of Consumer Research,* 1977, **4**(3), 131–140.

Pinneau, S. R., and Newhouse, A. Measures of invariance and comparability in factor analysis for fixed variables. *Psychometrika,* 1964, **29,** 271–281.

Price, B. Ridge regression: Application to nonexperimental data. *Psychological Bulletin,* 1977, **84**(4), 759–766.

Rahe, R. H. The pathway between subjects' recent life changes and their near-future illness reports: Representative results and methodological issues. In Dohrenwend, B. S. and Dohrenwend, B. P., eds. *Stressful Life Events: Their Nature and Effects.* New York: Wiley, 1974.

Rao, C. R. *Advanced Statistical Methods in Biometric Research.* New York: Wiley, 1952.

Rollins, B., and Feldman, H. Marital satisfaction over the family life cycle. *Journal of Marriage and Family,* 1970, **32,** 29–38.

Rosenberg, M. *Society and the Adolescent Self-Image.* Princeton, N.J.: Princeton University Press, 1965.

Rotter, J. B. Generalized expectancies for internal versus external control of reinforcement. *Psychological Monographs,* 1966, **80**(1, Whole No. 609).

Rozeboom, W. W. Ridge regression: Bonanza or beguilement? *Psychological Bulletin,* 1979, **86**(2), 242–249.

Rummel, R. J. *Applied Factor Analysis.* Evanston: Northwestern University Press, 1970.

St. Pierre, R. G. Correcting covariables for unreliability. *Evaluation Quarterly,* 1978, **2**(3), 401–420.

Schall, J. J., and Pianka, E. R. Geographical trends in numbers of species. *Science,* 1978, **201**, 679–686.

Scheffé, H. A. A method of judging all contrasts in the analysis of variance. *Biometrika,* 1953, **40**, 87–104.

Simonton, D. K. Cross-sectional time-series experiments: Some suggested statistical analyses. *Psychological Bulletin,* 1977, **84**(3), 489–502.

Singh, B., Greer, P. R., and Hammond, R. An evaluation of the use of the law in a free society materials on "responsibility." *Evaluation Quarterly,* 1977, **1**(4), 621–628.

Spence, J., and Helmreich, R. The attitudes towards women scale: An objective instrument to measure attitudes towards rights and roles of women in contemporary society. *Journal Supplementary Abstract Service* (Catalogue of Selected Documents in Psychology), 1972, **2**, 66.

Steiger, J. H. Tests for comparing elements of a correlation matrix. *Psychological Bulletin,* 1980, **87**(2), 245–251.

Stevens, S. S. On the theory of scales of measurement. *Science,* 1946, **103**, 677–680.

Stewart, D., and Love, W. A general canonical correlation index. *Psychological Bulletin,* 1968, **70**, 160–163.

Strober, M. H., and Weinberg, C. B. Working wives and major family expenditures. *Journal of Consumer Research,* 1977, **4**(3), 141–147.

Tatsuoka, M. M. *Multivariate Analysis: Techniques for Educational and Psychological Research.* New York: Wiley, 1971.

Tatsuoka, M. M. Classification procedures. In Amick, D. J., and Walberg, H. J., eds., *Introductory Multivariate Analysis.* Berkeley: McCutchan, 1975.

Thurstone, L. L. *Multiple Factor Analysis.* University of Chicago Press, 1947.

Vaughn, G. M., and Corballis, M. C. Beyond tests of significance: Estimating strength of effects in selected ANOVA designs. *Psychological Bulletin,* 1969, **72**(3), 204–213.

Wade, T. C., and Baker, T. B. Opinions and use of psychological tests: A survey of clinical psychologists. *American Psychologist,* 1977, **32**(10), 874–882.

Wiener, Y., and Vaitenas, R. Personality correlates of voluntary midcareer change in enterprising occupations. *Journal of Applied Psychology,* 1977, **62**(6), 706–712.

Wernimont, P. F., and Fitzpatrick, S. The meaning of money. *Journal of Applied Psychology,* 1972, **56**(3), 218–226.

Wesolowsky, G. O. *Multiple Regression and Analysis of Variance.* New York: Wiley-Interscience, 1976.

Wherry, R. J., Sr. A new formula for predicting the shrinkage of the coefficient of multiple correlation. *Annals of Mathematical Statistics,* 1931, **2**, 440–457.

Wilkinson, L. Tests of significance in stepwise regression. *Psychological Bulletin,* 1979, **86**(1), 168–174.

Wilkinson, L., and Dallal, G. E. Tests of significance in forward selection regression with an F-to-enter stopping rule. 1980 (in preparation).

Willis, J. W., and Wortman, C. B. Some determinants of public acceptance of randomized control group experimental designs. *Sociometry,* 1976, **39**(2), 91–96.

Winer, B. J. *Statistical Principles in Experimental Design.* 2nd ed. New York: McGraw-Hill, 1971.

Wingard, J. A., Huba, G. J., and Bentler, P. M. The relationship of personality structure to patterns of adolescent substance use. *Multivariate Behavioral Research,* 1979, **14**, 131–143.

Wong, M. R., and Allen, T. A three dimensional structure of drug attitudes. *Journal of Drug Education,* 1976, **6**(2), 181–191.

Woodward, J. A., and Overall, J. E. Multivariate analysis of variance by multiple regression methods. *Psychological Bulletin,* 1975, **82**(1), 21–32.

Wrigley, C., and Neuhaus, J. The matching of two sets of factors. *American Psychologist,* 1955, **10**, 418–419.

Young, R. K., and Veldman, D. J. *Introductory Statistics for the Behavioral Sciences.* 3rd ed. New York: Holt, Rinehart and Winston, 1977.

Index